DEVICES FOR INTEGRATED CIRCUITS

DEVICES FOR INTEGRATED CIRCUITS

Silicon and III-V Compound Semiconductors

H. CRAIG CASEY, JR.

Department of Electrical and Computer Engineering
Duke University

JOHN WILEY & SONS, INC.

New York • Chichester • Weinheim • Brisbane • Singapore • Toronto

ACQUISITIONS EDITOR Bill Zobrist
MARKETING MANAGER Katherine Hepburn
SENIOR PRODUCTION MANAGER Lucille Buonocore
SENIOR PRODUCTION EDITOR Monique Calello
COVER DESIGNER Karin Kincheloe
ILLUSTRATION COORDINATOR Gene Aiello
ILLUSTRATION STUDIO Radiant Illustration and Design
TEXT DESIGNER Nancy Field
COVER ILLUSTRATION Copyright 1995 International Business Machines Corporation. Reprinted with permission of the IBM Journal of Research and Development, Volume 39, Number 4.
COVER ILLUSTRATOR Ken Batelman

This book was set in Times Roman by Publication Services and printed and bound by Malloy Lithographing, Inc. The cover was printed by Phoenix Color Corporation.

This book is printed on acid-free paper. ♾

The paper in this book was manufactured by a mill whose forest management programs include sustained yield harvesting of its timberlands. Sustained yield harvesting principles ensure that the numbers of trees cut each year does not exceed the amount of new growth.

Library of Congress Cataloging-in-Publication Data

Casey, H. Craig.
Devices for integrated circuits : silicon and III-V compound semiconductors / H. Craig Casey, Jr.
p. cm.
Includes index.
ISBN 0-471-17134-4 (alk. paper)
1. Compound semiconductors. 2. Integrated circuits—Computer simulation. I. Title.
TK7871.99.C65C37 1999
621.3815—dc21 98-6992
CIP

Printed in the United States of America

10 9 8 7 6 5 4 3 2 1

PREFACE

This text is based on notes developed for a first-year graduate student (and advanced seniors) course. Some of the unique features of this text are:

- The *physics of semiconductor devices* is related to the parameters used in *s*imulation *p*rogram with *i*ntegrated *c*ircuit *e*mphasis (SPICE), and the use of PSpice is introduced.
- Both *silicon* and *III-V (three-five) compound semiconductors* are covered, although the emphasis is on silicon.
- Inclusion of topics often not considered in other texts such as the inversion layer thickness, drain current flow in field-effect transistors after pinch off, and the subthreshold current in MOSFETs.
- Detailed derivations are given so that class time can be spent on concepts while the details of the derivations are intended, in many cases, for the students to be able to follow on their own.
- Circuit applications are given to illustrate the application of the devices and to prepare for subsequent courses in electronic circuits.
- References are given to original journal publications and other specialized texts to provide further reading on a particular subject and to acknowledge the authors' contributions.

Both Si and III-V compound semiconductors are considered because in a semiconductor-device course for first-year graduate students, some students are preparing for study in analog or digital Si integrated circuits as well as submicron feature-size Si devices, while others are interested in advanced topics in III-V compound semiconductors and optoelectronics. This text prepares students for more advanced texts in Si devices as well as the recent texts *Semiconductor Optoelectronic Devices,* by Pallab

Bhattacharya and *Compound Semiconductor Device Physics,* by Sandip Tiwari. Also, during a professional career, an engineer or scientist may work in several different areas of solid-state electronics. Therefore, both Si and III-V compound semiconductors are presented, but the emphasis is on Si devices because of the dominance of bipolar and MOS integrated circuits.

With the growing emphasis on custom integrated-circuit design, the use of computer simulation of integrated circuits has become an integral part of circuit design. The de facto simulator for integrated-circuit design is one of the versions of SPICE. Device models both from the device physics and empirical models provide a description of how a device will behave in a circuit. Some device-physics topics are included here because they relate to parameters used in SPICE. The book *Semiconductor Device Modeling with SPICE, 2nd Edition,* by Giuseppe Massobrio and Paolo Antognetti not only made it possible to relate device physics to the SPICE parameters, but also made it very timely to include SPICE. Recent books such as *MOSFET Modeling with SPICE,* by Daniel Foty also connect SPICE with device models. The inclusion of SPICE required the omission of the usual introduction to integrated-circuit processing which may be found in numerous texts. PSpice may be run on personal computers as well as workstations. Free evaluation versions of PSpice are available from OrCAD, which merged with MicroSim Corporation (originator of PSpice) in January 1998. This free limited capability evaluation version of PSpice (release 8.0) is available on the John Wiley web site: http://www.wiley.com/college/casey. The one constant with software is continuous change and this is also true with PSpice; however, the parts of PSpice introduced here are not expected to change significantly.

Chapters 2 and 3 introduce the physics of semiconductors needed in the remaining chapters. These two chapters are intended to be covered rapidly. The homework problems in Chapters 2 and 3 are intended to familiarize students with the units and concepts, but not be an in-depth treatment. When both Si and III-V compound semiconductors are considered, subjects such as indirect and direct energy gaps must be introduced. Therefore, it was necessary to devote more effort to the physics of semiconductors than the usual device text. Because of the dominance of MOS integrated circuits, added emphasis was given to the MOSFET. The bipolar transistor chapter was placed after the MOSFET chapter to permit more ready comparison between the two transistors. However, the bipolar transistor chapter can be covered after the *p-n* junction chapters if desired without interruption of the continuity of the text.

Two sections on junction field-effect transistors (JFETs) were deleted from Chapter 5 because interest in JFETs for integrated circuit applications has become limited. Section 5.7 gave the derivation of the current-voltage behavior of JFETs, and Sec. 5.8 gave the application of PSpice to JFETs. These two sections may be found on the web site: www.ee.duke.edu/~hcc/DEVICE/JFET.html. The analysis of the JFET is almost identical to that given in Sec. 6.7 for the MESFET.

ACKNOWLEDGMENTS

Several reviewers provided helpful suggestions, including, Jiann S. Yuan of the University of Central Florida, Gary H. Bernstein of the University of Notre Dame, Pallab Bhattacharya of the University of Michigan, Bahram Nabet of Drexel University, and Kevin Brennan of Georgia Institute of Technology. I would like to thank Simon Sze for permission to use material given in Sec. 2.4.1 and some parts of the sections in Chapter 3 from his book *Semiconductor Devices: Physics and Technology.* The assistance of John Brews is greatly appreciated for providing the derivation of the inversion-layer thickness, the generalized definition of the threshold condition for MOS devices, and for useful comments on the MOS chapters.

Discussions of SiO_2 and MOSFETs with Hisham Massoud were very helpful. Michael A. Littlejohn provided useful comments on the overall text. The manuscript was carefully read by Michael Bergmann and he made many helpful suggestions. An additional debt of gratitude is owed to my many students who have asked thoughtful questions and found numerous errors in the text.

H. Craig Casey, Jr.
hcc@ee.duke.edu
September 1998

CONTENTS

CHAPTER 5
p-n JUNCTIONS: REVERSE BREAKDOWN AND JUNCTION CAPACITANCE 181

CHAPTER 6
SCHOTTKY-BARRIER DEVICES 226

CHAPTER 8
MOS FIELD-EFFECT TRANSISTORS 342

CHAPTER 9
BIPOLAR TRANSISTORS 427

APPENDIX I
INTRODUCTION TO PSPICE 491

INDEX 503

CHAPTER 1

INTEGRATED-CIRCUIT FAMILY TREE

1.1 THE EARLY YEARS

The evolution of semiconductor physics followed a rather uncertain path. Many unrelated discoveries and inventions led to the microelectronic devices and circuits which have become today's technology.[1] These discoveries and inventions make up a "family tree" and are some of the topics covered by the chapters of this book. A very interesting account of the scientists and engineers associated with the integrated-circuit family tree has been given by Queisser[2] in his book, *The Conquest of the Microchip.*

The first semiconductor device may be traced to Ferdinand Braun who discovered the crystal detector. In 1826, Georg Ohm had demonstrated that current flow in metallic conductors was directly proportional to the voltage and independent of the direction of current flow. This conduction behavior became known as Ohm's law. In 1876 Braun presented a lecture on his careful experiments on the conduction of current through the mineral crystal galena (lead sulfide). One contact was a metal point, while the other was a large surface contact. In Braun's experiments, the resistance became smaller as the current became greater and was the demonstration of a rectifier. This observation clearly deviated from Ohm's law. Braun made this discovery while undertaking systematic measurements of the flow of current through solids with the large collection of minerals at Würzburg University in Germany. This "detector" was later used as the receiver by Marchese Marconi for wireless telegraphy, and they shared the Nobel Prize in physics

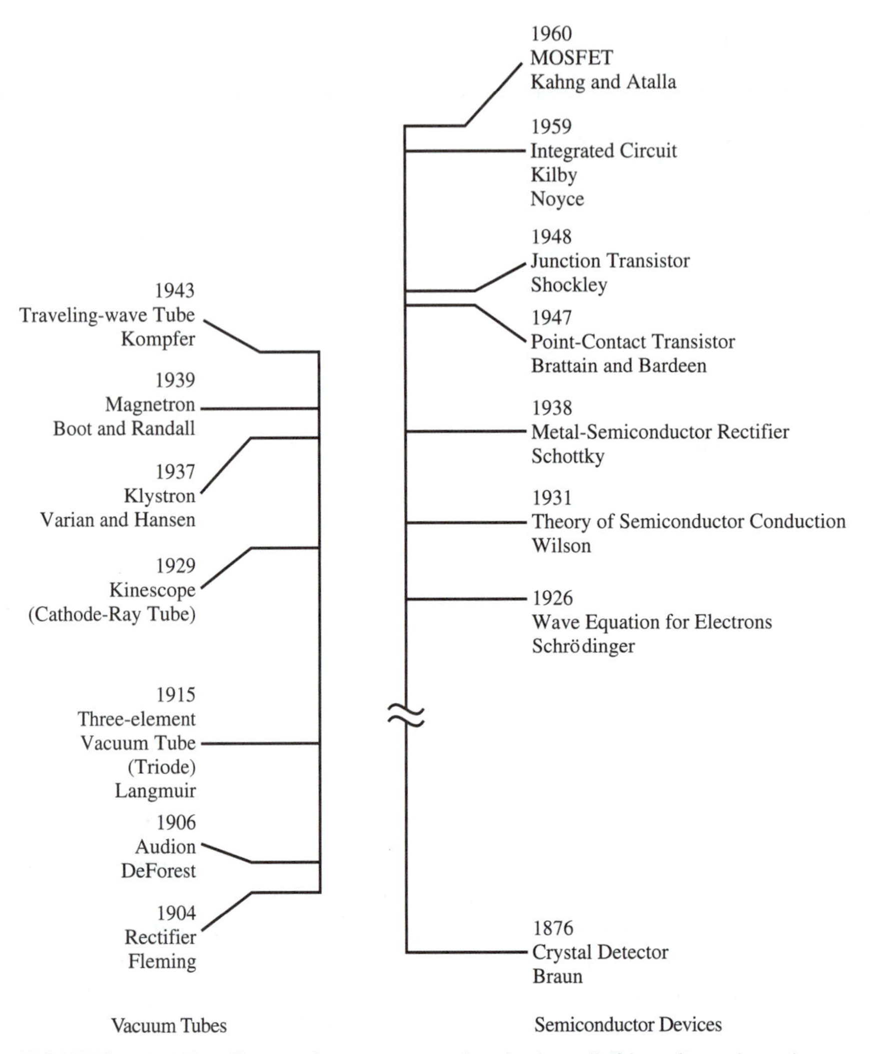

FIGURE 1.1 Family tree for vacuum tube devices (left) and semiconductor devices (right).

in 1909. Metal-semiconductor contacts are the subject of Chapter 6. The family tree for solid-state electronics is shown on the right in Fig. 1.1 and begins with Braun's crystal detector in 1876.

The crystal detectors were unreliable and were soon replaced by the vacuum tube. Thomas Edison's investigations of his carbon filament light bulbs led him in 1883 to add another electrode to the evacuated glass tube, and he found that he could control the current from the hot carbon filament by the voltage on this additional electrode. This behavior became known as the "Edison effect." The vacuum tube family tree begins with Fleming's vacuum tube rectifier in 1904, as shown on the left side of Fig. 1.1.

In 1906 Lee De Forest[3] inserted a third electrode between the cathode and anode to control the current, thus creating the *Audion.* Conduction was largely dependent on gas ionization. The three-terminal vacuum tube, later called the "triode," was largely the work of Irving Langmuir[4] and was well understood by 1915. This triode led to signal amplification and oscillators for signal sources. A variety of vacuum tubes provided the amplification devices upon which the electronics industry was built. The tubes still in use today, such as the cathode-ray tube, magnetron, klystron, and traveling-wave tube, are also shown in Fig. 1.1. Applications included radio, television, digital computers, radar (*ra*dio *d*etection *a*nd *r*anging), and undersea telephone cables from the United States to Europe. By comparison to solid-state devices now in use, vacuum tubes used considerable energy to heat the filament, were expensive to manufacture, and had a limited operating life.

During this almost one-half century when vacuum-tube electronics dominated the electronics industry, solid-state physicists were beginning to understand how electrons behaved in solids. The development of the understanding of electrons in solids, as needed to describe semiconductors, is given in Chapter 2. Many concepts and ideas led Erwin Schrödinger in 1926 to represent the wavelike nature of electrons in various force fields by what is now known as the *Schrödinger wave equation.* With the contributions of many solid-state physicists in the 1930s, concepts such as energy gaps and conduction by both holes and electrons were developed. Felix Bloch showed that impurities and lattice thermal vibrations resulted in the finite conductivity of metals. Alan Wilson[5–7] showed that there is a basic difference between a semiconductor, such as germanium, and a good conductor, such as silver, because impurities increase the resistivity of metals but decrease the resistivity of semiconductors. In 1938 Walter Schottky suggested that the rectification observed by Braun was the result of the formation of a stable space charge at the semiconductor surface. The semiconductors germanium and silicon were used during World War II with S-shaped metal "whiskers" as to give two terminal diode detectors for short wavelength radio and radar.

1.2 TRANSISTOR INVENTION

The steps that led to the invention of the transistor were described in a paper by William Shockley[8] which was entitled "The Path to the Conception of the Junction Transistor." The first steps that led to the transistor can be traced to 1936 when Shockley joined Bell Laboratories after receiving his Ph.D. at MIT. Mervin Kelley, Director of Research at Bell Laboratories, emphasized to Shockley his objective of introducing electronic switching in the telephone system. Shockley made a notebook entry in December 29, 1939 which describes a device now known as a field-effect transistor. A copper-oxide device was made, but it didn't work. His solid-state physics work was interrupted by World War II, and further work was not done on this copper-oxide device.

During World War II, semiconductor technology was stimulated by the need for point-contact detectors for radar, and silicon and germanium became the best-controlled semiconductors. In 1945, Shockley returned to Bell Laboratories and resumed his efforts to dream up a semiconductor amplifier. The first attempt was to place a metal plate above a thin layer of semiconductor which forms one plate of a parallel-plate capacitor. Charging the capacitor would alter the number of electrons on the semiconductor and modulate its conductance. No effect was observed. John Bardeen, who had joined Bell Laboratories in 1945, developed a theory for semiconductor surfaces that suggested the electrons were trapped at the surface.

This failure of the field effect led Bardeen, a theoretician, and Walter Brattain, an experimentalist, to study surface effects. Their experiments with metal probes on *n*-type germanium suggested that power amplification could be obtained if the metal contacts were spaced at distances of the order of 50 μm. Contacts were made by evaporating gold on a wedge and then separating the gold at the point of the wedge with a

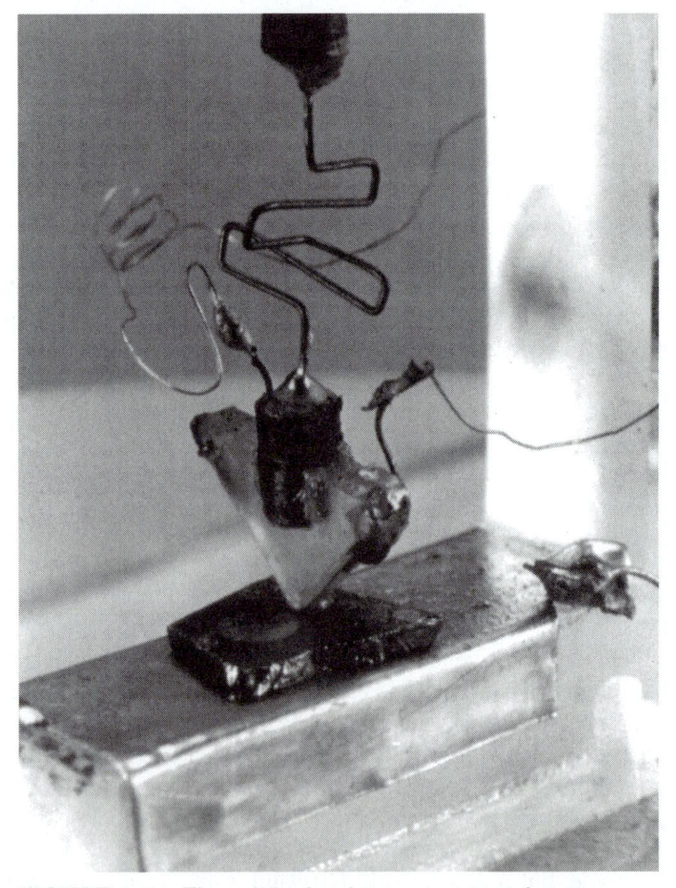

FIGURE 1.2 The original point-contact transistor structure comprising the plate of *n*-type germanium and two line-contacts of gold supported on a plastic wedge. The name "base," which arose from this structure, does not have functional significance as do "emitter" and "collector".[8] Photo courtesy of Lucent Technologies.

razor blade to make two closely spaced contacts.[8] This first point-contact transistor was demonstrated on December 23, 1947 and is shown in Fig. 1.2. John Pierce suggested the name *transistor* several months after the device was demonstrated. The prefix *trans-* emphasized that conduction traveled through the semiconductor, and the suffix *-istor* was often used to designate electronic elements. Like its predecessor, the point-contact detector, the point-contact transistor was unreliable.

Shockley's elation at Bardeen's and Brattain's success was mixed with the disappointment of not being a part of this success. He had continued to develop an understanding of how a *p-n* junction behaved, which is the subject of Chapters 4 and 5. Carrier transport and recombination, which are necessary to understand the *p-n* junction and the bipolar transistor, are given in Chapter 3. The final step for the conception of the junction transistor was the invention of minority-carrier injection, which was completed on January 23, 1948. A photograph of Shockley presenting the theory of the junction transistor is shown in Fig. 1.3. The junction transistor, now more commonly called the bipolar transistor, is the subject of Chapter 9.

Fabrication of a working junction transistor proved to be difficult. It was suggested that the name "persistor" be used for the junction transistor, because persistence was certainly needed to produce it. Working junction transistors were finally produced in the spring of 1950 by Morgan Sparks and Gordon Teal. Bardeen, Brattain, and Shockley were awarded the 1956 Nobel Prize in physics for inventing the transistor and for semiconductor research.

FIGURE 1.3 Shockley presenting the theory of the junction transistor probably during the "persistor" period of 1949 or early 1950.[8] Photo courtesy of Lucent Technologies.

The transistor was not met with enthusiasm by engineers familiar with vacuum-tube circuits. A miniaturized hearing aid was the first commercial product to use transistors. By 1952, only eight firms were producing germanium transistors.

1.3 THE INTEGRATED CIRCUIT

In the 1950s electronic circuits were limited because vacuum tubes were large and expensive, and their limited operating life required frequent replacement. Vacuum tubes cost more than the passive components such as resistors and capacitors, and cooling was often required. In 1958 Jack Kilby at Texas Instruments realized that only semiconductors were required to provide the active devices, resistors, and capacitors, and they all could be made from the same material. The resistors could be made from bulk semiconductors and capacitors could be made from *p-n* junctions. Kilby's first integrated circuits in 1958 used germanium with etched mesa structures to separate the components, which were electrically connected by bonded gold wires and formed a phase-shift oscillator.[9]

Several technological developments led to commercially practicable integrated circuits. The investigations of Cal Fuller and Jim Ditzenberger provided detailed information concerning the diffusion of group-III and group-V elements in silicon.[10] This understanding of diffusion in silicon was applied by Morris Tanenbaum and David Thomas to prepare *npn* silicon transistors with diffused emitter and base regions.[11] The next key step was the demonstration by Carl Frosch and Link Derick that SiO_2 could be used to mask the diffusion of the useful donors and acceptors in silicon.[12] Another significant step was the adaptation of photoengraving[13] by Andrus and Bond to form intricate patterns into the semiconductor surfaces, which is now known as photolithography. These four technological developments were produced at Bell Laboratories. At Fairchild Semiconductor in 1958, Jean Hoerni developed a transistor in which all the diffused *n* and *p* regions of the transistor were patterned by photolithography and were accessible for contacting on the top surface, which was protected by SiO_2 passivation.[14] This configuration, with all layers accessible on the top surface, is termed a *planar* structure and is used in all integrated circuits. Silicon became the preferred semiconductor material for transistors not only because of its decreased sensitivity to temperature but also because the SiO_2 native oxide had been shown to be a nearly ideal insulator and also prevented surface leakage currents.

Up to this time, the individual transistors on a wafer were separated by sawing or cleaving, and had tiny wires bonded to pins in the transistor package. In 1959 at Fairchild Semiconductor, Bob Noyce accomplished the separation of the circuit elements in the silicon wafer by using *p-n* junctions for isolation, and the circuit elements were interconnected by a conducting film of evaporated metal over an insulating SiO_2 layer.[15] The metal layer was patterned by photolithography and etched to give the desired interconnection pattern. These first Si integrated circuits were all based on planar bipolar transistors. It took almost 10 more years to develop techniques to produce stable SiO_2-Si interfaces to permit *m*etal-*o*xide *s*emiconductor (MOS) integrated circuits.

Although the early applications of SiO_2 were for patterning[14] and passivation of *p-n* junctions and bipolar transistors,[16] John Moll[17] proposed MOS capacitor structures in 1959. The basic behavior of a MOS device can be understood by considering the MOS capacitor which is the topic of Chapter 7. In 1960 Dwang Kahng and John Atalla[18] used a thermally oxidized Si structure to form a surface field-effect transistor which is now the *m*etal-*o*xide *s*emiconductor *f*ield-*e*ffect *t*ransistor (MOSFET). This initial MOSFET is shown in Fig. 1.4. The MOSFET is the subject of Chapter 8. A very detailed description of the evolution of the MOSFET was given by Sah.[19]

The next critical step was the concept of using a polycrystalline Si gate which serves as a self-aligning diffusion mask for the source and drain and subsequently as a self-aligned gate electrode. This self-aligned polycrystalline Si gate structure was developed by Bob Kerwin, Don Klein, and Jack Sarace[20–23] at Bell Laboratories in 1966. During the diffusion of source and drain regions, the polycrystalline Si also becomes sufficiently conductive that no gate metallization is required. Bob Noyce wrote in Bob Kerwin's laboratory notebook, "This (self-aligned gate) is the invention which made very large-scale integration (VLSI) a practical reality." The self-aligned gate greatly improved yield and reliability, which resulted in a rapid increase in the number of transistors on logic and memory chips.

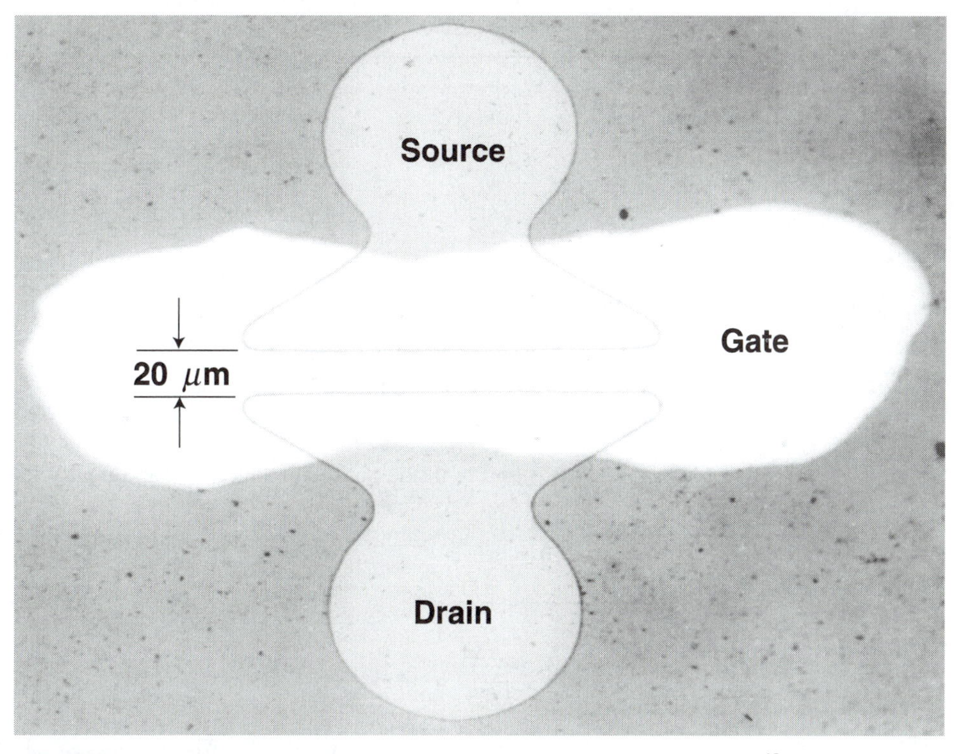

FIGURE 1.4 The first MOSFET demonstrated by Kahng and Atalla.[18] Photo courtesy of Lucent Technologies.

Integrated-circuit development proceeded at a very rapid pace, and feature sizes became smaller. This reduction in size of circuit elements reduces the cost and improves the basic performance of the device because delay times are proportional to dimensions of the circuit elements. During the late 1980s circuit feature sizes were reduced below 1 μm. Dynamic random-access memory (DRAM) development illustrates the continuing evolution of integrated-circuit technology. The 64-megabit (Mb) DRAM with minimum feature sizes of 0.35 μm became commercially available in 1994, while production of the 256-Mb DRAM with 0.25-μm minimum feature sizes began in 1996. The next generation, the 1-Gb DRAM with minimum feature sizes near 0.15 μm, can be expected to be available near the year 2000.

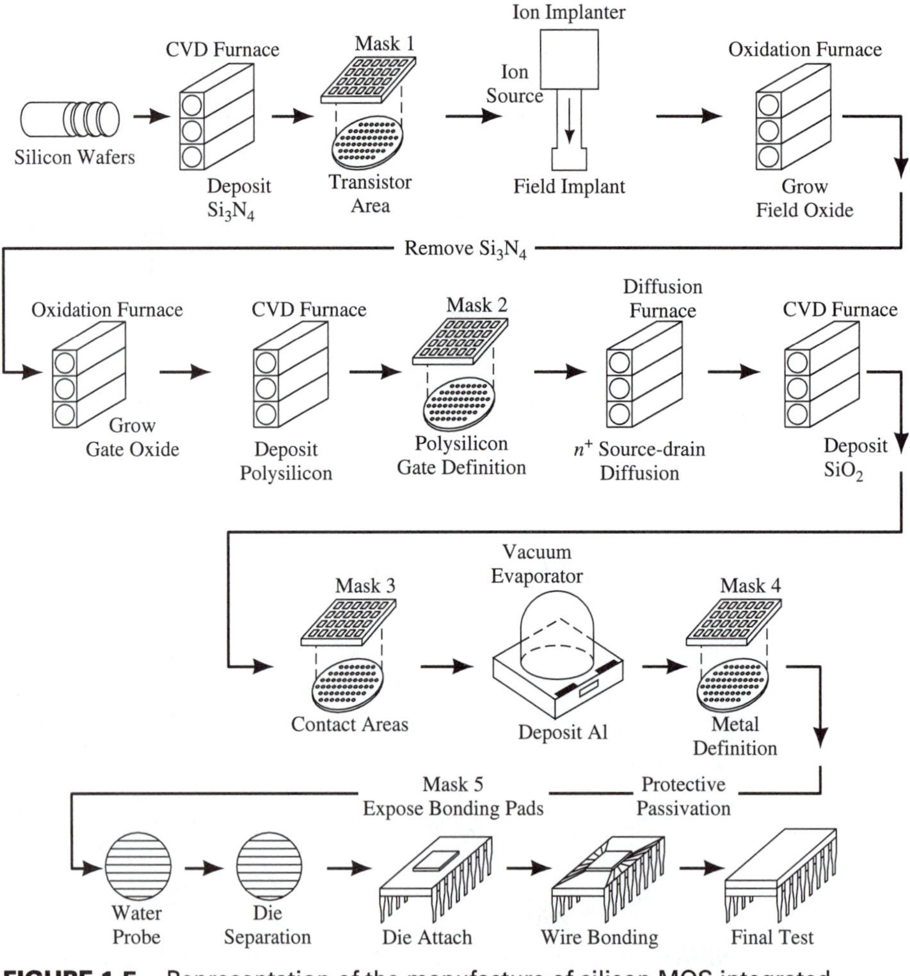

FIGURE 1.5 Representation of the manufacture of silicon MOS integrated circuits.[24]

The manufacturing process for Si MOS integrated circuits is outlined in Fig. 1.5. Design of the circuit is done with the aid of a workstation which also simulates the circuit to verify the design and to prepare a set of photolithographic masks. Silicon wafers up to 8 inches in diameter (next generation, 12 inches) and gallium arsenide (GaAs) wafers up to 6 inches in diameter are now used. Many texts are devoted to the detailed understanding of the numerous processes illustrated in Fig. 1.5, but space limitations prevent consideration of fabrication in this text.

The next figure shows an example of an integrated circuit with submicron feature sizes. Figure 1.6 is a micrograph of the metallization for connecting the components on the integrated circuit and was taken with a *s*canning *e*lectron *m*icroscope (SEM). This structure represents a complexity with small feature sizes never envisioned by the inventors of the transistor shown in Fig. 1.2.

Chapters 2 through 9 contain topics necessary for the understanding of the devices used in integrated circuits. Because students with diverse backgrounds, such as in materials science, physics, computer science, as well as electrical engineering, study devices for integrated circuits, all topics are derived from the relevant laws or postulates. These concepts are applied to realistic but one-dimensional derivations. Both ideal and nonideal phenomena are introduced. Most integrated circuits are produced with Si. However, combinations of group-III elements and group-V elements, which are called

SEM of Local Interconnects and First Level Aluminum in a CMOS SRAM Array

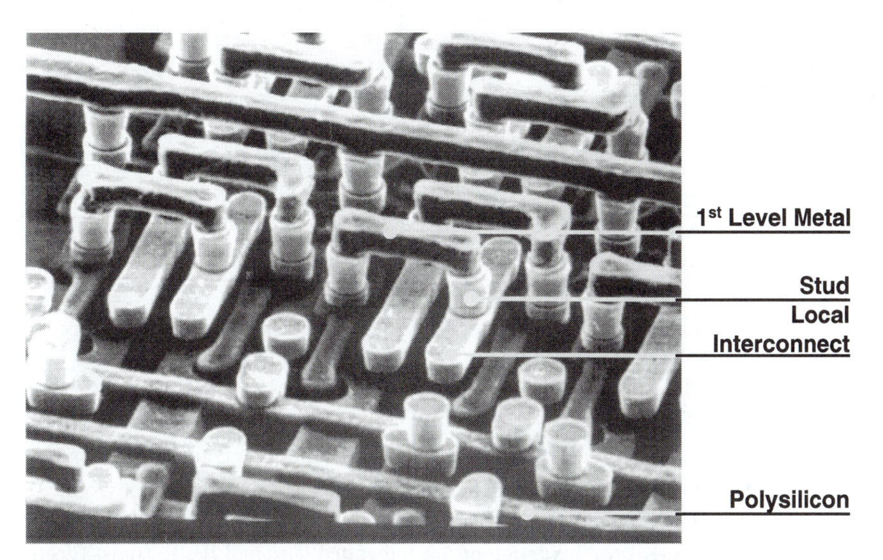

FIGURE 1.6 Scanning electron microscope (SEM) micrograph of the metallization on an integrated circuit with submicron feature sizes. The insulating layers of SiO_2 have been removed to reveal the metalization which connects the circuit elements.[25] Copyright 1995 International Business Machines Corporation. Reprinted with permission of the IBM Journal of Research and Development, Volume 39, Number 4.

the three-five (III-V) compound semiconductors, are used in specialized applications. The most common III-V compound semiconductors are based on GaAs and indium phosphide (InP) and are used in discrete devices as well as in *o*pto*e*lectronic *i*ntegrated *c*ircuits (OEICs) and *m*onolithic *mi*crowave *i*ntegrated *c*ircuits (MMICs). Because of the growing applications of devices based on semiconductors other than Si, III-V compound semiconductors are introduced in this textbook, although the emphasis is on Si-based devices for integrated circuits. The devices presented in this textbook are limited to those most commonly encountered in integrated circuits and are designated by the chapter titles. The book by Ng[26] provides an overview of 67 semiconductor devices, and includes the device characteristics, brief descriptions of how the device works, and examples of the device applications.

Representation of the devices introduced in Chapters 4 through 9 by the *s*imulation *p*rogram with *i*ntegrated-*c*ircuit *e*mphasis (SPICE) is given to relate the SPICE parameters to the device physics. Some topics are included or omitted based on the relevance to use with SPICE. A commercial version of SPICE named PSpice is used in this textbook and runs on personal computers. PSpice now has many enhanced features, such as the graphics output called probe, schematic capture, an extensive library of components, and models for short-channel MOSFETs. A free limited capability version is available from OrCAD.[27] MicroSim Corporation, the developer of PSpice, merged with OrCAD in January, 1998. The free version of PSpice (Release 8.0) is available on the Wiley World Wide Web site: http://www.wiley.com/college/casey. This free version is adequate for any tasks required in this text.

REFERENCES

1. M. Riordan and L. Hoddeson, *Crystal Fire: The Birth of the Information Age* (Horton, New York, 1997).
2. H. Queisser, *The Conquest of the Microchip* (Harvard University Press, Cambridge, 1988).
3. L. De Forest, "The Audion," Trans. AIEE **25**, 735 (1907).
4. I. Langmuir, "The Pure Electron Discharge," Proc. IRE **3**, 261 (1915).
5. A. H. Wilson, "The Theory of Electronic Semi-Conductors," Proc. R. Soc. Lond. **A133**, 458 (1931).
6. A. H. Wilson, "The Theory of Electronic Semi-Conductors—II," Proc. R. Soc. Lond. **A134**, 277 (1931).
7. A. H. Wilson, *The Theory of Metals* (Cambridge University Press, London, 1936).
8. W. Shockley, "The Path to the Conception of the Junction Transistor," IEEE Trans. Electron Dev. **ED-23**, 597 (1976).
9. J. S. Kilby, "Invention of the Integrated Circuit," IEEE Trans. Electron Devices **ED-23**, 648 (1976).
10. C. S. Fuller and J. A. Ditzenberger, "Diffusion of Boron and Phosphorus into Silicon," J. Appl. Phys. **25**, 1439 (1954).
11. M. Tanenbaum and D. E. Thomas, "Diffused Emitter and Base Silicon Transistor," Bell Syst. Tech. J. **35**, 1 (1956).

12. C. J. Frosch and L. Derick, "Surface Protection and Selective Masking during Diffusion in Silicon," J. Electrochem. Soc. **104**, 547 (1957).
13. J. Andrus and W. L. Bond, "Photoengraving in Transistor Fabrication," in *Transistor Technology, Vol. III,* ed. by F. J. Biondi (Van Nostrand, New York, 1958) pp. 151–162.
14. J. A. Hoerni, "Planar Silicon Transistors and Diodes," IRE Electron Devices Meeting, Washington, D.C. (1960).
15. R. N. Noyce, U.S. Patent No. 2,981,877, filed July 3, 1959.
16. M. M. Atalla, E. Tanenbaum, and E. J. Scheibner, "Stabilization of Silicon Surfaces by Thermally Grown Oxides," Bell Syst. Tech. J. **38**, 749 (1959).
17. J. L. Moll, "Variable Capacitance with Large Capacity Change," Wescon Convention Record, Part 3, p. 32 (1959).
18. D. Kahng and M. M. Atalla, "Silicon-Silicon Dioxide Field Induced Surface Devices," *IRE-AIEE Solid-State Device Res. Conf.* (Carnegie Inst. of Technol., Pittsburgh, 1960).
19. C. T. Sah, "Evolution of the MOS Transistor–From Conception to VLSI," Proc. IEEE **76**, 1280 (1988).
20. R. E. Kerwin, laboratory notebook dated May 27, 1966.
21. J. C. Sarace, R. E. Kerwin, D. L. Klein, and R. Edwards, Metallurgical Society (AIME) Meeting, New York (1967).
22. J. C. Sarace, R. E. Kerwin, D. L. Klein, and R. Edwards, "Metal-Nitride-Oxide-Silicon Field-Effect Transistors with Self-Aligned Gates," Solid-State Electron. **11**, 653 (1968).
23. R. E. Kerwin, D. L. Klein, and J. C. Sarace, "Method for Making MIS Structures," U.S. Patent No. 3,475,234, Oct. 28, 1969.
24. D. A. Hodges and H. G. Jackson, *Analysis and Design of Digital Integrated Circuits,* 2nd ed. (McGraw-Hill, New York, 1988), p. 16.
25. IBM J. Res. Devel. **39**, cover (July 1995).
26. K. K. Ng, *Complete Guide to Semiconductor Devices* (McGraw-Hill, New York, 1995).
27. OrCAD, 9300 S. W. Nimbus Avenue, Beaverton, OR 97008, USA. (503) 671-9500 and http://www.orcad.com

CHAPTER 2

ELECTRONS IN SOLIDS

2.1 INTRODUCTION

All semiconductor devices depend on the behavior of electrons in a solid. The behavior of electrons in free space as well as in solids is represented by the Schrödinger wave equation. The Schrödinger wave equation is not convenient for discussion of semiconductor devices, but it may be used to obtain the familiar quantities which adequately describe semiconductors, such as the energy gap and the intrinsic carrier concentration. If Si was the only commercially important semiconductor, then a table and brief description of the important properties would be an adequate introduction before a detailed consideration of semiconductor-device behavior. Specialized applications of many other semiconductor materials make a vast array of unique products possible. For example, the compact disk (CD) player uses a III-V (three-five) compound semiconductor laser to read the disk. Other semiconductor materials include the group-four (IV) compound semiconductor SiC, and II-VI (two-six) compound semiconductors such as zinc selenide (ZnSe). To be able to deal with this growing list of semiconductor materials, a more complete description of properties such as the energy gap is essential. These properties can be illustrated by considering both Si and the III-V compound gallium arsenide (GaAs).

The wave–particle duality for the electron is introduced in Sec. 2.2 by a discussion of the hydrogen atom. The orbit radius and ionization energy for the single electron

Schrödinger wave equation

effective mass

density of states

Fermi function

effective density of states

intrinsic carrier concentration

bound to the hydrogen nucleus are useful in describing the ionization energy of impurities in semiconductors. The Schrödinger wave equation is considered in Sec. 2.3 to permit the description of the motion of the electrons in a single-crystal solid. From the Schrödinger wave equation, the relation between the electron energy E and the wave vector k, which is also called the phase or propagation constant, will be established. The relation between E and k is shown in Sec. 2.4 to be represented by an *effective mass* for a free carrier in a semiconductor. In fact, many properties of real solids are summarized by this E versus k relationship. To represent the number of free electrons in a semiconductor, it is necessary to derive the concept of the *density of states,* which is given in Sec. 2.5. At a given temperature, the electrons are distributed among the allowed energy levels as described by the Fermi-Dirac distribution function. Section 2.6 introduces the Fermi-Dirac distribution function and illustrates the useful concept of the *effective density of states.* The electron concentration of a pure semiconductor, the intrinsic carrier concentration, is derived in Sec. 2.7. In Sec. 2.8, the control of the concentration of free carriers in the semiconductor by the introduction of small amounts of impurities into the semiconductor is presented. The significant concepts introduced in this chapter are summarized in Sec. 2.9, together with the physical constants, Si and GaAs parameters, and the expressions useful for representing free carriers in semiconductors.

The first part of Appendix A at the end of this chapter gives the expressions for the fraction of the donors or acceptors which are ionized as a function of the impurity ionization energy, the position of the Fermi level, and the temperature. In the second part of Appendix A, the concentration dependence of the ionization energy is presented. For devices considered in this text, the ionized impurity concentration will be given rather than the total impurity concentration which may exceed the ionized impurity concentration.

2.2 THE HYDROGEN ATOM

Gauss's law in integral form: $\epsilon_0 \oint \mathscr{E} \cdot d\mathbf{a} = \int \rho\, dv = q$ *and* $\mathscr{E} = q/4\pi\epsilon_0 r^2$. *Then,* $\mathscr{F} = -q\mathscr{E}$

Before the structure of the atom was well known, scattering experiments led Rutherford[1] to propose that the nucleus was the center of the atom. This nucleus was composed of positive protons and neutral neutrons, with electrons moving in orbits around the nucleus. For a neutral atom, the number of electrons equals the number of protons. This model for the hydrogen atom is shown in Fig. 2.1. For the electron in Fig. 2.1, the balance of the centrifugal force and the Coulomb force gives the following equation:

$$\frac{m_0 v^2}{r} = \frac{q^2}{4\pi\epsilon_0 r^2}, \tag{2.1}$$

where q is the charge, m_0 is the free electron mass, r is the orbit radius, and ϵ_0 is the dielectric constant of free space. From classical electromagnetic theory, charged particle acceleration (v^2/r) would result in light emission. With the loss of energy because of radiation, the electron radius would decrease and the electron would spiral into the nucleus. This behavior would predict a continuous range of frequency of the emitted

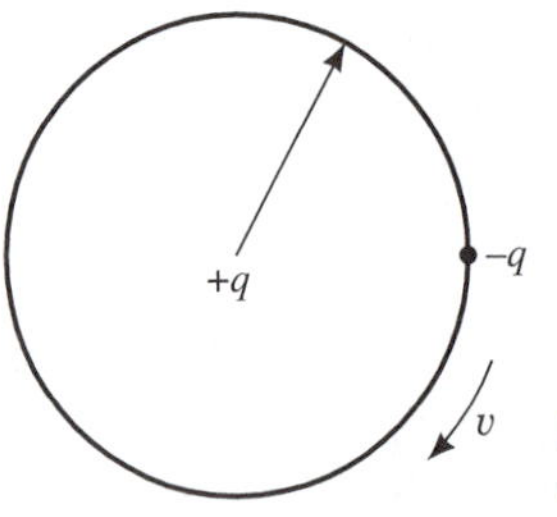

FIGURE 2.1 Illustration of an electron moving in a circular orbit around a proton.

light for an excited hydrogen gas. However, experimentally, the hydrogen emission spectrum was known to consist of discrete emission lines.[2]

To resolve this inconsistency, Bohr[3] proposed that the laws of classical electrodynamics must be modified by utilizing Planck's empirical formulation of quantized radiation.[4] He postulated that an electron can exist only in certain orbits at a definite energy and an electron can change from a higher energy orbit (E_2) to a lower energy orbit (E_1) by emitting a photon of energy $h\nu$, where h is Planck's constant (4.1354×10^{-15} eV-s) and ν is the frequency:

$$E_2 - E_1 = h\nu. \tag{2.2}$$

Another concept necessary for understanding the hydrogen atom was the proposal by de Broglie[5] that particles should be assigned a wavelength,

$$\boxed{\lambda = h/|\mathbf{p}|}\,, \tag{2.3}$$

where λ is the wavelength of the electron as a wave, and $|\mathbf{p}|$ is the magnitude of the electron momentum. Electrons were shown to undergo diffraction as expected for a wave in an experiment by Davisson and Germer.[6]

The orbits postulated by Bohr suggest that the orbit length should be an integer multiple of the wavelength:

$$2\pi r = n\lambda = nh/|\mathbf{p}| = nh/(m_0|\mathbf{v}|)\,, \tag{2.4}$$

where n is an integer number and Eq. (2.3) has been used to relate λ to $|\mathbf{p}|$. With

$$|\mathbf{v}| = \frac{nh}{2\pi m_0 r} \tag{2.5}$$

from Eq. (2.4), and substituting into Eq. (2.1),

$$\boxed{r = (n^2h^2\epsilon_0)/(\pi m_0 q^2) \simeq 0.529 \times 10^{-8} n^2 \text{ cm}}\,, \tag{2.6}$$

where r is the stable radius of the nth orbit. The radius of the ground state for $n = 1$ is the first Bohr orbit and is 0.529×10^{-8} cm.

Therefore, for the circular orbits described by Eq. (2.4), an integral number of whole wavelengths must be fitted around the circumference. Although an oversimplifi-

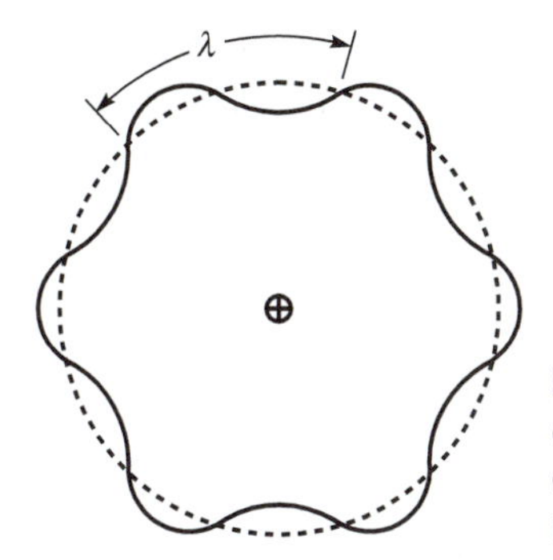

FIGURE 2.2 An electron orbit with a circumference of six wavelengths. The dashed line is the classical orbit and the solid line represents the quantum mechanical wave function with the wavelength λ.[7]

cation, an example of an orbit with an integral number of wavelengths is illustrated in Fig. 2.2.

The total energy of the electron is equal to the sum of its potential and kinetic energy,

$$E = \frac{-q^2}{4\pi\epsilon_0 r} + \frac{m_0 v^2}{2} = \frac{-q^2}{8\pi\epsilon_0 r}, \tag{2.7}$$

with Eq. (2.1) for $m_0 v^2$. With r from Eq. (2.6), the energy E_n of the nth orbit in Eq. (2.7) becomes

$$E_n = \frac{-q^4 m_0}{8\epsilon_0^2 h^2 n^2}. \tag{2.8}$$

Then, the change in energy corresponding to a transition from an initial orbit n_{in} to a final orbit n_f is

$$\Delta E = \frac{q^4 m_0}{8\epsilon_0^2 h^2}\left[\frac{1}{n_f^2} - \frac{1}{n_{in}^2}\right]. \tag{2.9}$$

The binding energy of the electron to the nucleus would be for $n_{in} = 1$ and $n_f = \infty$, which gives

$$\Delta E = -13.6 \text{ eV}. \tag{2.10}$$

The Bohr radius given by Eq. (2.6) and the binding energy given by Eq. (2.10) will be useful concepts when considering impurity elements in semiconductors.

EXAMPLE 2.1 To illustrate the conversion of eV to ergs, find the velocity of a free electron with a mass $m_0 = 9.1 \times 10^{-28}$ g and a kinetic energy of 1 eV. Note that the conversion of eV to joules (j) or ergs is given by 1 eV $= 1.6\times10^{-19}$ j $= 1.6\times10^{-12}$ erg.

Solution KE $= \frac{1}{2}m_0 v^2 = 1\text{ eV} \times (1.6 \times 10^{-12}\text{ erg})/1\text{ eV} = 1.6 \times 10^{-12}\text{ g cm}^2/\text{s}^2$. Therefore, $v = (2 \times 1.6 \times 10^{-12}\text{ g}/9.1 \times 10^{-28}\text{ g})^{1/2}$ cm/s $= 5.930 \times 10^7$ cm/s. In calculations related to semiconductors, three or four significant figures are sufficient. ■

2.3 SCHRÖDINGER WAVE EQUATION

2.3.1 The Wave Equation

Schrödinger[8] utilized de Broglie's ideas of the wavelike nature of electrons to understand effects that could not previously be understood in classical physics. This representation permitted a description of effects not previously explainable in classical terms. The Schrödinger wave equation in general depends on position and time. When the potential energy $U(x)$ depends only on position and not on time, the Schrödinger wave equation can be separated into time- and position-dependent equations. The position-dependent equation is a three-dimensional equation, but in one dimension, it can be written as

$$\frac{d^2\psi(x)}{dx^2} + \frac{2m_0}{h^2/4\pi^2}[E - U(x)]\psi(x) = 0\,, \tag{2.11}$$

where $\psi(x)$ is the wave function and E is the total electron energy. It is convenient to represent $h/2\pi$ by $\hbar$, which is called "*h bar.*"

The physical interpretation of the wave function was proposed by Born[9] to be the probability of finding the electron between x and $x + \Delta x$, and is given by $|\psi(x)|^2 \Delta x$. This probability density is given by the absolute value of $\psi(x)$ squared, and satisfies the normalization condition that the total integral has to be unity:

$$\int_{-\infty}^{\infty} |\psi(x)|^2\, dx = 1\,. \tag{2.12}$$

To solve the second-order differential equation given in Eq. (2.11), boundary conditions must be given to determine two arbitrary constants. In addition to the condition expressed by Eq. (2.12), $\psi(x)$ and $d\psi(x)/dx$ must be continuous, finite, and single-valued for all x. Determination of $\psi(x)$ gives the motion parameters such as energy and momentum.

2.3.2 Free-Electron Solution

The simplest solution to the Schrödinger wave equation is for an electron in free space where $U(x) = 0$. Equation (2.11) then reduces to

$$\frac{d^2\psi(x)}{dx^2} + \frac{2m_0}{\hbar^2}E\psi(x) = 0\,. \tag{2.13}$$

Assume a sinusoidal solution

$$\psi(x) = A\cos(-kx), \tag{2.14}$$

with the arbitrary constants A and k. Substitution of Eq. (2.14) back into Eq. (2.13) gives

$$k^2 = 2m_0E/\hbar^2, \tag{2.15}$$

or

$$\boxed{E = \hbar^2 k^2/2m_0}. \tag{2.16}$$

All the energy is kinetic energy, so that

$$E = \frac{\hbar^2 k^2}{2m_0} = \frac{m_0 v^2}{2} = \frac{p^2}{2m_0}, \tag{2.17}$$

and

$$\boxed{|\mathbf{p}| = \hbar|\mathbf{k}|}, \tag{2.18}$$

which relates k to the momentum p and energy E. Replacement of $|\mathbf{p}|$ from the de Broglie relation given in Eq. (2.3) gives

$$\boxed{|\mathbf{p}| = m_0|\mathbf{v}| = h/\lambda = \hbar|\mathbf{k}|}, \tag{2.19}$$

which shows that $h|\mathbf{k}|/2\pi = h/\lambda$ or $\boxed{|\mathbf{k}| = 2\pi/\lambda}$. The parabolic relationship between E and k is given by Eq. (2.17) and is shown by the plot in Fig. 2.3. This E vs. k diagram for the free electron, as represented by Eqs. (2.17) and (2.18), will be modified for the case of an electron in a crystalline solid.

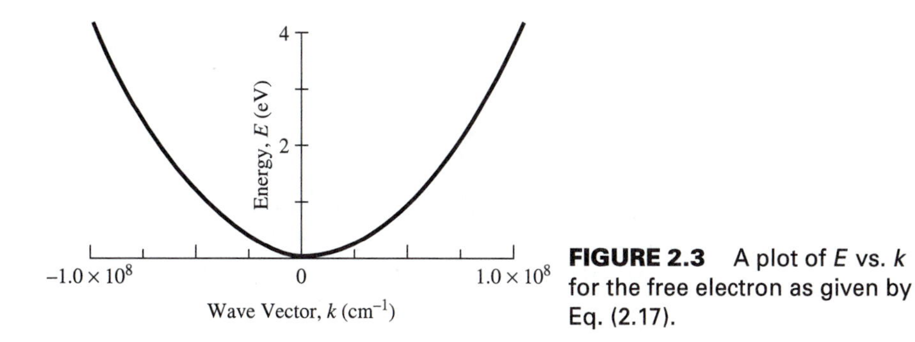

FIGURE 2.3 A plot of E vs. k for the free electron as given by Eq. (2.17).

Inclusion of time dependency in the sinusoidal solution of the Schrödinger wave equation, Eq. (2.14), gives

$$\psi(x, t) = A\cos(\omega t - kx)\,. \tag{2.20}$$

This equation represents a wave traveling in the $+x$ direction and can readily be illustrated by plotting Eq. (2.20). First, for $t = 0$, $\psi(x, 0)$ is simply $A\cos(-kx)$. With x in units of π/k, the resulting $\psi(x, 0)$ is plotted in Fig. 2.4(a). In Fig. 2.4(b), t has been taken as $(\pi/4\omega)$ to illustrate that the wave travels in the positive x direction. The solution becomes $A\cos(\pi/4 - kx)$, and the maximum in $\psi(x, t)$ has traveled in the x direction a distance of $(\pi/4k)$ in the time $(\pi/4\omega)$. Increasing t to $2\pi/\omega$ moves the wave a distance of one wavelength, as shown in Fig. 2.4(c).

electron wavelength

It is shown in Fig. 2.4 that the wavelength λ is given by $x = 2\pi/k$. Therefore,

$$\lambda = 2\pi/|\mathbf{k}| = h/(m_0|\mathbf{v}|)\,, \tag{2.21}$$

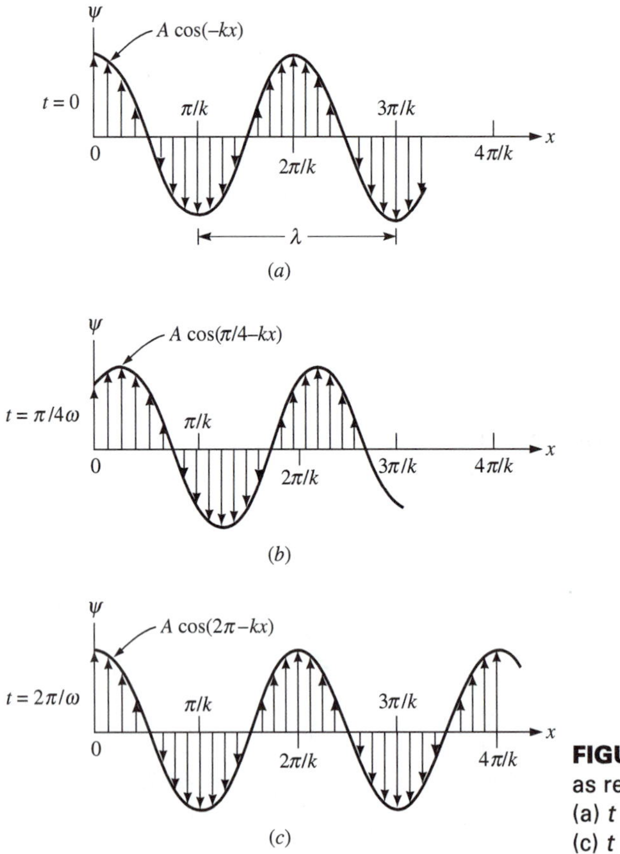

FIGURE 2.4 Traveling wave as represented by Eq. (2.20). (a) $t = 0$. (b) $t = \pi/4\omega$. (c) $t = 2\pi/\omega$.

or

$$\boxed{|\mathbf{k}| = 2\pi/\lambda}. \tag{2.22}$$

This relationship is also given by Eq. (2.19). The quantity k which appears in the traveling wave represented by Eq. (2.20) is called the wave vector and also the phase or propagation constant because it gives the change in phase per unit length, and is generally given in units of cm^{-1}.

By considering the free electron, the relation between the electron momentum as a particle, $m_0|\mathbf{v}|$, and the wave vector $|\mathbf{k}|$ as represented by Eq. (2.19) was found to be $|\mathbf{p}| = m_0|\mathbf{v}| = \hbar|\mathbf{k}|$. Although real crystals do not have the parabolic relationship between E and k as shown in Fig. 2.3, the most convenient summary of the optical and conduction properties of a real solid is a representation of E vs. k.

2.4 ELECTRONS IN CRYSTALS

2.4.1 Crystal Structure

To obtain the E vs. k diagram for electrons in semiconductor crystals, it is necessary to consider the arrangement of atoms which influences the electron motion. The semiconductor materials considered here are single crystals with the atoms arranged in a three-dimensional periodic fashion. The periodic arrangement of atoms in a crystal is called a *lattice*. For a given crystal, there is a *unit cell* that is representative of the entire lattice. By repeating the unit cell throughout the crystal, the entire lattice can be generated.

Figure 2.5 shows some basic cubic-crystal unit cells. Three basic cubic-crystal unit cells are shown in Fig. 2.5. In Fig. 2.5(a) a simple cubic crystal has each corner of

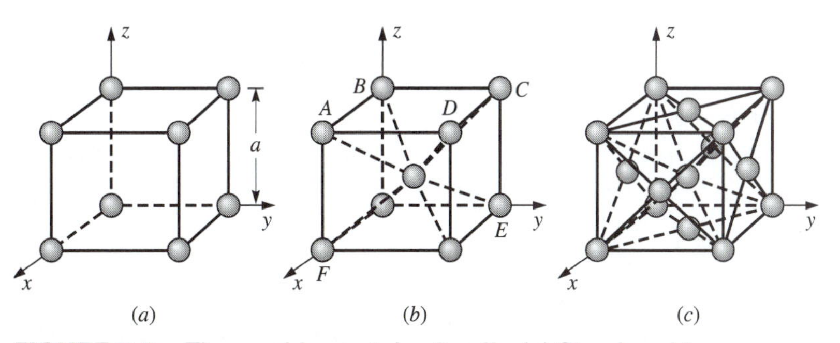

FIGURE 2.5 Three cubic-crystal unit cells. (a) Simple cubic. (b) Body-centered cubic. (c) Face-centered cubic.[10]

Section 2.4.1 is taken in part from S. M. Sze, *Semiconductor Devices: Physics and Technology* (Wiley, New York, 1985) pp. 3–7.

the cubic lattice occupied by an atom. The dimension a is called the *lattice constant.* The simple cubic lattice is not commonly encountered. Figure 2.5(b) is a body-centered cubic (bcc) crystal, where an atom is located at the center of the cube. Sodium and tungsten have bcc lattices. In the face-centered cubic (fcc) crystal shown in Fig. 2.5(c), there is an atom at the center of each cubic face. A large number of elements exhibit the fcc lattice, including aluminum, copper, gold, and platinum.

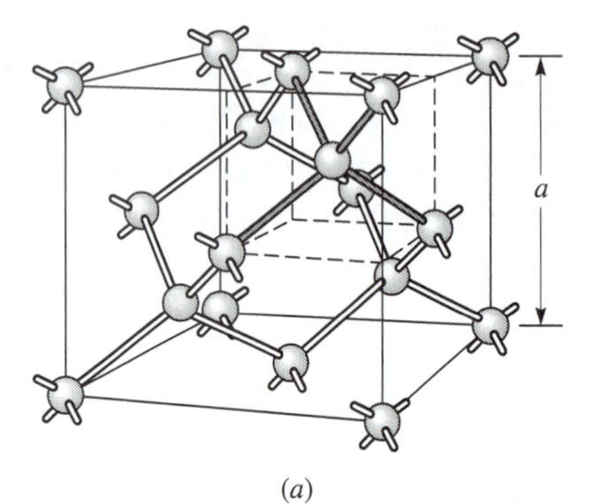

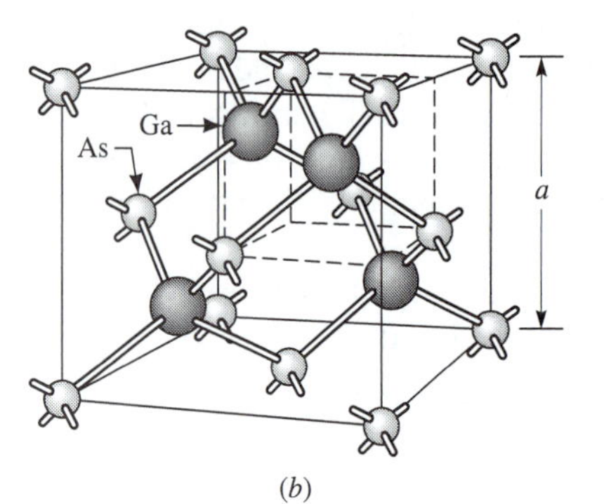

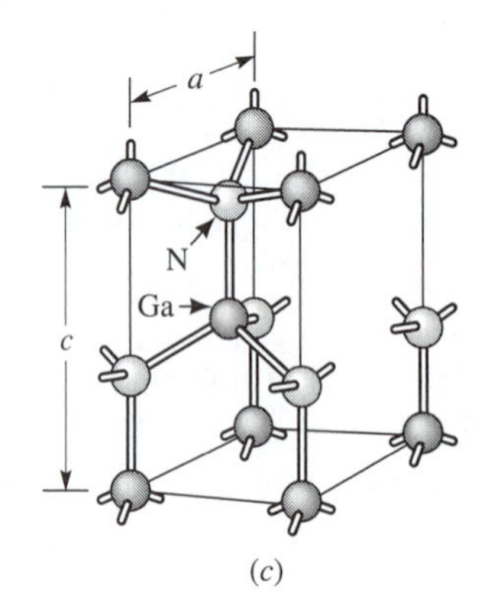

FIGURE 2.6 Tetrahedrally coordinated crystals (a) Diamond lattice with lattice constant *a* for Si or Ge, (b) Zinc-blende lattice with lattice constant *a* for GaAs (Ref. 10), and (c) Hexagonal wurtzite lattice with base a and height c for GaN.

The elemental semiconductors, silicon (Si) and germanium (Ge), have a *diamond-lattice* structure, as shown in Fig. 2.6(a). This structure also belongs to the cubic-crystal family and can be seen as two interpenetrating fcc sublattices with one sublattice displaced from the other by one quarter of the distance along a diagonal of the cube (i.e., a displacement of $a\sqrt{3}/4$). All atoms are identical in a diamond lattice, and each atom in the diamond lattice is surrounded by four equidistant nearest neighbors that lie at the corners of a tetrahedron [refer to the spheres connected by darkened bars in Fig. 2.6 (a)]. Most of the III-V compound semiconductors (e.g., GaAs and InP) have a *zinc-blende lattice*, shown in Fig. 2.6(b), which is identical to a diamond lattice except that one fcc sublattice has column III atoms (Al, Ga, or In) and the other has column V atoms (N, P, As, or Sb). The most extensively studied group-III nitrides have the hexagonal wurtzite lattice, as shown in Fig. 2.6(c) for GaN.

EXAMPLE 2.2 At 300 K, the lattice constant for Si is 5.430951 Å. Calculate the number of Si atoms per cubic centimeter and the density of Si at room temperature.

Solution Note that each corner atom is shared with eight adjoining unit cells, which gives a contribution of one atom for the eight shared atoms with eight adjoining unit cells at the corners 8(1/8), three atoms for the six shared atoms in the unit cell faces 6(1/2), and four atoms inside the unit cell for a total of eight atoms per unit cell:

$$\frac{8}{a^3} = \frac{8}{(5.431 \times 10^{-8})^3} = 5 \times 10^{22} \text{ atoms/cm}^3$$

$$\text{Density} = \frac{\text{no. atoms/cm}^3 \times \text{atomic weight}}{\text{Avogadro constant}}$$

$$\text{Density} = \frac{5 \times 10^{22}(\text{atoms/cm}^3) \times 28.09(\text{g/mole})}{6.02 \times 10^{23}(\text{atoms/mole})} = 2.33 \text{ g/cm}^3.$$

■

In Fig. 2.5(b), it may be noted that there are four atoms in the *ABCD* plane and five atoms in the *ACEF* plane (four atoms from the corners and one from the center), and that the atomic spacings are different for the two planes. Therefore, the crystal properties along different planes are different, and the electrical and other device characteristics are dependent on the crystal orientation. A convenient method of defining the various planes in a crystal is to use *Miller indices.*[11] These indices are obtained using the following steps:

1. Find the intercepts of the plane on the three Cartesian coordinates in terms of the lattice constant.
2. Take the reciprocals of these numbers and reduce them to the smallest three integers having the same ratio.
3. Enclose the result in parentheses (hkl) as the Miller indices for a single plane.

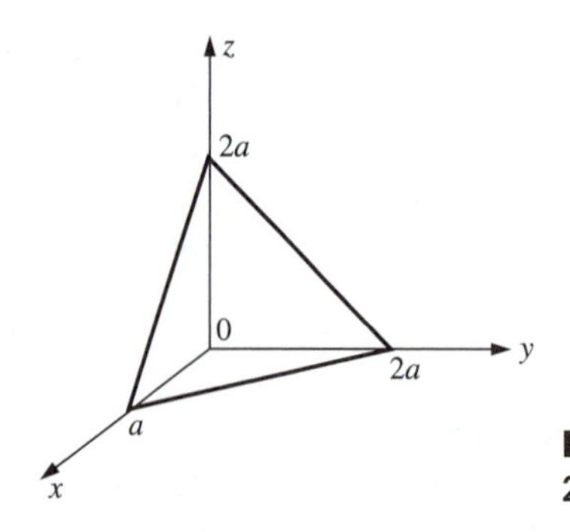

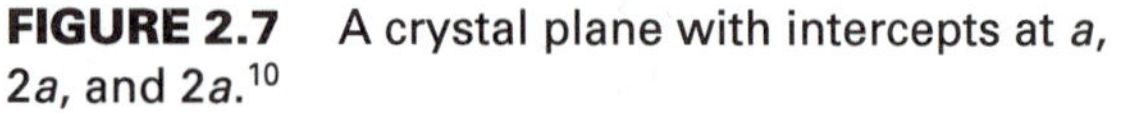
FIGURE 2.7 A crystal plane with intercepts at *a*, 2*a*, and 2*a*.[10]

EXAMPLE 2.3 Find the Miller indices for the plane shown in Fig. 2.7.

Solution The plane has intercepts at a, $2a$, and $2a$ along the three coordinates, and the reciprocals of these intercepts are 1, 1/2, and 1/2. The smallest three integers are 2, 1, and 1 (obtained by multiplying each fraction by 2). Thus, the plane is referred to as a (211) plane. ■

Figure 2.8 shows the Miller indices of important planes in a cubic crystal. Some other conventions are given as follows:

$(\bar{h}kl)$: For a plane that intercepts the x-axis on the negative side of the origin, such as $(\bar{1}00)$.

$\{hkl\}$: For planes of equivalent symmetry, such as {100}, for (100), (010), (001), $(\bar{1}00)$, $(0\bar{1}0)$, and $(00\bar{1})$.

$[hkl]$: For a crystal direction, such as [100] for the x-axis. Thus, the [100] direction is perpendicular to the (100) plane, and the [111] direction is perpendicular to the (111) plane.

$\langle hkl \rangle$: For a full set of equivalent directions, such as ⟨100⟩ for [100], [010], [001], $[\bar{1}00]$, $[0\bar{1}0]$, and $[00\bar{1}]$ in cubic symmetry.

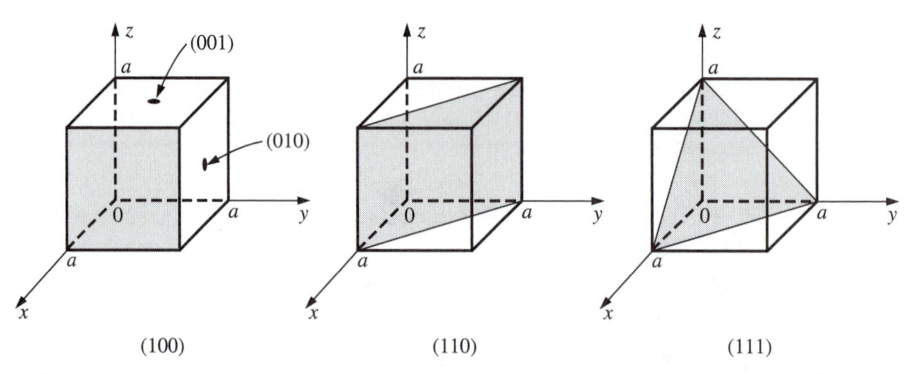

FIGURE 2.8 Miller indices of some important planes in a cubic crystal.[10]

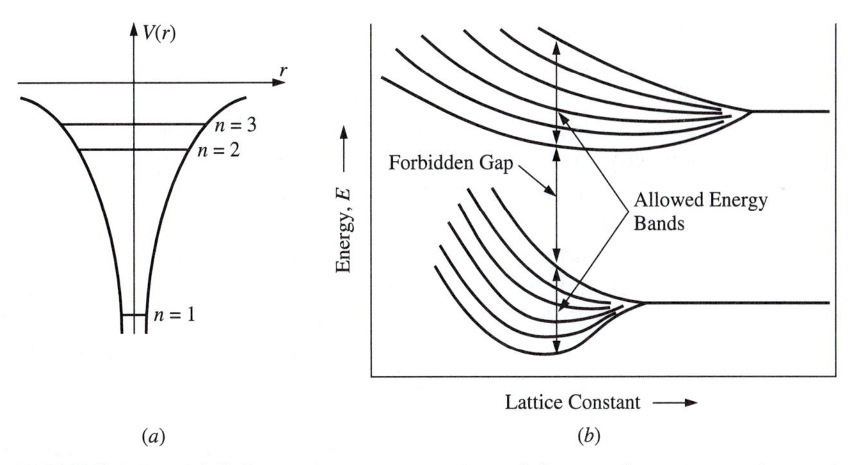

FIGURE 2.9 (a) Schematic representation of the nondegenerate electronic levels in an isolated atom with the potential $V(r)$ as a function of the radius r. (b) Energy levels for N atoms in a crystal as a function of atom spacing.

2.4.2 Energy Bands in Solids

The quantized electron energy levels of an isolated hydrogen atom were introduced in Sec. 2.2 and the related nondegenerate electron levels ($n = 1, n = 2, n = 3 \cdots$) in the atomic potential of an atom are shown in Fig. 2.9(a). Next, consider N atoms in a crystal with a very large lattice spacing so that the allowed energy levels will be the atomic energy levels. There will be N $n = 1$ levels, and therefore, each level will be N-fold degenerate. If the atoms are brought closer together, as in a real crystal, each level will split into N separate levels due to the influence of the potential of nearby atoms. These closely spaced levels will give a band of N degenerate levels instead of a single degenerate level. These levels are illustrated in Fig. 2.9(b). According to the Pauli exclusion principle, two electrons of opposite spin can occupy each level.

Inner-shell electrons are not significantly influenced by nearby atoms and are still tightly bound to a single nucleus and produce filled bands. However, the outermost or valence electrons bind to the crystal as a whole rather than to any particular atom. These valence electrons belong to bands which may be partially filled. Also, in correspondence to excited states of atoms, there will be additional empty higher energy bands. For lattice constants typical of crystalline solids, the energy bands may overlap as in a metal or have an energy gap as in a semiconductor or insulator. The detailed behavior of electrons in a crystal requires consideration of the Schrödinger wave equation, but these simple arguments suggest that energy bands and energy gaps should be obtained. The preceding arguments and those in Sec. 2.4.3 are meant to illustrate the physical origins of energy bands and energy gaps.

2.4.3 The *E* vs. *k* Diagram for a One-Dimensional Crystal

The crystal structure for Si was shown in Fig. 2.6(a). This diamond structure represents the atoms seen by a free electron in the solid. In this case, the potential U due to the nucleus of each atom would result in a very complex potential in the three-dimensional Schrödinger wave equation. Insight into the effect of a periodic potential on a conduction electron in a crystal can be obtained by considering a simple one-dimensional model first presented by Kronig and Penny.[12] This simple representation of the real crystal is only intended to emphasize that the regular arrangement of atoms in a crystal modifies the E vs. k diagram of the free electron and results in energy gaps between allowed bands.

The one-dimensional periodic potential is shown in Fig. 2.10. In part (b), the potential is simplified by the square-well approximation assumed by Kronig and Penny.[12,13] For the one-dimensional lattice, the total interatomic spacing is $(a + b)$, with the potential nonzero over the distance b. The complete solution for the Kronig-Penny model is somewhat lengthy and has been summarized elsewhere.[13,14] The resulting E vs. k diagram is shown in Fig. 2.11,[14] and is compared to the free-electron solution of $U = 0$. The solutions have sin and cos terms, and since k values that differ by $2\pi/(a + b)$ have the same sin and cos values, but differing values of E, the solution between $\pm 2\pi/(a + b)$ and $\pm 3\pi/(a + b)$ may be translated along the k-axis to give the dotted curves. It is readily seen that the solutions for the periodic potential have the greatest departure from the free-electron solution at low energies.

The k values associated with a given energy band form what is called a *Brillouin zone.* The first Brillouin zone is from $\pm\pi/(a + b)$, the second is from $\pm\pi/(a + b)$ to

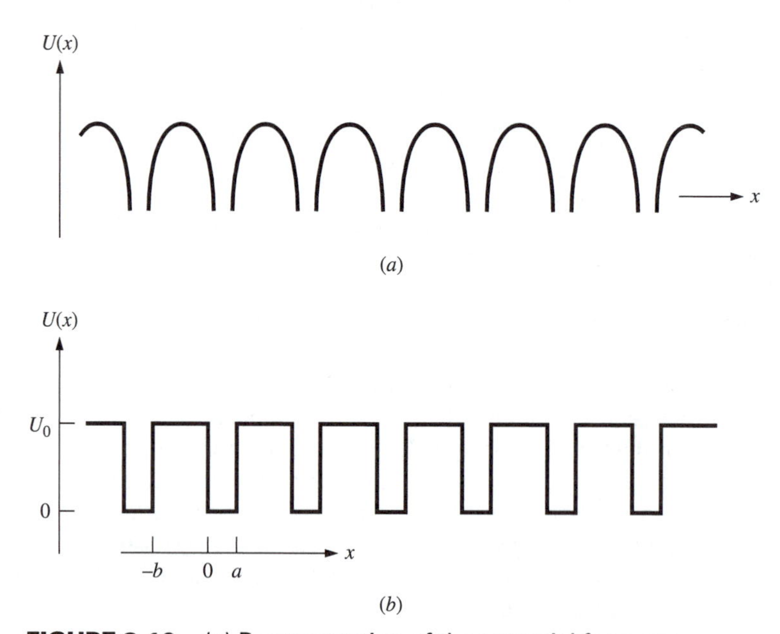

FIGURE 2.10 (a) Representation of the potential for a one-dimensional linear array. (b) Simplified representation of the periodic potential as represented in part (a) with the total interatomic spacing of $(a + b)$.[14]

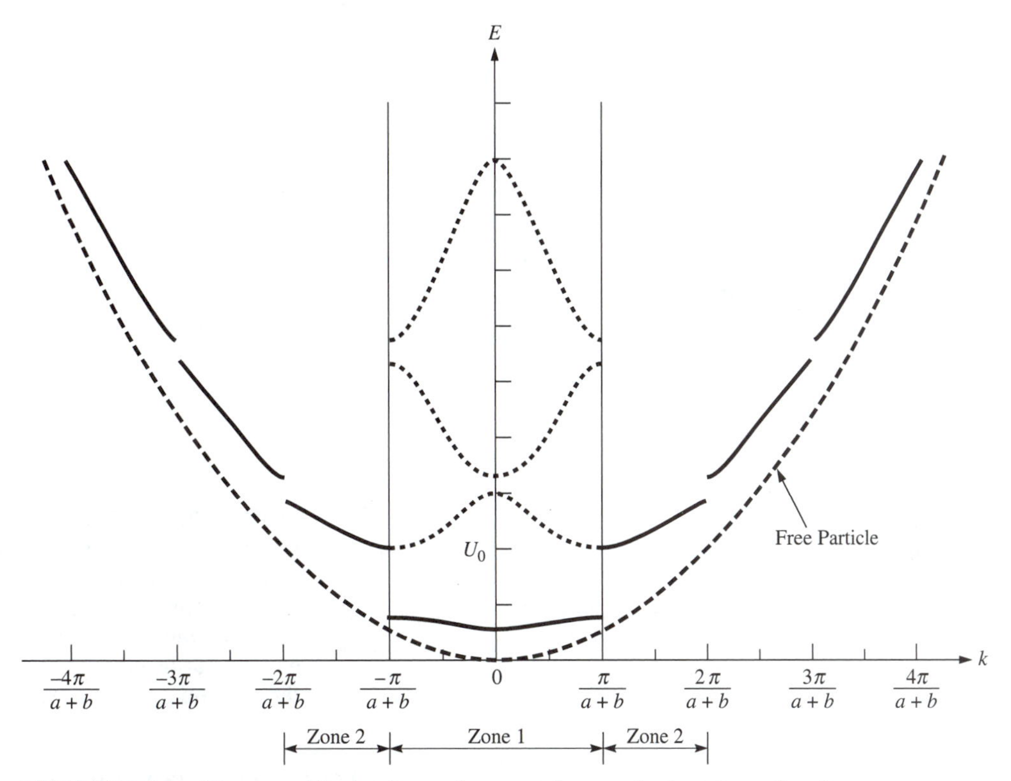

FIGURE 2.11 The permitted values of energy for an electron traveling in a one-dimensional periodic potential (solid lines) and are the results for the Kronig-Penny model. The *E* vs. *k* relationship for the free electron as given in Fig. 2.3 is shown by the dashed curve.[14]

$\pm 2\pi/(a + b)$, and so on, as shown in Fig. 2.11. The usual diagram for real crystals folds all the zones back into the first zone, as represented by the dotted curve and is called a *reduced-zone representation.* The reduced-zone representation is summarized in Fig. 2.12. The regions of allowed energy are called *energy bands,* while the excluded energy ranges are called *energy gaps* (or band gaps).

energy bands
energy gaps

It is helpful to note that the Brillouin zone boundaries are at $k = \pm n\pi/(a + b)$, where $n = 1, 2, 3, \ldots$. With the relationship for the free electron, Eq. (2.22) for $k = 2\pi/\lambda$, then the Brillouin zone boundaries occur when

$$n\lambda = 2(a + b)\,. \tag{2.23}$$

As shown in Fig. 2.11, the strongest interaction with the lattice occurs when the electron wavelength is twice the lattice spacing, the lattice spacing, half the lattice spacing, and so on. This condition is the same as the Bragg condition for the reflection of x rays.[15] Thus, *the electron wave cannot be propagated through the periodic structure when the*

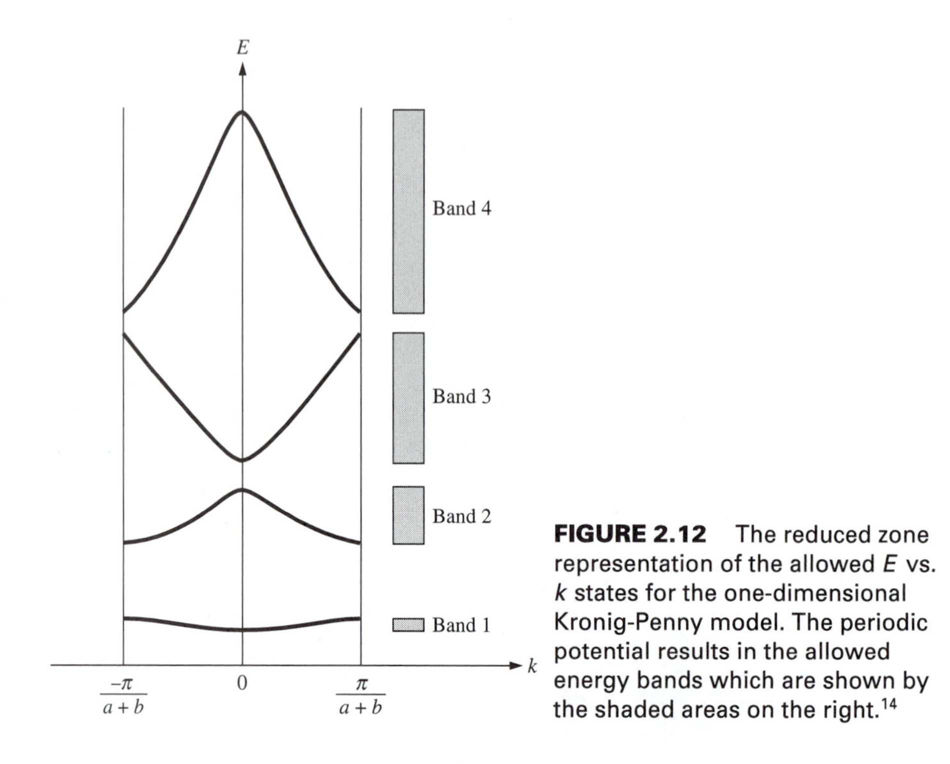

FIGURE 2.12 The reduced zone representation of the allowed *E* vs. *k* states for the one-dimensional Kronig-Penny model. The periodic potential results in the allowed energy bands which are shown by the shaded areas on the right.[14]

electron wavelength is an integer fraction of twice the lattice spacing, $\lambda = 2(a + b)/n$. For this condition, the propagating wave and the reflected wave interfere destructively and result in an energy gap between energy bands.

These simple models, which were illustrated in Figs. 2.11 and 2.12, allow a material to be classified as a metal, a semiconductor, or an insulator. In a metal, there is no energy gap. In a pure (no impurities) semiconductor, the energy gap between occupied and unoccupied states is generally less than ~1.5 eV. For energy gaps greater than ~1.5 eV, a crystal requires impurities to have semiconductor properties.

2.4.4 The *E* vs. *k* Diagram for Gallium Arsenide and Silicon

The simplified energy-band structures for the group-IV semiconductor Si and the III-V compound semiconductor gallium arsenide (GaAs) are shown in Fig. 2.13. The designations [111] and [100] refer to the Miller indices for particular directions in the crystal. The [100] direction is perpendicular to any of the six faces of the zinc-blende lattice cell shown in Fig 2.6(b). The [111] direction is, for example, perpendicular to the triangular plane connecting an atom at a corner on the top to two bottom atoms to the right and left of the top atom, as illustrated in Fig. 2.8. The carrier momentum p, often called the crystal momentum, was related to the wave vector k by Eq. (2.18) as $|\mathbf{p}| = \hbar|\mathbf{k}|$ for one dimension or $\mathbf{p} = \hbar\mathbf{k}$ in three dimensions. Carriers in crystals can exhibit only specific energies and momentum that are described by the E vs. k energy–momentum relationship.

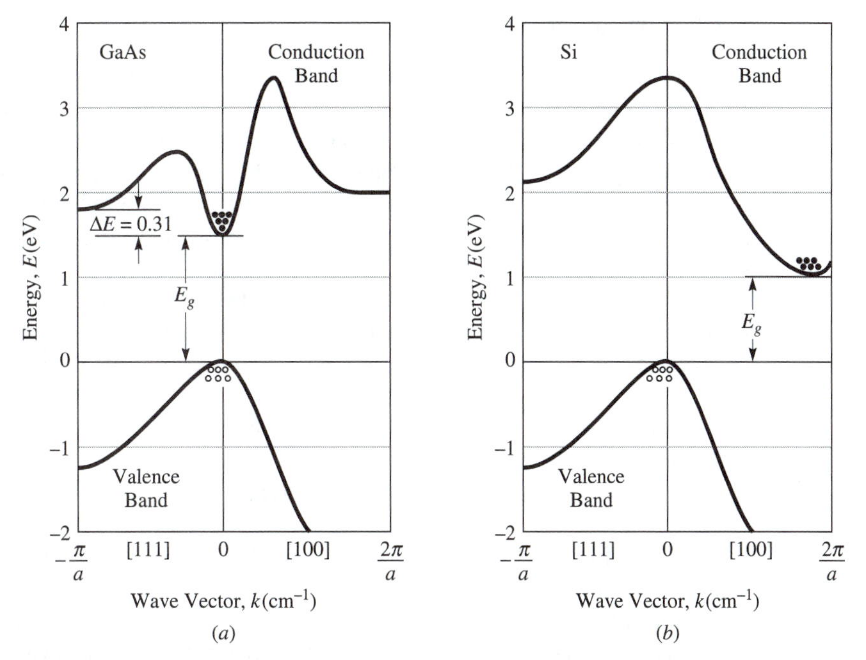

FIGURE 2.13 Energy-band structures of GaAs and Si. The lower band is the valence band and contains holes represented by ○. The upper band is the conduction band and contains conduction electrons represented by ●. The separation between the conduction and valence bands is the energy gap E_g.[16]

The energy-band structures shown in Fig. 2.13 are the result of theoretical calculations, but the significant parameters such as the energy gap are determined experimentally. The [100] and [111] are directions of crystal symmetry and the lowest conduction-band minima occur in these directions. The upper bands are called the *conduction bands,* and the lower bands are called the *valence bands. The separation between the energy of the lowest conduction band and the highest valence band is called the* energy gap E_g.

energy gap E_g

E_c and E_v

The bottom of the conduction band is designated by E_c *and the top of the valence band is designated by* E_v. The holes in the valence band will be defined in the last part of this section after the effective mass has been derived. In the energy-band diagram in Fig. 2.13, the electron energy is positive when measured upward. The hole energy is positive when measured downward, and the lowest energy holes are at the top of the valence band. Therefore, *electrons "sink" to the bottom of the conduction band and holes "float" to the top of the valence band.* For semiconductor devices, *most of the properties of the semiconductor given by the* E *vs.* k *diagrams can be summarized by the energy gap and the effective masses of the electrons and holes.* To define what is meant by a hole in the valence band and to derive the effective masses of the electron and hole, it is necessary to introduce the concept of wave packets and group velocity.

2.4.5 Wave Packets

In solutions of the Schrödinger wave equation for real crystals, values of k are restricted by boundary conditions to a series of discrete values, and the appropriate solution ψ is in the form of a sum (superposition) over all allowed values of k, $\psi(x, t) = \sum_k a_k \exp[j(\omega t - kx)]$. The example shown in Fig. 2.14 shows the superposition of seven sinusoidal waves. These waves have slightly different wavelengths, with phase and amplitudes which interfere constructively over a small region of space, outside of which they produce an amplitude that reduces to zero rapidly as a result of destructive interference. The sum of the waves is shown at the bottom of Fig. 2.14(a) for $t = 0$ and is known as a *wave packet.* This wave function represented by the wave packet gives the location of the electron as represented by the integral of the magnitude of ψ squared in Eq. (2.12).

Each of the seven waves travels with its own velocity [Eq. (2.19)],

$$v = \hbar k/m. \tag{2.24}$$

As shown in Fig. 2.14(b), all the waves are shifted by slightly different amounts of distance $d = vt = (\hbar k/m)t$. The sum of the waves shown at the bottom of part (b) has almost the same shape as for $t = 0$, except the wave packet has moved slightly more than twice as far as the average shift of the individual waves. The velocity of the wave packet is known as the *group velocity* v_g and can be shown to be[17]

group velocity

$$v_g = \frac{d\omega}{dk}, \tag{2.25}$$

and represents the velocity the electron moves in the crystal.

2.4.6 Effective Mass

Consideration of the effect of a force on an electron in a crystal results in a very useful concept for the behavior of electrons in semiconductors. This concept will be the effective mass. First, consider the electron as a particle according to Newton's laws, where the force F is related to the acceleration a and the mass m by

$$F = ma = m\frac{dv}{dt} = \frac{dp}{dt}. \tag{2.26}$$

Then, from Newton's laws, the acceleration can be written as

$$a = \frac{dv}{dt} = \frac{1}{m}\frac{dp}{dt}. \tag{2.27}$$

From the wave nature of the electron, the momentum was given by Eq. (2.18) as $|\mathbf{p}| = \hbar|\mathbf{k}| = \hbar k$, and taking the derivative of $\hbar k$ to obtain the acceleration from Eq. (2.27)

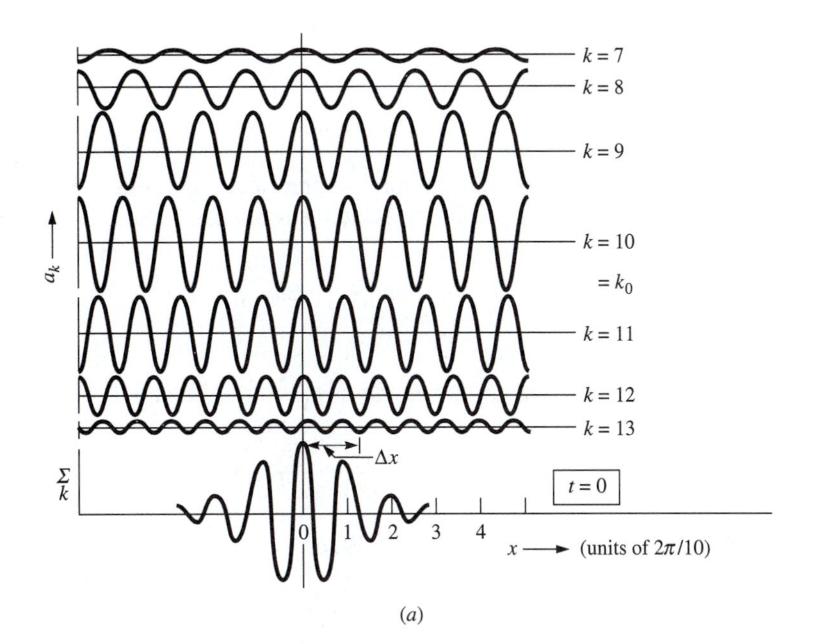

(a)

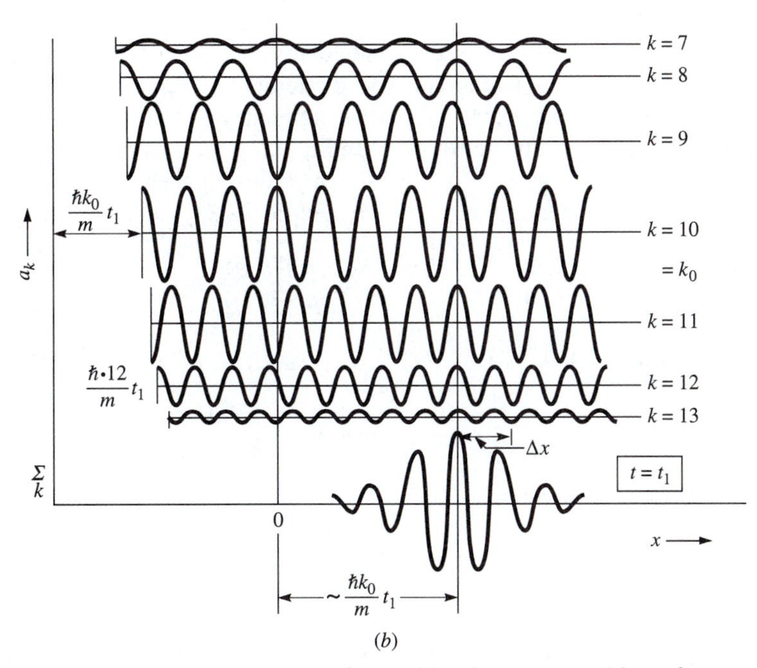

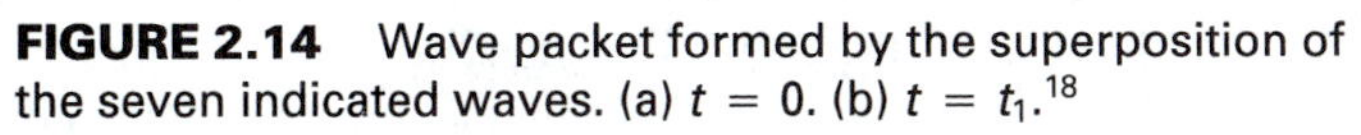

(b)

FIGURE 2.14 Wave packet formed by the superposition of the seven indicated waves. (a) $t = 0$. (b) $t = t_1$.[18]

gives

$$a = \frac{1}{m}\hbar\frac{dk}{dt}. \tag{2.28}$$

To account for the effect of a force on the electron, it is useful to write the acceleration in terms of the group velocity, which was defined in Sec. 2.4.5. The acceleration in terms of the group velocity is

$$a = \frac{dv_g}{dt} = \frac{\partial v_g}{\partial k}\frac{dk}{dt}. \tag{2.29}$$

But, from Eq. (2.25), the group velocity is $v_g = d\omega/dk$, and then,

$$\frac{\partial v_g}{\partial k} = \frac{\partial^2\omega}{\partial k^2}, \tag{2.30}$$

which gives the acceleration in Eq. (2.29) as

$$a = \frac{\partial^2\omega}{\partial k^2}\frac{dk}{dt}. \tag{2.31}$$

The objective of this section is to obtain a relationship for the carrier mass in the crystal from the E vs. k diagram. The acceleration in Eq. (2.31) can be written in terms of the energy E rather than the radian frequency ω by using the wave nature of the electron as proposed by de Broglie.[5] In Eq. (2.2), the energy E for a photon was related to the frequency ν as $E = h\nu = \hbar\omega$, which also applies to an electron of wavelength λ. Then, ω is related to E as $\omega = E/\hbar$, and Eq. (2.31) can be written in terms of E as

$$a = \frac{1}{\hbar}\frac{\partial^2 E}{\partial k^2}\frac{dk}{dt}. \tag{2.32}$$

With Eq. (2.28) and Eq. (2.32),

$$a = \frac{1}{m}\hbar\frac{dk}{dt} = \frac{1}{\hbar}\frac{\partial^2 E}{\partial k^2}\frac{dk}{dt}. \tag{2.33}$$

The mass can be written as

$$\boxed{m = m^* \equiv \hbar^2 / \frac{\partial^2 E}{\partial k^2}}, \tag{2.34}$$

effective mass

where m^* is called the *effective mass*. Note that $\partial^2 E/\partial k^2$ is the curvature of the E vs. k curve. A simplified E vs. k diagram is shown in Fig. 2.15. The upper band, the conduction band, is concave so that m^* is positive. In the lower band, the curvature is convex, and m^* is negative.

The *negative* mass is very inconvenient. Consider the behavior of the positive and negative masses in an electric field $\mathscr{E}$ as shown in Fig. 2.16. The force on a charged

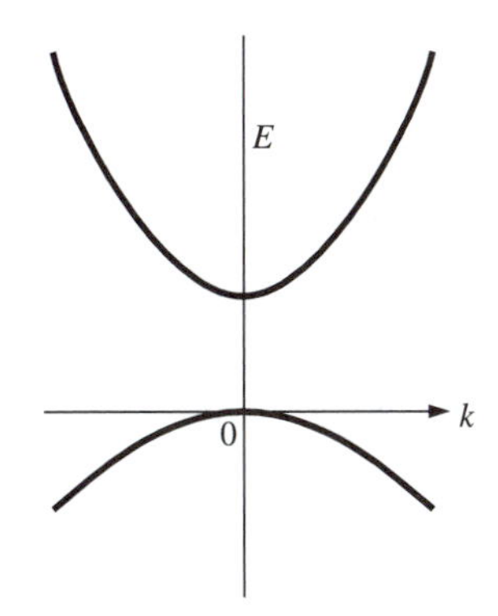

FIGURE 2.15 A simplified E vs. k diagram to illustrate the relationship of the effective mass to the curvature of the conduction and valence bands.

particle is $F = q\mathscr{E} = ma$, so that the acceleration is $a = q\mathscr{E}/m$. The acceleration of the negatively charged electron with a positive mass in Fig. 2.16(a) will be to the left, while the acceleration on the negatively charged particle with the negative mass in Fig. 2.16(b) will be to the right. In fact, this acceleration to the right is the same as the acceleration of a positively charged particle with a positive mass in Fig. 2.16(c). In the valence band, the unoccupied electron state will behave as if it had a positive charge and a positive mass and is called a *hole*. Therefore, in a crystal, the electron is treated like a free electron with a negative charge except that the mass is represented by the effective mass, and the hole in the valence band is taken to have a positive charge with a positive effective mass. This effective mass, which is related to the acceleration of an electron or hole in an electric field is defined as the *conductivity effective mass.*[19]

hole

When an electron is raised to the conduction band, the unoccupied electron state is designated as the hole. The motion of the hole is actually the motion of electrons from occupied to unoccupied states. This motion of holes can be illustrated by considering the movement of students from occupied seats (electrons) to empty seats (holes), as illustrated in Fig. 2.17. In Fig. 2.17(a), the empty seat in the middle of the group is one position to the right of its initial position when Brian takes the empty seat to the left as shown in Fig. 2.17(b). When Vanessa takes the now empty seat next to her, the empty seat is now at the end of the table as shown in Fig. 2.17(c). Thus, the movement of Brian and Vanessa to the left has the effect of the empty seat moving to the right. In the same manner, negatively charged electron motion in the valence band from occupied states to unoccupied states is the same as positively charged hole motion in the opposite direction. Therefore, it is a great convenience to take holes as positive-charge carriers rather than as unoccupied negative-charge states.

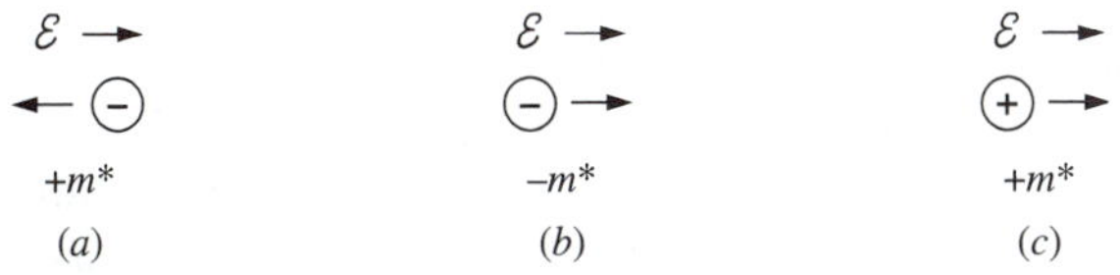

$\mathscr{E}$ → ← ⊖ $+m^*$ (a)

$\mathscr{E}$ → ⊖ → $-m^*$ (b)

$\mathscr{E}$ → ⊕ → $+m^*$ (c)

FIGURE 2.16 The direction of a force due to an electric field $\mathscr{E}$ on (a) a negative charge, positive mass, (b) a negative charge, negative mass, and (c) a positive charge, positive mass.

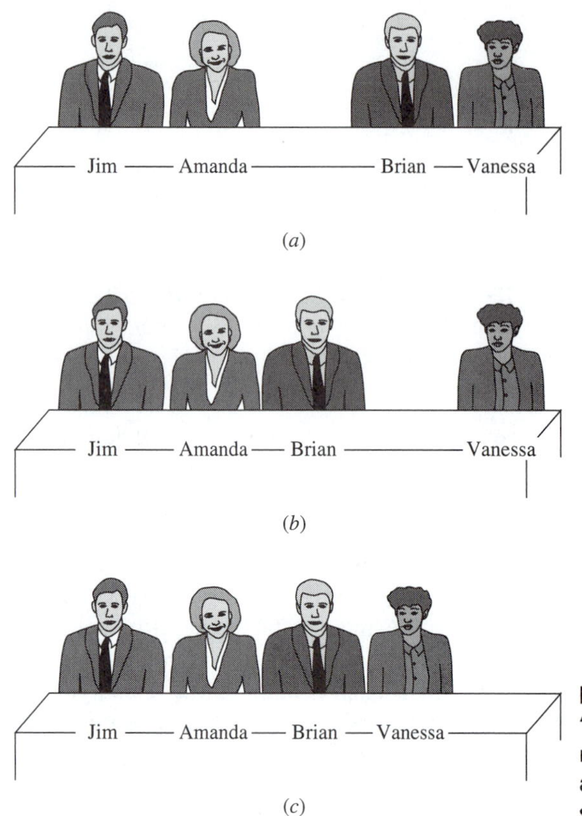

FIGURE 2.17 Representation of the 'movement' of the empty seat to the right by the movement of Brian and then Vanessa into the adjacent empty seat. The motion of holes is very similar.

2.5 DENSITY OF ELECTRON ENERGY STATES

Another useful concept which is necessary in order to deal with semiconductors is the density of states. Within the energy bands represented by the one-dimensional Kronig-Penny model or the actual E vs. k diagrams for real semiconductors, there are a finite number of energy levels which electrons can occupy. The origin of these levels was described in Sec. 2.4.2 and illustrated in Fig. 2.9. Because the separation of these numerous levels is very small, the number per electron volt of energy is near 10^{19} cm^{-3}.

Two quantities are necessary to express the electron (or hole) density distribution in the semiconductor crystal. One is the density of states or allowed number of electron states as obtained from the Schrödinger wave equation. The other is the probability that an electron will occupy that state and is given by the Fermi-Dirac distribution law, which is presented in Sec. 2.6.1. The density of states will be considered here.

Consider a cubic-shaped crystal with the dimensions $L = L_x = L_y = L_z$. Assume that the allowed solutions of the one-dimensional Schrödinger wave equation given by Eq. (2.13) with $U = 0$ are independent of the boundary conditions as long as the dimensions of the crystal are very large compared to the wavelengths of the electrons being considered. The simplest boundary conditions are periodic and require the minimum sample dimension to be many wavelengths. The wave traveling in the $+x$ direction was written in Eq. (2.20). The periodic boundary condition requires that

$$\psi(0, t) = \psi(L, t), \tag{2.35}$$

which restricts k to the discrete values of

$$k = 2\pi n/L, \tag{2.36}$$

where n is an integer $(0, \pm1, \pm2, \pm3, \ldots)$.

For a cubic-shaped crystal, the three-dimensional Schrödinger wave equation with periodic boundary conditions extends the one-dimensional case with the restriction

$$k_x = 2\pi n_x/L, \qquad k_y = 2\pi n_y/L, \qquad k_z = 2\pi n_z/L, \tag{2.37}$$

where n_x, n_y, and n_z are integers $(0, \pm1, \pm2, \pm3, \ldots)$. These discrete values of k give discrete values of $E_n = \hbar^2 k_n^2/2m^*$, which are called *states*. The problem now becomes one of counting the number of these states and deriving expressions for the number per unit volume which is the density of states.

The density of states is derived in terms of the propagation constant **k** in three dimensions, and then **k** is eliminated through its relationship to the energy E. It is useful to introduce the concept of **k**-space with

$$\mathbf{k} = \mathbf{a}_x k_x + \mathbf{a}_y k_y + \mathbf{a}_z k_z, \tag{2.38}$$

where $\mathbf{a}_x$, $\mathbf{a}_y$, and $\mathbf{a}_z$ are unit vectors in the x, y, and z directions. The propagation constant, commonly called the wave vector **k** in terms of k_x, k_y, and k_z given by Eq. (2.37) is shown in Fig. 2.18. The unit volume in **k**-space is indicated by the dotted cube in Fig. 2.18 and has a volume given by

$$\text{Unit volume in } \mathbf{k} \text{ space} = (2\pi/L)^3. \tag{2.39}$$

Only the values of **k** given by Eq. (2.37) are allowed, and it can then be concluded that the number of allowed values of **k** in any volume V_k in **k**-space is the number of cubes of side $2\pi/L$ in that volume.

The density of states may be found from the number of states in **k**-space between **k** and $(\mathbf{k} + d\mathbf{k})$. Let the volume of a thin spherical shell be $4\pi k^2 dk$, as illustrated in Fig. 2.19. The unit density in **k**-space is the reciprocal of the unit volume given in Eq. (2.39). The number of states in **k**-space is the unit density times the volume, which is

$$dN(k) = 2(L/2\pi)^3 4\pi k^2 dk, \tag{2.40}$$

where the factor of 2 accounts for the condition that two electrons with opposite spin can be accommodated for each allowed state. The density of states is the number of

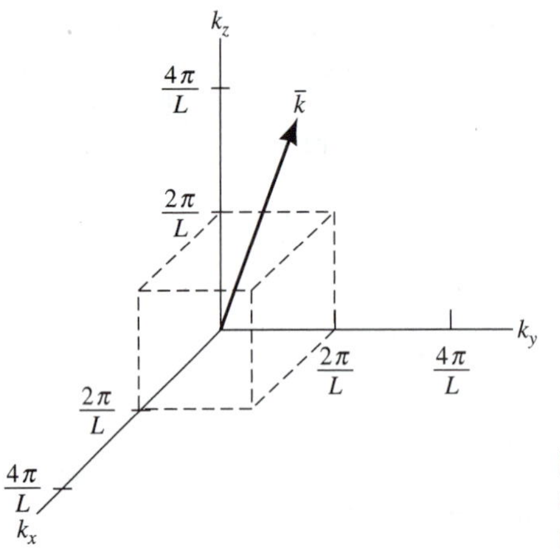

FIGURE 2.18 A plot of **k**-space in terms of k_x, k_y, and k_z given by Eq. (2.37) for periodic boundary conditions.

states per unit volume of the solid $V = L^3$ and is

$$dN(k) = (k^2/\pi^2)dk\,. \tag{2.41}$$

Next, use Eq. (2.15) with m_0 replaced by m^* to relate k to E, which gives

$$k^2 = 2m^*E/\hbar^2 \tag{2.42}$$

and

$$dk = (1/2)(2m^*/\hbar^2)^{1/2}E^{-1/2}dE. \tag{2.43}$$

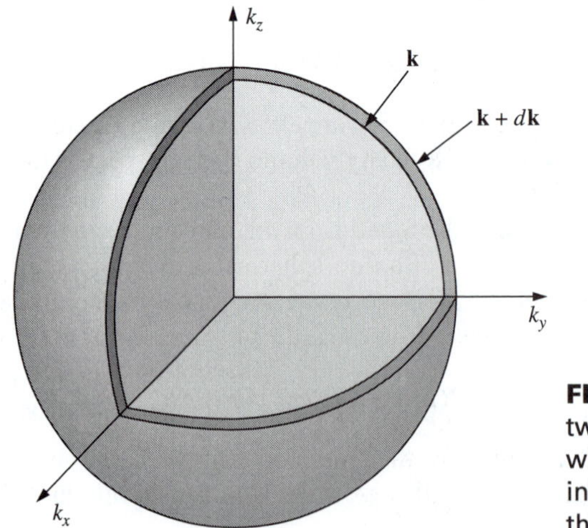

FIGURE 2.19 Representation of two spherical surfaces in **k**-space which are separated by the infinitesimally small radial thickness d**k**.

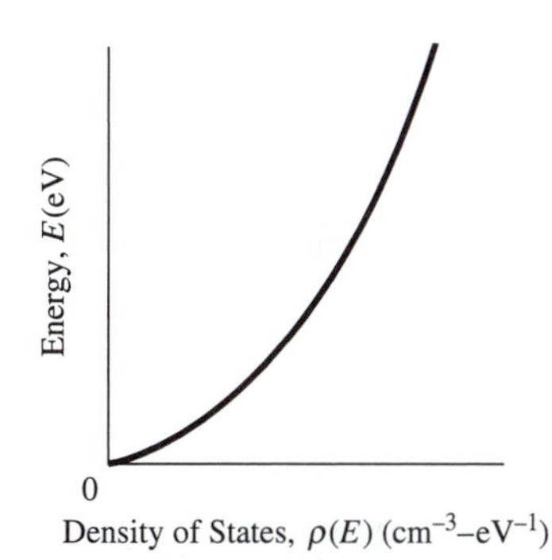

FIGURE 2.20 A representation of the density of states $\rho(E)$ as a function of the energy E. The shape of the curve only depends on the effective mass m^*.

Then Eq. (2.41) becomes

density of states

$$\boxed{dN(E)/dE = \rho(E) = (1/2\pi^2)(2m^*/\hbar^2)^{3/2}E^{1/2}}\,, \tag{2.44}$$

where the units for the density of states $\rho(E)$ are the number of states per cm^3 per eV. A representation of the density of states as a function of energy is shown in Fig. 2.20 to illustrate the square-root variation with energy.

2.6 THE EFFECTIVE DENSITY OF STATES

2.6.1 The Fermi-Dirac Distribution Function

Equation (2.44) gives the density of electronic states in a crystal. Another important property of electrons in a crystal is their distribution among the allowed states at thermal equilibrium. Electrons are indistinguishable, identical particles with half-integer spin that obey the Pauli exclusion principle. The occupation probability of an energy level E by an electron is given by the Fermi-Dirac distribution function, which is

Fermi-Dirac distribution function

$$\boxed{f(E) = 1/\{1 + \exp[(E - E_f)/kT]\}}\,, \tag{2.45}$$

where E_f is a reference energy called the *Fermi level.* Note that when $E = E_f$, $f(E) = \frac{1}{2}$. In Eq. (2.45), the temperature T is the absolute temperature in Kelvin, and k is Boltzmann's constant, which is 8.6164×10^{-5} eV-K^{-1}. This k will always appear with T as kT and should not be confused with the wave vector k, which is used as $\hbar k$. The Fermi-Dirac distribution function is generally called the Fermi function and is illustrated in Fig. 2.21 for temperatures between 0 K and 600 K. At 0°C, the absolute temperature is 273 K. Room temperature is generally taken as 27°C (80.6°F), which gives 300 K. For $E > E_f$, the occupation probability becomes very small, and for $E < E_f$, the occupation probability approaches unity.

From statistical mechanics, it is shown that the Fermi level is the chemical potential for the electrons,[21,22] and from thermodynamics, the chemical potential is constant

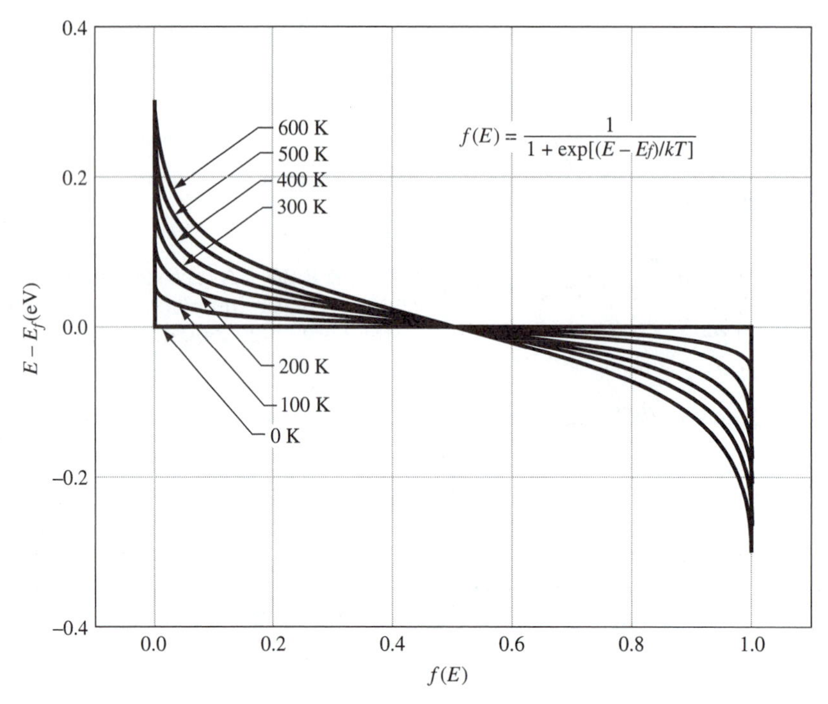

FIGURE 2.21 The Fermi function of Eq. (2.45) at the indicated temperatures.[20]

at thermal equilibrium[23] (no external excitation). In thermal equilibrium, there are no temperature gradients, no mechanical forces, and the rates of reactions in the forward and reverse directions are equal. Therefore, throughout any semiconductor structure, *the Fermi level is constant at thermal equilibrium,* which is generally expressed as

thermal equilibrium

$$\boxed{dE_f/dx = 0} \,. \tag{2.46}$$

The Fermi level is one of the principal quantities which is used to describe the behavior of semiconductor materials and devices. In the next part of this section, the relationship of the electron or hole concentration to the Fermi level will be demonstrated.

2.6.2 The Electron Concentration

The energy-band structure drawn as E vs. k diagrams in Fig. 2.13 can now be represented by plots of the density of states. The relevant features of the E vs. k diagram can be retained through the effective mass m_c^* of the lowest conduction band, E_c, the effective mass, m_v^* of the valence band, E_v, and the separation of E_c and E_v, which is the energy gap E_g. A representative density of states plot is shown in Fig. 2.22.

The concentration of electrons in the conduction band is found from the density of states in the conduction band and the Fermi function, which gives the occupation of

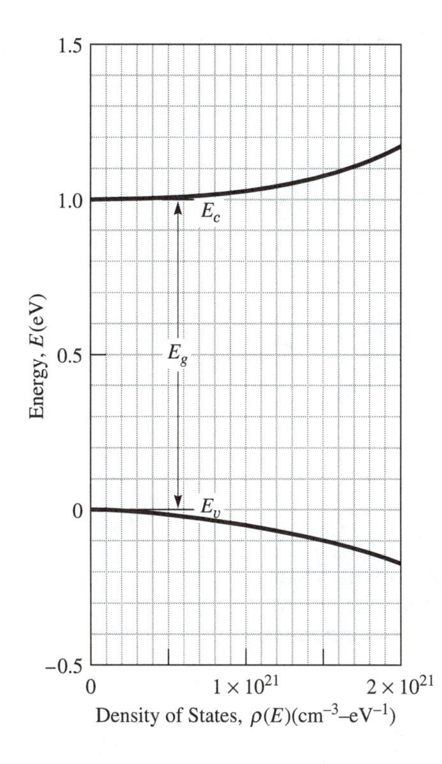

FIGURE 2.22 A plot of the density of states for a conduction band with $m_c^* = 0.5m_0$, a valence band with $m_v^* = 0.5m_0$, and an energy gap with $E_g = 1.0$ eV.

these states. Because the bottom of the conduction band occurs at $E = E_c$ as shown in Fig. 2.22, the density of states' expression becomes

$$\rho(E - E_c) = (1/2\pi^2)(2m_c^*/\hbar^2)^{3/2}(E - E_c)^{1/2}. \tag{2.47}$$

The electron concentration at a given energy is $n(E)$ and is given by the product of $\rho(E - E_c)f(E)$. By considering the conduction band shown in Fig. 2.22, this product becomes the shaded area in Fig. 2.23. The total electron concentration n is the area of the shaded area in Fig. 2.23, which is the integral of $n(E)$.

With Eq. (2.47) for the density of states and Eq. (2.45) for the Fermi function, the electron concentration in the conduction band may be written as

$$n = \frac{1}{2\pi^2}\left[\frac{2m_c^*}{\hbar^2}\right]^{3/2}\int_{E_c}^{\infty}\frac{(E - E_c)^{1/2}dE}{1 + \exp[(E - E_f)/kT]}. \tag{2.48}$$

Manipulation is enhanced by using the dimensionless notation,

$$\varepsilon = (E - E_c)/kT \tag{2.49}$$

and

$$\zeta = (E_f - E_c)/kT\,. \tag{2.50}$$

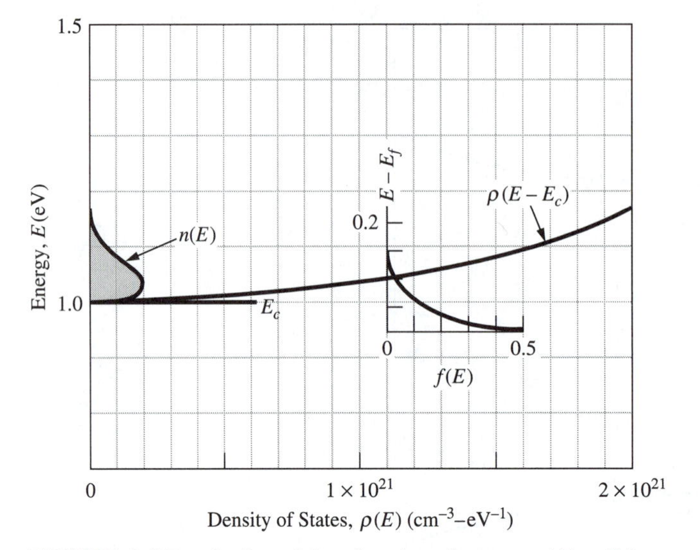

FIGURE 2.23 A plot of the density of states $\rho(E - E_c)$ and the Fermi function $f(E)$ (in the insert). The electron distribution $n(E)$ is indicated by the shaded area.

Equation (2.48) may be written as

$$n = \frac{1}{2\pi^2}\left[\frac{2m_c^* kT}{\hbar^2}\right]^{3/2}\int_0^\infty \frac{\varepsilon^{1/2}d\varepsilon}{1+\exp(\varepsilon-\zeta)}. \tag{2.51}$$

The integral in Eq. (2.51) may readily be evaluated when $(E - E_f)/kT$ or $(\varepsilon - \zeta) \gg 1$, so that the Fermi function $1/\{1 + [\exp(E - E_f)/kT]\}$ may be represented by $\exp[-(E - E_f)/kT]$ or $\exp\zeta\exp(-\varepsilon)$. *When the Fermi-Dirac distribution function is represented by the exponential approximation, this approximation is known as the* Boltzmann approximation. Then,

Boltzmann approximation

$$n = \frac{1}{2\pi^2}\left[\frac{2m_c^* kT}{\hbar^2}\right]^{3/2}\exp\zeta\int_0^\infty \varepsilon^{1/2}\exp(-\varepsilon)\,d\varepsilon\,, \tag{2.52}$$

and the integral is a standard form called the *gamma function* $\Gamma(3/2) = \pi^{1/2}/2$. The expression for the electron concentration is now

$$n = 2(2\pi m_c^* kT/h^2)^{3/2}\exp[-(E_c - E_f)/kT]. \tag{2.53}$$

The quantity multiplying the exponential occurs very frequently and is designated the conduction-band *effective density of states* N_c:

effective density of states

$$\boxed{N_c = 2(2\pi m_c^* kT/h^2)^{3/2}}. \tag{2.54}$$

Substitution of N_c into Eq. (2.53) gives

electron concentration

$$\boxed{n = N_c \exp[-(E_c - E_f)/kT]}\,, \tag{2.55}$$

or

Fermi level

$$\boxed{E_f = E_c + kT\ln(n/N_c)}\,, \tag{2.56}$$

where E_f is the Fermi level for electrons in the conduction band.

The effective density of states permits representing $\rho(E - E_c)$ in Eq. (2.47) by N_c located at $E = E_c$ as shown in Fig. 2.24. With $m^* = 0.5m_0$, the effective density of states $N_c = 8.84 \times 10^{18}\ \text{cm}^{-3}$. The exponential term in Eq. (2.53) is the Boltzmann distribution function $f(E) = \exp[-(E - E_f)/kT]$, evaluated at $E = E_c$. This representation of n by Eq. (2.55) is valid only for $(n/N_c) < 0.1$.

When the exponential (Boltzmann) approximation for the Fermi function may *not* be made, the integral may be written as

$$n = N_c \mathcal{F}_{1/2}[(E_f - E_c)/kT]\,, \tag{2.57}$$

where $\mathcal{F}_{1/2}$ is the Fermi-Dirac integral which requires evaluation with tables. The use of Fermi-Dirac integral tables is very inconvenient. Fortunately, when $(n/N_c) > 0.1$, several numerical representations are available. Joyce and Dixon[24] obtained an analytical approximation with a polynomial added to the Boltzmann expression given in

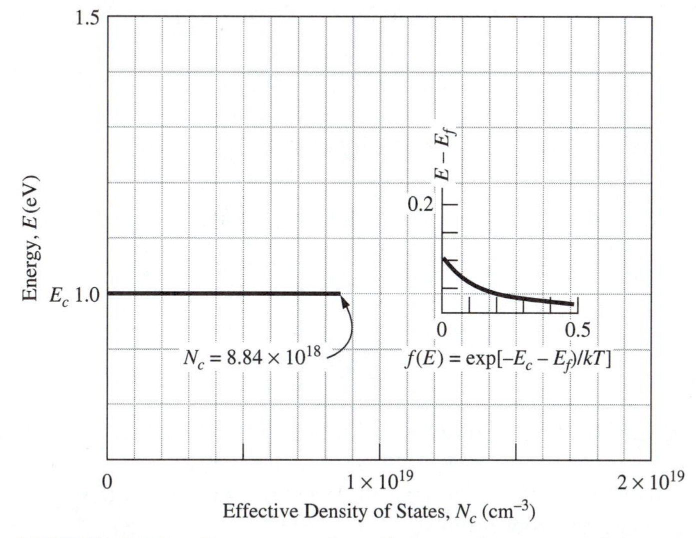

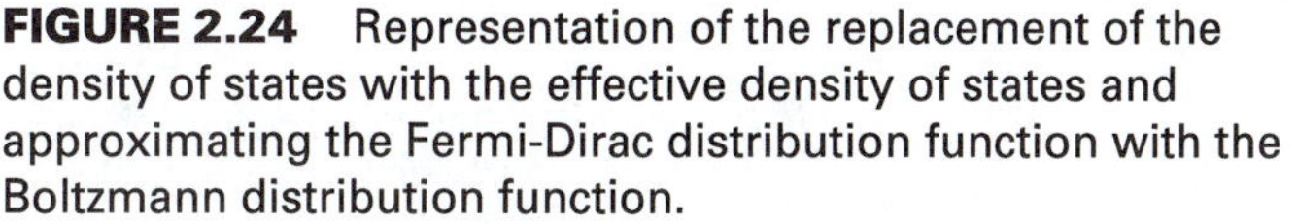
FIGURE 2.24 Representation of the replacement of the density of states with the effective density of states and approximating the Fermi-Dirac distribution function with the Boltzmann distribution function.

Eq. (2.56). Later, Nilsson[25] obtained a simple empirical expression which gives the Fermi level as

$$E_f = E_c + kT\left[\ln(n/N_c) + \frac{(n/N_c)}{\{64 + 0.05524(n/N_c)[64 + (n/N_c)^{1/2}]\}^{1/4}}\right]. \quad (2.58)$$

Next, consider the valence band. The probability that an electron state is occupied by an electron was given as $f(E)$ in Eq. (2.45). The probability that the electron state is not occupied by an electron (empty) is simply

$$1 - f(E) = \frac{1}{1 + \exp[(E_f - E)/kT]}, \quad (2.59)$$

and is the probability that a state in the valence band is occupied by a hole. The density of states for holes in the valence band is

$$\rho(E_v - E) = (1/2\pi^2)(2m_v^*/h^2)^{3/2}(E_v - E)^{1/2}, \quad (2.60)$$

where m_v^* is the effective mass in the valence band. The hole concentration in the valence band can be written as

$$p = \int_{-\infty}^{E_v} \rho(E_v - E)[1 - f(E)]\,dE. \quad (2.61)$$

With the Boltzmann approximation for the Fermi function when $(p/N_v) < 0.1$,

$$p = 2(2\pi m_v^* kT/h^2)^{3/2} \exp[-(E_f - E_v)/kT], \quad (2.62)$$

with the valence-band effective density of states given by

effective density of states

$$\boxed{N_v = 2(2\pi m_v^* kT/h^2)^{3/2}} \quad (2.63)$$

and

hole concentration

$$\boxed{p = N_v \exp[-(E_f - E_v)/kT]}. \quad (2.64)$$

Equation (2.64) is similar to Eq. (2.55) for the conduction band. The Boltzmann approximation also holds when $(p/N_v) < 0.1$, and an expression similar to Eq. (2.58) may be used when $(p/N_v) > 0.1$. When the Boltzmann approximation is appropriate, the Fermi level for holes may be written as

Fermi level

$$\boxed{E_f = E_v - kT\ln(p/N_v)}. \quad (2.65)$$

TABLE 2.1 Silicon and GaAs Effective Density of States at T = 300 K.

Semiconductor	N_c (cm^{-3})	N_v (cm^{-3})
Si	2.84×10^{19}	3.08×10^{19}
GaAs	4.34×10^{17}	8.06×10^{18}

It is convenient to write the expressions for the effective density of states with the numerical factors in units to give the electron concentration n and the hole concentration p in cm^{-3}. Equations (2.54) and (2.63) may be written as

$$\boxed{N_c = 2.5 \times 10^{19}(m_n^*/m_0)^{3/2}(T/300)^{3/2}\ \text{cm}^{-3}} \tag{2.66}$$

and

$$\boxed{N_v = 2.5 \times 10^{19}(m_p^*/m_0)^{3/2}(T/300)^{3/2}\ \text{cm}^{-3}}\,, \tag{2.67}$$

where $m_n^* = m_c^*$ is the *effective density of states mass* in the conduction band and $m_p^* = m_v^*$ is the *effective density of states mass* in the valence band. The density of states effective masses for Si and GaAs will be given in Sec. 2.7.3, and with these values, the room temperature effective density of states are summarized in Table 2.1.

2.7 THE INTRINSIC CARRIER CONCENTRATION

2.7.1 The Intrinsic Semiconductor

In a pure semiconductor, the thermal excitation will raise electrons from the valence band to the conduction band, which leaves an equal number of holes in the valence band. An *intrinsic semiconductor* contains relatively small amounts of impurities compared to the thermally generated electrons and holes. The thermal generation of an electron–hole pair may be written as a *chemical reaction:*

$$(\mathrm{e}^-\mathrm{e}^+) \leftrightharpoons \mathrm{e}^- + \mathrm{e}^+\,, \tag{2.68}$$

where $(\mathrm{e}^-\mathrm{e}^+)$ represents the recombined electron–hole pair in the valence band, e^- represents the free electron in the conduction band, and e^+ represents the free hole in the valence band. The arrow to the right represents the generation of an electron–hole pair, while the arrow to the left represents the recombination of a conduction band electron with a valence band hole. The equilibrium relationship for the reaction in Eq. (2.68) is

$$np = K(T)\,, \tag{2.69}$$

where n and p represent the electron and hole concentrations, respectively, and $K(T)$ is the reaction equilibrium constant, which is a function of temperature only. For the intrinsic semiconductor, $n = n_i$, $p = p_i$, and $E_f = E_i$, where *n_i is the intrinsic electron concentration,* and *E_i is the Fermi level for the intrinsic semiconductor.* From Eq. (2.55) for n and Eq. (2.64) for p, Eq. (2.69) may be written as

intrinsic Fermi level E_i

$$n_i^2 = N_c \exp[-(E_c - E_i)/kT] N_v \exp[-(E_i - E_v)/kT]. \tag{2.70}$$

intrinsic carrier concentration n_i

For the intrinsic semiconductor, $n_i = p_i$ and

$$\boxed{np = n_i^2 = N_c N_v \exp(-E_g/kT)}. \tag{2.71}$$

To evaluate n_i for a given semiconductor, it is necessary to have values for the energy gap E_g and the effective density of states masses m_n^* and m_p^*. *Much of the basic behavior of semiconductor devices requires relating* n *and* p *to* E_f *and to the addition of impurities to the crystal so that* n *or* p *exceeds* n_i.

2.7.2 The Energy Gap

In Sec. 2.4.4, the energy gap E_g was defined as the separation between the energy of the lowest conduction band and the highest valence band. The bottom of the conduction band, which was designated E_c, corresponds to the potential energy of an electron, that is, the energy of a conduction electron at rest. The kinetic energy of an electron is measured upward from E_c. Similarly, E_v corresponds to the potential energy of a hole. The kinetic energy of a hole is measured downward from E_v. When an electron is above E_c, the kinetic energy of the electron is increased by the amount it is above E_c. For a hole below E_v, its kinetic energy is the amount below E_v. These relationships for the electron and hole energies are summarized in the energy-band diagram shown in Fig. 2.25.

The energy-band structure for GaAs and Si were given in Fig. 2.13. In a semiconductor such as GaAs which has the lowest conduction band, E_c, at the same value

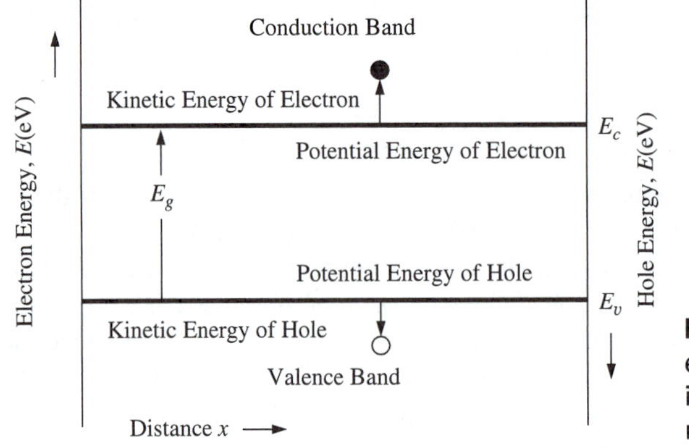

FIGURE 2.25 The potential energy and kinetic energy in an energy band diagram representation.[26]

direct- and indirect-energy gaps

of wave vector k as the highest valence band, E_v, the energy gap is designated a *direct-energy gap.* In other words, $k(\text{at } E_c) = k(\text{at } E_v)$. For GaAs, as shown in Fig. 2.13(a), the direct-energy gap is at $k = 0$. In Si, as shown in Fig. 2.13(b), the lowest E_c does not occur at the same wave vector k as the highest E_v, and Si is designated as an *indirect-energy gap* semiconductor. In the indirect-energy gap case, $k(\text{at } E_c) \neq k(\text{at } E_v)$. This difference between direct and indirect band structures is very important for light-emitting diodes (LED's) and semiconductor lasers. A direct-energy gap is required for efficient light emission by semiconductors.

EXAMPLE 2.4 The difference between direct- and indirect-energy gaps may be illustrated by the absorption or emission of light, which requires the conservation of both energy and momentum. As shown in Fig. 2.26 for Si, the difference in wave vector (Δk) between the top of the valence band, E_v, and the minimum in the conduction band, E_c, is approximately $0.85(2\pi/a)\ \text{cm}^{-1}$. For Si, the lattice constant a is 5.43×10^{-8} cm, which gives $\underline{\Delta k \approx 1 \times 10^8\ \text{cm}^{-1}}$. To create an electron–hole pair, absorption of a photon of light with an energy $h\nu \simeq E_g = 1.125$ eV would be required. The momentum of this photon is given by Eq. (2.19) as $\hbar k = h/\lambda$ with a frequency ν and velocity $c = 3 \times 10^{10}$ cm/s, which are related by $\lambda\nu = c$. Then, the value of the photon wave vector is

$$k = h/\hbar\lambda = h\nu/\hbar c = 1.125/6.579 \times 10^{-16} \times 3 \times 10^{10} = \underline{5.7 \times 10^4\ \text{cm}^{-1}},$$

which is three orders of magnitude smaller than the change in the wave vector between E_v and E_c.

The necessary wave vector and hence momentum between E_v and E_c for an indirect-energy gap semiconductor is provided by interaction with the thermal vibrational motions of the atoms. The allowed vibrational motions of the atoms are termed *phonons.* Thus, for the transition of an electron from the valence band to the conduction band in Si by the absorption of a photon, interaction with a phonon is also required to supply the necessary change in wave vector or momentum. With the

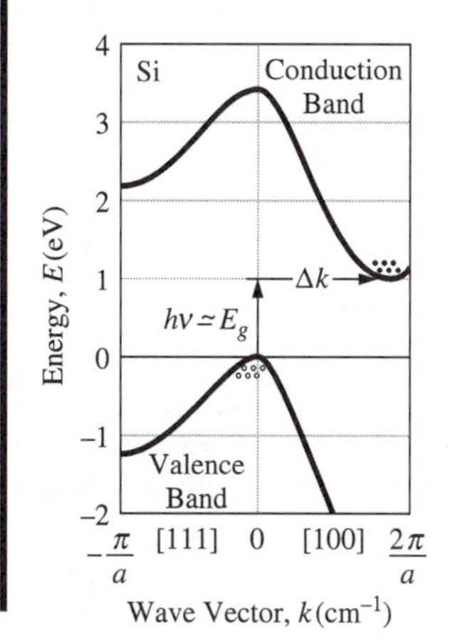

FIGURE 2.26 Illustration of the wave vector change Δk necessary for a transition from the valence band to the conduction band due to absorption of a photon of energy $h\nu \simeq 1.125$ eV.

necessary wave vector of $\Delta k \approx 1 \times 10^8$ cm^{-1}, the wavelength of the phonon would be $\lambda = 2\pi/\Delta k = 6.3 \times 10^{-8}$ cm. For an acoustic velocity v_a of 7.5×10^5 cm/s in Si, the phonon energy, E_p, would be

$$E_p = \frac{hv_a}{\lambda} = 4.135 \times 10^{-15} \times 7.5 \times 10^5/6.3 \times 10^{-8} \simeq 0.05 \text{ eV.}$$

The conservation of energy requires that $h\nu \pm E_p = E_g$, where $+E_p$ represents phonon absorption and $-E_p$ represents phonon emission. No phonon is required for this transition in a direct-energy gap semiconductor such as GaAs. ■

phonons

Sound waves in a crystal are composed of phonons. In analogy with the photon, which is the quantum of energy in an electromagnetic wave, the *phonon* is the quantum of energy in an elastic wave. Almost all of the concepts which apply to the photon, such as the particle–wave duality, apply equally well to the phonon. Thermal vibrations of the atoms in a crystal are excited phonons, and therefore, phonons are related to temperature.

Phonons are involved in the transfer of energy between electrons and the lattice (electron–phonon interaction). They can also interact with photons. For phonons, the movement of atoms can be in the direction of the phonon wave propagation (longitudinal mode) or in a direction transverse to the propagation (transverse mode). In Fig. 2.27(a), the lattice is represented by a simple linear chain of atoms. For the optical mode, the adjacent atoms move out of phase, as shown in the top diagram of Fig. 2.27(a). For the acoustical mode, the adjacent atoms move in phase, as shown in the bottom diagram

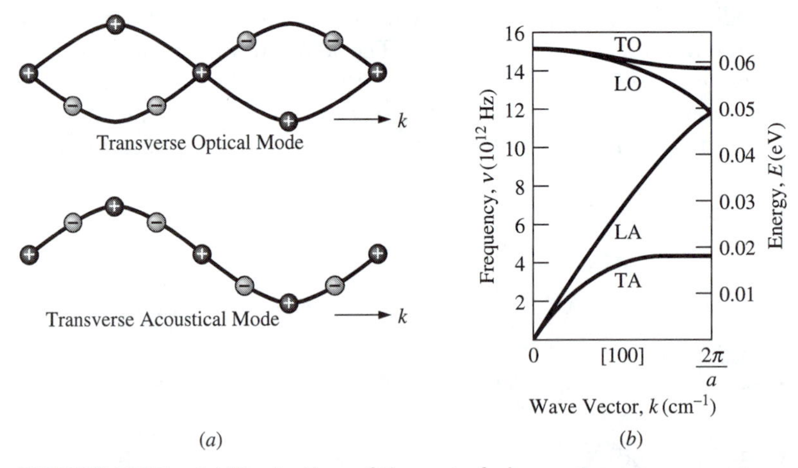

FIGURE 2.27 (a) Illustration of the out of phase atom movement for the optical phonons (top) and the in phase atom movement for the acoustical phonons (bottom). (b) E vs. k diagram for phonons in Si. The LO (TO) designation is the longitudinal (transverse) optical mode while the LA (TA) designation is the longitudinal (transverse) acoustical mode.[27]

TABLE 2.2 Room Temperature Energy Gaps of the Group IV Semiconductors and the Binary III-V Compound Semiconductors.

Semiconductor	Type of Energy Gap	Energy Gap E_g (eV)
Ge	indirect	0.663
Si	indirect	1.125
4H-SiC*	indirect	3.20
AlP	indirect	2.45
AlAs	indirect	2.163
AlSb	indirect	1.58
AlN†	direct	6.2
GaN†	direct	3.452
InN†	direct	1.95
GaP	indirect	2.261
GaAs	direct	1.424
GaSb	direct	0.726
InP	direct	1.351
InAs	direct	0.360
InSb	direct	0.172

*SiC exists in the cubic, hexagonal, and rhombohedral polytypes; however, most device applications use one of the hexagonal polytypes denoted as 4H-SiC or 6H-SiC.
†AlN, GaN, and InN exist both as zinc blende and the hexagonal wurtzite. E_g is given for the wurtzite structure which is generally used.

of Fig. 2.27(a). Only transverse modes are shown in Fig. 2.27(a). Motion of the atoms as "balls on springs" vibrating along a line in the direction of the propagation would be the longitudinal modes. The measured E vs. k diagram for phonons in Si is shown in Fig. 2.27(b).[27] It should be noted that the range of the wave vector for phonons in Fig. 2.27(b) is the same as for the Si-band structure shown in Fig. 2.26. As shown in Fig. 2.27(b), the optical phonon branch has both the energy and momentum needed for optical absorption in Si, as given in Example 2.4.

The type of energy gap and the room temperature energy gaps for the two group IV semiconductors Ge and Si, the group IV compound semiconductor SiC, and the binary III-V compound semiconductors are summarized in Table 2.2. The energy gap also varies with temperature. The experimental variation of the energy gaps with temperature may be expressed within ± 0.003 eV by[28]

$$\boxed{E_g(T) = 1.170 - 4.73 \times 10^{-4} T^2/(T + 636) \text{ eV} \quad \text{for Si}} \tag{2.72}$$

and

$$\boxed{E_g(T) = 1.519 - 5.405 \times 10^{-4} T^2/(T + 204) \text{ eV} \quad \text{for GaAs}}\,. \tag{2.73}$$

The experimental energy gaps for Si and GaAs are shown in Fig. 2.28.

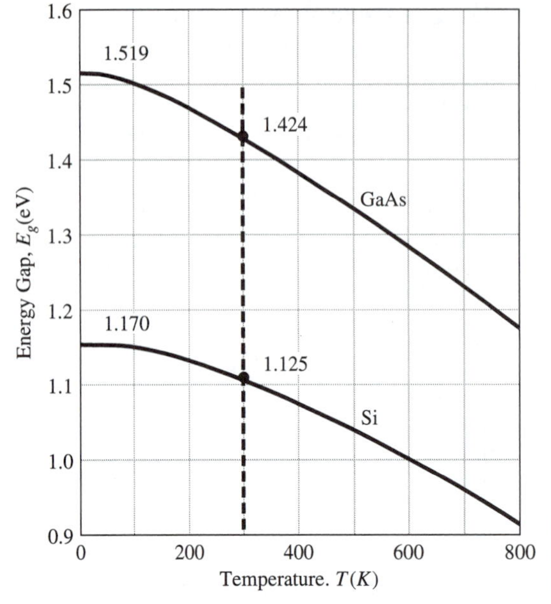

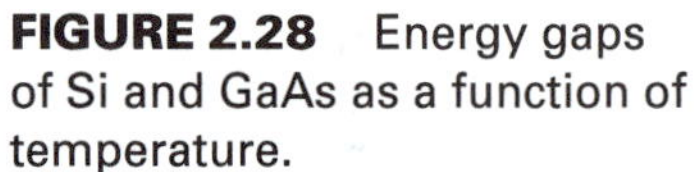
FIGURE 2.28 Energy gaps of Si and GaAs as a function of temperature.

2.7.3 Effective Mass Values

In Sec. 2.4.6 it was shown by Eq. (2.34) that the effective mass is given by the curvature of the E vs. k diagram, and the narrower the parabola, the smaller the effective mass. To obtain the density of states effective masses, the more complete energy-band structures are required instead of the simplified structures in Fig. 2.13 which were used to introduce the energy gap. The more complete band structure[29] for GaAs is shown in Fig. 2.29. The band structure has the notation from group theory. The conduction-band minimum at $k = 0$ is the Γ-point, while the minimum in the [100] direction is designated the X-point and the minimum in the [111] direction is the L-point. The valence band at $k = 0$ has the heavy-hole and light-hole bands, as well as the *split-off band.* The split-off band comes from the perturbation of the magnetic energy associated with the spin and orbit of the electron. The split-off band is separated by 0.35 eV from the heavy- and light-hole bands and can generally be ignored except for consideration of optical transitions.

When compounds are formed that have more than one group-III element distributed randomly on group-III lattice sites or more than one group-V element distributed randomly on group-V lattice sites, those compounds are *crystalline solid solutions*. The most commonly encountered III-V ternary solid solution is $Al_xGa_{1-x}As$, where x can vary from 0 to 1.0 and represents the fraction of the Al atoms on group-III lattice sites. The importance of $Al_xGa_{1-x}As$ is based on the close lattice match to GaAs for all values of x. For $x = 0$, $Al_xGa_{1-x}As$ reduces to GaAs and has the band structure shown in Fig. 2.29 with Γ, L, and X conduction-band minima. The energy-gap variation with composition for the Γ-direct conduction band at room temperature

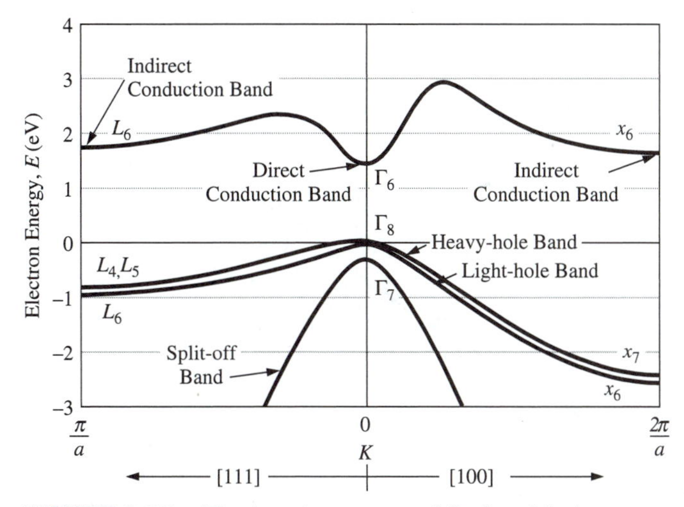

FIGURE 2.29 The band structure of GaAs with the energy E plotted as a function of $\mathbf{k}$ along the [100] and [111] directions. The Γ, X, and L designation identify the interband energy gaps.[29]

may be represented by[30]

$$E_g^{\Gamma} = 1.424 + 1.594x + x(1 - x)(0.127 - 1.310x) \text{ eV}. \tag{2.74}$$

The energy-gap variation of the L-indirect conduction band may be represented by[31]

$$E_g^{L} = 1.734 + 0.574x + 0.055x^2 \text{ eV}, \tag{2.75}$$

and the energy-gap variation of the X-indirect conduction band may be represented by[30]

$$E_g^{X} = 1.911 + 0.005x + 0.245x^2 \text{ eV}. \tag{2.76}$$

The valence-band maximum, E_v, is always at $k = 0$ for all compositions. Equations (2.74) through (2.76) for the room temperature compositional dependence of $Al_xGa_{1-x}As$ are plotted in Fig. 2.30 and show that the lowest conduction band changes from the Γ-direct conduction band to the X indirect-conduction band at $x = 0.4$. Because the energy gap for $Al_xGa_{1-x}As$ is larger than for GaAs, layers of $Al_xGa_{1-x}As$ are grown on GaAs to form *heterojunctions.* A heterojunction is the junction between two dissimilar semiconductors, and the most significant property is generally the energy-gap difference. The application of $Al_xGa_{1-x}As$/GaAs heterojunctions will be presented in Chapters 5, 6, and 9. Several other III-V ternary solid solutions have useful device applications such as $Ga_xIn_{1-x}P$, $Al_xIn_{1-x}As$, $Ga_xIn_{1-x}As$, and $Al_xGa_{1-x}N$. The quaternary solutions $Ga_xIn_{1-x}As_yP_{1-y}$ and $(Al_xGa_{1-x})_yIn_{1-y}P$ are also extensively used.

For GaAs, the effective masses have been evaluated by measuring the optical absorption of polarized light in a magnetic field (magnetooptical absorption). At 77 K,

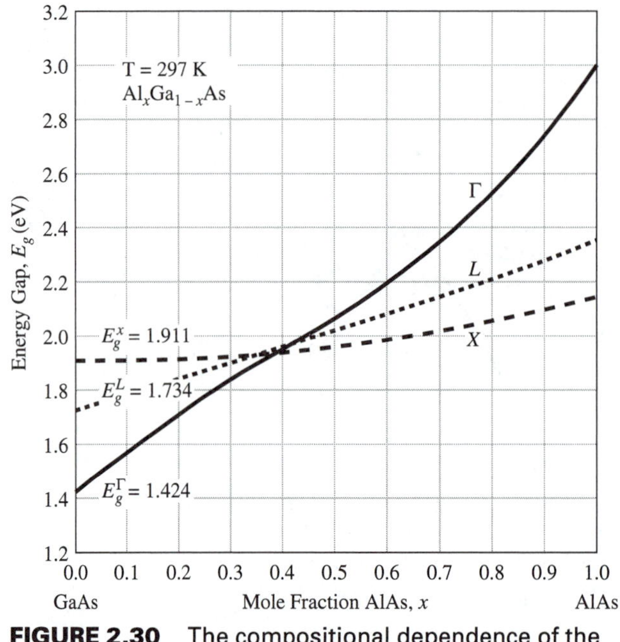

FIGURE 2.30 The compositional dependence of the direct $\Gamma_8 \rightarrow \Gamma_6$ and the indirect $\Gamma_8 \rightarrow X_6$ and $\Gamma_8 \rightarrow L_6$ energy gaps for $Al_xGa_{1-x}As$.

Vrehen[32] obtained $m_n^* = 0.067m_0$ for the Γ-direct conduction band. The density of states effective mass in the valence band must include the holes in both the heavy-hole band and the light-hole band. For the valence band,[32] the heavy-hole effective mass was $m_{hp}^* = 0.45m_0$ and the light-hole mass was $m_{lp}^* = 0.082m_0$. The hole concentration p given in Eq. (2.64) must be written as the sum of the concentration of holes in the heavy-hole valence band and the light-hole valence band with N_v given by Eq. (2.67):

$$\begin{aligned} p &= 2.5 \times 10^{19}(m_p^*/m_0)^{3/2}(T/300)^{3/2}\exp[-(E_f - E_v)/kT] \\ &= 2.5 \times 10^{19}(T/300)^{3/2}\exp[-(E_f - E_v)/kT] \\ &\quad \times [(m_{hp}^*/m_0)^{3/2} + (m_{lp}^*/m_0)^{3/2}]\,, \end{aligned} \tag{2.77}$$

which results in an effective hole mass given by

$$m_p^*/m_0 = [(m_{hp}^*/m_0)^{3/2} + (m_{lp}^*/m_0)^{3/2}]^{2/3}. \tag{2.78}$$

With the values for GaAs, the density of states effective mass for holes is $m_p^* = 0.47m_0$. In summary, for GaAs

GaAs

$$\boxed{m_n^* = 0.067m_0} \tag{2.79}$$

and

$$\boxed{m_p^* = 0.47m_0} \,. \tag{2.80}$$

conductivity effective mass $m_{X1}^* = m_c^*$

The more complete band structure[33] for Si is shown in Fig. 2.31. The indirect energy gap is between the $\Gamma_{25'}$-point and the X_1-point in the [100] direction. The separation of the split-off valence band is 0.044 eV. For Si, the effective masses have been evaluated at low temperature by measuring the microwave absorption in a magnetic field (cyclotron resonance).[34] The effective (conductivity) mass in the X-indirect conduction band is $0.32m_0$. As shown in Fig. 2.8, there are six {100} planes. Therefore, there are six equivalent X-conduction-band minima which contain electrons. The effective mass of a single minimum will be designated $m_{X1}^* = 0.32m_0$, and the density of states effective mass will have to include electrons in all six minima. Summing the electron concentrations in all six X-conduction bands by Eq. (2.55) gives

$$\begin{aligned} n = N_c \exp[-(E_c - E_f)/kT] = {} & N_{c1} \exp[-(E_c - E_f)/kT] \\ & + N_{c2} \exp[-(E_c - E_f)/kT] \\ & + \cdots + N_{c6} \exp[-(E_c - E_f)/kT] \,, \end{aligned} \tag{2.81}$$

or by Eq. (2.66) for N_c,

$$(m_n^*/m_0)^{3/2} = (m_{X1}^*/m_0)^{3/2} + (m_{X2}^*/m_0)^{3/2} + \cdots + (m_{X6}^*/m_0)^{3/2} \,. \tag{2.82}$$

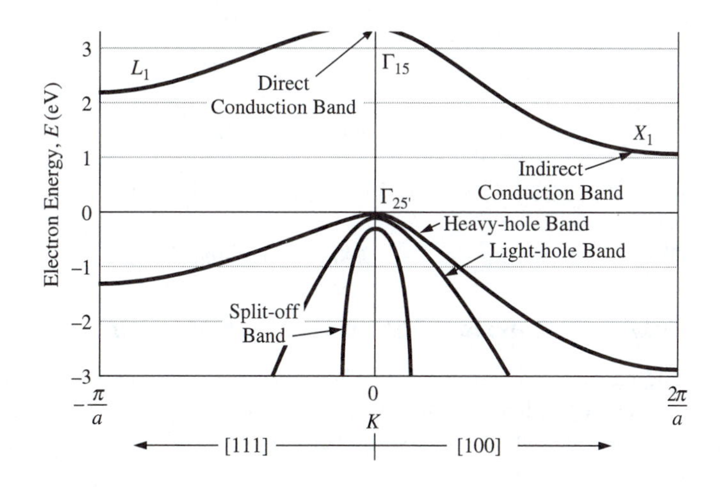

FIGURE 2.31 The band structure of Si with the energy E plotted as a function of **k** along the [100] and [111] directions. The Γ, X, and L designation identify the interband energy gaps.

With

$$m_{X1}^*/m_0 = m_{X2}^*/m_0 = \cdots = m_{X6}^*/m_0. \tag{2.83}$$

Eq. (2.82) becomes

$$(m_n^*/m_0)^{3/2} = 6(m_{X1}^*/m_0)^{3/2}\,, \tag{2.84}$$

density of states effective mass m_n^* for Si

or

$$m_n^* = 6^{2/3} m_{X1}^* = 1.06 m_0 \tag{2.85}$$

at low temperature for the density of states effective mass in the conduction band. *The density of states effective mass* m_n^* *given by Eq. (2.85) is only used to evaluate* N_c, *and the effective mass for a single minimum* m_{X1}^*, *the conductivity effective mass, is used for other evaluations such as drift in an electric field.* Green[35] made a careful evaluation of the literature to resolve several inconsistencies in the density of states effective masses and the intrinsic carrier concentration. Those results demonstrated a significant temperature dependence for the hole density of states effective mass with an increase from $0.59m_0$ at 4.2 K to $1.15m_0$ at 300 K. A summary of Green's results for m_n^*, m_v^*, and n_i from 4.2 K to 500 K are given in the next part of this section.

The effective mass for $Al_xGa_{1-x}As$ may be represented by a linear interpolation of the density of states effective mass between GaAs and AlAs. The electron effective mass for the indicated conduction band would be[36]

$$m_n^\Gamma = (0.067 + 0.083x)m_0\,, \tag{2.86}$$

$$m_n^L = (0.55 + 0.12x)m_0\,, \tag{2.87}$$

and,

$$m_n^X = (0.85 - 0.07x)m_0. \tag{2.88}$$

The density of states hole mass at Γ for $Al_xGa_{1-x}As$ would be[36]

$$m_p = (0.47 + 0.32x)m_0. \tag{2.89}$$

2.7.4 Intrinsic Carrier Concentration Values

GaAs

The intrinsic carrier concentration for GaAs can be calculated from Eq. (2.71) with the temperature dependence of the energy gap given by Eq. (2.73) and Eqs. (2.79) and (2.80) for the density of states effective masses. At 300 K, the intrinsic carrier concentration for GaAs is

$$\boxed{n_i = 2.4 \times 10^6\ \mathrm{cm}^{-3}}\,. \tag{2.90}$$

No information is presently available in regard to the possible temperature dependence of the density of states effective masses.

For Si, the generally accepted value for n_i at 300 K has been 1.45×10^{10} cm^{-3}. This value cannot be reconciled with the value calculated from the experimentally measured energy gap and the effective masses. *Green*[35] *resolved this discrepancy and showed that the hole density of states effective mass has a significant temperature dependence due to the anisotropy and nonparabolicity of the valence bands.* The temperature dependence of the energy gap was given by Eq. (2.72). Green's[35] values for E_g, m_n^*, m_p^*, and n_i between 4.2 K and 500 K are given in Table 2.3. For higher temperatures, Eq. (2.72) may still be used for E_g. The value for m_n^* at 500 K, $m_n^* = 1.13m_0$, may be used for $T > 500$ K, while a simple linear representation may be used for m_p^*:

$$m_p^*/m_0 = 1.29 + 1.80 \times 10^{-4}(T - 500)\,. \tag{2.91}$$

Values for n_i at higher temperature are used to determine impurity incorporation and diffusion during high-temperature processing.

More recent measurements of n_i for Si at 300 K gave $n_i = 1.0 \times 10^{10}$ cm^{-3}.[37,38] If n_i is calculated from Eq. (2.71) with $E_g(300\text{ K}) = 1.125$ eV from Eq. (2.72) and with m_n^* and m_p^* from Table 2.3, then $n_i = 1.06 \times 10^{10}$ cm^{-3}, which is within 6% of the most recent values for n_i. Therefore, in this textbook at 300 K n_i for Si will be taken as 1.0×10^{10} cm^{-3} with the energy gap given by Eq. (2.72) as 1.125 eV. For other temperatures, n_i will be given by $E_g(T)$ from Eq. (2.72), and m_n^* and m_p^* interpolated from the values given in Table 2.3.

$n_i^{Si}(300\text{ K}) = 1.0 \times 10^{10}$ cm^{-3}
$E_g^{Si}(300\text{ K}) = 1.125$ eV

The basic concepts which permit evaluation of n_i have been presented here because numerical values often are not available for other semiconductors and n_i is needed for numerical expressions used in the analysis and design of both bipolar and field-effect devices.

TABLE 2.3 Temperature Dependence for the Si Energy Gap, E_g, the Density of States Effective Masses, m_n^* and m_p^*, and the Intrinsic Carrier Concentration, n_i.

T (K)	E_g (eV)	m_n^*/m_0	m_p^*/m_0	n_i(cm^{-3})
4.2	1.170	1.06	0.59	3.14×10^{-686}
50	1.169	1.06	0.69	1.64×10^{-41}
100	1.165	1.06	0.83	1.95×10^{-11}
150	1.158	1.07	0.95	3.16×10^{-1}
200	1.148	1.08	1.03	5.03×10^{4}
250	1.137	1.08	1.10	7.59×10^{7}
300	1.124	1.09	1.15	1.07×10^{10}
350	1.110	1.10	1.19	3.92×10^{11}
400	1.097	1.11	1.23	6.00×10^{12}
450	1.083	1.12	1.29	5.11×10^{13}
500	1.070	1.13	1.29	2.89×10^{14}

Source: Green, J. Appl. Phys. **67**, 2944 (1990).[35]

2.8 DONORS AND ACCEPTORS

2.8.1 Donor and Acceptor Ionization Energies

Although Si has 14 electrons, only the four valence electrons are able to enter chemical reactions, and the remaining electrons are closely bound to the nucleus, producing a stable ionic core which is inactive. These four valence electrons form electron-pair bonds with the four neighboring atoms. Instead of the three-dimensional representation of the Si bonding, as shown in Fig. 2.6, it is convenient to use a two-dimensional representation, as shown in Fig. 2.32. The thermal excitation breaks these covalent bonds to form electron–hole pairs in the intrinsic semiconductor. Intrinsic Si at room temperature has a relatively high resistivity of approximately 1×10^5 ohm-cm. The addition of certain elements, often called *impurities* or *dopants,* to a semiconductor crystal causes the electron or hole concentration to exceed the intrinsic carrier concentration, and the

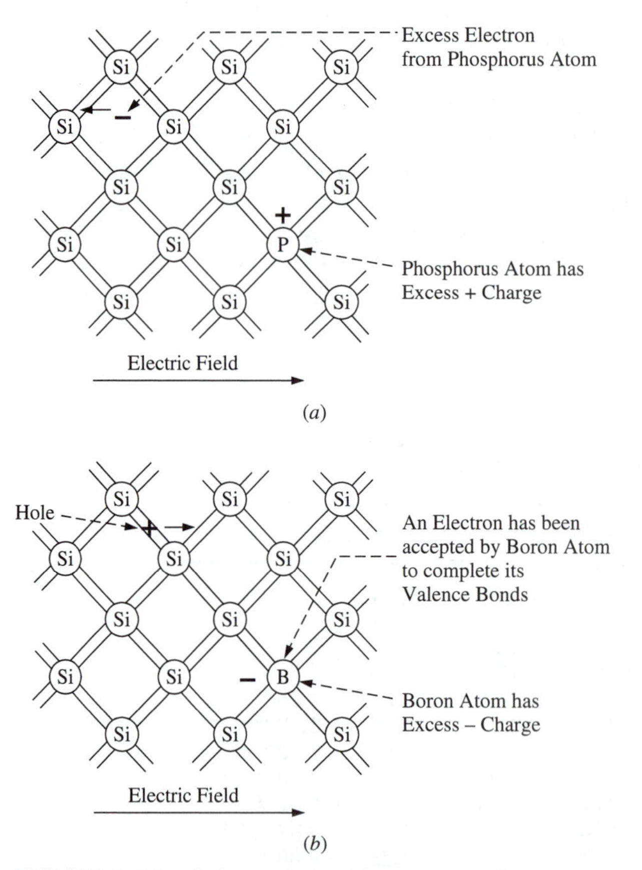

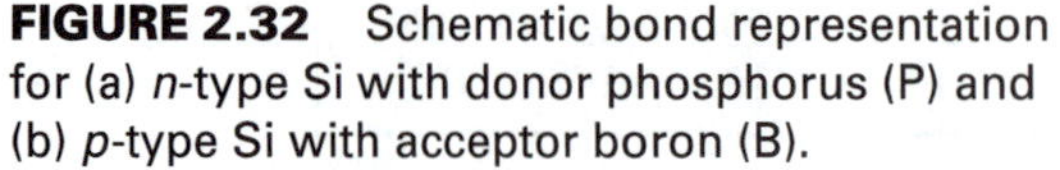

FIGURE 2.32 Schematic bond representation for (a) *n*-type Si with donor phosphorus (P) and (b) *p*-type Si with acceptor boron (B).

semiconductor becomes *extrinsic*. The intentional introduction of impurities into the semiconductor crystal is commonly called *doping*.

If an element from column V of the periodic table, such as phosphorus (P), substitutes for a Si lattice atom, as shown in Fig. 2.32(a), it forms four covalent bonds with the neighboring Si atoms. The fifth electron is considered to be similar to the hydrogen atom except that it is embedded in a dielectric material with a permittivity ϵ_s, and the electron has the effective mass (conductivity effective mass) of a carrier electron in the crystal. This electron, when bound to the impurity atom, has an orbit radius given by Eq. (2.6) as

$$r = 0.529 \times 10^{-8}(\epsilon_s/\epsilon_0)/(m^*/m_0) \text{ cm}, \tag{2.92}$$

which is a larger radius than for hydrogen when $m^* < m_0$. The energy necessary to free or *ionize* this fifth electron may be estimated from Eq. (2.9) for the ionization of hydrogen by replacing m_0 with the electron effective mass m^* and replacing the permittivity ϵ_0 with the permittivity of the semiconductor ϵ_s:

$$E_d = 13.6\frac{m^*/m_0}{(\epsilon_s/\epsilon_0)^2} \text{ eV}, \tag{2.93}$$

where E_d is called the *ionization energy.* For Si, the effective mass in Eqs. (2.92) and (2.93) is for a *single* conduction band minimum, $m^*_{X1} = 0.32m_0$, and the permittivity is $11.7\epsilon_0$.

donors

Impurities which have an extra electron that can easily be excited into the conduction band are called *donors.* The hydrogenic ionization energy for donors in Si (measured from the conduction-band edge) is given by Eq. (2.93) as 0.032 eV. For GaAs, the permittivity is $13.1\epsilon_0$ and the hydrogenic donor ionization energy is 0.005 eV. A donor atom whose electron has been excited into the conduction band has become *ionized* and the donor atom then has a net charge of $+q$ and is represented by N_d^+. A semiconductor with excess electrons due to donors is called *n-type.* This extra electron, when ionized to the conduction band, can drift in an electric field as shown in Fig. 2.32(a).

ionized donor N_d^+

Similarly, Fig. 2.32(b) shows schematically the replacement of a Si atom by an element from column III of the periodic table such as boron (B). The B atom would have one less outer electron than the four needed for bonding. An electron from the valence band can get enough energy to form the missing bond and leave a hole in the valence band. An element which can accept a bonding electron from the valence band is called an *acceptor.* The acceptor with the additional electron from the valence band has become *ionized* and the acceptor atom then has a net charge of $-q$ and is represented as N_a^-. A semiconductor with excess holes due to acceptors is called *p-type.* The hydrogenic ionization energy for acceptors in Si (measured from the top of the valence band) is given by Eq. (2.93) as 0.059 eV, and the hydrogenic acceptor ionization energy is 0.037 eV for GaAs. This hole can drift in an electric field as shown in Fig. 2.32(b).

acceptors

ionized acceptor N_a^-

An abbreviated periodic table is given in Table 2.4 and includes elements important for Si and III-V compound semiconductors. Elements of similar chemical

TABLE 2.4 Partial Periodic Table for the Elements Important for Si and the III-V Compound Semiconductors.

Column Number				
II	III	IV	V	VI
4* Beryllium Be	5 Boron B	6 Carbon C	7 Nitrogen N	8 Oxygen O
12 Magnesium Mg	13 Aluminum Al	14 Silicon Si	15 Phosphorus P	16 Sulfur S
30 Zinc Zn	31 Gallium Ga	32 Germanium Ge	33 Arsenic As	34 Selenium Se
48 Cadmium Cd	49 Indium In	50 Tin Sn	51 Antimony Sb	52 Tellurium Te

*Atomic number of the element which is the number of positive charges in the nucleus.

properties are in the same column and are known as a *group*. The group-IV elements have already been described with four valence electrons. Elements with consecutive atomic numbers are in the same row.

shallow impurities

deep levels

Impurities with ionization energies near the hydrogenic value are labeled the shallow impurities, while those with larger ionization energy values are designated deep levels. This simple hydrogen atom model cannot account for many of the details of ionization energy, particularly for deep impurity levels. However, the calculated values do predict the correct approximate value of the experimentally measured ionization energies for shallow impurity levels. Figure 2.33 summarizes the measured ionization energies for various impurities in Si and GaAs. Note that it is possible for some elements to have more than one level. For example, oxygen (O) in Si has two donor levels and two acceptor levels in the energy gap. An important crystal defect which behaves as a deep donor is the As-antisite defect shown by EL2 in Fig. 2.33. This defect is an As atom on a Ga site and has a donor energy 0.77 eV below E_c. A more thorough treatment of impurities in the III-V compound semiconductors is given in the book by Schubert.[39]

2.8.2 Majority and Minority Carriers

If both donor and acceptor impurities are present simultaneously, the impurity that is present in a greater concentration determines the type of conductivity in the semiconductor. The Fermi level must adjust itself to preserve charge neutrality in a *homoge-*

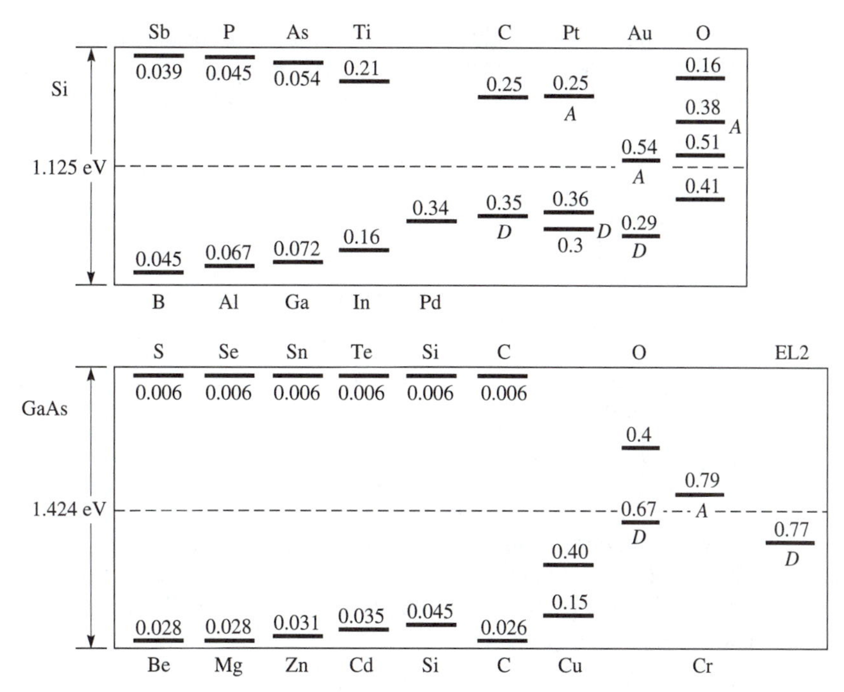

FIGURE 2.33 Measured ionization energies for various impurities in Si and GaAs. The levels below the energy gap center are measured from the top of the valence band and are acceptor levels unless indicated by D for a donor level. The levels above the energy gap center are measured from the bottom of the conduction band and are donor levels unless indicated by an A for acceptor level.

electrical neutrality

neous semiconductor. For charge neutrality, the total negative charges (electrons and ionized acceptors) must equal the total positive charges (holes and ionized donors):

$$\boxed{n + \sum N_a^- = p + \sum N_d^+} \,. \tag{2.94}$$

Electrical neutrality as expressed in Eq. (2.94) applies to both thermal equilibrium and steady-state conditions (nonthermal equilibrium).

As shown in Appendix A.1 at the end of this chapter, the fraction of the donors or acceptors which are ionized depends on the impurity ionization energy, the position of the Fermi level, and the temperature. To concentrate on device behavior in this text, the ionized impurity concentration will be given rather than the total concentration. Also, as the impurity concentration is increased, the ionization energy will become smaller and go to zero. This concentration dependence of the ionization energy is presented in Appendix A.2. Consideration of the behavior of actual semiconductor devices may require application of the concepts presented in the Appendix, especially at high impurity concentrations and at low temperatures.

When the semiconductor is in thermal equilibrium, the electron and hole concentrations are related by Eq. (2.71) as $np = n_i^2$. For the case when $N_d^+ > N_a^-$, p in Eq. (2.94) may be replaced by $p = n_i^2/n$ which gives the quadratic equation:

$$n^2 - (N_d^+ - N_a^-)n - n_i^2 = 0\,, \tag{2.95}$$

and

$$n = \frac{1}{2}\left[(N_d^+ - N_a^-) + \sqrt{(N_d^+ - N_a^-)^2 + 4n_i^2}\right]. \tag{2.96}$$

Typical net impurity concentrations $|N_d^+ - N_a^-|$ generally are much greater than n_i, and then Eq. (2.96) with $N_d^+ > N_a^-$ becomes

$$n_{no} = N_d^+ - N_a^-. \tag{2.97}$$

majority and minority carriers

The subscript n refers to the n-type semiconductor, and the subscript o represents thermal equilibrium. Because the electron is the dominant carrier, it is called the *majority carrier.* The hole in the n-type semiconductor is called the *minority carrier* and at thermal equilibrium is given by

$$p_{no} = n_i^2/n_{no}. \tag{2.98}$$

Similarly, the hole concentration (majority carrier) and the electron concentration (minority carrier) in a p-type semiconductor is given by

$$p_{po} = \frac{1}{2}\left[(N_a^- - N_d^+) + \sqrt{(N_a^- - N_d^+)^2 + 4n_i^2}\right] \tag{2.99}$$

and

$$n_{po} = n_i^2/p_{po}. \tag{2.100}$$

The subscript p refers to the p-type semiconductor, and the subscript o represents thermal equilibrium. Generally, the preceding relationships simplify to

$$n_{no} = N_d^+ - N_a^- \quad \text{or} \quad n_{no} = N_d^+ \quad \text{if} \quad N_d^+ > N_a^- \tag{2.101}$$

and

$$p_{po} = N_a^- - N_d^+ \quad \text{or} \quad p_{po} = N_a^- \quad \text{if} \quad N_a^- > N_d^+\,. \tag{2.102}$$

It is important to note that electrical neutrality given by Eq. (2.94) holds for *thermal equilibrium* and *non-thermal equilibrium.* The equilibrium relationship $np = n_i^2$ given in Eq. 2.71 only holds for *thermal equilibrium.*

2.8.3 Intrinsic Fermi Level

It is convenient to express electron and hole concentrations in terms of the intrinsic carrier concentration n_i and the Fermi level of the intrinsic semiconductor E_i. The intrinsic Fermi level E_i is frequently used as a reference level when discussing extrinsic semiconductors. First it is useful to obtain an expression for E_i. From Eqs. (2.55) and (2.64),

$$n_i = N_c \exp[-(E_c - E_i)/kT] = p_i = N_v \exp[-(E_i - E_v)/kT]. \quad (2.103)$$

Then,

$$N_c \exp[-(E_c - E_i)/kT] = N_v \exp[-(E_i - E_v)/kT], \quad (2.104)$$

or

$$N_c/N_v = (m_n^*/m_p^*)^{3/2} = \exp\{-[(E_i - E_v)/kT] + [(E_c - E_i)/kT]\}. \quad (2.105)$$

For $E_v = 0$ and $E_c = E_g$, Eq. (2.105) becomes

$$(m_n^*/m_p^*)^{3/2} = \exp[-(2E_i - E_g)/kT], \quad (2.106)$$

or

$$\boxed{E_i = E_g/2 - (3kT/4)\ln(m_n^*/m_p^*)}. \quad (2.107)$$

intrinsic Fermi level

For Si at 300 K, $m_n^*/m_p^* \simeq 1.0$ and $E_i \simeq E_g/2$, which is 0.562 eV. For GaAs at 300 K, $m_n^*/m_p^* = 0.143$ and $E_i = 0.712 + 0.038 = 0.750$ eV. The electron concentration may be written in terms of n_i and E_i by obtaining N_c from Eq. (2.55) as

$$N_c = n_i \exp[(E_c - E_i)/kT]; \quad (2.108)$$

then

n in terms of n_i and E_f

$$\boxed{n = n_i \exp[(E_f - E_i)/kT]}. \quad (2.109)$$

Similarly from Eq. (2.64),

p in terms of n_i and E_f

$$\boxed{p = n_i \exp[(E_i - E_f)/kT]}. \quad (2.110)$$

EXAMPLE 2.5 A Si wafer is doped with the donor phosphorus (P) to give $n_{no} = 1\times10^{16}$ cm^{-3}. Find the Fermi level and hole concentration at room temperature (300 K).

Solution With $n_i = 1.0 \times 10^{10}$ cm^{-3} and by Eq. (2.71)

$$p_{no} = n_i^2/n_{no} = (1.0 \times 10^{10})^2/1 \times 10^{16} = 1.0 \times 10^4 \text{ cm}^{-3}.$$

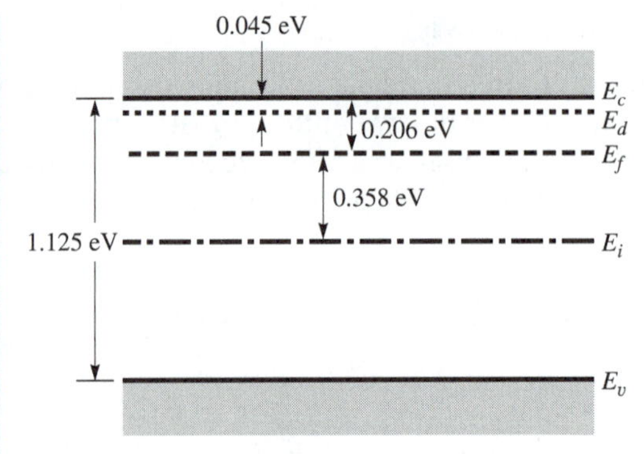

FIGURE 2.34 Energy band diagram showing the Fermi level E_f and the intrinsic Fermi level E_i.

The Fermi level measured from the bottom of the conduction band as given by Eq. (2.56) is

$$E_c - E_f = -kT \ln(n/N_c) = -0.0259 \ln\left[\frac{1 \times 10^{16}}{2.84 \times 10^{19}}\right] = 0.206 \text{ eV}.$$

The Fermi level measured from the intrinsic Fermi level is given by Eq. (2.109) as

$$E_f - E_i = kT \ln(n/n_i) = 0.0259 \ln\left[\frac{1 \times 10^{16}}{1.0 \times 10^{10}}\right] = 0.358 \text{ eV}.$$

These results are shown graphically in Fig. 2.34. ■

2.9 SUMMARY AND USEFUL EXPRESSIONS

The following significant concepts were introduced in this chapter:

- Particle–wave duality of the electron.
- Much of the quantum mechanical behavior of the free electrons and holes in the crystal are summarized by the energy gap and effective masses for the conduction and valence bands.
- Counting the allowed energies (states) for electrons in a crystal per cm^3 per eV, which are the allowed solutions to the Schrödinger equation, gives the density of states as $\rho(E) = (1/2\pi)^2(2m^*/\hbar^2)^{3/2}E^{1/2}$.
- The probability that a state within the conduction or valence band contains an electron is given by the Fermi-Dirac distribution function.
- When the Fermi-Dirac distribution function can be represented by the Boltzmann approximation, the free electron concentration n can be represented by $n = N_c \exp[-(E_c - E_f)/kT]$, where N_c is the effective density of states in the conduction band, E_c is the conduction band edge, and E_f is the Fermi level.

- For Si, the single-conduction-band effective mass, called the conductivity effective mass $m^*_{X1} = m^*_c$, is used in calculations of the motion of electrons in the crystal, while the density of states effective mass m^*_n combines the six equivalent conduction-band minima as $m^*_n = 6^{2/3} m^*_{X1}$ and is used in N_c for the number of free carriers.
- The carrier concentration due to thermal excitation alone is called the intrinsic carrier concentration n_i, and at thermal equilibrium, $n_o p_o = n_i^2$.
- A semiconductor is extrinsic when the presence of impurities gives an electron or hole concentration greater than n_i.
- Donor impurities, such as phosphorus in Si, make the semiconductor *n*-type, with the electrons as the majority carriers and the holes as the minority carriers.
- Acceptor impurities, such a boron in Si, make the semiconductor *p*-type, with the holes as the majority carriers and the electrons as the minority carriers.
- In a uniformly doped semiconductor at thermal equilibrium, the Fermi level adjusts to a position to ensure equality of positive and negative charges; that is, electrical neutrality is maintained.

From the numerous equations given in this chapter, the following parameters and expressions summarize the most important results for electrons in solids.

Physical Constants	**Units**
$m_0 = 9.1 \times 10^{-28}$ g	1 eV $= 1.6 \times 10^{-19}$ joules
$q = 1.6 \times 10^{-19}$ coulombs	1 eV $= 1.6 \times 10^{-12}$ ergs
$k = 8.616 \times 10^{-5}$ eV/K	volt = joule/coulomb
$k = 1.38 \times 10^{-16}$ erg/K	ampere = coulomb/s
$h = 4.135 \times 10^{-15}$ eV-s	farad = coulomb/volt
$h = 6.616 \times 10^{-27}$ ergs-s	
$\hbar = 1.054 \times 10^{-27}$ ergs-s	
$\epsilon_0 = 8.85 \times 10^{-14}$ F/cm	

Parameter	Si Value	GaAs Value
m^*_{x1}(300 K)	$0.32m_0$	—
m^*_n(300 K)	$1.09m_0$	$0.067m_0$
m^*_p(300 K)	$1.15m_0$	$0.47m_0$
ϵ	$11.7\epsilon_0$	$13.1\epsilon_0$
E_g(eV)	$1.170 - \dfrac{4.73 \times 10^{-4} T^2}{(T + 636)}$	$1.519 - \dfrac{5.405 \times 10^{-4} T^2}{(T + 204)}$
E_g(300 K)	1.125 eV	1.424 eV
N_c(300 K)	2.84×10^{19} cm^{-3}	4.34×10^{17} cm^{-3}
N_v(300 K)	3.08×10^{19} cm^{-3}	8.06×10^{18} cm^{-3}
n_i(300 K)	1.0×10^{10} cm^{-3}	2.4×10^{6} cm^{-3}

Electrons in Solids

$$|\mathbf{p}| = h/\lambda = |\hbar \mathbf{k}|$$

$$|\mathbf{k}| = 2\pi/\lambda$$

$$f(E) = 1/[1 + \exp(E - E_f)/kT]$$

$$N_{c \text{ or } v} = 2.5 \times 10^{19}(m^*/m_0)^{3/2}(T/300)^{3/2}$$

$$E_f = E_c + kT \ln(n/N_c) = E_v - kT \ln(p/N_v) \quad \text{for} \quad n/N_c < 0.1 \quad \text{or} \quad p/N_v < 0.1$$

$$E_f = E_c + kT \left[\ln(n/N_c) + \frac{(n/N_c)}{\{64 + 0.05524(n/N_c)[64 + (n/N_c)^{1/2}]\}^{1/4}} \right] \quad \text{for} \quad n/N_c > 0.1$$

$$np = n_i^2$$

$$E_i = E_g/2 - (3kT/4)\ln(m_n^*/m_p^*)$$

$$n + N_a^- = p + N_d^+$$

REFERENCES

1. E. Rutherford, "The Scattering of α and β Particles by Matter and the Structure of the Atom," Philos. Mag. **21**, 669 (1911). (For an introductory presentation of the development of quantum theory, see R. B. Leighton, *Principles of Quantum Theory* (McGraw-Hill, New York, 1959).
2. J. J. Balmer, "Notiz über die Spectrallinien des Wasserstoffs," Ann. Phys. **25**, 80 (1885).
3. N. Bohr, "On the Constitution of Atoms and Molecules," Philos. Mag. **26**, 1 (1913).
4. M. Planck, "Distribution of Energy in the Spectrum," Ann. Phys. **4**, 553 (1901).
5. L. de Broglie, "A Tentative Theory of Light Quanta," Philos. Mag. **47**, 446 (1924).
6. C. Davisson and L. H. Germer, "Diffraction of Electrons by a Crystal of Nickel," Phys. Rev. **30**, 705 (1927).
7. R. T. Weidner and R. L. Sells, *Elements of Modern Physics* (Allyn & Bacon, Boston, 1960), p. 184.
8. E. Schrödinger, "Quantization as an Eigenvalue Problem," Ann. Phys. **79**, 489 (1926).
9. M. Born, "Quantum Mechanics of Collisions," Z. Phys. **37**, 863 (1926).
10. S. M. Sze, *Semiconductor Devices: Physics and Technology* (Wiley, New York, 1985), pp. 3-7.
11. C. Kittel, *Introduction to Solid State Physics,* 7th ed. (Wiley, New York, 1996), p. 12.
12. R. de L. Kronig and W. G. Penny, "Quantum Mechanics of Electrons in Crystal Lattices," Proc. R. Soc. (Lond.), **A130**, 499 (1930).
13. For a detailed solution, see S. Wang, *Fundamentals of Semiconductor Theory and Device Physics* (Prentice-Hall, Englewood Cliffs, N.J., 1989), pp. 146-154.
14. An additional useful description is given by R. F. Pierret, *Advanced Semiconductor Fundamentals,* Vol. VI, *Modular Series on Solid State Devices,* R. F. Pierret and G. W. Neudeck, Eds. (Addison-Wesley, Reading, Mass., 1987), pp. 53–55.
15. Weidner and Sells, *Elements of Modern Physics,* p. 134.

16. Sze, *Semiconductor Devices,* p. 14.

17. W. B. Jones, Jr., *Introduction to Optical Fiber Communication Systems* (Holt, Rinehart & Winston, New York, 1988), pp. 44–46.

18. C. W. Sherwin, *Introduction to Quantum Mechanics* (Holt, Rinehart & Winston, New York, 1959), p. 132.

19. C. M. Wolfe, N. Holonyak, Jr., and G. E. Stillman, *Physical Properties of Semiconductors* (Prentice Hall, New York, 1989), p. 155.

20. Sze, *Semiconductor Devices,* p. 17.

21. C. Kittel and H. Kroemer, *Thermal Physics* (Freeman, San Francisco, 1980), p. 154.

22. R. A. Swalin, *Thermodynamics of Solids* (Wiley, New York, 1962), p. 253.

23. L. S. Darken and R. W. Gurry, *Physical Chemistry of Metals* (McGraw-Hill, New York, 1953), p. 145.

24. W. B. Joyce and R. W. Dixon, "Analytic Approximation for the Fermi Energy of an Ideal Fermi Gas," Appl. Phys. Lett. **31**, 354 (1977).

25. N. G. Nilsson, Appl. Phys. Lett. **33**, 653 (1978).

26. A. S. Grove, *Physics and Technology of Semiconductor Devices* (Wiley, New York, 1967) p. 94.

27. B. N. Brockhouse, "Lattice Vibrations in Silicon and Germanium," Phys. Rev. Lett. **2**, 256 (1959).

28. C. D. Thurmond, "The Standard Thermodynamic Function of the Formation of Electrons and Holes in Ge, Si, GaAs, and GaP," J. Electrochem. Soc. **122**, 1133 (1975).

29. F. H. Pollak, C. W. Higginbotham, and M. Cardona, "Band Structure of GaAs, InP and AlSb: The kp Method," *Proc. Int. Conf. Phys. Semiconductors* (Kyoto, Japan, 1966) (J. Phys. Soc. Jpn. **21**, Supplement, 1966).

30. D. E. Aspnes, S. M. Kelso, R. A. Logan, and R. Bhat, "Optical Properties of $Al_xGa_{1-x}As$," J. Appl. Phys. **60**, 754 (1986).

31. H. J. Lee, L. Y. Juravel, J. C. Woolley, and A. J. SpringThrope, "Electron Transport and Band Structure of $Ga_{1-x}Al_xAs$ Alloys," Phys. Rev. B **21**, 659 (1980).

32. Q. H. F. Vrehen,"Interband Magneto-Optical Absorption in Gallium Arsenide," J. Chem. Phys. Solids **29**, 129 (1968).

33. J. R. Chelikowsky and M. L. Cohen, "Nonlocal Pseudopotential Calculation for the Electronic Structure of Eleven Diamond and Zinc-Blende Semiconductors," Phys. Rev. **B14**, 556 (1976).

34. J. C. Hensel, H. Hasegawa, and M. Nakayama, "Cyclotron Resonance in Uniaxially Stressed Silicon. II. Nature of the Covalent Bond," Phys. Rev. **138**, A225 (1965).

35. M. A. Green, "Intrinsic Concentration, Effective Density of States, and Effective Mass in Silicon," J. Appl. Phys. **67**, 2944 (1990).

36. H. C. Casey, Jr. and M. B. Panish, *Heterostructure Lasers, Part A: Fundamental Principles* (Academic Press, New York, 1978) p. 194.

37. A. B. Sproul and M. A. Green, "Intrinsic Carrier Concentration and Minority-Carrier Mobility of Silicon from 77 to 300 K," J. Appl. Phys. **73**, 1214 (1993).

38. K. Misiakos, "Accurate Measurement of the Silicon Intrinsic Carrier Density from 78 to 340 K," J. Appl. Phys. **74**, 3293 (1993).

39. E. F. Schubert, *Doping in III-V Semiconductors* (Cambridge University Press, Cambridge, 1993).

40. G. L. Pearson and J. Bardeen, "Electrical Properties of Pure Silicon and Silicon Alloys Containing Boron and Phosphorus," Phys. Rev. **75**, 865 (1949).

41. O. V. Emeĺyanenko, T. S. Lagunova, D. N. Nasledov, and G. N. Talalakin, "Formation and Properties of an Impurity Band in *n*-Type GaAs," Sov. Phys. Solid State **7**, 1063 (1965) [Translated from Fiz. Tverd. Tela. **7**, 1315 (1965)].

42. F. Ermanis and W. Wolfstirn, "Hall Effect and Resistivity of Zn-Doped GaAs," J. Appl. Phys. **37**, 1963 (1966).

43. N. F. Mott and W. D. Twose, "The Theory of Impurity Conduction," Adv. Phys. **10**, 107 (1961).

PROBLEMS

All problems are for room temperature unless another temperature is specified.

2.1. What is the de Broglie wavelength for a beam of electrons at room temperature ($T = 300$ K) whose kinetic energy is $(3/2)kT$?

2.2. What is the de Broglie wavelength for a beam of electrons whose kinetic energy is 100 eV?

2.3. What is the relationship between a free electron's velocity and its wavelength?

2.4. For GaAs, the electron mass m^* is 6.1×10^{-29} g. With the thermal velocity v_{th} given by $\sqrt{8kT/\pi m^*}$, what is the electron wavelength at 300 K?

2.5. To measure feature sizes which require greater magnification than possible with an optical microscope (greater than 2000×), a scanning-electron microscope is used. Electrons are accelerated to give a kinetic energy E.

Derive a relationship for electron *wavelength* in terms of the kinetic energy E and free electron mass m_0.

2.6. For InP, the electron mass m^* is 7.0×10^{-29} g. With the thermal velocity v_{th} given by $\sqrt{8kT/\pi m^*}$, what is the electron wavelength at 300 K?

2.7. Plot E vs. k for a free electron from $E = 0$ to $E = 0.25$ eV.

2.8. Plot E vs. k for a free electron from $E = 0$ to $E = 6kT$ at $T = 300$ K.

2.9. How far does the plane wave $\mathscr{E}_x(z, t) = A\cos(\omega t - kz)$ travel in the time $2\pi/\omega$?

2.10. How far does the plane wave $\mathscr{E}_x(z, t) = A\cos(\omega t - kz)$ travel in the time $5\pi/\omega$?

2.11. Plot the plane wave $\mathscr{E}_x(z, t) = A\cos(\omega t - kz)$ for $A = 1$ and for $t = 0$ and $t = \pi/4\omega$.

2.12. Plot the plane wave $\mathscr{E}_x(z, t) = A\cos(\omega t - kz)$ for $A = 1$ and for $t = 0$ and $t = \pi/\omega$.

2.13. **(a)** What is the distance between nearest neighbors in Si?

(b) For the diamond lattice, there are eight atoms at the corners and each atom is shared by eight adjoining unit cells to give $8 \times \frac{1}{8} = 1$ atom per unit cell due to the atoms at the the corners. On the six cube faces, each atom is shared by two adjoining unit cells. Four atoms are entirely within the unit cell and not shared. In a plane such as the (100), an atom at a corner occupies 90° out of 360° in the plane to give $\frac{1}{4}$ atom. *Find the number of atoms per cm*^{2} *in Si* in the (100), (110), and (111) planes. It is helpful to make a copy of Fig. 2.6(a) and draw the (100), (110), and (111) planes on the unit cell. Note that only the atoms at the corners and in the cube faces are in these planes.

2.14. If a plane has intercepts at $2a$, $3a$, and $4a$ along the three Cartesian coordinates where a is the lattice constant, find the Miller indices of the plane.

2.15. For a crystal with $E = \hbar^2 k^2 / 0.5 m_0$, what is the effective mass?

2.16. Consider two semiconductors with E vs. k diagrams for the conduction band as shown. Both have $n = 1 \times 10^{16}$ cm^{-3} in the conduction band at $T = 300$ K.

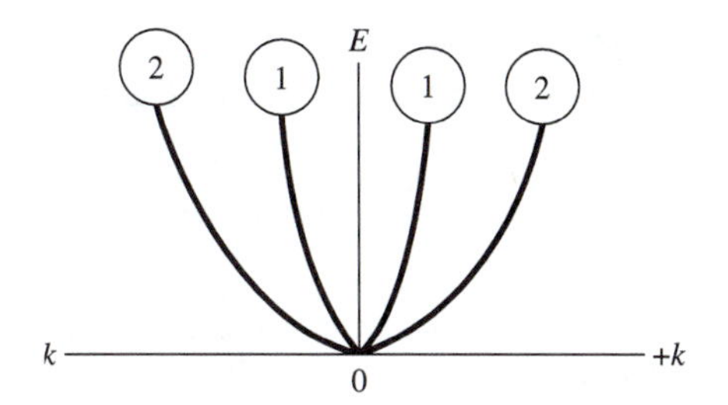

(a) In which case (1, 2, or both) will the effective density of states N_c in the conduction band be the greatest? Also, give the reason for your choice.

(b) In which case will the Fermi level be located closest to the conduction band? Also, give the reason for your choice.

2.17. Find the intrinsic carrier concentration n_i in GaAs at 900°C.

2.18. Consider a photon with an energy of 1.5 eV and a phonon with an energy of 0.026 eV. If the velocity of sound in Si is 7.5×10^5 cm/s, what is the ratio of the phonon momentum in Si to the photon momentum?

2.19. The Example in Sec. 2.7.2 showed that the wave vector difference (Δk) between E_v and E_c was $\Delta k \approx 1 \times 10^8$ cm^{-1}. For a phonon with energy of 0.058 eV and a velocity of 9.2×10^5 cm/s, show that the phonon momentum is approximately equal to Δk between E_v and E_c.

2.20. Silicon is an indirect energy gap semiconductor. What is meant by an indirect-energy gap?

2.21. GaAs is a direct-energy-gap semiconductor. What is meant by a direct-energy gap?

2.22. Calculate the density of states for a two-dimensional electron gas. Take the electron degeneracy as two.

2.23. Show that $2(2\pi m_n^* kT/h^2)^{3/2}$ reduces to $2.5 \times 10^{19} (m_n^*/m_0)^{3/2} (T/300)^{3/2}$ cm^{-3}.

2.24. Consider an indirect-energy-gap semiconductor with four equivalent minima in the conduction band. With an effective mass of $0.22 m_0$ for a single-conduction-band minimum, what is the density of states effective mass for electrons?

2.25. Plot the Fermi-Dirac distribution function at 27°C and for $E - E_f = 0.0$ eV, ± 0.030 eV, ± 0.060 eV, and ± 0.090 eV. Repeat for $T = 50$ K and 100°C. Plot all the curves on the same figure.

2.26. For the case of the Boltzmann approximation in Si, (a) what is the Fermi level measured from E_c at 27°C, and for an electron concentration of 3×10^{17} cm^{-3}? (b) What is the Fermi level measured from E_v for a hole concentration of 1×10^{16} cm^{-3}?

2.27. The III-V compound semiconductor GaP has a density of states effective mass for electrons of $0.82m_0$ and a density of states effective mass for holes of $0.60m_0$. The energy gap is 2.261 eV. For an ionized donor concentration $n = N_d^+$ of 5.0×10^{16} cm^{-3}, find the Fermi level at room temperature with respect to the conduction band edge.

2.28. Plot the Fermi level in Si measured from E_c vs. the log of concentration for $n = 1 \times 10^{16}$ to $n = 1 \times 10^{20}$ at 27°C for E_f obtained by the Boltzmann approximation and compare with E_f obtained from the Nilsson expression.

2.29. What is the Boltzmann approximation?

2.30. The III-V compound semiconductor InP has a density of states effective mass for electrons of $0.077m_0$ and a density of states effective mass for holes of $0.64m_0$. The energy gap is 1.351 eV. Find the intrinsic carrier concentration at room temperature.

2.31. The group-IV element germanium (Ge) is substitutional on Ga sites in GaAs. Is Ge a donor or an acceptor?

2.32. For Si at 27°C, calculate the Fermi level by the Boltzmann approximation expression and compare with the Fermi level obtained with the correct formula. Tabulate the results for the following electron concentrations: $n = 1 \times 10^{16}$ cm^{-3}, $n = 1 \times 10^{17}$ cm^{-3}, $n = 1 \times 10^{18}$ cm^{-3}, $n = 1 \times 10^{19}$ cm^{-3}, $n = 3 \times 10^{19}$ cm^{-3}, $n = 6 \times 10^{19}$ cm^{-3}, and $n = 1 \times 10^{20}$ cm^{-3}.

2.33. Repeat Problem 2.32 with $n = 1 \times 10^{16}$, 1×10^{17}, 1×10^{18}, 2×10^{18}, 3×10^{18}, and 6×10^{18} cm^{-3} for GaAs.

2.34. For Si at 300 K, plot E_f vs. n for $n = 2.8\times10^{17}, 2.8\times10^{18}, 1\times10^{19}, 2\times10^{19}, 3\times10^{19}, 5\times10^{19}$, and 1×10^{20} cm^{-3}. Plot on a semi-log scale and compare the result for both the Boltzmann approximation and the expression by Nilsson.

2.35. At 300 K and thermal equilibrium in Si, the ionized phosphorus concentration is 7×10^{16} cm^{-3} and the ionized boron concentration is 4×10^{16} cm^{-3}.

(a) What is the majority carrier concentration in cm^{-3}?

(b) What is the minority carrier concentration in cm^{-3}?

2.36. A Si sample at thermal equilibrium contains 3×10^{16} cm^{-3} ionized phosphorus atoms, 2×10^{16} cm ionized boron atoms, and 1.5×10^{16} cm ionized arsenic atoms.

(a) Find the majority carrier concentration and whether the majority carrier concentration is electrons or holes.

(b) What is the minority carrier concentration?

2.37. Consider the bond model for GaAs. If an As atom is replaced by a group-VI element such as Te, would Te be expected to be a donor or an acceptor and why?

2.38. **(a)** GaAs has a lattice constant of 5.65325 Å. What is the density?

(b) A GaAs sample is doped with tin. If tin substitutes for a gallium atom in the lattice, are donors or acceptors formed? Why? Is the semiconductor n- or p-type?

2.39. Calculate the location of E_i in Si at liquid nitrogen temperature (77 K), and at 100°C. Is it reasonable to assume that E_i is in the center of the energy gap?

2.40. Sketch an energy-band diagram for Si doped to give an electron concentration of 1×10^{16} cm^{-3} at (a) 77 K, (b) 300 K, and (c) 500 K. Show the Fermi level and use the intrinsic Fermi level as the energy reference.

2.41. Repeat Problem 2.40 for GaAs.

2.42. For n-type Si at 27°C, $n = 2 \times 10^{17}$ cm^{-3}, $N_d^+ = 3 \times 10^{17}$ cm^{-3}, what is the value of p?

2.43. Repeat Problem 2.42 for $T = 100$°C.

2.44. Find the electron and hole concentrations and Fermi level in Si at 300 K (a) for 1×10^{15} cm^{-3} electrons/cm^3 and (b) for 3×10^{16} ionized acceptors and 2.9×10^{16} ionized donors/cm^3.

2.45. Consider Si at 300 K doped with phosphorus to a concentration of 1×10^{19} cm^{-3}.

(a) Ignore the concentration dependence of the donor ionization energy. Calculate the concentration of neutral phosphorus and ionized phosphorus. What is the free-electron concentration?

(b) At what phosphorus concentration should the ionization energy go to zero?

(c) If E_d goes to zero, what is the free-electron concentration?

2.46. Repeat Problem 2.45 for GaAs with tellurium at a concentration of 1×10^{18} cm^{-3}.

2.47. Consider GaAs at 300 K doped with zinc to a concentration of 1×10^{19} cm^{-3}.

(a) Ignore the concentration dependence of the acceptor ionization energy. Calculate the concentration of neutral zinc and ionized zinc. What is the free hole concentration?

(b) At what zinc concentration should the ionization energy go to zero?

(c) If E_a goes to zero, what is the free hole concentration?

2.48. A GaAs wafer at 300 K contains 6×10^{17} cm^{-3} ionized Zn atoms and 3×10^{17} cm^{-3} ionized Te atoms, and 4×10^{17} cm^{-3} ionized Be atoms.

(a) Find the majority carrier concentration.

(b) Find the minority carrier concentration.

2.49. The III-V compound semiconductor GaP has a density of states effective mass for electrons of $0.82m_0$ and a density of states effective mass for holes of $0.60m_0$. The energy gap is 2.261 eV. For an ionized donor concentration of 5.0×10^{16} cm^{-3}, find the Fermi level with respect to the conduction-band edge at room temperature.

2.50. From Eq. (A.5) for the relation between the total acceptor concentration N_a and the ionized acceptor concentration N_a^- and the electrical neutrality condition taken as $p = N_a^-$ (neglect n and N_d^+), show that the free hole concentration is given by $p = \sqrt{\frac{1}{4}N_a N_v}\exp(-E_a/2kT)$ at low temperature or large E_a so that $N_a N_v >> pN_v$.

APPENDIX A

A.1 DONOR AND ACCEPTOR IONIZATION

For donor levels, the probability $P(E)$ of an electron of either possible spin orientation occupying the level has a spin degeneracy of two and is given by

$$P(E) = 1/\{1 + (1/2)\exp[(E - E_f)/kT]\}. \tag{A.1}$$

If there are N_d donors, the number of neutral donors (donors with electrons) is given by

$$N_d^o = N_d/\{1 + (1/2)\exp[(E_d' - E_f)/kT]\}, \tag{A.2}$$

ionized donors

where E_d' is the donor level measured from the valence-band edge. Note that the previous donor ionization energy $E_d = E_g - E_d'$. The number of ionized donors N_d^+ is simply $(N_d - N_d^o)$, which gives

$$\boxed{N_d^+ = N_d/\{1 + 2\exp[(E_f - E_d')/kT]\}}. \tag{A.3}$$

For acceptors, when an extra electron is attached (an ionized acceptor), the electron accepted by the acceptor must have the correct spin to correctly pair with the other covalent electron. The probability of electron occupancy is

$$P(E) = 1/\{1 + 2\exp[(E_f - E_a)/kT]\}. \tag{A.4}$$

ionized acceptors

Because electrons can come from the heavy- or light-hole bands, the 2 is replaced by a 4, and the ionized acceptor concentration N_a^- becomes

$$\boxed{N_a^- = N_a/\{1 + 4\exp[(E_a - E_f)/kT]\}} \tag{A.5}$$

where E_a is the previously defined acceptor level. The neutral acceptor with a bound hole is given by $N_a^\circ = N_a - N_a^-$. In Eqs. (A.1) through (A.5), E_d', E_a, and E_f are measured as a function of increasing electron energy. These designations are illustrated in

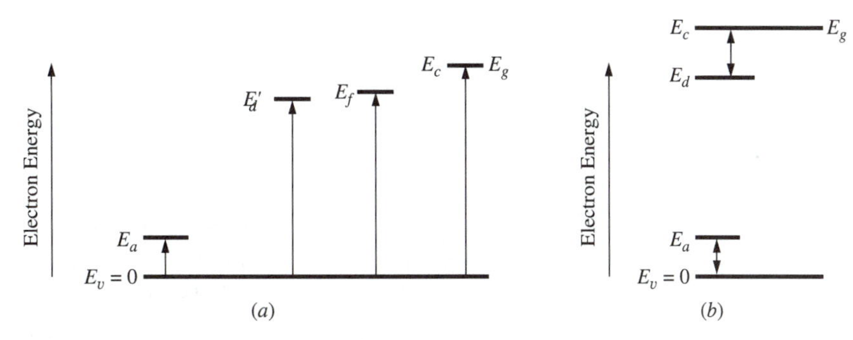

FIGURE A.1 (a) Illustration of the designation of E_a, E_d', E_f, and $E_c = E_g$ in Eqs. (A.1)–(A.5). All quantities are measured from $E_v = 0$. (b) Illustration of the designation of the acceptor ionization energy as $E_a - E_v$, and the donor ionization energy as $E_c - E_d'$.

Fig. A.1 (a). It is convenient to assign E_v as $E = 0$ so that these quantities are related to their position relative to the valence band. As previously described, it has become customary to give the acceptor ionization energy as $E_a - E_v$, which is also the acceptor level, and to give the donor ionization energy as $E_c - E_d$. The values given in Fig. 2.33 have been defined in this manner and are illustrated in Fig. A.1 (b). These relationships for the fraction of the donor or acceptor impurity ionized are necessary to understand the effect of deep level impurities such as chromium (Cr) or O in GaAs, but are not generally encountered in consideration of Si-based devices.

A.2 CONCENTRATION DEPENDENCE OF IONIZATION ENERGY

Impurities in semiconductors are generally represented as a localized level at a fixed energy with respect to the valence or conduction bands. Pearson and Bardeen[40] showed that the ionization energy of B in Si decreased as the impurity concentration was increased. Similar observations have been well documented for both donors and acceptors in GaAs. Shallow donors in GaAs such as S, Se, Te, Sn, and C have a donor ionization energy E_d of 0.006 eV at low concentrations. Measurements[41] on n-type GaAs showed that the donor ionization energy goes to zero for free-electron concentrations near 2×10^{16} cm^{-3} as shown in Fig. A.2. For the shallow acceptor zinc in GaAs, the acceptor ionization energy E_a at low concentration is 0.031 eV,[42] but E_a for Zn goes to zero somewhere between a hole concentration of 1 and 5×10^{18} cm^{-3}. These results emphasize the necessity of recognizing that impurities in semiconductors cannot be treated by the usual localized level concepts unless the impurity concentrations are very low.

When the donor or acceptor ionization energy goes to zero, the carrier concentration becomes temperature independent,[41,42] and all the substitutional donor or acceptor atoms are ionized. Therefore, the free-electron concentration n will equal the donor concentration, and the free hole p concentration will equal the acceptor

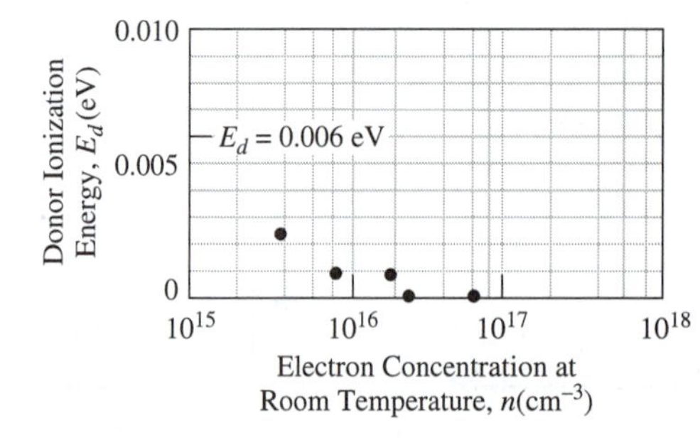

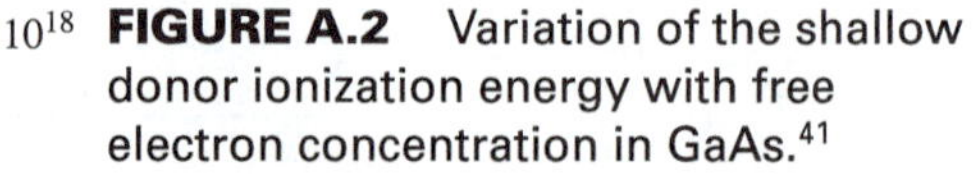

FIGURE A.2 Variation of the shallow donor ionization energy with free electron concentration in GaAs.[41]

zero ionization energy

concentration. The best criterion for the concentration where the ionization energy goes to zero is the impurity concentration at which metallic impurity conduction occurs. This transition has been predicted[43] to occur when the ratio of the average separation r of the impurity atoms to the radius of the hydrogenic impurity a^* is about three:

$$r/a^* \approx 3.0. \tag{A.6}$$

In Eq. (A.6), r is $(3/4\pi N)^{1/3}$, where N is the impurity concentration, a^* is the hydrogenic radius given in Eq. (A.6) with the appropriate semiconductor effective mass and dielectric constant. Then a^* becomes

$$(\epsilon_s/\epsilon_0)(m_0/m^*)0.5 \times 10^{-8} \text{ cm}.$$

For GaAs, Eq. (A.6) has been solved for N and plotted as a function of m^*/m_0 in Fig. A.3. Most commonly encountered semiconductors have about the same ϵ_s/ϵ_0 but different effective masses. For p-type GaAs with $m_p^*/m_0 = 0.47$, N in Fig. A.3 is $\sim 4 \times 10^{18}$ cm^{-3}, while for n-type GaAs with $m_n^*/m_0 = 0.067$, N is $\sim 1 \times 10^{16}$ cm^{-3}. These values are in reasonable agreement with the concentrations in Fig. A.3 where E_d and E_a go to zero.

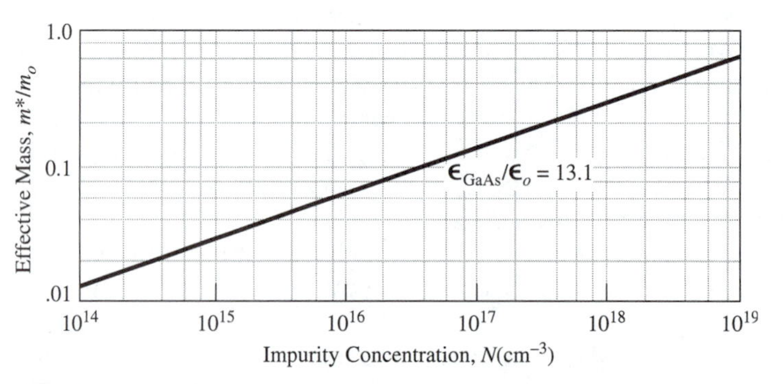

FIGURE A.3 Concentration given by Eq. (A.6) for the transition to metallic impurity conduction as a function of effective mass ratio.

CHAPTER 3

CARRIER TRANSPORT AND RECOMBINATION

3.1 INTRODUCTION

This chapter presents the various transport phenomena that arise from the motion of charge carriers (electrons and holes) in semiconductors such as *drift* under the influence of an electric field and *diffusion* that arise from a carrier concentration gradient. The carrier flow into and out of an infinitesimal volume is shown to be represented by the *continuity equation,* and this equation is used to analyze current flow in semiconductor materials and devices. In addition, the mechanisms by which carrier concentrations in excess of the thermal equilibrium concentrations recombine, which is called *carrier recombination,* are introduced. These concepts provide the necessary background to understand the behavior of the semiconductor devices presented in the remaining chapters.

carrier drift and mobility

resistivity

SPICE

The drift of electrons (or holes) in an electric field is described by introducing carrier mobility in Sec. 3.2. Then, in the next part of that section, consideration of conduction in the semiconductor due to an applied bias voltage leads to expressions for the resistivity, which is shown to depend on the carrier mobility and concentration. Section 3.2 concludes with the introduction of the *s*imulation *p*rogram with *i*ntegrated *c*ircuit *e*mphasis (SPICE), which is the most common circuit simulation program. SPICE is illustrated with a resistor formed in Si.

carrier diffusion

excess carrier concentrations

continuity equation

recombination, lifetime, and diffusion length

high electric fields

The motion of carriers from a region of high concentration to a region of low concentration, which is driven by the concentration gradient, is called *diffusion* and is presented in Sec. 3.3. Also in Sec. 3.3, the process of introducing carrier concentrations in excess of the thermal equilibrium concentrations is described. This process is called *carrier injection* and the concentrations in excess of the thermal equilibrium concentrations are called *excess* carrier concentrations. The overall effect of drift, diffusion, generation, and recombination in a semiconductor material is governed by the *continuity equation,* which is derived in Sec. 3.4. The mechanisms for decay of injected excess carrier concentrations are called *recombination,* and the various recombination processes are described in Sec. 3.5. The recombination process is characterized by the minority carrier *lifetime* and *diffusion length.* The basic equations which are introduced in this chapter and will be used to derive the behavior of the semiconductor devices considered in Chapters 4 through 9 are summarized in Table 3.2 in Sec. 3.6. Examples of the application of the continuity equation to nonequilibrium concentrations of excess carriers in bulk semiconductors are given in Sec. 3.7. *Generation* of electron–hole pairs by absorption of light is also introduced in this section. In Sec. 3.8, the effects of large electric fields on the drift mobility are described. It will be shown that the drift velocity is no longer proportional to the electric field when the electric field exceeds approximately 2×10^3 V/cm. The significant concepts introduced in this chapter are summarized in Sec. 3.9 together with the expressions useful for representing carrier transport and recombination in semiconductors.

An appendix is included at the end of the chapter to provide a detailed derivation of the theory for nonradiative recombination and radiative recombination.

3.2 CARRIER DRIFT

3.2.1 Mobility

Consider an n-type semiconductor sample with uniform donor concentration in thermal equilibrium. As discussed in Chapter 2, the conduction electrons in the semiconductor conduction band are essentially free particles, since they are not associated with any particular lattice or atom site. The influence of the crystal lattice is incorporated in the effective mass of conduction electrons, which differs from the mass of free electrons. The average thermal velocity v_{th} of carriers is defined as

$$v_{th} = \frac{\int_0^\infty v(E)\rho(E)\exp[-(E/kT)]\,dE}{\int_0^\infty \rho(E)\exp[-(E/kT)]\,dE}, \tag{3.1}$$

where $v(E)$ is the velocity of the carrier of energy E when all of the energy is kinetic energy so that $v(E)$ is $\sqrt{2E/m^*}$. In Eq. (3.1), $\rho(E)$ is the density of states as given

This subsection of Sec. 3.2 is taken in part from S. M. Sze, *Semiconductor Devices: Physics and Technology* (Wiley, New York, 1985), pp. 30–34.

in Eq. (2.44), while $\exp[-(E/kT)]$ is the Boltzmann representation for the probability an electron occupies the energy E. The integral in the numerator of Eq. (3.1) is the gamma function $\Gamma(2) = 1$, and the integral in the denominator is the gamma function $\Gamma(3/2) = \pi^{1/2}/2$, which gives

thermal velocity

$$\boxed{v_{th} = \sqrt{8kT/\pi m^*}}\,. \tag{3.2}$$

At room temperature, v_{th} is approximately 1×10^7 cm/s for most semiconductors.

The electrons in the semiconductor are therefore moving rapidly in all directions. The thermal motion of an individual electron may be visualized as a succession of random scattering from collisions with lattice atoms, impurity atoms, and other scattering centers, as illustrated in Fig. 3.1(a). The random motion of electrons leads to a zero net displacement of an electron over a sufficiently long period of time. The average distance between collisions is called the *mean free path,* and the average time between collisions is called the *mean free time* $\langle\tau_c\rangle$. For a typical value of 1×10^{-5} cm for the mean free path, $\langle\tau_c\rangle$ is about 1 ps (i.e., $10^{-5}/v_{th} \simeq 10^{-12}$ s).

When a small electric field $\mathscr{E}$ is applied to the semiconductor sample, each electron will experience a force $-q\mathscr{E}$ from the field and will be accelerated along the field (in the opposite direction to the field) during the time between collisions. Therefore, an additional velocity component will be superimposed upon the thermal motion of electrons. This additional component is called the *drift velocity.* The combined displacement of an electron due to the random thermal motion and the drift component is illustrated in Fig. 3.1(b). Note that there is a net displacement of the electron in the direction opposite to the applied field.

The mean drift velocity v_n may be obtained by equating the momentum (force $\times$ time) applied to an electron during the free flight between collisions to the momentum gained by the electron in the same period. The equality is valid because in steady state

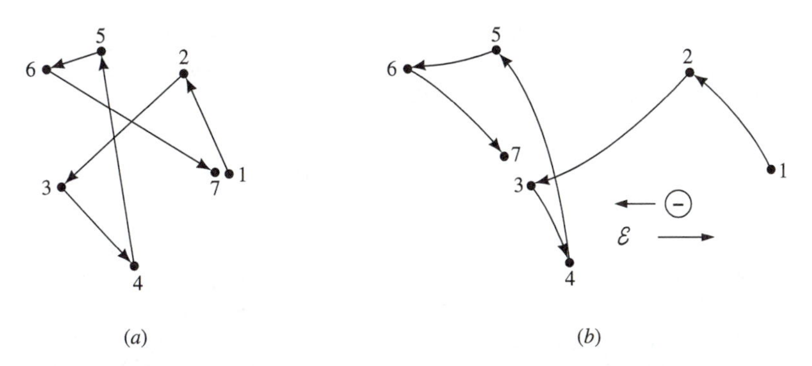

FIGURE 3.1 Schematic representation of the path of an electron in a semiconductor. (a) Random thermal motion of electron. (b) Combined electron motion due to random thermal motion and an applied electric field.[1]

all momentum gained between collisions is lost to the lattice in the collision. The momentum applied to an electron is given by $-q\mathscr{E}\langle\tau_c\rangle$, and the momentum gained is $m^* v_n$, which gives

$$-q\mathscr{E}\langle\tau_c\rangle = m^* v_n\,, \tag{3.3}$$

or

$$v_n = -\left[\frac{q\langle\tau_c\rangle}{m^*}\right]\mathscr{E}. \tag{3.4}$$

EXAMPLE 3.1 The expression on the right-hand side of Eq. (3.4) has units of:

$$\frac{\text{coulomb-s}}{\text{g}}\frac{\text{V}}{\text{cm}}\frac{\text{joule/coulomb}}{\text{V}}\frac{\text{erg}}{10^{-7}\,\text{joule}}\frac{\text{g-cm}^2/\text{s}^2}{\text{erg}} = \frac{\text{cm}}{\text{s}}.$$

■

Equation (3.4) states that the electron drift velocity is proportional to the applied electric field. The proportionality factor depends on the mean free time and the effective mass. The proportionality factor is called the *electron mobility* μ_n in units of cm^2/V-s, or

mobility

$$\mu_n \equiv \frac{q\langle\tau_c\rangle}{m^*}. \tag{3.5}$$

Thus,

$$\boxed{v_n = -\mu_n\mathscr{E}}\,. \tag{3.6}$$

Mobility is an important parameter for carrier transport because it describes how strongly the motion of an electron is influenced by an applied electric field. A similar expression can be written for holes in the valence band:

$$\boxed{v_p = \mu_p\mathscr{E}}\,, \tag{3.7}$$

where v_p is the hole drift velocity and μ_p is the hole mobility. The negative sign is removed in Eq. (3.7) because holes drift in the same direction as the electric field.

In Eq. (3.5), the mobility is related directly to the mean free time between collisions, which in turn is determined by the various scattering mechanisms. The two most important mechanisms are lattice scattering and impurity scattering. Lattice scattering results from thermal vibrations of the lattice atoms at any temperature above absolute zero. These vibrations disturb the lattice periodic potential and allow energy to be transferred between the carriers and the lattice. Since lattice vibration increases with increasing temperature, lattice scattering becomes dominant at high temperatures; therefore, the mobility decreases with increasing temperature ($T\uparrow$, $\mu\downarrow$). As introduced in Sec. 2.7.2, the allowed vibrational motions which interact with the free electrons are termed phonons. Scattering by acoustic phonons has often been found to limit mobility in semiconductors at room temperature. The acoustic phonons have energies of approximately 0.05 eV (see Example 2.4). The mobility due to acoustical phonon scattering μ_L decreases with temperature as $T^{-3/2}$.[2]

Also in Eq. (3.5), the mobility varies inversely with the effective mass (conductivity mass for electrons in Si). Therefore, a smaller effective mass is expected to give a larger mobility. Generally, the electron effective mass is smaller than the hole effective mass, $m_n^* < m_p^*$. At a given impurity concentration, the electron mobility is found to exceed the hole mobility, $\mu_n > \mu_p$, as expected from the difference in effective mass.

Ionized impurity scattering results when a charge carrier travels past an ionized dopant impurity (donor or acceptor). The charge carrier path will be deflected due to Coulomb force interaction. The probability of impurity scattering depends on the *total* concentration of ionized impurities, that is, the *sum* of the concentration of ionized negatively and positively charged ions. However, unlike lattice scattering, impurity scattering becomes less significant at higher temperatures. Because the carriers move faster at higher temperature, they remain near the impurity center for a shorter time and are therefore less effectively scattered. The mobility near room temperature due to ionized impurity scattering varies with temperature and ionized impurity concentration N_I as $T^{3/2}/N_I$.[3]

The probability of a collision taking place in unit time, $1/\langle\tau_c\rangle$, is the sum of the probabilities of collisions due to the various scattering mechanisms:

$$\frac{1}{\langle\tau_c\rangle} = \frac{1}{\langle\tau_{c,lattice}\rangle} + \frac{1}{\langle\tau_{c,impurity}\rangle} \tag{3.8}$$

or

$$\frac{1}{\mu} = \frac{1}{\mu_L} + \frac{1}{\mu_I}. \tag{3.9}$$

Figure 3.2 shows the measured electron mobility for phosphorus-doped Si as a function of temperature with the indicated net donor concentrations.[4] For the lightly doped samples with $n = 1.0 \times 10^{17}$ cm^{-3} and $n = 1.2 \times 10^{18}$ cm^{-3}, lattice scattering dominates, and the mobility decreases as the temperature increases. For the other two more heavily doped samples, the mobility is lower due to impurity scattering. The expressions for mobility due to these scattering mechanisms, as well as several less important mechanisms, have been summarized for p-type GaP and may be applied to any semiconductor.[5]

The electron and hole mobilities in Si at 297 K are shown in Fig. 3.3. Generally, the plots of mobility for Si are given as a function of impurity concentration, which refers to uncompensated Si with $N_d^+ = n$ and $N_a^- = p$. For *compensated* samples

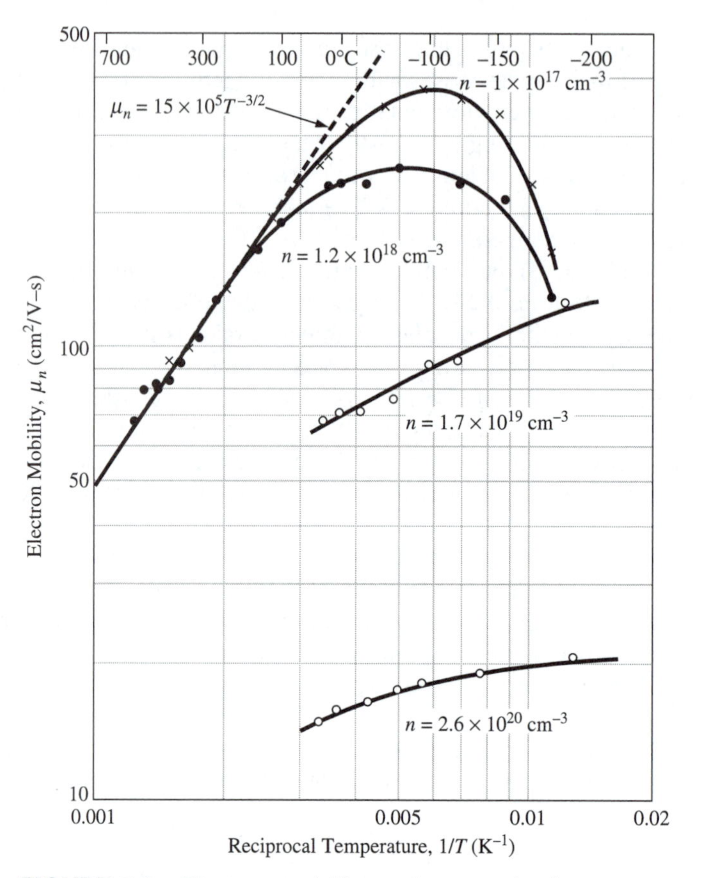

FIGURE 3.2 Electron mobility vs. inverse absolute temperature for phosphorus-doped Si.[4] Carrier concentrations are for $T = 300$ K.

which contain both donors and acceptors, the mobility will be decreased at a given electron or hole concentration due to increased impurity scattering. The mobility of electrons in Fig. 3.3 is represented by an expression which was given by Baccarani and Ostoja as[6]

$$\mu_n = \frac{1{,}360 - 92}{1 + (N/1.3 \times 10^{17})^{0.91}} + 92\,, \tag{3.10}$$

while the mobility of holes is represented by the expression given by Antoniadis *et al.*,[7]

$$\mu_p = \frac{468 - 49.7}{1 + (N/1.6 \times 10^{17})^{0.7}} + 49.7. \tag{3.11}$$

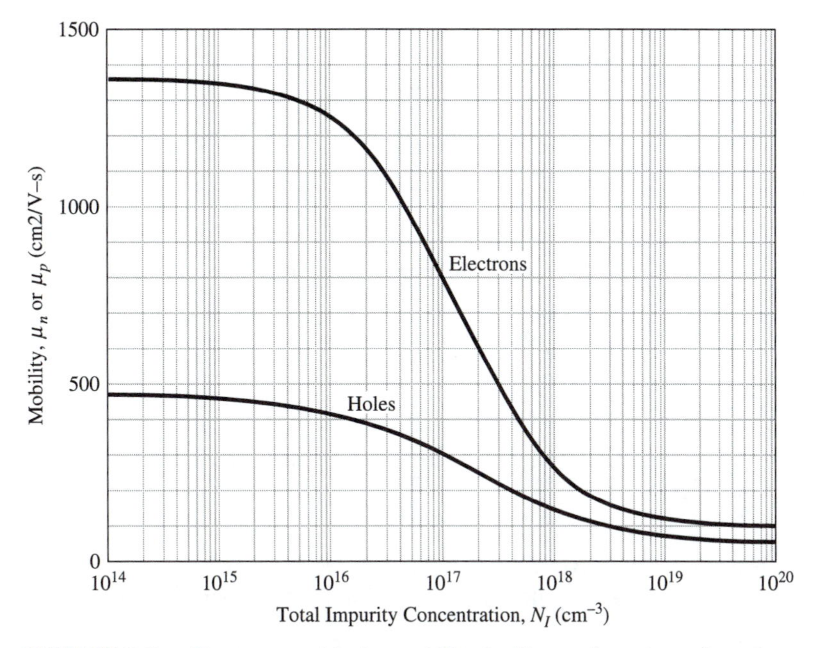

FIGURE 3.3 Electron and hole mobility in Si as a function of total impurity concentration at room temperature.

Arora *et al.*[8] obtained empirical expressions for the electron and hole mobilities which were based on experimental data over the temperature range of 250 to 500 K. Their mobility relationship as a function of temperature and carrier concentration for electrons was

$$\mu_n = 88T_n^{-0.57} + \frac{7.4 \times 10^8\, T^{-2.33}}{1 + [N/(1.26 \times 10^{17}\, T_n^{2.4})]\, 0.88\, T_n^{-0.146}}, \tag{3.12}$$

and for holes was

$$\mu_p = 54.3\, T_n^{-0.57} + \frac{1.36 \times 10^8\, T^{-2.23}}{1 + [N/(2.35 \times 10^{17}\, T_n^{2.4})]\, 0.88 T_n^{-0.146}}, \tag{3.13}$$

where N is the electron or hole concentration and T_n is (T/300 K). These numerical representations are useful for computer-device modeling.

The electron and hole mobilities as a function of carrier concentration in GaAs at 297 K are shown in Fig 3.4.[9] The effect of compensation due to the presence of acceptors in n-type samples has been summarized by Rode and Knight[10] and illustrates the decrease in mobility due to compensation. By comparing Figs. 3.3 and 3.4, it may readily be seen that the mobility of electrons is about six times larger in lightly doped

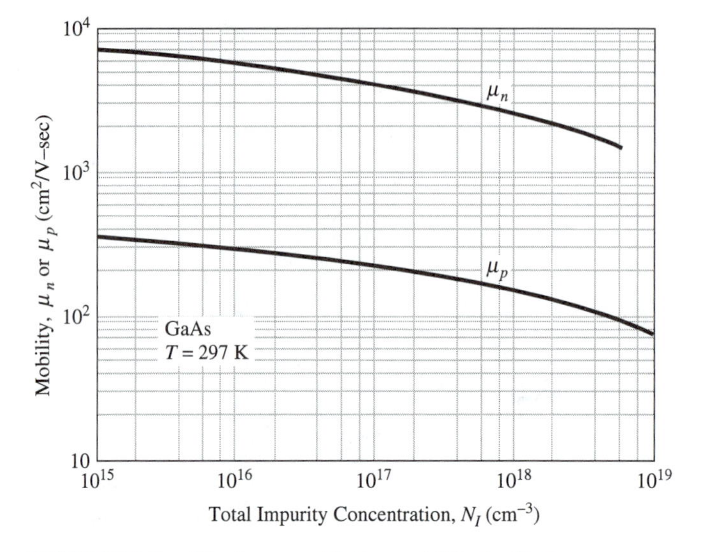

FIGURE 3.4 Electron and hole mobility in GaAs as a function of total impurity concentration at room temperature.[9]

GaAs as compared to Si while the hole mobilities are about the same. This greater electron mobility in GaAs is one characteristic which makes GaAs appealing for high-speed applications. However, as shown in Sec. 3.7, the mobility at high electric fields is a function of the electric field, and the limiting velocity at high electric fields must be considered in the analysis of high-speed devices.

minority-carrier mobility

For a Si sample with $n = N_d^+ = 1 \times 10^{17}$ cm^{-3}, the majority carrier electron mobility μ_n from Fig. 3.3 would be 800 cm^2/V-s. The *minority-carrier hole mobility* μ_p in this n-type sample is determined by the ionized donor concentration of 1×10^{17} cm^{-3} and is found in Fig. 3.3 to be 300 cm^2/V-s. Further discussion of minority-carrier mobility is given in Sec. 4.7.4.

3.2.2 Resistivity

Next, consider conduction in a homogeneous semiconductor. Figure 3.5(a) shows an n-type semiconductor and its energy-band diagram at thermal equilibrium. Figure 3.5(b) shows the corresponding band diagram when a biasing voltage is applied to the right-hand terminal. It is assumed that the contacts at the left-hand and right-hand terminals are ohmic, which means that the voltage drop at each of the contacts is negligible. The behavior of ohmic contacts is considered in Sec. 7.6. As mentioned previously, when an electric field $\mathscr{E}$ is applied to a semiconductor, each electron in the conduction band will experience a force $-q\mathscr{E}$. The electrons in the conduction band will accelerate as the result of the applied field $\mathscr{E}$, and they will neither gain nor lose a significant amount of their total energy as represented by the horizontal path in Fig. 3.5(b). As previously presented in Sec. 2.7.2, the conduction band edge E_c corresponds to the potential

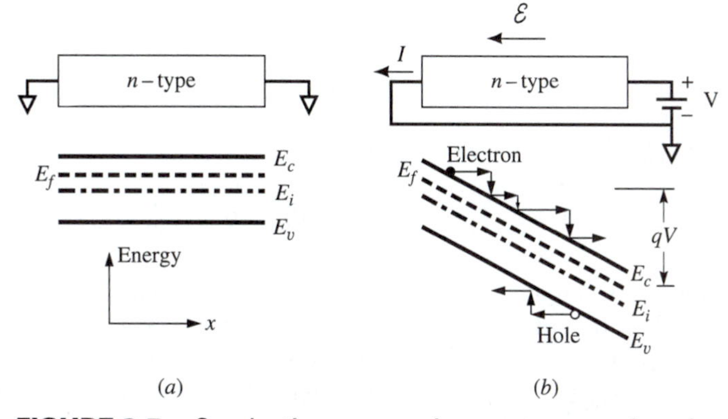

FIGURE 3.5 Conduction process in an *n*-type semiconductor (*a*) at thermal equilibrium, and (*b*) under a biasing condition.[11,12]

energy of the electron at rest. The kinetic energy of an electron is measured upward from E_c (distance from E_c). Although the total energy of electrons does not change significantly during acceleration [hence the horizontal arrow at the beginning of acceleration in Fig. 3.5(*b*)], they will increase their kinetic energy and lose a corresponding amount of potential energy (given by E_c). When an electron is scattered, it loses some or all of its kinetic energy, which is given to the lattice as heat. Then, the electron can begin to accelerate and the process is repeated many times. This process shown in Fig. 3.5(*b*) is the representation of conduction in the energy-band diagram. Conduction by holes is described by a similar but opposite process also shown in Fig. 3.5(*b*) (recall that holes "float").

The force on the electron due to the electric field is equal to the negative gradient of potential energy:[11,12]

$$\text{Force} = (-q)\mathcal{E} = -(\text{gradient of potential energy}). \tag{3.14}$$

As just mentioned and in Sec. 2.7.2, the bottom of the conduction band E_c corresponds to the potential energy of an electron and the gradient of the potential energy is dE_c/dx. Since only the gradient of potential energy enters the relationship in Eq. (3.14), any part of the energy-band diagram that is parallel to E_c such as E_f, E_i, or E_v, as shown in Fig. 3.5(*b*), can be used interchangeably. It is convenient to use the intrinsic Fermi level E_i because E_i will be used when *p-n* junctions are considered in Chapter 4. Therefore, Eq. (3.14) may be written as,[11,12]

electric field $\mathcal{E}$

$$\boxed{\mathcal{E}(x) = \frac{1}{q}\frac{dE_c(x)}{dx} = \frac{1}{q}\frac{dE_i(x)}{dx}}\,. \tag{3.15}$$

A related quantity ψ may be defined as the *electrostatic potential* whose negative gradient equals the electric field:

electrostatic potential

$$\boxed{\mathscr{E}(x) \equiv -\frac{d\psi(x)}{dx}} \,. \tag{3.16}$$

Comparison of Eqs. (3.15) and (3.16) gives

$$\boxed{\psi(x) = -E_i(x)/q} \,, \tag{3.17}$$

where a nonzero reference potential is often selected, but the reference potential is set to zero here. Equation (3.17) provides a relationship between the electrostatic potential, often just called *potential,* and the potential energy of an electron. For the homogeneous semiconductor shown in Fig. 3.5(*b*), the potential energy and E_i decrease linearly with distance; thus, the electric field is a constant in the negative x direction. The magnitude of the electric field equals the applied voltage divided by the sample length.

The transport of carriers under the influence of an applied electric field produces a current called the *drift current.* Consider a semiconductor sample shown in Fig. 3.6, which has a cross-sectional area A, a length L, and a carrier concentration n electrons/cm^3. When an electric field $\mathscr{E}$ is applied along the length L, the electron *current density* J_n flowing in the sample can be found by summing the product of the charge $(-q)$ on each electron times the electron's velocity over all electrons per unit volume n:

electron-drift current density

$$\boxed{J_n = \frac{I_n}{A} = \sum_{i=0}^{n}(-qv_i) = -qnv_n = qn\mu_n\mathscr{E}} \,, \tag{3.18}$$

where I_n is the electron current. Equation (3.6) was used for the relationship between v_n and $\mathscr{E}$.

A similar argument applies to holes. By taking the charge on the hole to be positive, the hole-drift current density may be written as

hole-drift current density

$$\boxed{J_p = qpv_p = qp\mu_p\mathscr{E}} \,. \tag{3.19}$$

The total current flowing in the semiconductor sample due to the applied field $\mathscr{E}$ can be written as the sum of the electron and hole-current densities:

$$\boxed{J = J_n + J_p = (qn\mu_n + qp\mu_p)\mathscr{E}} \,. \tag{3.20}$$

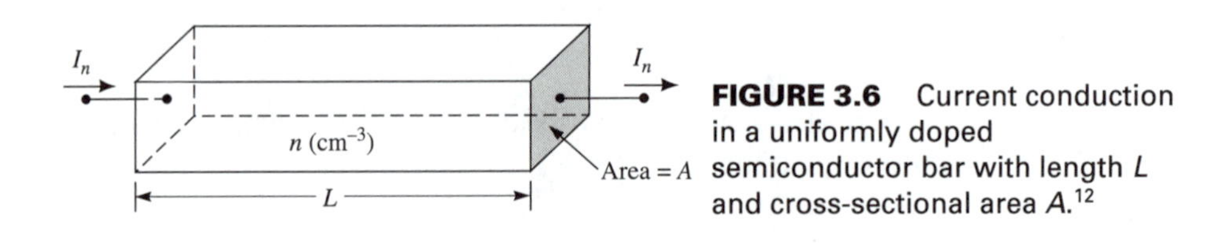

FIGURE 3.6 Current conduction in a uniformly doped semiconductor bar with length L and cross-sectional area A.[12]

conductivity

The quantity in parentheses is known as the *conductivity:*

$$\boxed{\sigma = (qn\mu_n + qp\mu_p)} \ . \tag{3.21}$$

The electron and hole contributions to the conductivity are simply additive. The corresponding resistivity of the semiconductor, which is the reciprocal of σ, is given by

resistivity

$$\boxed{\rho \equiv 1/\sigma = 1/(q\mu_n n + q\mu_p p)} \ . \tag{3.22}$$

Generally, in extrinsic semiconductors, only one of the components in Eq. (3.21) or Eq. (3.22) is significant because of the many orders of magnitude difference between the two carrier densities. Therefore, Eq. (3.22) reduces to

$$\boxed{\rho = 1/(q\mu_n n)} \tag{3.23}$$

for an n-type semiconductor and to

$$\boxed{\rho = 1/(q\mu_p p)} \tag{3.24}$$

for a p-type semiconductor.

Figure 3.7 shows the measured resistivity for Si and GaAs at 300 K as a function of the electron or hole concentration. The data for Si are from Beadle *et al.*[13] The data for GaAs were calculated by using Eqs. (3.23) and (3.24) for the resistivity and with the mobility given in Fig. 3.4.

3.2.3 Integrated-circuit Resistors

When resistors are required for integrated circuits, two types of resistors are used. One type is fabricated by the deposition of thin films such as Nichrome (NiCr) or tantalum on the surface of an insulating layer of SiO_2 on the semiconductor. These materials have sheet resistances of 40 to 4,000 ohms/sq and have been summarized by Hamilton and Howard,[14] and will not be considered here. The other type of resistor is formed by the introduction of impurities into the semiconductor surface by techniques such as diffusion or ion implantation which are used to form the semiconductor devices. These layers have sheet resistances in the 1,000-ohms/sq range. A representation of a diffused resistor is shown in Fig. 3.8. Calculation of the resistance for the integrated-circuit resistor will illustrate the definition of sheet resistance. The analysis of field-effect transistors (FETs) which are "resistors" whose conducting channel thickness is controlled by the gate electrode is very similar to the analysis for the integrated-circuit resistor.

The geometry for this analysis is shown in Fig. 3.9 with a length L, width W, and thickness x_j in the x direction. The resistance R of a rectangular resistor of length L and

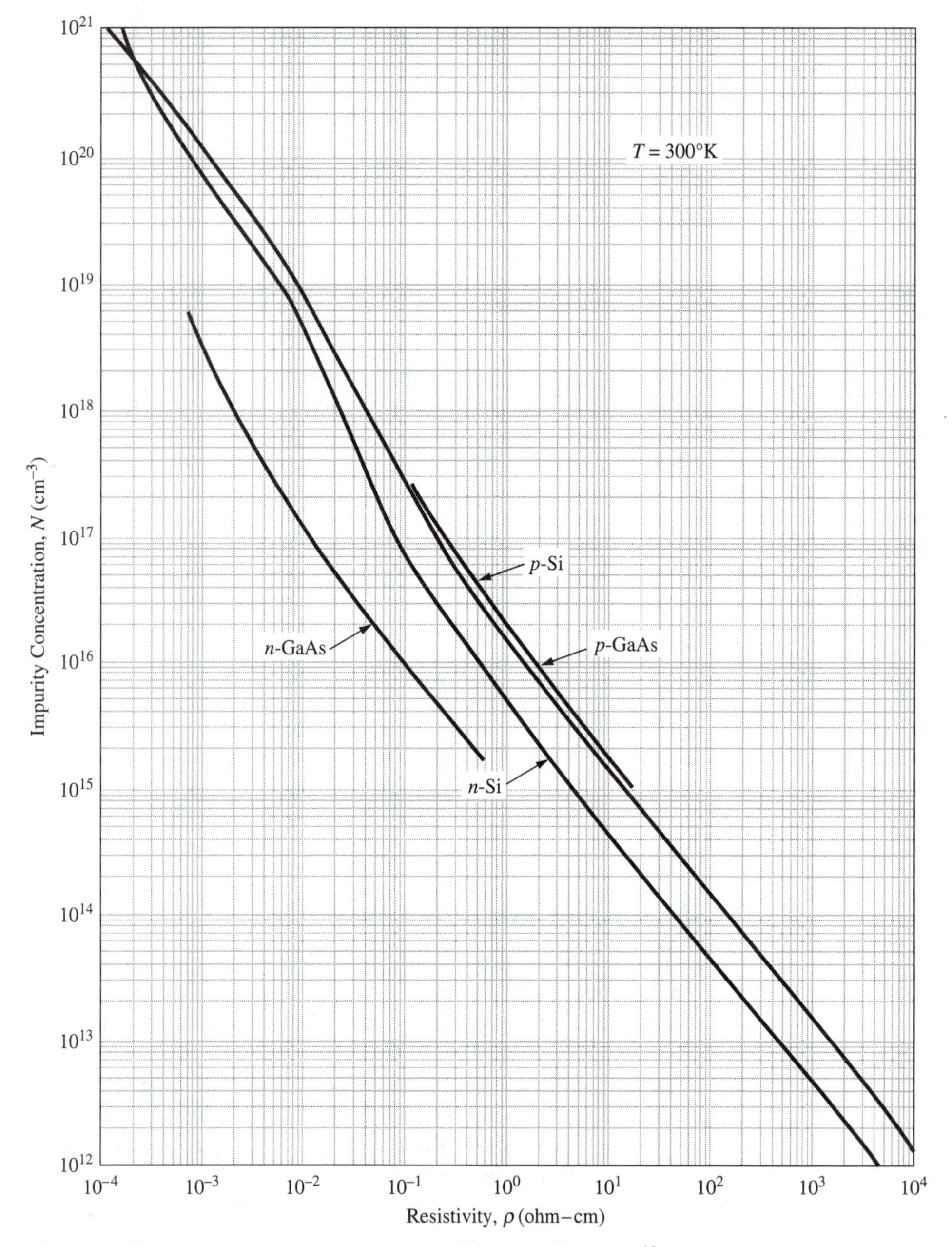

FIGURE 3.7 Resistivity vs. impurity concentration for Si[13] and GaAs.

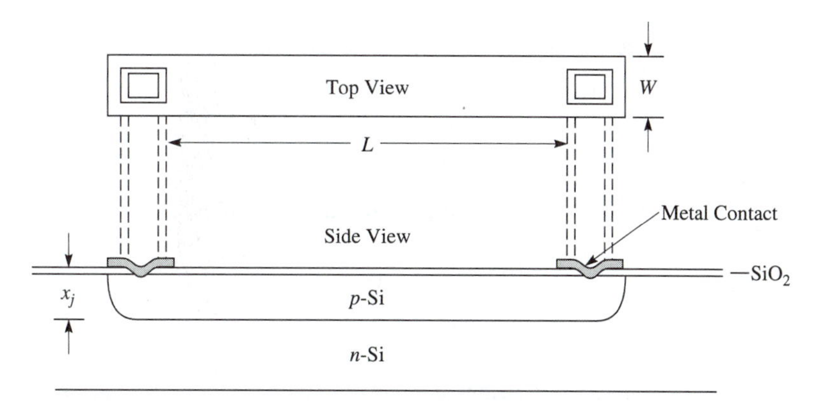

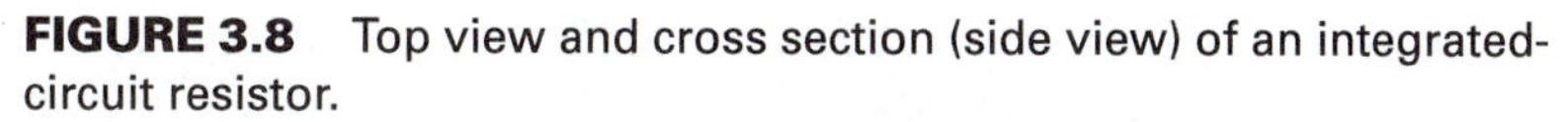

FIGURE 3.8 Top view and cross section (side view) of an integrated-circuit resistor.

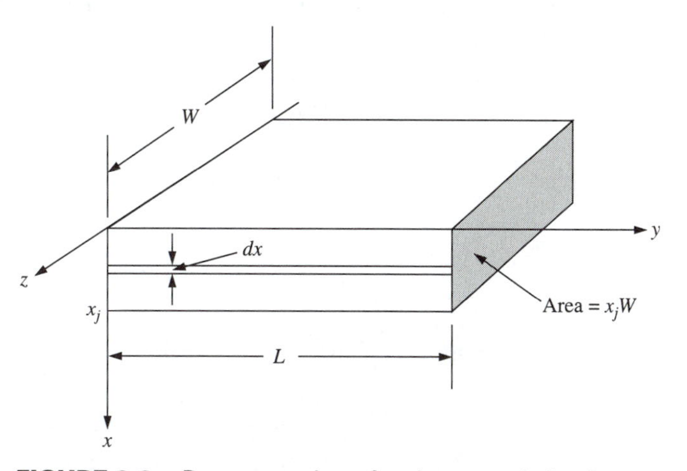

FIGURE 3.9 Representation of an integrated-circuit resistor for the calculation of the resistance.

cross-section area A is

$$\boxed{R = \rho(L/A)}\ . \tag{3.25}$$

It is more convenient to consider the conductance

$$G = 1/R. \tag{3.26}$$

Then, the differential conductance dG of a thin layer of thickness dx parallel to the surface at depth x is

$$dG = \frac{W\,dx}{\rho(x)L}, \tag{3.27}$$

where $\rho(x)$ varies with x. With Eq. (3.24) for ρ,

$$dG = q\mu_p(x)p(x)\frac{W\,dx}{L}. \tag{3.28}$$

The total conductance G may be obtained by the integral

$$G = \frac{qW}{L}\int_0^{x_j} \mu_p(x)p(x)\,dx, \tag{3.29}$$

where the hole concentration $p(x)$ varies with x depending on the method of forming the p-layer, and as shown in Figs. 3.3 and 3.4, the mobility varies with p and hence x. Correct evaluation would require numerical evaluation of the integral in Eq. (3.29). However, for our purposes, $\mu_p(x)p(x)$ will be taken as constant from $x = 0$ to $x = x_j$. The conductance becomes

$$G = q\mu_p p x_j \frac{W}{L}, \tag{3.30}$$

and the resistance is

$$R = \frac{1}{G} = \frac{1}{q\mu_p p\, x_j}\frac{L}{W} = \frac{\rho}{x_j}\frac{L}{W}. \tag{3.31}$$

Let $L = W$ for a square, and then

sheet resistance

$$R_\square = \frac{\rho}{x_j}\frac{\text{ohm-cm}}{\text{cm}} = \frac{\rho}{x_j}\ \text{ohms/square}, \tag{3.32}$$

which is defined as the *sheet resistance*. Therefore, for a resistor with length L and width W, the resistance is

$$\boxed{R = R_\square(L/W)}\ , \tag{3.33}$$

where the ratio L/W is the number of squares as shown in Fig. 3.10.

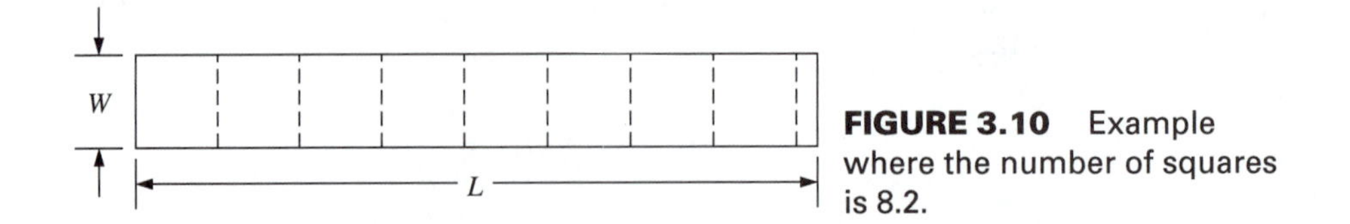

FIGURE 3.10 Example where the number of squares is 8.2.

A more complex analysis is given by Glasser and Subak-Sharpe.[15] A meander pattern is used to minimize the surface area because cost is related to the chip area used. Correction factors must be used for corners and bends.[16] Additional description may also be found in Muller and Kamins.[16]

3.2.4 Application of SPICE to Integrated-circuit Resistors

SPICE is intended for circuit analysis and integrated-circuit (IC) design. PSpice was initially written for personal computers (PCs), but now also runs on Sun workstations and includes many features not available in SPICE2. These features include extensive component libraries, schematic capture, and a graphical waveform output called *probe.* Other high-performance commercial circuit-analysis programs such as HSPICE are based on SPICE and require a prior knowledge of SPICE. Complete descriptions of SPICE and PSpice may be found in several books.[17–21] Free versions of PSpice, the Evaluation version, are available from OrCAD, 9300 S. W. Nimbus Avenue, Beaverton, OR 97008, USA. (503) 671-9500 and http://www.orcad.com. MicroSim Corporation, the developer of PSpice, merged with OrCAD in January, 1998. The free version of PSpice (Release 8.0) is available on the Wiley World Wide Web site: http://www.wiley.com/college/casey. The Evaluation version is limited to circuits containing no more than 10 transistors and 20 electrical nodes, which is adequate for the Examples and Problems considered here. Complete on-line manuals are included in the Evaluation version. A brief introduction to PSpice is given in Appendix I at the end of this text. PSpice, which is written in the C programming language, will be used in the Examples and run on a Sun workstation or on a PC with Windows 95. *The emphasis on PSpice in this book will be to relate the device physics to the PSpice parameters.*

PSpice

To analyze an electronic circuit, a schematic circuit diagram is prepared with the necessary notation for PSpice and each component is assigned a unique name (see Appendix I). A resistor must start with `R` and can contain up to a total of seven alphanumeric characters. Capacitors begin with `C`, inductors begin with `L`, and voltage sources begin with `V`. Diodes and transistors will be introduced in the next chapters. Nodes in the circuit must be numbered. The ground node is zero and each node must be assigned a positive integer other than zero. Each node must be connected to at least two elements and have a DC path to ground. An input file is created with the computer system text editor (some systems require capital letters). The name of the PSpice file must be of the form *name*.cir. (The suffix .cir is required in PSpice only.) The first line of the input file must be a title line, and the last line of the input file must be `.END` or `.end` with the "." as the first character. Between the `TITLE` line and the `.END` line, the lines in the file contain three sections: the *Circuit Description,* the *Analysis Description,* and the *Output Description* and they may be in any order. When PSpice is run, output files are created which will permit display of the results for the analysis as a numerical tabulation and as a graphical representation. PSpice is run with the command *pspice filename* with or without the suffix .cir. The results from PSpice are sent to the *filename.out* file and to the graphical output program named *probe.*

PSpice is now intended to be run with the schematic capture front end, Schematics. It is helpful to learn the line command version of PSpice in which the file.cir is created directly before using Schematics. When an error is generated during execution in Schematics, a reference is given to the netlist. The user must be able to understand the netlist in order to correct the error. The netlist is composed of line commands that specify element locations and values similar to the element lines in the .cir file. The debugging process is greatly assisted with knowledge of the line command version of PSpice. The relation of the device physics to the PSpice parameters is more easily illustrated with the line command version.

The general form of the *element line* for a resistor is

```
RXXXXXX N1 N2 VALUE<TC=TC1<,TC2>>.
```

In many PSpice texts, the element line is called the General Form; however, the notation of Banzhaf[17] as the element line is more descriptive of the purpose for the line in the Circuit Description. In the resistor element line, XXXXXX is an alphanumeric string with up to six characters, N1 and N2 represent the nodes at each end of the resistor, and VALUE gives the value of the resistor. Data contained within < . . . > represents optional data, but punctuation such as commas, equals signs, and parentheses must be included with data. To continue an element line on an additional line, use a "+" in the first column of the next line. The value of a resistor as a function of temperature is given by

$$\mathrm{VALUE} = \mathrm{VALUE}(T_o)[1 + \mathrm{TC1}(T - T_o) + \mathrm{TC2}(T - T_o)^2],$$

where T_o is the temperature of the reference value of the resistance, T is the temperature at which the resistance is being determined, TC1 is the linear resistance temperature coefficient, and TC2 is used when the resistance temperature dependence is nonlinear. If the temperature dependence is neglected, the VALUE for a 1.5 kΩ resistor is written as 1.5E+3. An example of an input file is given in Example 3.2.

The general form of the *element line* for a voltage source must be specified. Only a simple DC source will be introduced here as

```
VXXXXXXX N1 N2 DC VALUE,
```

where VXXXXXXX is the voltage source name, N1 and N2 are the nodes of the source, and VALUE is the DC source voltage. To measure a particular current, a dead voltage source, which has zero volts across it, could be used as an ammeter. For example, the *element line* for the current through the dead voltage source,

```
VDEAD 10 16 0,
```

would give the current flowing through nodes 10 and 16. Node currents are generated in PSpice without dead voltage sources, but sometimes they are convenient to keep track of a particular current without remembering the node number.

Before an example is given for the use of PSpice, a *control line* for the temperature will be introduced. The control line for temperature is given by

```
.TEMP T1 < T2 < T3 ...>> .
```

The .TEMP line sets the temperatures(s) for the SPICE simulation, and T1, T2, T3 ... are the temperature in degrees Celsius. Analysis is performed for each temper-

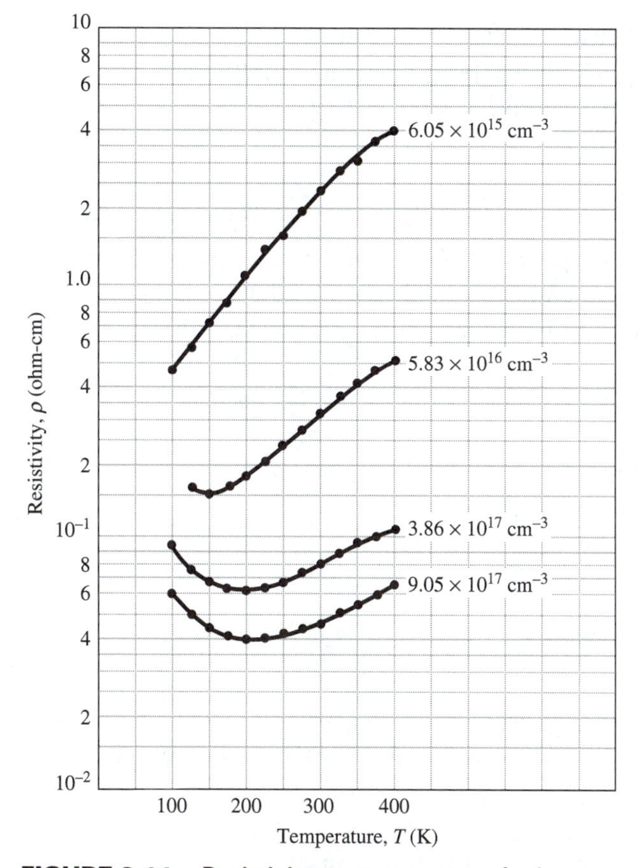

FIGURE 3.11 Resistivity vs. temperature for boron-doped Si.[22]

ature specified. If a temperature control line is not given, PSpice will assume that the temperature is 27°C.

For application of PSpice to integrated-circuit resistors, it is helpful to have the variation of resistivity for Si with temperature. The variation of resistivity for boron-doped Si at several room temperature hole concentrations is given in Fig. 3.11. The temperature coefficients for the variation of resistivity with temperature may be determined from these data.

EXAMPLE 3.2 To illustrate the use of SPICE for a resistor, consider a p-type Si integrated-circuit resistor with a resistivity of 0.1 ohm-cm at 300 K and 0.2 ohm-cm at 400 K. The resistor has a thickness x_j of 3.0 μm, a width W of 4.0 μm, and a length L of 14.0 μm.

(a) What is the sheet resistance at 300 K?

(b) What is the resistance of the resistor at 300 K?

(c) What is the temperature coefficient `TC1`?

(d) Use PSpice to determine the current at 27°C and at 100°C for the resistor in parts (a) to (c) when connected to a 5-V battery.

Solution

(a) From Eq. (3.32), $R_\square = 0.1/3 \times 10^{-4} = 3.33 \times 10^2$ ohms/square.

(b) Number of squares $= L/W = 14/4 = 3.5$ squares. Then, $R = R_\square \times$ number of squares $= 3.33 \times 10^2 \times 3.5 = 1.17 \times 10^3$ ohms.

(c) Let `TC2` $= 0$. Then, $0.2 = 0.1[1 +$ `TC1`$(400 - 300)]$, which gives `TC1` $= 1.0 \times 10^{-2}$.

(d) First, draw a circuit diagram and number the nodes.

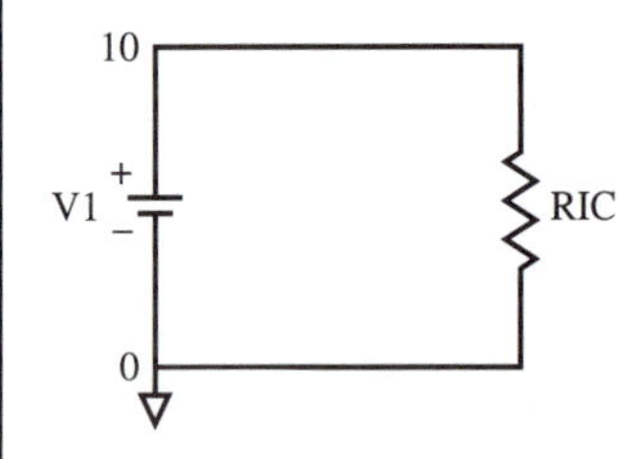

Then create a file named *icr.cir* with the title `icresistor icr.cir`. Note that for temperature analysis, a control line `.TEMP` gives the desired temperatures.

```
icresistor icr.cir
V1 10 0 dc 5
RIC 10 0 1.17E+3 TC=1.0E-2, 0
.TEMP 27 100
.END
```

Run PSpice as described in Appendix I at the end of the text. From the data given in the file *icr.out,* the current at 27 K is 4.27E-03 A and the power dissipation is 2.14E-02 W. At 100°C, the current is 2.47E-3 A and the power dissipation is 1.24E-02 W. ■

3.2.5 Resistivity Temperature Dependence

In a Hall measurement,[23] the carrier concentration is measured with the sample in a magnetic field and the resistivity is measured to permit determination of the mobility. The temperature dependence of the resistivity for a p-type GaP sample[5] doped with the acceptor Zn is compared to the temperature dependence of the resistivity for Al over the temperature range from 10 K to room temperature in Fig. 3.12.[24] The Zn acceptor concentration in the GaP is 6.7×10^{16} cm^{-3}. The energy gap for GaP at room temperature is 2.261 eV and the Zn ionization energy E_a is 0.060 eV.

The resistivity of the GaP rapidly decreases with temperature ($T\uparrow, \rho\downarrow$) because a larger fraction of the Zn acceptors become ionized, which increases the number of free holes ($p = N_a^-$) as represented by Eq. (A.5) in the appendix at the end of Chapter 2.

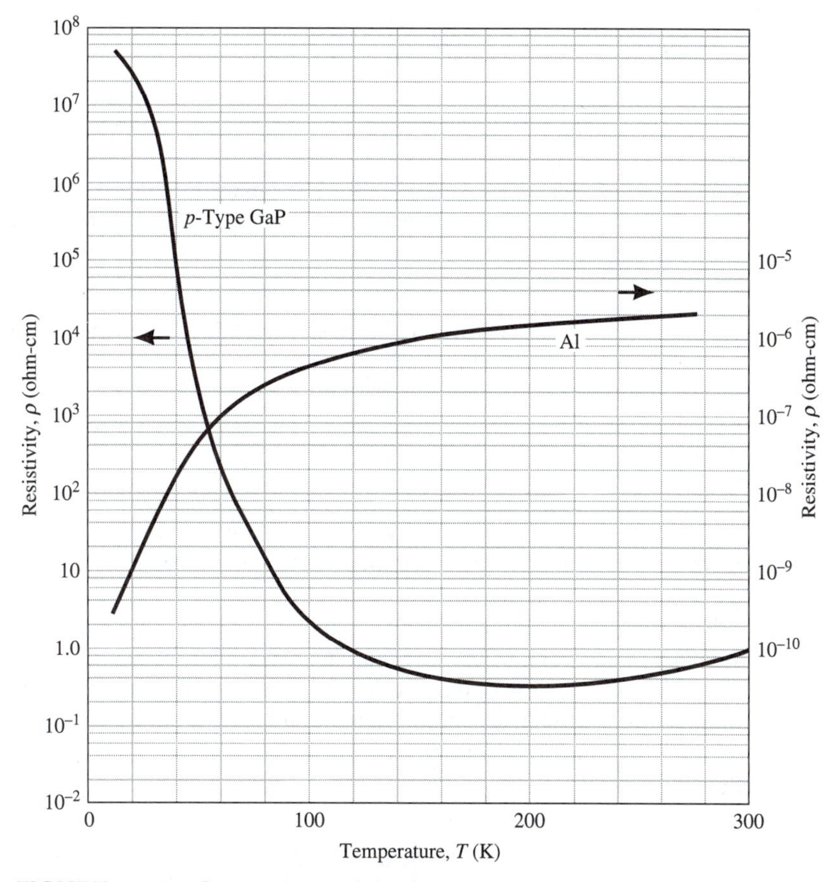

FIGURE 3.12 Comparison of the temperature dependence of *p*-type GaP[5] and Al.[24]

As the temperature is increased, the Al resistivity increases ($T\uparrow$, $\rho\uparrow$) due to increased phonon scattering at higher temperatures. The opposite temperature dependence of the resistivity is one major difference between semiconductors and metals.

3.3 CARRIER DIFFUSION

3.3.1 The Diffusion Process

In Sec. 3.2.2, the drift current, which is the transport of carriers under the influence of an electric field, was introduced. Another important current component can exist if there is a spatial variation of carrier concentration in the semiconductor material, that is, the carriers tend to move from a region of high concentration to a region of low concentration. This current component is called *diffusion current.*

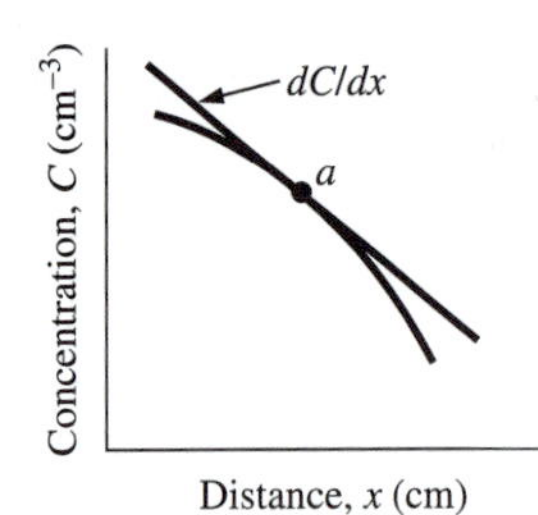

FIGURE 3.13 Illustration of a spatially varying concentration with the distance which has a negative concentration gradient dC/dx at the point a.

For the hole concentration profile shown in Fig. 3.13, the flux of holes F_p across a given plane is given by Fick's First Law as

$$F_p = -D_p \frac{dp}{dx}, \tag{3.34}$$

where D_p is the hole-diffusion coefficient or hole diffusivity. Units for diffusivity are cm^2/s. Note that in Fig. 3.13 the gradient dp/dx is negative so that when multiplied by $-D_p$ the flux is positive. Fick's First Law applies to electrons and holes as well as to the impurities in semiconductors at the high temperatures used for fabrication, and it applies to any case where an inhomogeneous distribution exists. The mathematical form of Fick's First Law given in Eq. (3.34) is analogous to the equation for heat flow, where the heat-flow flux is proportional to the temperature gradient.

The hole-diffusion current density J_p is obtained by multiplying the flux by the charge $+q$ to give

hole-diffusion current density

$$\boxed{J_p = -qD_p\, dp/dx}, \tag{3.35}$$

while the electron current density J_n is obtained by multiplying the flux by the charge $-q$ to give

electron-diffusion current density

$$\boxed{J_n = qD_n\, dn/dx}. \tag{3.36}$$

3.3.2 The Current-density Equations

When an electric field is present in addition to a carrier concentration gradient, both drift and diffusion current will flow. For electrons, the drift current density was given by Eq. (3.18) and the diffusion current density was given by Eq. (3.36). The total electron current density is the sum of the drift and diffusion current densities:

electron current density

$$\boxed{J_n = q\mu_n n\mathscr{E} + qD_n\, dn/dx}, \tag{3.37}$$

where $\mathscr{E}$ is the electric field in the x direction.

By the addition of Eq. (3.19) for the hole-drift current density to Eq. (3.35) for the hole-diffusion current density, the hole current density is

hole current density

$$\boxed{J_p = q\mu_p p\mathscr{E} - qD_p\, dp/dx}\,. \tag{3.38}$$

The negative sign is used in Eq. (3.38) because, for a negative hole gradient, the holes will diffuse in a positive x direction and give a positive current.

The total conduction current density is given by the sum of Eqs. (3.37) and (3.38):

total current density

$$\boxed{J_{cond} = J_n + J_p}\,. \tag{3.39}$$

The three expressions given by Eqs. (3.37), (3.38), and (3.39) constitute the current-density equations. These equations are very important for analyzing device operations under low electric fields. However, at sufficiently high electric fields, the terms $\mu_n\mathscr{E}$ and $\mu_p\mathscr{E}$ should be replaced by a field-dependent carrier velocity.

3.3.3 The Einstein Relation

Determination of the relationship between the diffusivity D and the mobility μ provides a good application of several concepts already presented. Begin with the equation for the electron current density as given by Eq. (3.37) at thermal equilibrium so that $J_n = 0$:

$$J_n = q\mu_n n\mathscr{E} + qD_n\, dn/dx = 0. \tag{3.40}$$

The electron concentration in this equation may be represented by Eq. (2.109), as $n = n_i \exp[(E_f - E_i)/kT)]$, and the electric field was given in Eqs. (3.15) and (3.16), as $\mathscr{E} = (1/q)(dE_i/dx) = -(d\psi/dx)$. Substitution for n and $\mathscr{E}$ in Eq. (3.40) gives

$$q\mu_n n_i \exp\left[\frac{E_f - E_i}{kT}\right]\frac{1}{q}\frac{dE_i}{dx} = -\frac{q}{kT}D_n n_i \exp\left[\frac{E_f - E_i}{kT}\right]\left[\frac{dE_f}{dx} - \frac{dE_i}{dx}\right]. \tag{3.41}$$

For either a homogeneous or inhomogeneous semiconductor at thermal equilibrium,

$$\frac{dE_f}{dx} = 0, \tag{3.42}$$

but for an inhomogenous semiconductor, $dE_i/dx \neq 0$ (see Fig. 4.7). Then by canceling common terms,

Einstein relation

$$\boxed{D_n = \mu_n kT/q}\,, \tag{3.43}$$

which is termed the *Einstein relation.* It relates two important quantities, diffusivity and mobility, which characterize carrier transport by diffusion and drift in a semiconductor. The Einstein relation also applies between D_p and μ_p. Values for mobilities for Si were given in Fig. 3.3 and for GaAs in Fig. 3.4.

3.3.4 Excess-carrier Concentrations

In thermal equilibrium, the electron–hole density product was given by Eq. (2.71) as $np = n_i^2$. When excess carriers are introduced into a semiconductor so that $np > n_i^2$, nonequilibrium exists. The process of introducing excess carriers is called *carrier injection.* Carriers can be injected by various methods such as the absorption of light with energy $h\nu > E_g$, or as presented in the next chapter, by forward biasing a *p-n* junction. Injection increases the electron and hole concentrations above their thermal-equilibrium values. The excess-electron concentration Δn is given by

excess-electron concentration

$$\boxed{\Delta n = n - n_o}\,, \tag{3.44}$$

where n is the nonequilibrium electron concentration and n_o is the thermal-equilibrium electron concentration. Equation (3.44) also may be written as

$$n = n_o + \Delta n. \tag{3.45}$$

In a similar manner, the excess-hole concentration Δp is given by

excess-hole concentration

$$\boxed{\Delta p = p - p_o}\,, \tag{3.46}$$

where p is the nonequilibrium hole concentration and p_o is the thermal-equilibrium hole concentration. Equation (3.46) may be written as

$$p = p_o + \Delta p. \tag{3.47}$$

The magnitude of the excess-carrier concentration relative to the thermal-equilibrium majority carrier concentration determines the injection level. An example will be used to clarify the meaning of injection level. Consider *n*-type Si with a donor concentration which gives a thermal-equilibrium electron concentration of $n_{no} = 1 \times 10^{15}$ cm^{-3}. The minority carrier concentration is given by $p_{no} = n_i^2/n_{no} = 1.0 \times 10^5$ cm^{-3}. In this notation, the first subscript, n or p, refers to the type of semiconductor and the subscript o refers to the thermal-equilibrium condition. Thus, n_{no} and p_{no} denote the electron and hole concentrations, respectively, in an *n*-type semiconductor at thermal equilibrium.

When excess carriers are introduced by a technique such as optical absorption, the excess-electron concentration Δn must equal the excess-hole concentration Δp, because electrons and holes are produced in pairs. *Charge neutrality as given by Eq. (2.94) must also hold.* If 1×10^{12} cm^{-3} minority carrier holes are introduced into the *n*-type Si in the preceding example, the hole concentration is

$$p_n = p_{no} + \Delta p = 1.0 \times 10^5 + 1 \times 10^{12} = 1 \times 10^{12}\ \text{cm}^{-3},$$

which is an increase by seven orders of magnitude (from 10^5 to 10^{12}). At the same time, the addition of $1 \times 10^{12}\ \text{cm}^{-3}$ majority carrier electrons to the $1 \times 10^{15}\ \text{cm}^{-3}$ thermal-equilibrium concentration gives

$$n_n = n_{no} + \Delta n = 1 \times 10^{15} + 1 \times 10^{12} = 1.001 \times 10^{15}\ \text{cm}^{-3}.$$

The percentage change in the majority carrier concentration is only 0.1% ($10^{12}/10^{15}$). *This condition, in which the excess-carrier concentration is small in comparison to the thermal-equilibrium majority carrier concentration, that is,* $\Delta n = \Delta p \ll n_{no}$, *is referred to as low-level injection.* High-level injection is when the injected excess-carrier concentration is comparable to or larger than the thermal-equilibrium majority carrier concentration. High-level injection is sometimes encountered in device operation. However, because of the complexities involved in the treatment of high-level injection, only low-level injection will be considered unless otherwise stated.

low-level injection

3.4 THE CONTINUITY EQUATION

To analyze carrier and current flow in semiconductor materials and devices, an equation may be written which accounts for the flux of free carriers into and out of an infinitesimal volume as represented in Fig. 3.14. The governing equation is called the *continuity equation,* and it applies to *both majority* and *minority carriers.* To derive the one-dimensional continuity equation for electrons, consider an infinitesimal slice

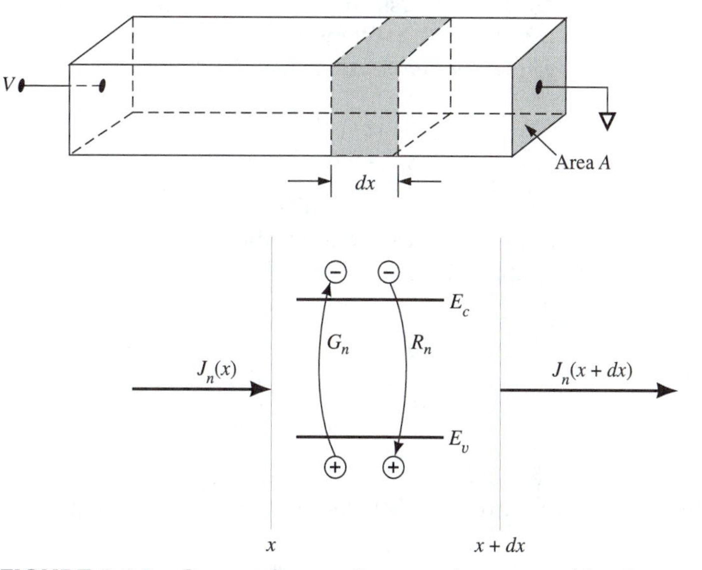

FIGURE 3.14 Current flow and generation–recombination processes in an infinitesimal slice of thickness *dx*.

with a thickness dx located at x, as shown in Fig. 3.14. The number of electrons in the slice may increase due to the net flow into the slice and the net carrier generation in the slice. The overall rate of electron increase is the algebraic sum of four components: the number of electrons flowing into the slice at x, minus the number of electrons flowing out at $x + dx$, plus the rate at which electrons are generated, minus the rate at which they combine with holes in the slice. The figure schematically represents the generation of a hole in the valence band E_v and an electron in the conduction band E_c with the generation rate G_n, which has units of $\mathrm{cm}^{-3}\mathrm{s}^{-1}$, and the recombination of a conduction-band electron with a valence-band hole with the recombination rate R_n, which also has units of $\mathrm{cm}^{-3}\mathrm{s}^{-1}$. The detailed nature of G_n and R_n is not required to obtain the continuity equation.

The first two components are found by dividing the currents at each side of the slice by the charge of an electron. The generation and recombination rates were designated in Fig. 3.14 as G_n and R_n, respectively. The overall rate of change in the number of electrons in the slice is then

$$\frac{\partial n}{\partial t} A\,dx = \left[\frac{J_n(x)A}{-q} - \frac{J_n(x+dx)A}{-q}\right] + (G_n - R_n)A\,dx, \tag{3.48}$$

where A is the cross-sectional area and $A\,dx$ is the volume of the slice. Expanding the expression for the current at $x + dx$ in a Taylor series yields

$$J_n(x+dx) = J_n(x) + \frac{1}{1!}\frac{\partial J_n}{\partial x}dx + \frac{1}{2!}\frac{\partial^2 J_n}{\partial x^2}(dx)^2 + \cdots. \tag{3.49}$$

Because dx is small, only the first two terms in the series are retained. With this approximation, the $J_n(x)$ term in Eq. (3.49) for $J_n(x+dx)$ cancels the $J_n(x)$ term in Eq. (3.48) and the *continuity equation* for electrons may be written as

continuity equation for electrons

$$\boxed{\frac{\partial n}{\partial t} = \frac{1}{q}\frac{\partial J_n}{\partial x} + (G_n - R_n)}\ . \tag{3.50}$$

A similar continuity equation can be derived for holes, except that the sign of the first term on the right-hand side of Eq. (3.50) is changed because of the positive charge associated with a hole:

continuity equation for holes

$$\boxed{\frac{\partial p}{\partial t} = -\frac{1}{q}\frac{\partial J_p}{\partial x} + (G_p - R_p)}\ . \tag{3.51}$$

The current density given by Eq. (3.37) may be used for J_n in Eq. (3.50) to represent the minority carrier electron concentration n_p in a p-type semiconductor as

$$\frac{\partial n_p}{\partial t} = n_p\mu_n\frac{\partial \mathscr{E}}{\partial x} + \mu_n\mathscr{E}\frac{\partial n_p}{\partial x} + D_n\frac{\partial^2 n_p}{\partial x^2} + (G_n - R_n). \tag{3.52}$$

Similarly, the current density given by Eq. (3.38) may be used for J_p in Eq. (3.51) to represent the minority carrier hole concentration p_n in an n-type semiconductor as

$$\frac{\partial p_n}{\partial t} = -p_n \mu_p \frac{\partial \mathscr{E}}{\partial x} - \mu_p \mathscr{E} \frac{\partial p_n}{\partial x} + D_p \frac{\partial^2 p_n}{\partial x^2} + (G_p - R_p). \tag{3.53}$$

The form in which Eqs. (3.52) and (3.53) have been written assumes that the mobility μ and the diffusivity D are not functions of x.

For Eqs. (3.52) and (3.53) at steady state, $\partial/\partial t = 0$ and the partial derivative becomes an ordinary differential: $\partial \mathscr{E}/\partial x = d\mathscr{E}/dx$. The derivative of the electric field $d\mathscr{E}/dx$ appears frequently as Gauss' law in the analysis of semiconductor devices and is given by

Gauss's law in differential form from Gauss's theorem: $\int \rho\, dv = \epsilon \oint \mathscr{E} \cdot d\mathbf{a} = \epsilon \int \nabla \cdot \mathscr{E}\, dv.$

$$\boxed{\frac{d\mathscr{E}(x)}{dx} = \frac{\rho(x)}{\epsilon}}, \tag{3.54}$$

where $\rho(x)$ is the net charge per unit volume and ϵ is the dielectric constant. For charge neutrality in the semiconductor bulk, $\rho(x) = 0$ and the $\partial \mathscr{E}/\partial x$ term in Eqs. (3.52) and (3.53) is zero. For the p-n junctions and the bipolar junction transistors considered in the Chapters 4 through 6, the electric field $\mathscr{E}$ is zero in the neutral semiconductor bulk where minority carrier diffusion current dominates drift current, and the term with $\mathscr{E}$ in Eqs. (3.52) and (3.53) may be neglected, and these equations reduce to

continuity equation for electrons

$$\boxed{\frac{\partial n_p}{\partial t} = D_n \frac{\partial^2 n_p}{\partial x^2} + (G_n - R_n)} \tag{3.55}$$

and

continuity equation for holes

$$\boxed{\frac{\partial p_n}{\partial t} = D_p \frac{\partial^2 p_n}{\partial x^2} + (G_p - R_p)}\ . \tag{3.56}$$

For steady-state conditions, $\partial n_p/\partial t$ and $\partial p_n/\partial t$ are zero and the previous two equations reduce to

continuity equations for steady state

$$D_n \frac{\partial^2 n_p}{\partial x^2} + (G_n - R_n) = 0\,, \tag{3.57}$$

and

$$D_p \frac{\partial^2 p_n}{\partial x^2} + (G_p - R_p) = 0. \tag{3.58}$$

Another frequent application of the continuity equations is when the carrier concentration distribution is uniform so that $\partial^2 n_p/\partial x^2$ and $\partial^2 p_n/\partial x^2$ are zero. In this case, the continuity equations reduce to

continuity equations for uniform carrier distribution

$$\frac{\partial n_p}{\partial t} = (G_n - R_n) \tag{3.59}$$

and

$$\frac{\partial p_n}{\partial t} = (G_p - R_p). \tag{3.60}$$

3.5 CARRIER RECOMBINATION

3.5.1 Introductory Comments

Whenever the thermal-equilibrium condition is disturbed so that $np \neq n_i^2$, processes exist to restore the system to equilibrium with $np = n_i^2$. In the case of the injection of excess carriers, the mechanism that restores equilibrium is recombination of the injected minority carriers with the majority carriers. Depending on the nature of the recombination process, the energy released by the electron–hole recombination can be dissipated as heat to the lattice or emitted as a photon. When the recombination energy is dissipated as heat, the recombination process is called *nonradiative* recombination, while the recombination process is called *radiative* recombination when a photon is emitted.

The nonradiative recombination process via recombination centers within the energy gap will be presented in Sec. 3.5.2. The radiative recombination process which is important in III-V compound semiconductors for light-emitting diodes (LEDs) and semiconductor lasers is described in Sec. 3.5.3. Another nonradiative recombination process, called Auger (pronounced OH-JAY) recombination, is given in Sec. 3.5.4. Auger recombination is important for small-energy-gap semiconductors and at high carrier concentrations. In Auger recombination, either two electrons and one hole or two holes and one electron are involved in electron–hole recombination. The recombination energy is given to the remaining electron or hole, which dissipates the excess energy to the lattice as heat. Auger recombination is significant in semiconductor lasers which emit at wavelengths of 1.3 or 1.55 μm. These wavelengths are selected to match the minimum optical-fiber attenuation and the smallest optical-pulse spreading.

3.5.2 Shockley-Hall-Read Recombination

The nonradiative recombination process via energy levels within the energy gap is common to all semiconductors. These energy levels within the energy gap are due to chemical impurities as well as crystalline defects such as dislocations. The ionization energy of these levels E_t exceeds the ionization E_d and E_a of the shallow donors and acceptors. Gold in Si is an example of an intentionally introduced impurity to increase recombination. The theory for nonradiative recombination was first published by Hall[25] and Shockley and Read.[26] This theory is referred to as Shockley-Hall-Read (SHR) recombination.

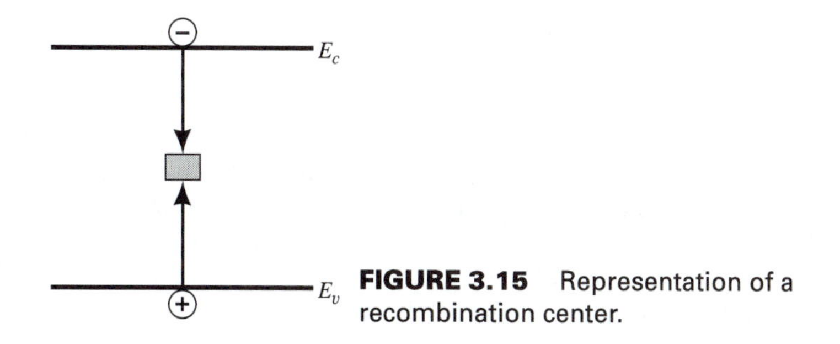

FIGURE 3.15 Representation of a recombination center.

The terminology used for the energy levels within the energy gap is often not well defined. The energy levels which determine the free carrier concentrations are the *shallow levels* near the conduction or valence bands and are called donors and acceptors, respectively. The energy levels with greater ionization energies than the shallow donors and acceptors can serve as recombination centers. A recombination center is illustrated in Fig. 3.15. In this process, the *recombination center* first captures an electron (or hole) and then eliminates an electron–hole pair by capturing a hole (or electron).

recombination center

Consider a p-type semiconductor with $N_a^- > N_d^+$. The nonradiative recombination centers have a concentration N_t and are located at an energy E_t above the valence band. The electron capture probability is given by the product of the capture cross section σ_n and the thermal velocity v_{th}, which was given in Eq. (3.2) as $\sqrt{8kT/\pi m^*}$ and is approximately 1×10^7 cm/s for most semiconductors. A capture cross section σ_n represents the effectiveness of the center to capture an electron and is a measure of how close the electron has to come to the center to be captured. It could be expected that the capture cross section would be of the order of atomic dimensions, that is, of the order of 10^{-15} cm^2. The capture rate is also proportional to the number of recombination centers N_t. The quantity $1/(\sigma_n v_{th} N_t)$ has units of 1/(cm^2 cm/s cm^{-3}) or seconds, and is designated the nonradiative carrier lifetime τ_{nr}:

nonradiative carrier lifetime

$$\boxed{\tau_{nr} \equiv \frac{1}{\sigma_n v_{th} N_t}} \, . \tag{3.61}$$

For nonequilibrium but steady state, the net rate of capture of electrons must equal that of holes. As shown in Appendix B.1 at the end of this chapter, the net nonradiative recombination rate U may be written as

net nonradiative recombination rate

$$\boxed{U = \frac{np - n_i^2}{\tau_{nr}(n + n_1) + \tau_{nr}(p + p_1)}} \, , \tag{3.62}$$

where n_1 is the electron concentration if $E_f = E_t$, and p_1 is the hole concentration if $E_f = E_t$. From Eq. (2.109), n_1 is given as

$$n_1 \equiv n_i \exp[(E_t - E_i)/kT] \, , \tag{3.63}$$

and from Eq. (2.110), p_1 is given as

$$p_1 \equiv n_i \exp[(E_i - E_t)/kT] \,. \tag{3.64}$$

Note that when $E_t = E_i$, $n_1 = p_1 = n_i$. Equation (3.62) is the general equation for the net recombination rate, but in many applications it can be simplified.

A more convenient form of Eq. (3.62) may be written for low-level injection when $\Delta n = \Delta p$, and both Δn and Δp are much less than the majority carrier concentration. For an n-type semiconductor, the net recombination rate for the minority carrier holes becomes

$$R_p = \frac{(n_{no} + \Delta n)(p_{no} + \Delta p) - n_i^2}{\tau_{nr}(n_{no} + \Delta n + n_1) + \tau_{nr}(p_{no} + \Delta n + p_1)} \,. \tag{3.65}$$

In the numerator, $n_{no} p_{no}$ cancels the $-n_i^2$ term, the $\Delta n \, \Delta p$ term may be neglected, and $n_{no} \, \Delta p \gg p_{no} \, \Delta n$ because $n_{no} \gg p_{no}$, which leaves only the $n_{no} \, \Delta p$ term. In the denominator, let $n_1 = p_1 = n_i$ (which has been found to be a reasonable approximation) and then n_1 and p_1 may be neglected, and for low-level injection, $n_{no} > \Delta n$ or Δp. The recombination rate for a minority carrier hole in an n-type semiconductor reduces to

hole recombination rate

$$\boxed{U(\text{cm}^{-3}\text{s}^{-1}) = R_p = \frac{n_{no} \, \Delta p}{n_{no} \tau_{nr}} = \frac{\Delta p}{\tau_{nr}} = \frac{p_n - p_{no}}{\tau_{nr}}} \,. \tag{3.66}$$

Equation (3.66) is the most frequently used form of the nonradiative Shockley-Hall-Read recombination rate for minority carriers. For an electron in a p-type semiconductor, the minority carrier electron recombination rate would be

electron recombination rate

$$\boxed{R_n = \frac{\Delta n}{\tau_{nr}} = \frac{n_p - n_{po}}{\tau_{nr}}} \,. \tag{3.67}$$

The lifetime as given in Eq. (3.61) is inversely proportional to N_t, the concentration of recombination centers per unit volume. For device applications that require long recombination lifetimes, the concentration of the recombination centers must be minimized. On the other hand, for high-speed switching operations, short recombination lifetimes are required and it is desirable that the semiconductor be heavily doped with recombination centers.

Several impurities have energy levels close to the middle of the energy gap. These impurities are efficient recombination centers. A typical example is gold in Si.[27] Gold has an acceptor state at $E_t - E_i = 0.02$ eV or $(E_t - E_i)/kT = 0.77$ at room temperature. The minority carrier lifetime decreases linearly with gold concentration as shown in Fig. 3.16. By increasing the gold concentration from 10^{14} cm^{-3} to 10^{18} cm^{-3}, the

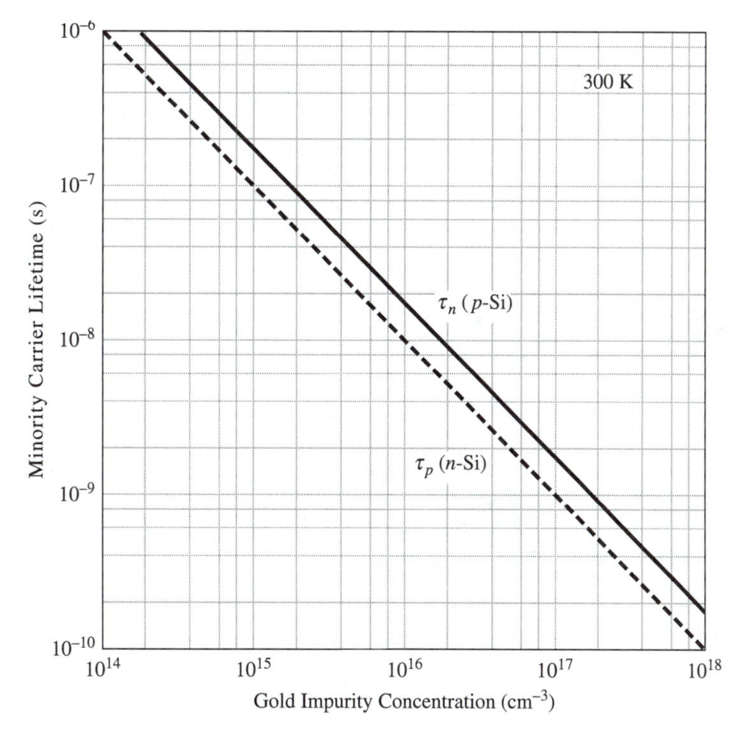

FIGURE 3.16 Recombination lifetime vs. gold impurity concentration in Si.[27]

minority carrier lifetime can be reduced from 1 μs to 0.1 ns. Another method of changing the minority carrier lifetime is by high-energy irradiation, which causes displacement of lattice atoms and introduces energy levels in the energy gap. For example, electron irradiation in Si results in an acceptor level at 0.4 eV above the valence band and a donor level at 0.36 eV below the conduction band. Irradiation with neutrons creates an acceptor level very near E_i.

3.5.3 Radiative Recombination

In direct-energy-gap semiconductors, the recombination of conduction-band electrons with valence-band holes with the same wave vector **k** results in the emission of the difference in energy as photons. The recombination of electrons and holes within the conduction and valence bands, which is illustrated in Fig. 3.17(*a*), results in the emission spectrum illustrated in Fig. 3.17(*b*) for *p*-type GaAs with $p_{po} = 1.2 \times 10^{18}$ cm^{-3}. The radiative recombination rate in semiconductors was derived by van Roosbroeck and Shockley.[28] It is shown in Appendix B.2 at the end of this chapter that for nonequilibrium with excess carriers, the total rate of radiative recombination R_r per unit volume in steady state is proportional to the product of the nonequilibrium electron–hole

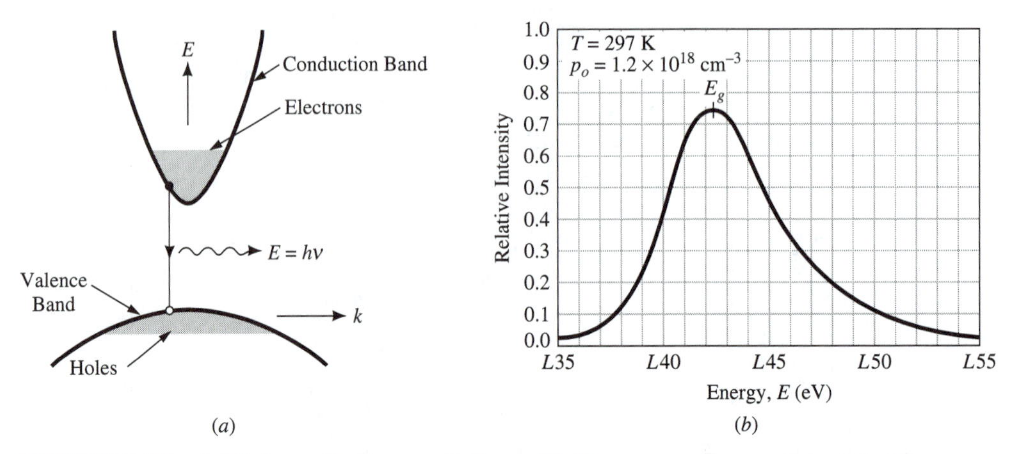

FIGURE 3.17 (*a*) Schematic representation with the energy E vs. wave vector k diagram for the direct electron–hole recombination with the emission of a photon of energy $E = h\nu$. (*b*) The emission spectrum of GaAs for the indicated temperature and free hole concentration. The energy gap E_g at this temperature is also indicated.

concentration, and without momentum conservation, is given by

radiative recombination rate

$$\boxed{R_r \equiv Bnp}, \tag{3.68}$$

where B is the radiative constant. When low-level excitation in an n-type semiconductor exists so that $n \approx n_{no}$, the radiative lifetime is given by

radiative lifetime

$$\boxed{\tau_r = \frac{\Delta p}{R_r} = \frac{p_n - p_{no}}{Bn_{no}p_n} = \frac{1}{Bn_{no}}}, \tag{3.69}$$

B for GaAs

with $p_n \gg p_{no}$ and $n_n = n_{no}$ for $n_{no} \gg \Delta n$. For GaAs at room temperature,[29] the radiative constant $B \simeq 2\times10^{-10}$ cm^3/s, which by Eq. (3.69) for $p_{po} = 1.2\times10^{18}$ cm^{-3} gives a radiative lifetime $\tau_r = 4.2 \times 10^{-9}$ s.

Several useful expressions may be obtained by combining the expressions for nonradiative lifetime τ_{nr} in Eq. (3.66) and the radiative lifetime τ_r in Eq. (3.69). The overall lifetime τ for the overall recombination rate R for holes in an n-type semiconductor is

$$R = R_{nr} + R_r = \frac{\Delta p}{\tau} = \frac{\Delta p}{\tau_{nr}} + \frac{\Delta p}{\tau_r}, \tag{3.70}$$

and then

overall lifetime

$$\tau = \frac{\tau_{nr}\tau_r}{\tau_{nr} + \tau_r}. \tag{3.71}$$

The quantum efficiency η is the fraction of electron–hole recombinations which result in photon emission and is given by

quantum efficiency

$$\eta = \frac{R_r}{R} = \frac{\Delta p/\tau_r}{\Delta p/\tau_{nr} + \Delta p/\tau_r} = \frac{\tau_{nr}}{\tau_{nr} + \tau_r}. \tag{3.72}$$

Therefore, high quantum efficiency requires $\tau_r \ll \tau_{nr}$; i.e., the radiative lifetime must be small compared to the nonradiative lifetime.

The room temperature experimental absorption coefficients for Si and GaAs are shown in Fig. 3.18. It is shown in Appendix B.2 that the radiative constant in Eq. (3.68) is proportional to the value of the absorption coefficient at the energy gap $\alpha(E_g)$:

$$B \propto \alpha(E_g). \tag{3.73}$$

As shown in Fig. 3.18, $\alpha(E_g) \approx 15\ \text{cm}^{-1}$ for Si, and $\alpha(E_g) \approx 10^4\ \text{cm}^{-1}$ for GaAs. This smaller radiative constant B for indirect-energy-gap semiconductors, such as Si, is a result of the requirement of electron interaction with both a photon and a phonon, as described in Sec. 2.7.2, which is a less probable recombination as compared to direct-energy-gap recombination, which does not require phonon interaction. Additional description of optical processes in semiconductors has been given by Bhattacharya.[32]

3.5.4 Auger Recombination

The final recombination process to be considered is Auger recombination which is non-radiative, but very different from SHR recombination. This recombination mechanism becomes important for semiconductors with E_g less than about 1.0 eV and at high carrier

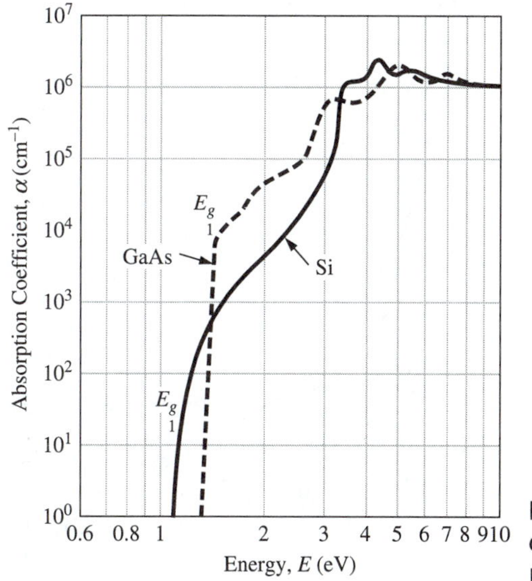

FIGURE 3.18 The absorption coefficients for Si[30] and GaAs[31] at room temperature.

concentrations. Auger recombination results from electron–electron and hole–hole collisions. In the electron process, two electrons collide. One electron drops to the empty hole state in the valence band while the second electron assumes this recombination energy which is the difference in the conduction and valence band, the energy gap E_g. The second electron moves to a higher energy state by the E_g recombination energy and then relaxes to the thermal distribution by processes such as carrier–lattice interactions by phonon scattering. This process is designated as CCCH. In the hole process, two holes collide. One hole recombines with an electron from the conduction band while the other is excited into the split-off valence band by the energy E_g, and this process is designated as CHHS. In this notation, C stands for a conduction-band electron, H stands for a valence-band heavy hole, and S stands for a hole in the split-off valence band. The heavy-hole valence band and the light-hole valence band were illustrated in Figs. 2.28 and 2.30. As mentioned in Sec. 3.5.1, Auger recombination is a detrimental nonradiative recombination process for semiconductor lasers used in optical-fiber communication systems. The CCCH and the CHHS Auger recombination processes are illustrated in Fig. 3.19.

The recombination rates for SHR recombination and radiative recombination were derived by application of the principle of detailed balance. Calculation of the Auger recombination rate requires representation of the electrons and holes by their wave functions and the application of quantum mechanics, which requires a more extensive physics background than for any other subject covered in this book. Therefore, only a description of the quantities which are necessary to represent the carrier lifetime due to Auger recombination will be presented here.

The Auger recombination rate was first derived by Beattie and Landsberg.[34] Numerous authors have made refinements to the theory with a more realistic band structure[33,35–38] and a useful summary was given by Agrawal and Dutta.[39] The transition probability for Auger recombination gives the Auger recombination rate R_A as

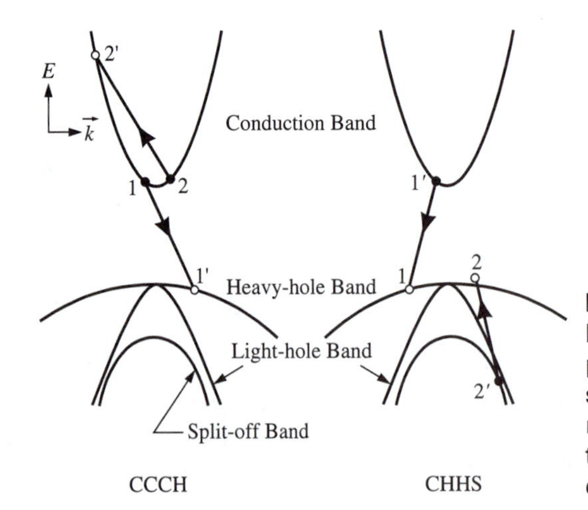

FIGURE 3.19 Two different band-to-band Auger recombination processes are represented schematically. The electrons are represented by closed circles and the holes are represented by open circles.[33]

Table 3.1 Representative Room Temperature Values for the Auger Coefficient C and the Auger Lifetime τ_a.

Semiconductor	Auger Coefficient (cm^6/s)	Reference	Lifetime (s) for $n_{no} = 2 \times 10^{18}\ cm^{-3}$
GaAs	7×10^{-30}	40	3.8×10^{-8}
p-type Si	9.9×10^{-32}	41	2.5×10^{-6}
n-type Si	2.8×10^{-31}	41	8.9×10^{-7}
$Ga_{0.28}In_{0.72}As_{0.6}P_{0.4}$	6×10^{-29}	42	4.2×10^{-9}

Auger recombination rate

$$\boxed{R_A = Cn^2p}\ , \tag{3.74}$$

where C is the Auger coefficient in units of cm^6/s. The Auger coefficient has terms which depend on the energy gap and only becomes important when $E_g \leq 1.0$ eV. When Auger recombination is significant, C is in the range of 10^{-28} to $10^{-29}\ cm^6/s$. For low-level excitation in an n-type semiconductor so that $n \approx n_{no}$, the Auger lifetime is given by

Auger lifetime

$$\boxed{\tau_A = \frac{\Delta p}{R_A} = \frac{p_n - p_{no}}{Cn_{no}^2(p_n - p_{no})} = \frac{1}{Cn_{no}^2}}\ , \tag{3.75}$$

which shows that Auger lifetime varies with the majority carrier concentration as $1/n_{no}^2$ while the radiative lifetime varies as $1/n_{no}$.

There are significant uncertainties in the measured and calculated values of the Auger lifetimes and therefore uncertainties in the assigned values of C. However, the Auger coefficient tends to be larger for narrow-energy-gap semiconductors. Table 3.1 gives representative values for C and τ_A for $n_{no} = 2 \times 10^{18}\ cm^{-3}$. The quaternary $Ga_{0.28}In_{0.72}As_{0.6}P_{0.4}$ has an energy gap at room temperature of 0.95 eV and is used for semiconductor lasers with an emission wavelength of 1.3 μm. The short Auger lifetime for $Ga_{0.28}In_{0.72}As_{0.6}P_{0.4}$ degrades the performance of semiconductor lasers with this material at high injection levels.

EXAMPLE 3.3 Find the steady-state minority carrier concentration for an n-type GaAs sample with $n_{no} = N_d^+ = 3 \times 10^{16}\ cm^{-3}$ and a lifetime $\tau = 3 \times 10^{-9}$ s. By the absorption of light, the uniform excitation rate is $2.6 \times 10^{21}\ cm^{-3}s^{-1}$.

Solution The continuity equation [Eq. (3.56)] for steady state and uniform excitation becomes

$$G_p = 2.6 \times 10^{21} = R_p = \Delta p/3 \times 10^{-9}$$

or

$$\Delta p = p_n - p_{no} \simeq p_n = 2.6 \times 10^{21}\ cm^{-3}\ s^{-1} \times 3 \times 10^{-9}\ s = 7.8 \times 10^{12}\ cm^{-3}. \ \blacksquare$$

Table 3.2 Basic Equations for Semiconductor Device Analysis.

Equation	Equation Name	Equation Number
$\mathscr{E}(x) = -\dfrac{d\psi(x)}{dx}$	Relation of $\mathscr{E}$ to ψ	(3.16)
$J_n = q\mu_n n\mathscr{E} + qD_n\dfrac{dn}{dx}$	Electron current density	(3.37)
$J_p = q\mu_p p\mathscr{E} - qD_p\dfrac{dp}{dx}$	Hole current density	(3.38)
$\dfrac{d\mathscr{E}}{dx} = \dfrac{\rho(x)}{\epsilon}$	Gauss's law	(3.54)
$\dfrac{\partial n_p}{\partial t} = D_n\dfrac{\partial^2 n_p}{\partial x^2} + (G_n - R_n)$	Electron continuity equation	(3.55)
$\dfrac{\partial p_n}{\partial t} = D_p\dfrac{\partial^2 p_n}{\partial x^2} + (G_p - R_p)$	Hole continuity equation	(3.56)
$R_p = \dfrac{p_n - p_{no}}{\tau_{nr}}$	Hole recombination rate	(3.66)
$R_n = \dfrac{n_p - n_{po}}{\tau_{nr}}$	Electron recombination rate	(3.67)

3.6 THE BASIC EQUATIONS

With 75 numbered equations in the preceding sections of this chapter, it is helpful to tabulate the basic equations which are used in the derivation of the behavior of semiconductor devices. These basic equations are summarized in Table 3.2. One additional basic equation, which is for the relationship between potential differences and minority carrier concentration ratios, is given in Sec. 4.5.

3.7 CONTINUITY EQUATION EXAMPLES

3.7.1 Transient Response

The transient response of a semiconductor slab after the removal of a light source which generates excess carriers illustrates the physical meaning of minority carrier lifetime.

These examples in this section are taken from S. M. Sze, *Semiconductor Devices, Physics and Technology* (Wiley, New York, 1985), pp. 47, 59–60, and A. S. Grove, *Physics and Technology of Semiconductor Devices* (Wiley, New York, 1967), pp. 120, 122, 126, and 136.

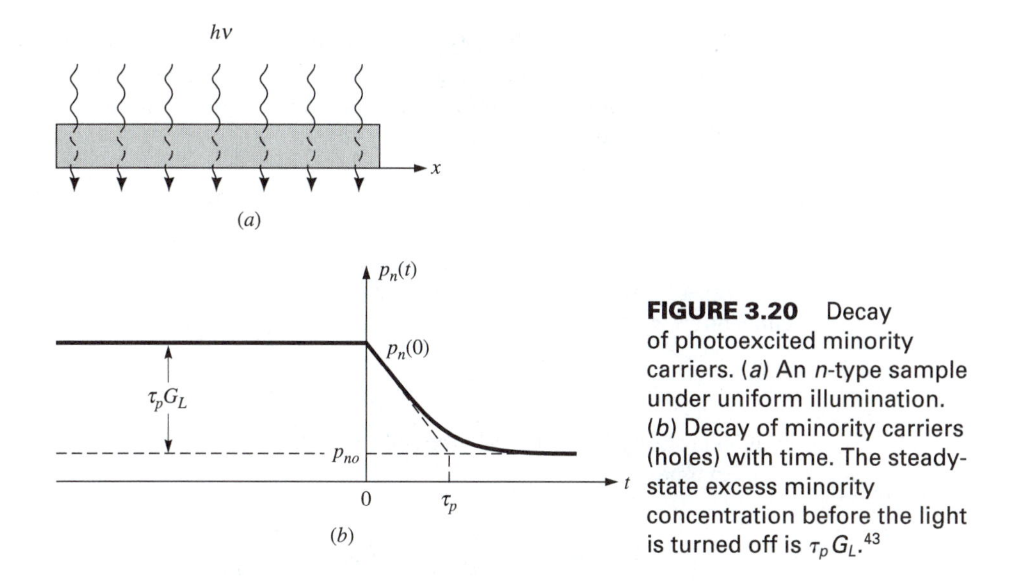

FIGURE 3.20 Decay of photoexcited minority carriers. (*a*) An *n*-type sample under uniform illumination. (*b*) Decay of minority carriers (holes) with time. The steady-state excess minority concentration before the light is turned off is $\tau_p G_L$.[43]

Consider an *n*-type sample, as shown in Fig. 3.20 (*a*), that is illuminated with light and in which the electron–hole pairs are generated uniformly throughout the sample with a generation rate G_L. For the generation of carriers by the absorption of light, the generation rate is governed by Lambert's law of photon absorption and the net generation rate may be written as

$$G(x) = \frac{(1 - R)N}{A}\alpha \exp(-\alpha x), \tag{3.76}$$

where R is the reflectivity, N the number of incident photons per unit time, A the cross-sectional area, and α the optical absorption coefficient with units of cm^{-1}. For most semiconductors, the reflectivity at photon energies near the energy gap is approximately 0.3.

EXAMPLE 3.4 The incident light is generally given in watts (W) and can be converted to the number of incident photons per second by noting that the incident photon energy is given by $Nh\nu$, where the photon energy $E = h\nu$ is generally near 2 eV. Units of eV are converted to joules (J) by multiplying by 1.6×10^{-19}:

$$1 \text{ eV} = 1.6 \times 10^{-19} \text{ J}.$$

Also, 1 W = 1 J/s, and the number of incident photons/s is given by

$$N(\text{number/s}) = W/h\nu = W/[h\nu\ (\text{eV}) \times (1.6 \times 10^{-19} \text{ J/eV})].$$

Therefore, 1 W at a photon energy of 2.0 eV would give 3.1×10^{18} photons/s. ■

It is assumed that the n-type sample in Fig. 3.20 is sufficiently thin and that α is small enough so that uniform generation can be assumed. In steady state ($\partial p_n/\partial t = 0$) and for a uniform distribution ($\partial^2 p_n/\partial x^2 = 0$), the continuity equation given by Eq. (3.56) becomes

$$G_L = R_p \qquad \text{or} \qquad G_L = \frac{p_n - p_{no}}{\tau_p}, \tag{3.77}$$

with Eq. (3.66) for R_p and the overall lifetime written as τ_p. The steady-state hole concentration becomes

$$p_n = p_{no} + \tau_p G_L. \tag{3.78}$$

If at an arbitrary time of $t = 0$ the light is suddenly turned off, the continuity equation then becomes

$$\frac{dp_n}{dt} = -R_p = -\frac{p_n - p_{no}}{\tau_p}, \tag{3.79}$$

with the initial condition of $p_n(t = 0) = p_{no} + \tau_p G_L$ as given by Eq. (3.78), and for the return to thermal equilibrium, $p_n(t \to \infty) = p_{no}$. The solution is

$$p_n(t) = p_{no} + \tau_p G_L \exp(-t/\tau_p). \tag{3.80}$$

Figure 3.20 (b) shows the variation of p_n with time. *The minority carriers recombine with majority carriers and decay exponentially with the time constant τ_p, which corresponds to the lifetime.*

3.7.2 Steady-state Injection from One Side

An n-type semiconductor is shown in Fig. 3.21 (a) where excess carriers are injected from one side as a result of illumination and absorption at the surface. At steady state ($\partial p_n/\partial t = 0$) there is a concentration gradient of carriers into the bulk. Equation (3.58) gives the differential equation for minority carriers inside the semiconductor where the effect of the illumination is included as a boundary condition at $x = 0$. In the bulk, $G_L = 0$ and with R_p given by Eq. (3.66) as $R_p = (p_n - p_{no})/\tau_p$. The differential equation for the spatial variation of the minority carriers is

$$\frac{d^2 p_n}{dx^2} - \frac{p_n}{D_p \tau_p} = -\frac{p_{no}}{D_p \tau_p}. \tag{3.81}$$

Equation (3.81) may readily be solved by assuming a solution which can be verified by substituting the assumed solution back into the equation. The solution to Eq. (3.81) is

$$p_n(x) = C_1 \exp(-x/\sqrt{D_p \tau_p}) + C_2 \exp(x/\sqrt{D_p \tau_p}) + p_{no}, \tag{3.82}$$

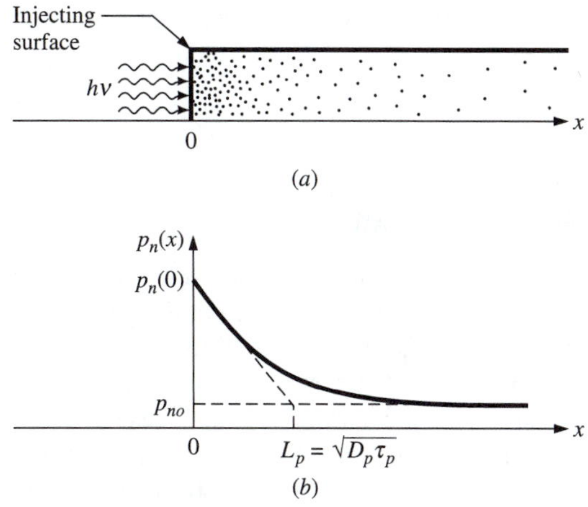

FIGURE 3.21 Steady-state carrier injection from one side. (*a*) Schematic representation. (*b*) Variation of the minority carrier concentration p_n with distance.[44]

where C_1 and C_2 are arbitrary constants determined by the boundary conditions. The boundary conditions are $p_n(x = 0) = p_n(0)$ and $p_n(x \to \infty) = p_{no}$. For $x = \infty$, Eq. (3.82) becomes

$$p_n(\infty) = p_{no} = C_1 \exp(-\infty) + C_2 \exp(\infty) + p_{no}. \tag{3.83}$$

The first term is zero because $\exp(-\infty) \to 0$, and C_2 must be zero because $\exp(\infty)$ would violate the boundary condition. For the other boundary condition at $x = 0$, Eq. (3.83) with $C_2 = 0$ becomes

$$p_n(0) = C_1 + p_{no}, \tag{3.84}$$

which gives

$$C_1 = p_n(0) - p_{no}. \tag{3.85}$$

The solution of $p_n(x)$ is

$$p_n(x) = [p_n(0) - p_{no}] \exp(-x/L_p) + p_{no}. \tag{3.86}$$

The length L_p is equal to $\sqrt{D_p \tau_p}$ and is called the *diffusion length*:

hole-diffusion length

$$\boxed{L_p \equiv \sqrt{D_p \tau_p}}\ . \tag{3.87}$$

Figure 3.21 (*b*) shows the variation of the minority carrier concentration, which decays with a characteristic length L_p. At $x = L_p$, the excess carrier concentration has decayed by exp(−1) of the initial value, which is 36.8%. For a p-type semiconductor, the electron diffusion length L_n is given by

electron-diffusion length

$$\boxed{L_n \equiv \sqrt{D_n \tau_n}} \ . \tag{3.88}$$

3.7.3 Surface Recombination

For the examples given in Secs. 3.7.1 and 3.7.2, any recombination at the surface was ignored. For any real semiconductor, recombination at the surface must be considered. The dangling bonds at a semiconductor surface are illustrated in Fig. 3.22.[45] Because of the abrupt discontinuity of the lattice structure at the surface, a large number of localized energy states may be introduced at the surface region. These energy states may greatly enhance the recombination rate at the surface region. An understanding of the surface recombination process is important because it has a strong effect on the characteristics of many semiconductor devices and the first applications of SiO_2 grown on Si were to reduce surface recombination for Si bipolar transistors.

The kinetics of surface recombination are similar to those considered before for recombination centers in the semiconductor bulk. The total number of carriers recombining at the surface per unit area and unit time can be expressed in a form analogous to Eq. (3.62):

$$U_s = (\sigma v_{th} N_s) \left[\frac{n_s p_s - n_i^2}{(n_s + n_1) + (p_s + p_1)} \right], \tag{3.89}$$

where n_s and p_s denote the electron and hole concentrations at the surface and N_s is the recombination center density per unit area in the surface region. For low excess

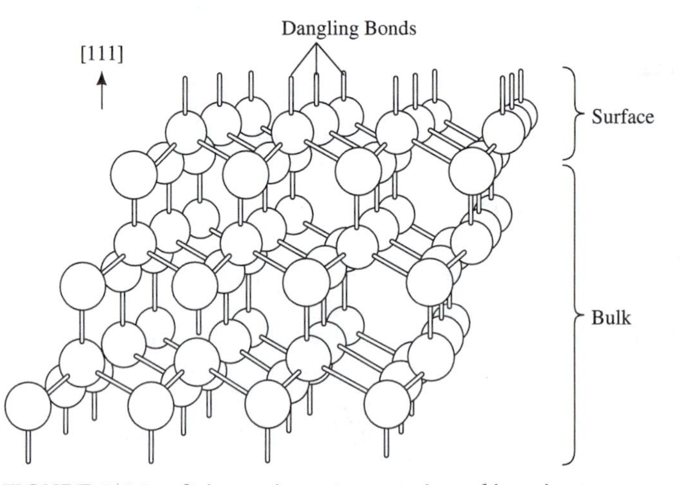

FIGURE 3.22 Schematic representation of bonds at a clean semiconductor surface. The bonds are anisotropic and differ from those in the bulk.[45]

carriers, and for the limiting case where n_s is essentially equal to the bulk majority carrier concentration so that in the numerator $n_s \simeq n_{no}$, $p_s > p_{no}$ and in the denominator $n_s \gg p_s$ and $n_s \gg n_1$ or p_1, Eq. (3.89) simplifies to

$$U_s = v_{th}\sigma N_s(p_s - p_{no}). \tag{3.90}$$

Since the product $v_{th}\sigma N_s$ has units of centimeters per second, it is called the *surface recombination velocity* S:

surface recombination velocity

$$S = v_{th}\sigma N_s. \tag{3.91}$$

When there is surface recombination at one end of a semiconductor sample under uniform illumination as shown in Fig. 3.23, the hole current density flowing into the surface from the bulk of the semiconductor is given by qU_s, where U_s is given by Eq. (3.90). The surface recombination reduces the minority concentration at the surface. This gradient of hole concentration results in a diffusion current density that is equal to the surface recombination current. Therefore, the boundary condition at $x = 0$ is

$$qD_p \left.\frac{dp_n}{dx}\right|_{x=0} = qU_s = qS[p_n(0) - p_{no}]. \tag{3.92}$$

The boundary condition at $x = \infty$ is given by Eq. (3.78). At steady state, the continuity equation given by Eq. (3.58) becomes

$$D_p\frac{\partial^2 p_n}{\partial x^2} + G_L - \frac{p_n - p_{no}}{\tau_p} = 0. \tag{3.93}$$

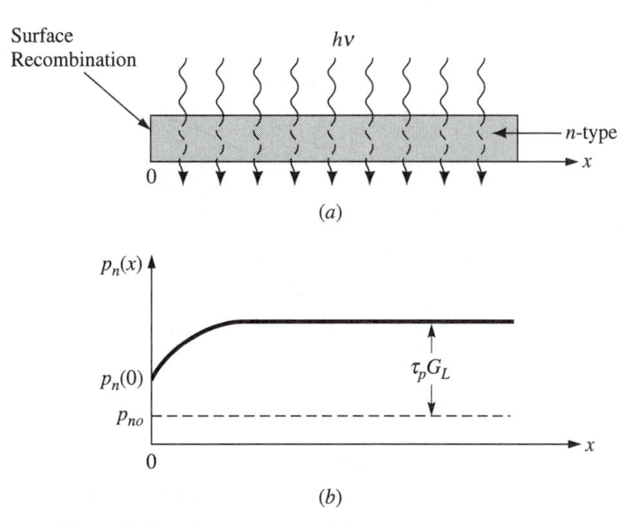

FIGURE 3.23 Illustration of the steady-state minority carrier distribution in a uniformly illuminated sample with surface recombination.[47]

The solution of the equation, subject to the boundary conditions above, is[46,47]

$$p_n(x) = p_{no} + \tau_p G_L \left[1 - \frac{\tau_p S \exp(-x/L_p)}{L_p + \tau_p S} \right]. \quad (3.94)$$

A plot of this equation for a finite S is shown in Fig. 3.23. When $S \rightarrow 0$, then $p_n(x) \rightarrow p_{no} + \tau_p G_L$, as obtained previously in Eq. (3.78). When S is large so that $\tau_p S \gg L_p$, then

$$p_n(x) = p_{no} + \tau_p G_L[1 - \exp(-x/L_p)]. \quad (3.95)$$

From Eq. (3.95), it may be seen that at the surface the minority carrier density approaches its thermal equilibrium value p_{no}.

3.8 HIGH-FIELD EFFECTS

At low electric fields, the drift velocity v_d is linearly proportional to the applied field. It is assumed that the time interval between collisions, $< \tau_c >$, is independent of the applied field. This assumption is reasonable as long as the drift velocity is small compared to the thermal velocity of carriers, which is about 10^7 cm/s for most semiconductors at room temperature.

As the drift velocity approaches the thermal velocity, its field dependence on the electric field will begin to depart from the linear relationship given in Sec. 3.2. Figure 3.24 shows the measured drift velocities of electrons and holes in high-purity Si as a function of electric field $\mathscr{E}$.[48] Initially, the field dependence is linear, corresponding to a

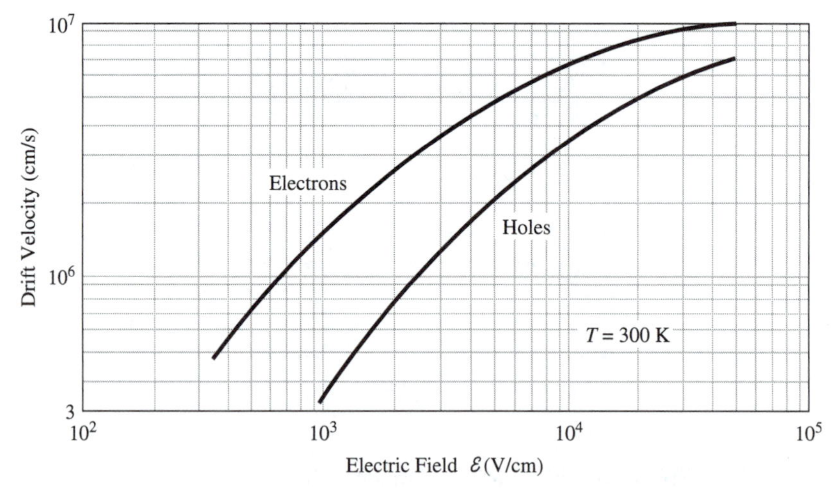

FIGURE 3.24 Drift velocity vs. electric field in Si.[48]

Table 3.3 Parameters for Electric-field Dependence of Drift in Si at 300 K.

Parameter	Electrons	Holes
v_m (cm/s)	1.07×10^7	8.34×10^6
$\mathscr{E}_c$ (V/cm)	6.98×10^3	1.80×10^4
β	1.11	1.21

SOURCE: Jacoboni *et al.*[49]

constant mobility. As the electric field is further increased, the drift velocity increases less rapidly. At sufficiently large fields, the drift velocity approaches a saturation velocity. The experimental results can be approximated by the empirical expression[49]

$$v_d = v_m \frac{\mathscr{E}/\mathscr{E}_c}{[1 + (\mathscr{E}/\mathscr{E}_c)^\beta]^{1/\beta}}. \tag{3.96}$$

The parameters v_m, $\mathscr{E}_c$, and β are given for electrons and holes at room temperature in Table 3.3. It should be noted that v_d also depends on the impurity concentration and decreases for a given $\mathscr{E}$ when the impurity concentration exceeds 10^{16} cm^{-3}.

The high-field transport in n-type GaAs is quite different from that of Si. Figure 3.25 shows the measured drift velocity for n-type high-purity GaAs at room temperature.[50] The drift velocity for p-type GaAs is very similar to Si. Note that for n-type GaAs, the drift velocity reaches a maximum, then decreases as the field further

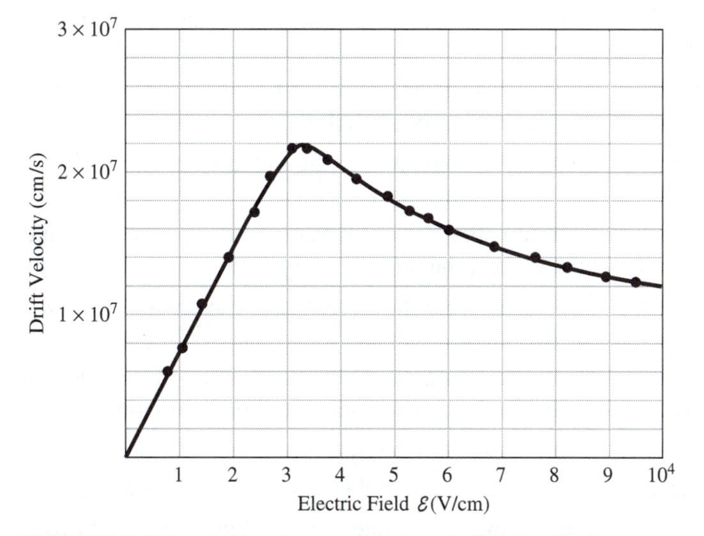

FIGURE 3.25 Drift velocity vs. electric field in GaAs at 300 K.[50]

increases. This phenomenon is due to the energy-band structure of GaAs that allows the transfer of conduction electrons from a high-mobility conduction-band minimum (called a valley) to low-mobility, higher conduction-band minima. The band structure for GaAs was given in Fig. 2.28. The lowest conduction-band minimum is the direct conduction band at $k = 0$, while the next conduction-band minima is the indirect conduction band at $k = 2\pi/a$. The energy separation between the two valleys is $\Delta E = 0.284$ eV. As the electric field is increased, electrons in the lower high-mobility valley at $k = 0$ obtain enough energy to begin to populate the lower mobility valley at $k = 2\pi/a$. As this transfer takes place, the mobility peaks and then decreases as shown in Fig. 3.25. This negative differential mobility in GaAs results in microwave oscillations in devices called Gunn diodes.[52]

The low-field electron and hole mobility were shown in Fig. 3.3 for Si and in Fig. 3.4 for GaAs. In both cases, the mobility decreases for impurity concentrations greater than 10^{16} cm^{-3}. This decrease in mobility is due to greater ionized impurity scattering. The electron drift velocity versus electric field as a function of impurity concentration at room temperature is shown in Fig. 3.26 (*a*) for Si and in Fig. 3.26 (*b*) for GaAs. This figure illustrates the effect of impurities on the drift velocity at high electric fields and shows that the peak in the drift velocity for GaAs is eliminated as the carrier concentration reaches 1×10^{18} cm^{-3}.

3.9 SUMMARY AND USEFUL EXPRESSIONS

The following significant concepts were introduced in this chapter:

- Sheet resistance $R_\square$ is used to give the resistance of integrated circuit resistors as $R = R_\square(L/W)$.
- Current flow due to free electrons and holes is by drift in an electric field or diffusion due to a concentration gradient.
- Carrier diffusivity D is related to the mobility μ by the Einstein relation: $D = \mu(kT/q)$.
- The SPICE circuit simulation program used here is PSpice and it requires specification of the circuit, the desired analysis and the desired output data.
- When excess carriers are introduced into a semiconductor, $np > n_i^2$ with $p_n = p_{no} + \Delta p$ or $n_p = n_{po} + \Delta n$.
- The continuity equation governs the flow of carriers into and out of an infinitesimal volume and includes generation and recombination in that volume.
- Electric field is related to charge by Gauss's law.
- Recombination restores thermal equilibrium to excess carrier concentrations.
- Nonradiative recombination due to recombination centers within the energy gap is called Shockley-Hall-Read (SHR) recombination and is important in all semiconductors.

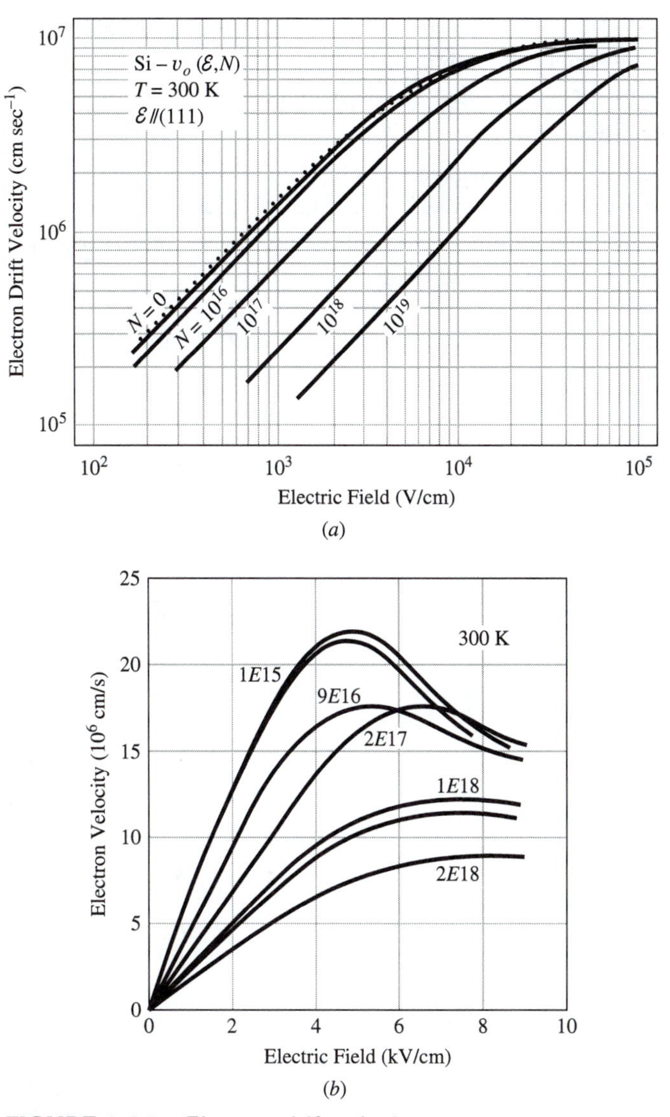

FIGURE 3.26 Electron drift velocity at room temperature as a function of impurity concentration and electric field strength. (*a*) Si.[49] (*b*) GaAs.[51]

- Radiative recombination gives the emission of a photon and is important in the direct-energy-gap semiconductors.
- Auger recombination results from electron–electron or hole–hole collisions, is nonradiative, and is important at high carrier concentrations and for small energy-gap ($E_g \leq 1$ eV) semiconductors.

- Recombination at the surface is similar to SHR recombination and is represented by the surface recombination velocity S.
- Carrier lifetime is the time at which the excess concentration decays to $\exp^{-1}$ of the initial value.
- Minority carrier diffusion length is the distance at which the excess carrier concentration decays to $\exp^{-1}$ of the initial value.
- When the electric field exceeds about 10^3 V/cm, the drift velocity is no longer linear with the electric field.

The following expressions summarize the most important results for carrier transport and recombination.

Carrier Transport

$$v_d = \mu \mathscr{E}$$

Electric Field and Potential

$$\mathscr{E} = \frac{1}{q}\frac{dE_c}{dx} = \frac{1}{q}\frac{dE_i}{dx} = -\frac{d\psi(x)}{dx}$$

$$\psi = -E_i/q$$

Diffusion Current Density

$$J_n = qD_n dn/dx$$

$$J_p = -qD_p dp/dx$$

Einstein Relation

$$D = \mu kT/q$$

Excess Carrier Concentrations

$$\Delta n = n_p - n_{po}$$

$$\Delta p = p_n - p_{no}$$

Gauss's Law

$$\frac{d\mathscr{E}}{dx} = \frac{\rho(x)}{\epsilon}$$

Drift Current Density

$$J_n = q\mu_n n\mathscr{E}$$

$$J_p = q\mu_p p\mathscr{E}$$

$$\rho = 1/(q\mu_n n)$$

$$\rho = 1/(q\mu_p p)$$

Continuity Equation

$$\frac{\partial n_p}{\partial t} = D_n\frac{\partial^2 n_p}{\partial x^2} + (G_n - R_n)$$

$$\frac{\partial p_n}{\partial t} = D_p\frac{\partial^2 p_n}{\partial x^2} + (G_p - R_p)$$

Carrier Recombination

$$R_{nr} = \frac{np - n_i^2}{\tau_{nr}(n + n_1) + \tau_{nr}(p + p_1)}$$

$$R_p = \Delta p/\tau_{nr}$$

$$R_n = \Delta n/\tau_{nr}$$

$$\tau_r = 1/(Bn_{no})$$

$$\tau_A = 1/(Cn_{no}^2)$$

$$L_n = \sqrt{D_n\tau_n}$$

$$L_p = \sqrt{D_p\tau_p}$$

REFERENCES

1. A. S. Grove, *Physics and Technology of Semiconductor Devices* (Wiley, New York, 1967), p. 107.
2. J. Bardeen and W. Shockley, "Deformation Potentials and Mobilities in Non-Polar Crystals," Phys. Rev. **80**, 72 (1950).
3. F. J. Blatt, "Theory of Mobility of Electrons in Solids," *Solid State Physics,* Vol. 4, F. Seitz and D. Turnbull, Eds. (Academic Press, New York, 1957), p. 200.

4. G. L. Pearson and J. Bardeen, "Electrical Properties of Pure Silicon and Silicon Alloys Containing Boron and Phosphorus," Phys. Rev. **75**, 865 (1949).
5. H. C. Casey, Jr., F. Ermanis, and K. B. Wolfstirn, "Variation of Electrical Properties with Zn Concentration in GaP," J. Appl. Phys. **40**, 2945 (1969).
6. G. Baccarani and P. Ostoja, "Electron Mobility Empirically Related to the Phosphorus Concentration in Silicon," Solid-State Elec. **18**, 579 (1975).
7. D. A. Antoniadis, A. G. Gonzalez, and R. W. Dutton, "Boron in Near-intrinsic $< 100 >$ and $< 111 >$ Silicon Under Inert and Oxidizing Ambients," J. Electrochem. Soc. **125**, 813 (1978).
8. N. D. Arora, J. R. Hauser, and D. J. Roulston,"Electron and Hole Mobilities in Silicon as a Function of Concentration and Temperature," IEEE Trans. Electron Devices, **ED-29**, 292 (1982).
9. H. C. Casey, Jr., and M. B. Panish, *Heterostructure Lasers, Part B: Materials and Operating Characteristics* (Academic Press, New York, 1978), p. 170.
10. D. L. Rode and S. Knight, "Electron Transport in GaAs," Phys. Rev. B **3**, 2534 (1971).
11. Grove, *Physics and Technology,* p. 111.
12. S. M. Sze, *Semiconductor Devices: Physics and Technology* (Wiley, New York, 1986), pp. 30–37.
13. W. E. Beadle, J. C. C. Tsai, and R. D. Plummer, *Quick Reference Manual for Silicon Integrated Circuit Technology* (Wiley, New York, 1985), pp. 2–37.
14. D. J. Hamilton and W. G. Howard, *Basic Integrated Circuit Engineering* (McGraw-Hill, New York, 1975), p. 97.
15. A. B. Glaser and G. E. Subak-Sharpe, *Integrated Circuit Engineering: Design, Fabrication, and Applications* (Addison-Wesley, Reading, Mass., 1977), p. 122.
16. R. S. Muller and T. I. Kamins, *Device Electronics for Integrated Circuits,* 2nd ed. (Wiley, New York, 1986), p. 110.
17. W. Banzhaf, *Computer-aided Circuit Analysis Using PSPICE,* 2nd ed. (Prentice Hall, Englewood Cliffs, N.J., 1992).
18. M. H. Rashid, *SPICE for Circuits and Electronics Using PSpice,* 2nd ed. (Prentice Hall, Englewood Cliffs, N.J., 1992).
19. A. Vladimirescu, *The SPICE Book* (Wiley, New York, 1994).
20. P. W. Tuinenga, *SPICE: A Guide to Circuit Simulation and Analysis Using PSpice,* 3rd ed. (Prentice Hall, New York, 1995).
21. G. W. Roberts and A. S. Sedra, *SPICE,* 2nd ed. (Oxford University Press, New York, 1997).
22. Beadle *et al., Quick Reference Manual,* pp. 2–49.
23. Sze, *Semiconductor Devices,* p. 38.
24. J. E. Hatch, ed., *Aluminum* (American Society for Metals, Metals Park, Ohio, 1984), p. 9.
25. R. N. Hall,"Electron-Hole Recombination in Germanium," Phys. Rev. **87**, 387 (1952).
26. W. Shockley and W. T. Read, "Statistics of the Recombinations of Holes and Electrons," Phys. Rev. **87**, 835 (1952).
27. H. F. Wolf, *Semiconductors* (Wiley, New York, 1971), p. 326.
28. W. van Roosbroeck and W. Shockley, "Photon-Radiative Recombination of Electrons and Holes in Germanium," Phys. Rev. **94**, 1558 (1954).
29. H. C. Casey, Jr., and M. B. Panish, *Heterostructure Lasers, Part A: Fundamental Principles* (Academic Press, New York, 1978), p. 160.

30. H. R. Phillip and E. A. Taft, "Optical Constants of Silicon and Germanium in the Region 1 to 10 eV," Phys. Rev. **113**, 1002 (1959).

31. H. C. Casey, Jr., D. D. Sell, and K. W. Wecht, "Concentration Dependence of the Absorption Coefficient for *n*- and *p*-type GaAs Between 1.3 and 1.6 eV," J. Appl. Phys. **46**, 250 (1975).

32. P. Bhattacharya, *Semiconductor Optoelectronic Devices,* 2nd ed. (Prentice-Hall, Saddle River, N.J., 1997).

33. N. K. Dutta and R. J. Nelson, "The Case for Auger Recombination in $In_{1-x}Ga_xAs_yP_{1-y}$," J. Appl. Phys. **53**, 74 (1982).

34. A. B. Beattie and P. T. Landsberg, "Auger Effect in Semiconductors," Proc. R. Soc. London Ser. A, **249**, 16 (1959).

35. A. R. Beattie and G. Smith, "Recombination in Semiconductors by a Light Hole Auger Transition," Phys. Status Solidi **19**, 577 (1967).

36. M. Takeshima, "Auger Recombination in InAs, GaSb, InP, and GaAs," J. Appl. Phys. **43**, 4114 (1972).

37. A. Sugimura, "Band-to-Band Auger Recombination Effect on InGaAsP Laser Threshold," IEEE J. Quantum Electron. **QE-17**, 627 (1981).

38. A. Haug, "Recombination in InGaAsP," Appl. Phys. Lett. **42**, 512 (1983).

39. G. P. Agrawal and N. K. Dutta, *Long-Wavelength Semiconductor Lasers* (Van Nostrand Reinhold, New York, 1986), pp. 95–141.

40. U. Strauss, W. W. Rühle, and K. Köhler, "Auger Recombination in GaAs," Appl. Phys. Lett. **62**, 55 (1993).

41. J. Dziewior and W. Schmid, "Auger Coefficients for Highly Doped and Highly Excited Silicon," Appl. Phys. Lett. **31**, 346 (1977).

42. H. C. Casey, Jr., "Temperature Dependence of the Threshold Current Density in InP-$Ga_{0.28}In_{0.72}As_{0.6}P_{0.4}$ ($\lambda = 1.3\ \mu$m) Double Heterostructure Lasers," J. Appl. Phys. **56**, 1959 (1984).

43. Sze, *Semiconductor Devices,* p. 47.

44. Ibid., p. 58.

45. M. Prutton, *Surface Physics, 2nd Edition* (Oxford University Press (Clarendon), London, 1983), p. 88.

46. Grove, *Physics and Technology,* p. 124.

47. Sze, *Semiconductor Devices,* p. 59.

48. C. Canali, G. Ottaviani, and A. A. Quaranta, "Drift Velocity of Electrons and Holes and Associated Anisotropic Effects in Silicon," J. Phys. Chem. Solids **32**, 1707 (1971).

49. C. Jacoboni, C. Canali, G. Ottaviani, and A. A. Quaranta, "A Review of Some Charge Transport Properties of Silicon," Solid-State Electron. **20**, 77 (1977).

50. J. G. Ruch and G. S. Kino, "Measurement of the Velocity-Field Characteristics of Gallium Arsenide," Appl. Phys. Lett. **10**, 40 (1967).

51. W. T. Masselink and T. F. Kuech, "Velocity-Field Characteristics of Electrons in Doped GaAs," J. Electron. Mat. **18**, 579 (1989).

52. K. K. Ng, *Complete Guide to Semiconductor Devices* (McGraw-Hill, New York, 1995), p. 63.

53. B. Sermage, H. J. Eichler, J. P. Heritage, R. J. Nelson, and N. K. Dutta, "Photoexcited Carrier Lifetime and Auger Recombination in 1.3 μm InGaAsP," Appl. Phys. Lett. **42**, 259 (1983).

54. J. Kestin and J. R. Dorfman, *A Course in Statistical Thermodynamics* (Wiley, New York, 1971), p. 426.

PROBLEMS

All problems are for room temperature unless another temperature is specified.

3.1. **(a)** Calculate the mean free time of an electron in Si at room temperature which has a mobility of 1,000 cm^2/V-s at 300 K.

(b) Calculate the mean free path (i.e., the distance traveled by an electron between collisions). Use the conduction-band effective mass m^*_{X1}.

3.2. **(a)** Calculate the mean free time of a hole in Si at room temperature which has a mobility of 400 cm^2/V-s at 300 K.

(b) Calculate the mean free path (i.e., the distance traveled by a hole between collisions).

3.3. **(a)** Compare the room temperature mobility μ_n obtained from Fig. 3.7 at $n = 1 \times 10^{16}$ cm^{-3} for Si and GaAs with the mobility given in Figs. 3.3 and 3.4.

(b) Repeat part (a) for μ_n with $n = 1 \times 10^{17}$ cm^{-3}.

(c) Repeat part (a) for μ_p with $p = 1 \times 10^{16}$ cm^{-3}.

(d) Repeat part (a) for μ_p with $p = 1 \times 10^{17}$ cm^{-3}.

3.4. If Si at room temperature contains 2×10^{16} cm^{-3} ionized P atoms and 5×10^{16} cm^{-3} ionized B atoms, and the mobility is measured to be 320 cm^2/V-s, what is the resistivity?

3.5. Show that the units for $1/(q\mu_n n)$ are ohm-cm.

3.6. For a Si sample at room temperature, the intrinsic Fermi level decreases linearly by 0.3 eV in a distance of 0.15 μm. Find the electric field.

3.7. A current of 10 mA flows through a GaAs bar 500 μm long. The voltage across the bar is 0.34 V and the ends of the bar are 20 μm by 5 μm. What is the resistivity?

3.8. A Si wafer contains 4.8×10^{16} cm^{-3} ionized B atoms and 5.2×10^{16} cm^{-3} ionized P atoms.

(a) Is the wafer *n*- or *p*-type and what is the majority carrier concentration?

(b) Find the resistivity.

3.9. For Si with $n = N_d^+ = 5 \times 10^{16}$ cm^{-3}, find the mean free time $\langle \tau_c \rangle$ (the average time between collisions).

3.10. Compare the mean free time $< \tau_c >$ (the average time between collisions) for *n*- and *p*-type GaAs with a carrier concentration of 1×10^{16} cm^{-3}.

3.11. For Si, sketch the variation of the electron diffusivity D_n and the hole diffusivity D_p on a semi-logarithmic plot for carrier concentrations from 1.0×10^{15} cm^{-3} to 1.0×10^{19} cm^{-3}.

3.12. Why is the resistivity in Fig. 3.7 at a given impurity concentration larger for *n*-type Si than for *n*-type GaAs?

3.13. Consider a Si integrated-circuit resistor doped with boron to give $p = 5.8 \times 10^{16}$ cm^{-3} (see Fig. 3.11) which gives a resistivity of 0.3 ohm-cm at 27°C. The thickness of the layer is 4.0 μm and the width is 10.0 μm.

(a) What is the length to give a resistance of 1.5×10^3 ohms at 27°C?

(b) What is the temperature coefficient between 27 and 100°C?

(c) For a DC 5-V source, draw and label the circuit diagram to give the current at 27 and 100°C.

(d) Run PSpice and give the current at 27 and 100°C.

3.14. Consider a Si integrated-circuit resistor doped with boron to give $p = 6.0 \times 10^{15}$ cm^{-3} (see Fig. 3.11) which gives a resistivity of 2.4 ohm-cm at 27°C. The thickness of the layer is 6.0 μm and the width is 7.0 μm.

(a) What is the length to give a resistance of 2.5×10^4 ohms at 27°C?

(b) What is the temperature coefficient between 27 and 100°C?

(c) For a DC 5-V source, draw and label the circuit diagram to give the current at 27 and 100°C.

(d) Run PSpice and give the current at 27 and 100°C.

3.15. At steady state, but nonequilibrium, a uniformly doped Si sample contains 2.6×10^{16} cm^{-3} ionized P atoms and 3.5×10^{16} cm^{-3} ionized B atoms. When the hole concentration is 1.2×10^{16} cm^{-3}, what is the electron concentration?

3.16. Find the resistivities of intrinsic Si and GaAs at room temperature.

3.17. At thermal equilibrium, $np = n_i^2$ and the resistivity is given as $\rho = 1/q(\mu_n n + \mu_p p) = 1/\sigma$, where σ is the conductivity.

(a) Find the ratio n/p for the maximum resistivity by using $d\sigma/dn = 0$ at minimum conductivity (maximum resistivity) and then obtain n and p.

(b) Write an expression for the maximum resistivity in terms of the mobilities and n_i.

3.18. With the result from Problem 3.17, find the maximum resistivity for Si and GaAs at room temperature.

3.19. In *n*-type Si at room temperature, the minority carrier hole concentration varies as $p(x) = 3 \times 10^{18} \exp(-x/L_p)$, where L_p is 3×10^{-4} cm and D_p is 12 cm^2/s. What is the diffusion current density at $x = 4 \times 10^{-4}$ cm?

3.20. **(a)** For *n*-type Si at room temperature with $n = 2.5 \times 10^{16}$ cm^{-3} at thermal equilibrium, what is the hole concentration at nonequilibrium when the excess electron concentration is 1×10^4 cm^{-3}?

(b) What is the hole concentration for the *n*-type Si in part (a) when the excess electron concentration is 5×10^{12} cm^{-3}?

3.21. Find the current density for an electric field of 4.5×10^3 V/cm in an *n*-type Si sample with $n = 8.5 \times 10^{16}$ cm^{-3} and a mobility of 1.15×10^3 cm^2/V-s.

3.22. A Si sample at room temperature has 7.5×10^{16} cm^{-3} ionized P atoms and a resistivity of 0.1 ohm-cm. Find the mobility and diffusivity (include the correct units).

3.23. **(a)** What is a carrier recombination center?

(b) What is a carrier trap?

3.24. Show that p_1 is the value of p for $E_f = E_t$ and that $p_1 = n_i$ for $E_t = E_i$.

3.25. For a carrier lifetime of 1×10^{-8} s, what is the recombination rate for an excess carrier density of 5×10^{13} cm^{-3}?

3.26. **(a)** For n-type Si at room temperature, what gold concentration is required to give τ_p of 5×10^{-8} s?

(b) What is the capture cross section?

3.27. An n-type sample has a thermal equilibrium electron concentration $n = N_d^+ - N_a^- = 1 \times 10^{16}$ cm^{-3} and a hole lifetime of 1×10^{-8} s. The uniform generation rate due to excitation with light is 1×10^{21} cm^{-3}s^{-1}.

(a) What is the steady-state minority carrier concentration?

(b) What is the steady-state electron concentration?

3.28. For the lifetime versus carrier concentration[53] shown in the figure, what are the radiative constant B and the Auger coefficient C?

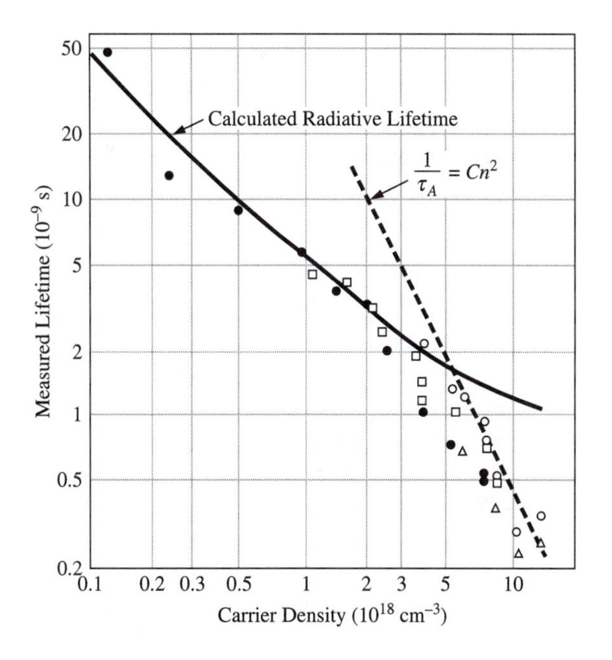

3.29. A p-type Si sample has a resistivity of 2.4 ohm-cm and a lifetime of 5×10^{-8} s. The hole mobility is 430 cm^2/V-s and the electron mobility is 1,320 cm^2/V-s.

(a) What is the thermal equilibrium majority carrier concentration?

(b) What is the minority carrier concentration?

(c) What is the minority carrier diffusion length?

3.30. A Si sample has the following properties: $n = N_d^+ = 4.0 \times 10^{16}$ cm^{-3}; $\mu_n = 1.0 \times 10^3$ cm^2/V-s; $\mu_p = 350$ cm^2/V-s; and $\tau_p = 2 \times 10^{-8}$ s. For parts (a) through (d), please include the units.

(a) Find the resistivity.

(b) Find the minority carrier concentration.

(c) Find the minority carrier diffusion length.

(d) What is the recombination rate for holes when the excess hole concentration is 3×10^{13} cm^{-3}?

3.31. If the radiative constant $B = 2.0 \times 10^{-10}$ cm^3/s for p-type GaAs at room temperature and the hole concentration p_{po} is 1.0×10^{18} cm^{-3} and the overall lifetime is 3.33×10^{-9} s, what is the nonradiative lifetime?

3.32. For p-type GaAs, at what carrier concentration are the radiative and Auger lifetimes equal?

3.33. Compare the Auger lifetimes at a carrier concentration of $p = 3 \times 10^{18}$ cm^{-3} for GaAs, Si, and $Ga_{0.28}In_{0.72}As_{0.6}P_{0.4}$.

3.34. Show that $kT/q = 0.026$ V at room temperature.

3.35. Consider a uniformly excited p-type Si sample at steady state with a minority carrier lifetime $\tau_o = 3 \times 10^{-9}$ s. What generation rate is necessary to give an excess electron concentration of 1×10^{15} cm^{-3}?

3.36. In uniformly illuminated Si at $T = 27°$C, $n = 6 \times 10^{17}$ cm^{-3}, $N_d^+ = 4 \times 10^{17}$ cm^{-3}, $N_a^- = 3 \times 10^{17}$ cm^{-3}. What is the value of the hole concentration p?

3.37. A uniform Si wafer is measured and found to have a resistivity of 0.01 ohm-cm and a mobility of 200 cm^2/V-s. What is the carrier concentration?

3.38. For the Si sample in Problem 3.35, what is the recombination rate for an excess carrier concentration of 3×10^{15} cm^{-3}?

3.39. In an $Al_xGa_{1-x}As$ light-emitting diode, the nonradiative lifetime is 5×10^{-8} s and the radiative lifetime is 1.6×10^{-8} s. Find the overall lifetime and the internal quantum efficiency.

3.40. Consider a Si sample with $N_d^+ = 3.5 \times 10^{16}$ cm^{-3} and $N_a^- = 1.7 \times 10^{16}$ cm^{-3} at room temperature. Ignore surface recombination.

(a) For thermal equilibrium, what is the majority carrier concentration?

(b) For thermal equilibrium, what is the minority carrier concentration?

(c) For a uniform excitation with a generation rate of 3×10^{22} cm^{-3} s^{-1}, and a hole lifetime of 2×10^{-8} s, what is the steady-state hole concentration?

(d) When the excitation is turned off, how long does it take for the hole concentration to decay to 10% of its steady-state value?

3.41. For an n-type Si sample at room temperature with 5×10^{16} cm^{-3} impurities and a lifetime of 5×10^{-8} s, what is the minority carrier diffusion length if the mobility is taken as the same as the hole mobility when the hole is a majority carrier?

3.42. Consider a Si sample at room temperature with $n = N_d^+ = 3.5 \times 10^{16}$ cm^{-3}.

(a) For a uniform excitation with a generation rate of 4.5×10^{22} cm^{-3}s^{-1} and a hole lifetime of 2×10^{-8} s, what is the steady-state hole concentration?

(b) When the excitation is turned off, how long does it take for the hole concentration to decay to 10% of its steady-state value?

3.43. At steady state, but nonequilibrium, a uniformly doped Si sample contains 2.6×10^{16} cm^{-3} ionized P atoms and 3.5×10^{16} cm^{-3} ionized B atoms. When the hole concentration is 1.2×10^{16} cm^{-3}, what is the electron concentration?

3.44. A Si sample has the following properties: $N_d^+ = 4.0 \times 10^{16}$ cm^{-3}; $N_a^- = 3.5 \times 10^{16}$ cm^{-3}; $\mu_n = 870$ cm^2/V-s; $\mu_p = 310$ cm^2/V-s; and $\tau_p = 2 \times 10^{-8}$ s. For parts (a) through (d), please include the units.

(a) Find the resistivity.

(b) Find the minority carrier concentration.

(c) Find the minority carrier diffusion length.

(d) What is the recombination rate for holes when the excess hole concentration is 3×10^{13} cm^{-3}?

3.45. A Si sample at thermal equilibrium contains 3×10^{16} cm^{-3} ionized phosphorus atoms, 2×10^{16} cm^{-3} ionized boron atoms, and 2×10^{16} cm^{-3} ionized arsenic atoms. The minority carrier lifetime is 6×10^{-8} s.

(a) Find the majority carrier concentration. Is the majority carrier concentration electrons or holes?

(b) What is the resistivity?

(c) What is the minority carrier concentration?

(d) Find the minority carrier diffusion length.

3.46. Consider a semiconductor region which is completely depleted of carriers so that n and $p \ll n_i$. In this region, electron–hole pairs are generated by the recombination–generation centers. Derive an expression for the generation rate of carriers with $E_t = E_i$.

3.47. At what electric field does the maximum drift velocity occur in lightly doped GaAs?

3.48. At an electron concentration of 1×10^{17} cm^{-3}, compare the room temperature drift velocities of Si and GaAs at small and large electric fields.

3.49. An n-type Si sample with uniform illumination as shown in Fig. 3.23 has a minority carrier lifetime of 2.6×10^{-8} s, a diffusion length of 3.6 μm, and an electron concentration of 6.25×10^{18} cm^{-3} at room temperature. For a generation rate of 3.2×10^{21} cm^{-3}-s^{-1}, plot $p_n(x)$ on a linear scale from 0 to 1×10^{14} cm^{-3} for x between 0 and 25 μm for $S = 10^2$ cm/s, 10^3 cm/s, 10^4 cm/s, 10^5 cm/s, and 10^6 cm/s.

3.50. Emission at a wavelength of 0.6328 μm from a He-Ne laser is used to generate photoluminescence (radiative recombination due to photoexcitation) in a GaAs sample at room temperature. The incident emission energy is measured to be 25 mW and the light is focused to a circular spot 200 μm in diameter. The wavelength may be converted to photon energy by the relation E_g (eV) = 1.24/wavelength (μm). The reflectivity is 0.34 and the sample is sufficiently thick to absorb all the light not reflected.

(a) Find the absorption coefficient.

(b) Find the net generation rate.

APPENDIX B

B.1 SHOCKLEY-HALL-READ RECOMBINATION

The theory for nonradiative recombination was first published by Hall[25] and Shockley and Read.[26] This theory is referred to as Shockley-Hall-Read (SHR) recombination and is based on the *principle of detailed balance* where every process and its inverse must balance. This principle states that under equilibrium conditions every process and its inverse must proceed at exactly equal rates.

The four processes which occur in the recombination process are illustrated in Fig. B.1. In this case, $N_d^+ > N_a^-$ so that the semiconductor is *n*-type. The nonradiative recombination centers have a concentration N_t and are located at an energy E_t above the valence band. In the first process r_n^c, *electron capture*, an electron in the conduction band is captured by the nonradiative recombination center N_t. In the second process, the rate of emission of electrons r_n^e from the nonradiative recombination center is the inverse of the electron capture process. The transitions between holes in the valence band and the nonradiative recombination centers are analogous to those for the electron. The hole capture process is represented by r_p^c and the hole emission process is represented by r_p^e. To determine the recombination rate R_n, each of these four processes, r_n^c, r_n^e, r_p^c, and r_p^e must be specified. The quantities describing these processes are not all independent, and a relationship will be derived relating these four processes.

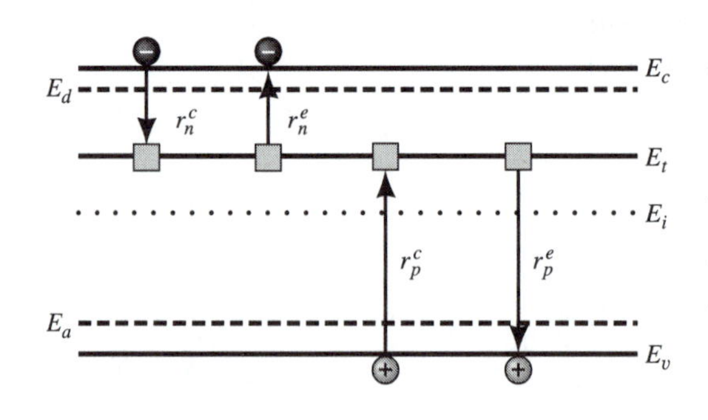

FIGURE B.1 Illustration of the four processes involved in Shockley-Hall-Read recombination. The process r_n^c is electron capture, while r_n^e is electron emission. Similarly, r_p^c is hole capture and r_p^e is hole emission.

The electron capture rate r_n^c is given by the product of the free-electron concentration, the concentration of empty nonradiative recombination centers, and the electron capture probability. The occupation probability of a level E was given by the Fermi-Dirac distribution function in Eq. (2.45). The probability that N_t is occupied is

$$f(E_t) = \frac{1}{1 + \exp[(E_t - E_f)/kT]}, \tag{B.1}$$

and the probability that N_t is empty (unoccupied) is $[1 - f(E_t)]$. The electron capture probability is given by the product of the capture cross section σ_n and the thermal velocity v_{th} as given in Eq. (3.2). The quantity σ_n represents the effectiveness of the center to capture an electron and is a measure of how close the electron has to come to the center to be captured. It could be expected that the capture cross section would be of the order of atomic dimensions, that is, of the order of 10^{-15} cm^2. The product $\sigma_n v_{th}$ may be visualized as the volume swept out per unit time by an electron with cross section σ_n. If the center lies within this volume, the electron will be captured by it. The electron capture rate is given by

$$r_n^c = nN_t[1 - f(E_t)]\sigma_n v_{th}. \tag{B.2}$$

The rate of electron emission r_n^e is the inverse of the electron capture process. The rate is proportional to the concentration of centers occupied by electrons, $N_t f(E_t)$. The electron emission rate is given by

$$r_n^e = e_n N_t f(E_t). \tag{B.3}$$

The proportionality constant e_n is called the electron emission probability.

At thermal equilibrium, detailed balance requires that the rates of capture and emission of electrons must be equal; therefore,

$$r_n^c = r_n^e = nN_t[1 - f(E_t)]\sigma_n v_{th} = e_n N_t f(E_t). \tag{B.4}$$

The emission probability can be represented by

$$e_n = \frac{n[1 - f(E_t)]\sigma_n v_{th}}{f(E_t)}. \tag{B.5}$$

With Eq. (B.1) for $f(E_t)$,

$$\frac{1 - f(E_t)}{f(E_t)} = \left[1 - \frac{1}{1 + \exp(\cdots)}\right][1 + \exp(\cdots)] = \exp[(E_t - E_f)/kT], \tag{B.6}$$

and with n represented by Eq. (2.109) as $n = n_i \exp[(E_f - E_i)/kT]$, the electron emission probability may be written as

$$e_n = n_i \exp[(E_f - E_i)/kT] \exp[(E_t - E_f)/kT]\sigma_n v_{th}, \tag{B.7}$$

which reduces to

$$e_n = n_i \exp[(E_t - E_i)/kT]\sigma_n v_{th}. \tag{B.8}$$

From Eq. (2.109), the quantity $n_i \exp[(E_t - E_i)/kT]$ would be the electron concentration if $E_f = E_t$. It is convenient to define

$$n_1 \equiv n_i \exp[(E_t - E_i)/kT], \tag{B.9}$$

and then

$$e_n = n_1 \sigma_n v_{th}. \tag{B.10}$$

Similar expressions may be written for holes. The hole capture rate r_p^c is the product of the hole concentration, the concentration of filled nonradiative recombination centers (filled with electrons), and the hole capture probability. The hole capture rate is given by

$$r_p^c = pN_t f(E_t)\sigma_p v_{th}. \tag{B.11}$$

The hole emission rate is proportional to the centers filled with holes (unoccupied by electrons), and the hole emission probability. The hole emission rate is given by

$$r_p^e = e_p N_t[1 - f(E_t)], \tag{B.12}$$

where e_p is the emission probability. At thermal equilibrium, detailed balance requires that $r_p^c = r_p^e$, which gives the emission probability as

$$e_p = \frac{pf(E_t)\sigma_p v_{th}}{[1 - f(E_t)]}. \tag{B.13}$$

The quantity $f(E_t)/[1 - f(E - t)]$ is the reciprocal of the expression given in Eq. (B.6). With p represented by Eq. (2.110) as $n_i \exp[(E_i - E_f)/kT]$, the hole emission probability may be written as

$$e_p = n_i \exp[(E_i - E_f)/kT] \exp[-(E_t - E_f)/kT]\sigma_p v_{th}, \tag{B.14}$$

which reduces to

$$e_p = n_i \exp[(E_i - E_t)/kT]\sigma_p v_{th}. \tag{B.15}$$

From Eq. (2.110), the quantity $n_i \exp[(E_i - E_t)/kT]$ would be the hole concentration if $E_f = E_t$. In the same manner as for electrons, it is convenient to define

$$p_1 \equiv n_i \exp[(E_i - E_t)/kT], \tag{B.16}$$

and then

$$e_p = p_1 \sigma_p v_{th}. \tag{B.17}$$

For nonequilibrium and steady state, the net rate of capture of electrons $(r_n^c - r_n^e)$ must equal that of holes $(r_p^c - r_p^e)$. The first step is to use this equality to obtain $f(E_t)$ and $[1 - f(E_t)]$. It is customary to represent the net recombination by U:

$$U \equiv r_n^c - r_n^e = r_p^c - r_p^e. \tag{B.18}$$

Then, with Eqs. (B.2), (B.3), and (B.10) for electrons and Eqs. (B.11), (B.12), and (B.17) for holes gives

$$\begin{aligned} U &= n\sigma_n v_{th} N_t [1 - f(E_t)] - n_1 \sigma_n v_{th} N_t f(E_t) \\ &= p\sigma_p v_{th} N_t f(E_t) - p_1 \sigma_p v_{th} N_t [1 - f(E_t)]. \end{aligned} \tag{B.19}$$

From Eq. (B.19), $f(E_t)$ is

$$f(E_t) = \frac{\sigma_n v_{th} N_t n + p_1 \sigma_p v_{th} N_t}{\sigma_n v_{th} N_t (n + n_1) + \sigma_p v_{th} N_t (p + p_1)}. \tag{B.20}$$

The most useful expressions for U are obtained with the assumption that $\sigma_n = \sigma_p = \sigma$, and Eq. (B.20) reduces to

$$f(E_t) = \frac{n + p_1}{(n + n_1) + (p + p_1)}, \tag{B.21}$$

and then

$$1 - f(E_t) = \frac{p + n_1}{(n + n_1) + (p + p_1)}. \tag{B.22}$$

With these expressions for $f(E_t)$ and $[1 - f(E_t)]$,

$$U = r_n^c - r_n^e = \frac{(\sigma v_{th} N_t) n (p + n_1)}{(n + n_1) + (p + p_1)} - \frac{(\sigma v_{th} N_t) n_1 (n + p_1)}{(n + n_1) + (p + p_1)}, \tag{B.23}$$

which becomes

$$U = (\sigma v_{th} N_t) \left[\frac{np + nn_1 - nn_1 - n_1 p_1}{(n + n_1) + (p + p_1)} \right]. \tag{B.24}$$

With Eq. (B.9) for n_1 and Eq. (B.16) for p_1,

$$n_1 p_1 = n_i^2, \tag{B.25}$$

and Eq. (B.24) becomes

$$U = (\sigma v_{th} N_t) \left[\frac{np - n_i^2}{(n + n_1) + (p + p_1)} \right]. \tag{B.26}$$

The quantity $1/(\sigma v_{th} N_t)$ has units of $1/(\text{cm}^2\ \text{cm/s}\ \text{cm}^{-3})$ or seconds, and is designated the nonradiative carrier lifetime τ_{nr}:

$$\tau_{nr} \equiv \frac{1}{\sigma v_{th} N_t}. \tag{B.27}$$

The net recombination rate may now be written as

$$U = \frac{np - n_i^2}{\tau_{nr}(n + n_1) + \tau_{nr}(p + p_1)}. \tag{B.28}$$

Equation (B.28) is the general equation for the net recombination rate, and in many applications, it can be simplified.

Equation (B.28) may be written with Eq. (B.9) for n_1 and Eq. (B.16) for p_1, which gives

$$U = \frac{1}{\tau_{nr}} \left[\frac{np - n_i^2}{n + p + n_i \exp[(E_t - E_i)/kT] + n_i \exp[-(E_t - E_i)/kT]} \right] \tag{B.29}$$

or

$$U = \frac{1}{\tau_{nr}} \left[\frac{np - n_i^2}{n + p + 2n_i \cosh[(E_t - E_i)/kT]} \right]. \tag{B.30}$$

The dependence of the recombination rate on the recombination center level is contained in the hyperbolic cosh term. The quantities n_1 and p_1 will be n_i for $E_i = E_t$, and n_1 will exceed n_i for $E_t > E_i$ and will be less than n_i for $E_t < E_i$.

A more convenient form of Eq. (B.28) was given in Sec. 3.5.2 where $U = R$, so that

$$R_n = \frac{\Delta n}{\tau_{nr}}. \tag{B.31}$$

B.2 RADIATIVE RECOMBINATION

The radiative recombination rate in semiconductors was derived by van Roosbroeck and Shockley[28] by the application of the principle of detailed balance to the thermal equilibrium rate of photon absorption. In Sec. B.1, detailed balance was used in the derivation of SHR recombination.

From the principle of detailed balance, the rate of radiative recombination at *thermal equilibrium* for an incremental photon energy interval $d(h\nu)$ at the photon energy $E = h\nu$ is equal to the corresponding rate of generation of electron–hole pairs by thermal radiation. This radiative recombination rate is given by the product of the probability per unit time that a photon is absorbed $P(h\nu)$ and the photon density distribution. The photon density distribution is the density of states for photons $\rho(h\nu)$ times the photon distribution function $f(h\nu)$. At thermal equilibrium, the total generation rate rate per unit volume for photons is

$$G = \int_0^\infty P(h\nu)\rho(h\nu)f(h\nu)\, d(h\nu). \tag{B.32}$$

The probability of absorption of a photon per unit time $P(h\nu)$ is given by the product of the absorption coefficient $\alpha(h\nu)$ times the photon velocity v in the semiconductor:

$$P(h\nu) = \alpha(h\nu)v. \tag{B.33}$$

Since the frequency ν is the same in a dielectric medium such as a semiconductor and free space,

$$\nu = \frac{v}{\lambda} = \frac{c}{\lambda_o}, \tag{B.34}$$

where λ is the wavelength in the dielectric medium, λ_o is the wavelength in free space, and c is the velocity of light in free space. Also, the ratio of c/v is the refractive index $\overline{n}$:

$$\overline{n} = \frac{c}{v}, \tag{B.35}$$

and $P(h\nu)$ becomes

$$P(h\nu) = \alpha(h\nu)c/\overline{n}. \tag{B.36}$$

The photon density of states $\rho(h\nu)$ is obtained in the same manner as the electron density of states. The wave equation for the photon electric field is the same as the Schrödinger wave equation of Eq. (2.13), with ψ replaced by $\mathscr{E}$ and $2m_o/\hbar^2$ replaced by k^2 because photons are massless particles. As in Eq. (2.41), the density of states per unit volume in **k**-space is

$$dN(k) = \frac{k^2}{\pi^2}\, dk. \tag{B.37}$$

Rather than using $E = \hbar^2/2m_o$ to eliminate k, Eq. (2.22) gave $k = 2\pi/\lambda$, and Eqs. (B.34) and (B.35) are used to write λ in terms of ν:

$$k = \frac{2\pi}{\lambda} = \frac{2\pi}{v/\nu} = \frac{2\pi\bar{n}\nu}{c} = \frac{2\pi\bar{n}h\nu}{hc}. \tag{B.38}$$

Now,

$$dk = \frac{2\pi\bar{n}}{hc}d(h\nu), \tag{B.39}$$

where the variation of $\bar{n}$ with $h\nu$, called dispersion, has been neglected. Equation (B.37) becomes

$$\rho(h\nu) = \frac{dN(h\nu)}{d(h\nu)} = \frac{8\pi\bar{n}^3 h^2 \nu^2}{h^3 c^3}. \tag{B.40}$$

The photon distribution function $f(h\nu)$ is given by the Bose-Einstein distribution law:

$$f(h\nu) = \frac{1}{\exp(h\nu/kT) - 1}. \tag{B.41}$$

The product $\rho(h\nu)f(h\nu)$ is called the spectral density at a specific energy $h\nu$ and is

$$\rho(h\nu)f(h\nu) = \frac{8\pi\bar{n}^3 h^2 \nu^2}{h^3 c^3[\exp(h\nu/kT) - 1]}. \tag{B.42}$$

The spectral density given by Eq. (B.42) is commonly called *blackbody radiation,* and a descriptive treatment is given by Kestin and Dorfman.[54] For example, the emission spectra of incandescent light bulbs are the blackbody emission of the heated filament. The generation rate given by Eq. (B.32) may now be written with Eqs. (B.36) and (B.42) to give

$$G = \frac{8\pi\bar{n}^2}{h^3 c^2} \int_0^\infty \frac{\alpha(h\nu) h^2 \nu^2 \, d(h\nu)}{[\exp(h\nu/kT) - 1]}. \tag{B.43}$$

For a nonequilibrium state with excess carriers, the total rate of radiative recombination R_r per unit volume in steady state will be proportional to the departure from equlibrium given by the np product, and R_r is given by

$$R_r = G\frac{np}{n_i^2} \equiv Bnp, \tag{B.44}$$

since R_r must be proportional to the product of the nonequilibrium electron–hole concentration and be equal to G when $np = n_i^2$ at thermal equilibrium. In Eq. (B.44), the radiative constant B has been defined as G/n_i^2. When low-level excitation in an n-type semiconductor exists so that $n \approx n_{no}$, the radiative lifetime is given by

$$\tau_r = \frac{\Delta p}{R_r} = \frac{p_n - p_{no}}{B n_{no} p_n} = \frac{1}{B n_{no}}, \tag{B.45}$$

with $p_n \gg p_{no}$ and $n_{no} \gg \Delta n$.

The room temperature experimental absorption coefficients for Si and GaAs were shown in Fig. 3.18. These absorption coefficients permit illustration of the effect of the absorption coefficient for indirect- and direct-energy-gap semiconductors on the radiative constant B. In the integral of Eq. (B.43), the term $1/[\exp(h\nu/kT) - 1]$ reduces to $\exp(-h\nu/kT)$ so that the quantity in the integral becomes $\alpha(h\nu)\exp(-h\nu/kT)$. Since $\alpha(h\nu) \to 0$ for $E < E_g$ and $\exp(-h\nu/kT) \to 0$ for $E > E_g$, then with Eq. (2.71) for $n_i^2 \propto \exp(-E_g/kT)$, the value of B for $h\nu$ near E_g is

$$B = \frac{G}{n_i^2} \propto \alpha(h\nu)\exp(-h\nu/kT)\exp(E_g/kT). \tag{B.46}$$

In the preceding equation, the exponential terms cancel for $h\nu \simeq E_g$ and

$$B \propto \alpha(E_g). \tag{B.47}$$

CHAPTER 4

p-n JUNCTIONS: I–V BEHAVIOR

4.1 INTRODUCTION

p-n junction

p-n junction diode

bipolar devices

The previous two chapters introduced the carrier concentrations and transport in uniform semiconductors. This chapter will apply these concepts to single-crystal material which contains *p*- and *n*-type regions to form a *p-n* junction. The term *junction* refers to the boundary between two semiconductor regions, while a *p-n junction diode* usually means a structure with metal contacts or a packaged device. Semiconductor devices in which both electrons and holes participate in the conduction process are termed bipolar devices. Both the *p-n* junctions presented in Chapters 4 and 5, and the *npn* and *pnp* transistors presented in Chapter 9 are *bipolar devices.* The most important characteristic of a *p-n* junction is that it can conduct large currents for one polarity of voltage, while for the other polarity of voltage, the current through the *p-n* junction is very small. The current–voltage (*I–V*) behavior for a representative Si *p-n* junction is shown in Fig. 4.1. The forward conduction is considered in this chapter, and the reverse breakdown and the junction capacitance will be considered in the next chapter.

Rather than the metal-semiconductor rectifier observed by Braun in 1876 (see Chapter 1), a *p-n* junction rectifier occurs when a piece of semiconductor has a variable concentration of donors or acceptors so that a transition from *p*-type to *n*-type occurs in a continuous solid specimen. As summarized by Moll,[1] early theories did not recognize

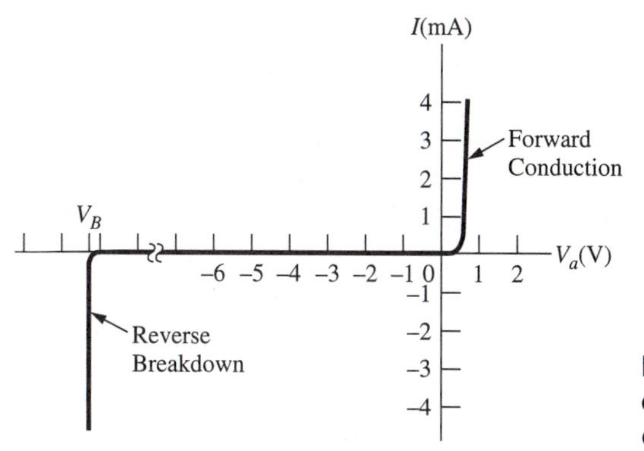

FIGURE 4.1 Current-voltage characteristic of a Si *p-n* junction diode.[3]

the role of minority carriers which was presented by Shockley[2] in 1949. In this definitive theory, infinitely long pieces of *p*-type and *n*-type semiconductors are divided into three separate regions: (1) the neutral *p*-type region, (2) the neutral *n*-type region, and (3) the interface, or depletion region, on both sides of the *p-n* boundary. It is assumed that essentially all the majority carriers have been swept out of the interface region, which results in a space charge due to the ionized acceptors and donors on the *p*- and *n*-sides, respectively. This "depletion theory" is presented in this chapter.

Early *p-n* junctions were found to occur naturally in crystals of relatively pure Si because of segregation of impurities upon solidification. Grown *p-n* junctions were prepared by dropping acceptors into the melt of a growing semiconductor crystal to form a *p*-region and then dropping larger amounts of a donor into the melt to form the *n*-region. Also, alloyed junctions were formed by alloying an Al wire into a piece of *n*-type Si or Ge. Currently, *p-n* junctions in Si are formed by the controlled introduction of impurities by diffusion[4] or ion implantation[5] processes. For the III-V compound semiconductors, *p-n* junctions are generally formed by growth of the *n*- and *p*-layers by one of the epitaxial layer growth processes which are described by Bhattachaya.[6]

The applications of *p-n* junctions are too numerous to attempt a comprehensive list. These applications range from power rectifiers to high-frequency detectors and light-emitting diodes. Also, *p-n* junctions are the basic building blocks for most other devices used in integrated circuits, such as bipolar transistors and the source and drain regions of metal-oxide semiconductor field-effect transistors (MOSFETs). Electrical isolation for the individual devices in bipolar integrated circuits is provided by *p-n* junctions. A thorough understanding of the *p-n* junction is essential to the understanding of other semiconductor devices.

depletion approximation

The theory of *p-n* junctions begins with the energy-band diagram in Sec. 4.2 which illustrates the built-in potential. In Sec. 4.3, the depletion approximation is made which assumes that essentially all the majority carriers have been swept out of the interface region of the *p-n* junction. This depletion region has a negative space charge due to ionized acceptors N_a^- on the *p*-side which are not neutralized by holes and a positive space charge due to ionized donors N_d^+ on the *n*-side which are not neutralized by electrons. Gauss's law is applied to the depletion region space charge to obtain the electric field. Integration of the electric field gives the junction potential as a function of distance. Then, the width of the depletion region is obtained in Sec. 4.4 from the expressions for the electric field and potential given in Sec. 4.3. In Sec. 4.5, the relationship of the junction potential variation with carrier concentration is established and shows that a carrier concentration difference is related to a potential difference.

I–V behavior

The *p-n* junction with an applied bias is considered in Sec. 4.6, which leads to the derivation of the *I–V* behavior in Sec. 4.7 for an abrupt (step) junction where the transition from *p*- to *n*-type occurs abruptly. Current flow by diffusion due to a concentration gradient is the first current mechanism considered, then current flow due to recombination of carriers in the depletion region is derived. The derived expressions are compared with experimental *I–V* measurements in the last part of Sec. 4.7. The significant concepts introduced in this chapter are summarized in Sec. 4.8 together with the expressions useful for representation of *p-n* junctions.

Chapter 5 continues the analysis of the *p-n* junction with consideration of the inherently nondestructive (if the current is limited) junction breakdown in reverse bias. The capacitance associated with the *p-n* junction is obtained in Chapter 5 from expressions derived in this chapter. The use of PSpice for the *p-n* junctions as well as applications of *p-n* junctions is given in Chapter 5. The *p-n* junction may be used as the gate electrode to control the current flow between the source and drain of the three-terminal junction field-effect transistor (JFET). A section on the JFET is available on the website listed in the Preface.

4.2 ENERGY-BAND DIAGRAM

A schematic representation of a *p-n* junction diode is shown in Fig. 4.2(*a*). To fabricate the device, a relatively thin and lightly doped *n*-type (or *p*-type) epitaxial layer is grown

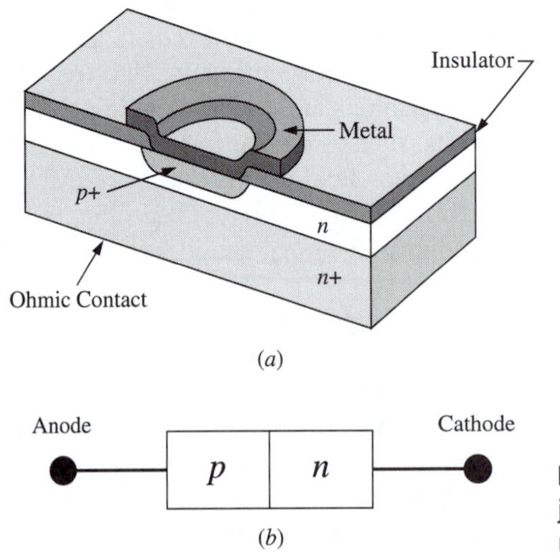

FIGURE 4.2 (*a*) Illustration of a *p-n* junction diode. (*b*) One-dimensional representation of a *p-n* junction diode.

on a much thicker and heavily doped n^+-substrate (or p^+-substrate) to minimize series resistance. A window is opened in the silicon dioxide (SiO_2) insulator and the *p-n* junction is formed by acceptor (or donor) diffusion or ion implantation into the region where the SiO_2 has been removed. This process gives the *p-n* junction illustrated in the figure. A metal layer is deposited in a vacuum system and the circular contact is defined by a lithographic step. Analysis of the *p-n* junction will consider a one-dimensional structure as represented in Fig. 4.2(*b*).

Figure 4.3(*a*) shows the energy-band diagram of an isolated *p*-type semiconductor adjacent to an isolated *n*-type semiconductor and represents the allowed energies E

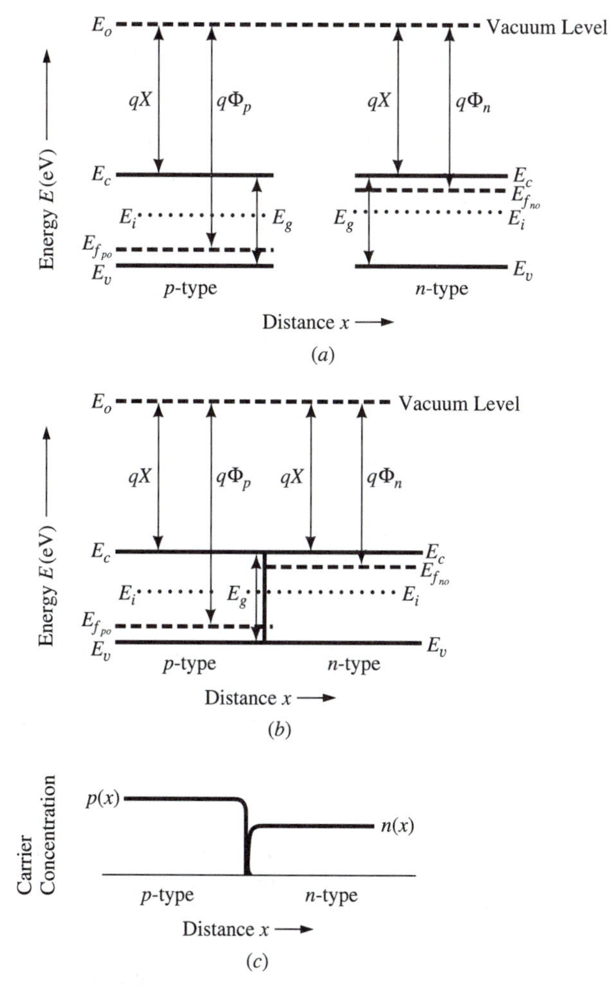

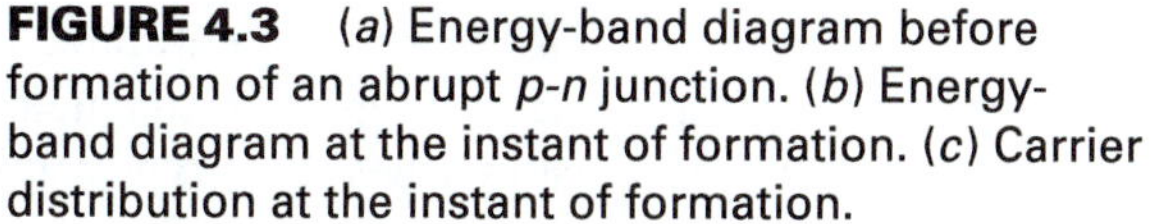

FIGURE 4.3 (*a*) Energy-band diagram before formation of an abrupt *p-n* junction. (*b*) Energy-band diagram at the instant of formation. (*c*) Carrier distribution at the instant of formation.

in the bulk of the material versus position which is given by the distance x. The semiconductor *work function* $q\Phi$ is the difference between the vacuum or free-electron energy E_o and the Fermi level E_f and depends on the impurity concentration in the semiconductor. The difference between E_o and the conduction band edge E_c is a constant for a particular semiconductor. This difference between E_o and E_c is called the *electron affinity* and is represented by qX. The intrinsic Fermi level is represented by E_i. The electron affinities of Si and GaAs have been measured and within the experimental uncertainties have the same value: $\boxed{qX_{\mathrm{Si}} = qX_{\mathrm{GaAs}} = 4.05\ \mathrm{eV}}$. The work function in the p-type semiconductor is designated by $q\Phi_p$ and as shown in Fig. 4.3(a) is given by

work function

electron affinity

$$\boxed{q\Phi_p = qX + E_g - (E_{f_{po}} - E_v)}, \tag{4.1}$$

where the value of the thermal equilibrium Fermi level in the p-type semiconductor $E_{f_{po}}$ will depend on the hole concentration. The work function in the n-type semiconductor is designated as $q\Phi_n$ and is also shown in Fig. 4.3 (a). For the n-type semiconductor, the work function is given by

$$\boxed{q\Phi_n = qX + (E_c - E_{f_{no}})}, \tag{4.2}$$

where the value of the thermal equilibrium Fermi level in the n-type semiconductor $E_{f_{no}}$ will depend on the electron concentration. The built-in potential V_{bi} is the difference in the work functions,

$$qV_{bi} \equiv q\Phi_p - q\Phi_n = E_{f_{no}} - E_{f_{po}}, \tag{4.3}$$

which becomes the difference in the Fermi levels.

Figure 4.3(b) shows the energy-band diagram for the p- and n-type regions at initial contact before equilibrium can be established. The majority carrier concentrations are shown in Fig. 4.3(c) at initial contact to illustrate the high concentration gradient of holes and electrons at the junction interface. Instantaneously, holes will be lost from the p-region at the interface and give a space charge due to unneutralized negatively ionized acceptors N_a^-, while electrons will be lost from the n-region at the interface and give a space charge due to unneutralized positively ionized donors N_d^+. This space charge is illustrated in Fig. 4.4(a) and is called the *depletion region,* with the edge of the depletion region given by $-x_p$ on the p-side and by x_n on the n-side. In the depletion region, few carriers are present, but there are ionized donors and acceptors. In the p-type region, the ionized acceptors are negatively charged, while in the n-type region, the ionized donors are positively charged. These ionized acceptors and donors are located in substitutional lattice sites and cannot move in the electric field. The concentrations of these donors and acceptors are selected to give the p-n junction desired device properties.

depletion region

As given in Eq. (3.42), $dE_f/dx = 0$ for thermal equilibrium. Therefore, at thermal equilibrium, the Fermi levels in the p- and n-type semiconductors must be equal. The requirement for the constant Fermi level pushes the n-type semiconductor Fermi

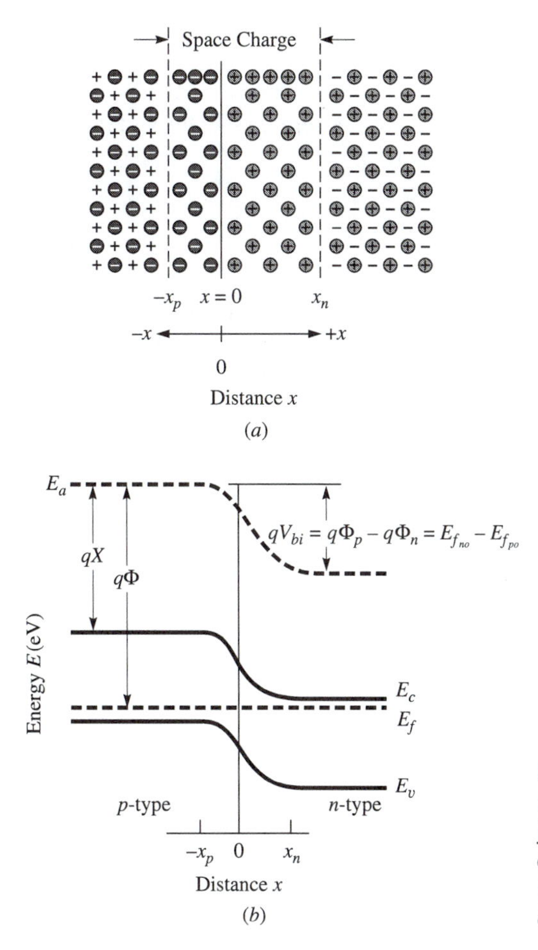

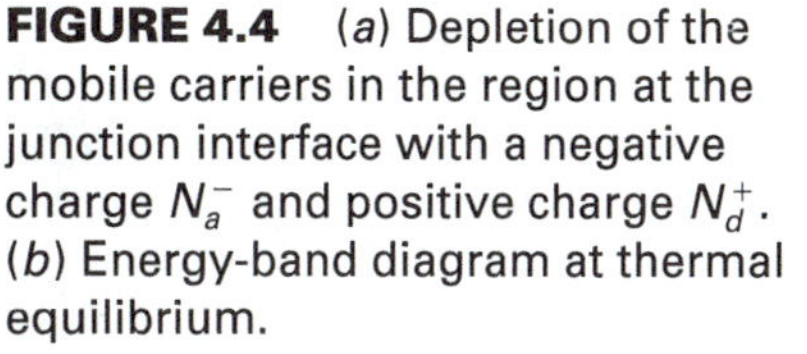
FIGURE 4.4 (*a*) Depletion of the mobile carriers in the region at the junction interface with a negative charge N_a^- and positive charge N_d^+. (*b*) Energy-band diagram at thermal equilibrium.

level down to be constant with the p-type semiconductor Fermi level as shown in Fig. 4.4(*b*), which is the thermal equilibrium energy-band diagram. The amount the bands are bent is the difference in work functions which was given in Eq. (4.3) as the difference in the Fermi levels. To specify the distance over which the bands bend requires determination of the electric field as a function of distance $\mathscr{E}(x)$ and then the potential as a function of distance $\psi(x)$ and is the subject of the next section. The energy-band diagram for thermal equilibrium, as well as for nonthermal equilibrium with an applied bias, is a useful representation for the derivation of the behavior of p-n junctions.

4.3 JUNCTION POTENTIAL

Detailed analysis of the p-n junction begins with determination of the electric field $\mathscr{E}(x)$, which is related to the space charge by Gauss's law, which was given in Eq. (3.54) as

$$d\mathcal{E}(x)/dx = \rho(x)/\epsilon, \tag{4.4}$$

where ϵ is the dielectric constant for the semiconductor and the charge density $\rho(x)$ is the sum of all of the charged species:

$$\rho(x) = q(p - n + N_d^+ - N_a^-). \tag{4.5}$$

depletion approximation

The solution to Eq. (4.4) is generally obtained by assuming that the mobile carrier concentrations can be neglected in the space-charge region and is called the *depletion approximation.* The resulting space-charge region is represented in Fig. 4.5(*a*) with the

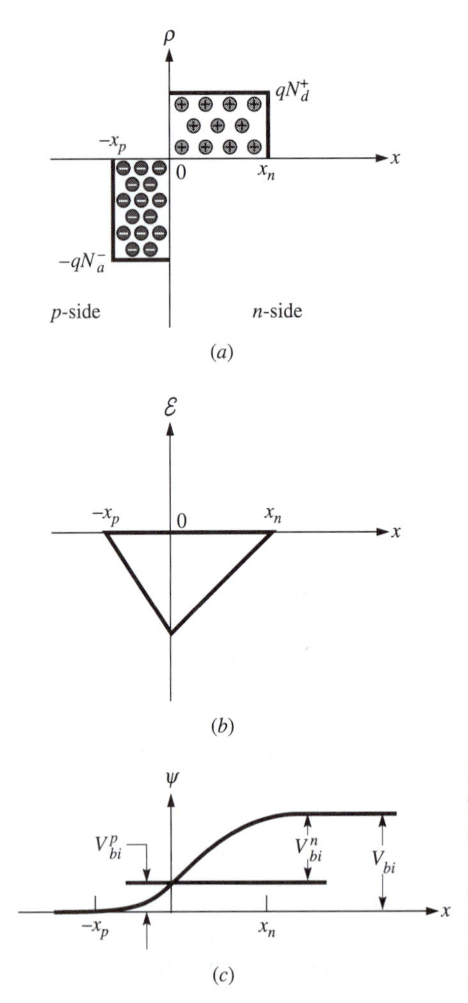

FIGURE 4.5 Abrupt *p-n* junction at thermal equilibrium. The edge of the space-charge region on the *p*-side is $-x_p$ and on the *n*-side is x_n. (*a*) Impurity distribution illustrating the space-charge region. (*b*) Electric-field variation with distance *x*. (c) Potential variation with distance *x*.

edge of the depletion region given by $-x_p$ on the p-side and the edge of the depletion region given by x_n on the n-side.

Gauss's law may be written as

$$d\mathcal{E}(x)/dx = 0 \qquad \text{for} \qquad -\infty < x < -x_p,\ x_n < x < \infty, \tag{4.6}$$

$$d\mathcal{E}(x)/dx = (-qN_a^-)/\epsilon \qquad \text{for} \qquad -x_p \le x < 0, \tag{4.7}$$

and

$$d\mathcal{E}(x)/dx = (qN_d^+)/\epsilon \qquad \text{for} \qquad 0 < x \le x_n. \tag{4.8}$$

dielectric constant of free space ϵ_0

The dielectric constants for commonly encountered semiconductors are summarized in Table 4.1. The value of the dielectric constant of free space ϵ_0 is 8.85×10^{-14} F/cm.

Integration of Eqs. (4.7) and (4.8) gives

$$\mathcal{E}(x) = (-qN_a^-/\epsilon)x + A_1 \qquad \text{for} \qquad -x_p \le x < 0, \tag{4.9}$$

and

$$\mathcal{E}(x) = (qN_d^+/\epsilon)x + A_2 \qquad \text{for} \qquad 0 < x \le x_n, \tag{4.10}$$

where A_1 and A_2 are arbitrary integration constants to be determined by the boundary conditions. At the edges of the space-charge region, the field must go to zero: $\mathcal{E}(-x_p) = 0$ and $\mathcal{E}(x_n) = 0$. This condition gives

$$A_1 = -(q/\epsilon)N_a^- x_p, \tag{4.11}$$

and

$$A_2 = -(q/\epsilon)N_d^+ x_n. \tag{4.12}$$

TABLE 4.1 Semiconductor Dielectric Constants.

Semiconductor	Dielectric Constant
Si	$11.7\epsilon_0$
Ge	$16.0\epsilon_0$
AlAs	$10.1\epsilon_0$
GaN	$9.5\epsilon_0$
GaP	$11.1\epsilon_0$
GaAs	$13.1\epsilon_0$
InP	$12.4\epsilon_0$
$Al_xGa_{1-x}As$	$(13.1 - 3.0x)\epsilon_0$

Equations (4.9) and (4.10) may be written as

$$\boxed{\mathscr{E}(x) = -(qN_a^-/\epsilon)(x + x_p) \qquad \text{for} \qquad -x_p \le x < 0}, \tag{4.13}$$

electric field

and

$$\boxed{\mathscr{E}(x) = -(qN_d^+/\epsilon)(x_n - x) \qquad \text{for} \qquad 0 < x \le x_n}. \tag{4.14}$$

Note that for $x = 0$,

$$\mathscr{E}(0) = -\frac{qN_a^-}{\epsilon}x_p = -\frac{qN_d^+}{\epsilon}x_n. \tag{4.15}$$

For a static electric field, the potential $\psi(x)$ is related to the electric field $\mathscr{E}(x)$ by

$$\boxed{\mathscr{E}(x) = -d\psi(x)/dx}, \tag{4.16}$$

which gives $\psi(x) = -\int \mathscr{E}(x)\,dx$. Integration of Eqs. (4.13) and (4.14) gives

$$\psi(x) = (qN_a^-/\epsilon)(x^2/2 + x_p x) + B_1 \qquad \text{for} \qquad -x_p \le x < 0 \tag{4.17}$$

and

$$\psi(x) = (qN_d^+/\epsilon)(x_n x - x^2/2) + B_2 \qquad \text{for} \qquad 0 < x \le x_n. \tag{4.18}$$

Let $\psi(x) = 0$ at $x = -x_p$, and then

$$B_1 = (q/\epsilon)N_a^-(x_p^2/2) \tag{4.19}$$

and the potential on the p-side is given by

potential on p-side

$$\boxed{\psi(x) = (qN_a^-/2\epsilon)(x_p + x)^2 \qquad \text{for} \qquad -x_p \le x < 0}. \tag{4.20}$$

The built-in potential V_{bi}^p on the p-side is now obtained from Eq. (4.20) with $x = 0$ as

built-in potential on p-side

$$\boxed{V_{bi}^p = (qN_a^-/2\epsilon)x_p^2}. \tag{4.21}$$

On the n-side, $\psi(x)$ at $x = x_n$ is the total built-in potential V_{bi}. Therefore, from Eq. (4.18),

$$B_2 = V_{bi} - (qN_d^+/\epsilon)(x_n^2/2) \tag{4.22}$$

and

potential on n-side

$$\boxed{\psi(x) = V_{bi} - (qN_d^+/2\epsilon)(x_n - x)^2 \qquad \text{for} \qquad 0 < x \le x_n}. \tag{4.23}$$

At $x = 0$ where $\psi(x) = V_{bi}^p$, Eq. (4.23) gives

$$\psi(0) = V_{bi}^p = V_{bi} - (qN_d^+/2\epsilon)x_n^2, \tag{4.24}$$

and with V_{bi}^n designated as the built-in potential on the n-side,

built-in potential on n-side

$$\boxed{V_{bi} = V_{bi}^p + V_{bi}^n}. \tag{4.25}$$

Then, by Eqs. (4.24) and (4.25),

$$\boxed{V_{bi}^n = V_{bi} - V_{bi}^p = (qN_d^+/2\epsilon)x_n^2}. \tag{4.26}$$

If Eq. (4.23) is written with Eqs. (4.25) and (4.26),

$$\psi(x) = V_{bi}^p + (qN_d^+/2\epsilon)x_n^2 - (qN_d^+/2\epsilon)(x_n - x)^2, \tag{4.27}$$

then a convenient form for $\psi(x)$ on the n-side is

$$\psi(x) = V_{bi}^p + (qN_d^+/2\epsilon)[x_n^2 - (x_n - x)^2] \qquad \text{for} \qquad 0 < x \le x_n. \tag{4.28}$$

Note that for Eq. (4.27) at $x = x_n$,

$$\psi(x_n) = V_{bi}^p + \underbrace{(qN_d^+/2\epsilon)x_n^2}_{V_{bi}^n} = V_{bi}. \tag{4.29}$$

A schematic representation of the space charge, electric field, and potential are given in Fig. 4.5. By Eqs. (4.13) and (4.14), it may readily be seen that the electric field is negative and varies linearly with distance, as illustrated in Fig. 4.5(b). The potential was given by Eqs. (4.20) and (4.28). It is positive and varies quadratically with distance as shown in Fig. 4.5(c).

By equating Eqs. (3.15) and (4.16),

$$\mathscr{E}(x) = (1/q)\,dE_c(x)/dx = -d\psi/dx, \tag{4.30}$$

then,

$$(1/q)\,dE_c(x)/dx = -d\psi(x)/dx. \tag{4.31}$$

Equations (4.30) and (4.31) show that the conduction-band edge bends in the opposite direction to the potential [compare Figs. 4.4 (*b*) and 4.5 (*c*)].

EXAMPLE 4.1 Show how the relationship given by Eq. (4.31) may be used with Eqs. (4.20) and (4.28) to obtain the energy-band diagram.

Solution Because $E_c(x) - E_g = E_v(x)$, Eq. (4.31) may be written as

$$(1/q)\, dE_c(x)/dx = (1/q)\, dE_v(x)/dx = -d\psi(x)/dx.$$

Integration gives

$$(1/q)E_v(x) = -\psi(x) + B,$$

where B is an integration constant. With $E_v(-x_p)$ selected as qV_{bi}, $B = V_{bi}$ because $\psi(-x_p) = 0$. With $\psi(x)$ given by Eq. (4.20) for the p-side,

$$E_v(x) = q[V_{bi} - (qN_a^-/2\epsilon)(x_p + x)^2] \qquad \text{for} \qquad -x_p \le x < 0.$$

Then $\psi(x)$ given by Eq. (4.28) for the n-side with $(V_{bi} - V_{bi}^p) = V_{bi}^n$,

$$E_v(x) = q\{V_{bi}^n - (qN_d^+/2\epsilon)[x_n^2 - (x_n - x)^2]\} \qquad \text{for} \qquad 0 < x \le x_n.$$

The potential and the corresponding energy-band diagram are plotted in Fig. 4.6 to demonstrate that $E_v(x)$ and $E_c(x)$ are the mirror images of $\psi(x)$. ■

4.4 DEPLETION WIDTH

The expressions for the electric field and potential for the p-n junction depend on the depletion widths $-x_p$ and x_n on the p- and n-sides, respectively. Therefore, $-x_p$ and x_n,

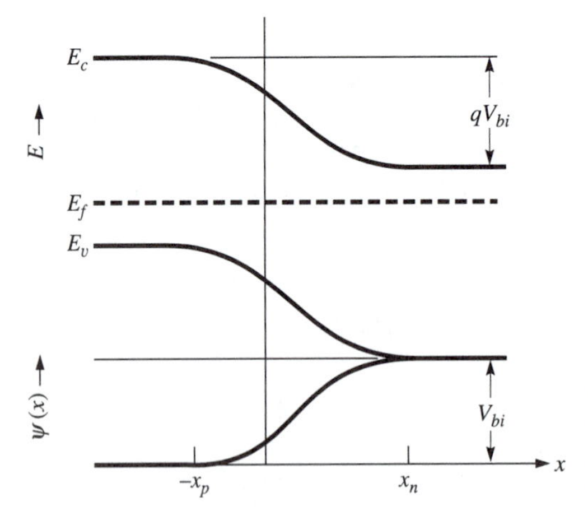

FIGURE 4.6 Comparison of the energy-band diagram for an abrupt p-n junction with the potential variation.

together with their sum $x_d = |x_p| + |x_n|$, are fundamental parameters in the description of *p-n* junctions. The depletion width on the *p*-side is given by Eq. (4.21) as

p-side depletion width

$$\boxed{x_p = \sqrt{2\epsilon V_{bi}^p / qN_a^-}}, \tag{4.32}$$

and the depletion width on the *n*-side is given by Eq. (4.26) as

n-side depletion width

$$\boxed{x_n = \sqrt{2\epsilon V_{bi}^n / qN_d^+}}. \tag{4.33}$$

The expressions in Eqs. (4.32) and (4.33) have a diverse combination of units:

units example

$$\frac{\epsilon\, V_{bi}^{n \text{ or } p}}{qN_a^- \text{ or } qN_d^+} = \frac{\text{F/cm V}}{\text{coulombs-cm}^{-3}} = \frac{\text{(coulombs/V)/cm V}}{\text{coulombs-cm}^{-3}} = \frac{1}{\text{cm}^{-2}} = \text{cm}^2,$$

which demonstrates that the depletion width has units of cm.

The expressions for the depletion width on the *p*- and *n*-sides of the *p-n* junction require V_{bi}^p and V_{bi}^n. The ratio formed by dividing V_{bi}^n by V_{bi}^p from Eqs. (4.21) and (4.26) gives

$$\frac{V_{bi}^n}{V_{bi}^p} = \frac{(q/2\epsilon)N_d^+ x_n^2}{(q/2\epsilon)N_a^- x_p^2}. \tag{4.34}$$

The depletion widths may be eliminated from Eq. (4.34) by setting $x = 0$ in Eqs. (4.13) and (4.14) for the electric field, which gave Eq. (4.15) as $(qN_a^-/\epsilon)x_p = (qN_d^+/\epsilon)x_n$. Then, with Eq. (4.15),

$$\frac{x_n^2}{x_p^2} = \left[\frac{N_a^-}{N_d^+}\right]^2. \tag{4.35}$$

The expression for (x_n^2/x_p^2) in Eq. (4.35) may be used to replace (x_n^2/x_p^2) in Eq. (4.34) to give

$$\frac{V_{bi}^n}{V_{bi}^p} = \frac{N_d^+}{N_a^-}\left[\frac{N_a^-}{N_d^+}\right]^2 = \frac{N_a^-}{N_d^+}. \tag{4.36}$$

With Eq. (4.25) for the built-in potential V_{bi} and Eq. (4.36) for V_{bi}^n:

$$V_{bi} = V_{bi}^n + V_{bi}^p = V_{bi}^p(N_a^-/N_d^+) + V_{bi}^p \tag{4.37}$$

or

$$\boxed{V_{bi}^{p} = V_{bi}/[1 + (N_a^-/N_d^+)]}. \tag{4.38}$$

In Eq. (4.38), evaluation of V_{bi} and V_{bi}^p only requires knowledge of the acceptor and donor concentration on each side of the junction. In a similar manner,

$$\boxed{V_{bi}^{n} = V_{bi}/[1 + (N_d^+/N_a^-)]}. \tag{4.39}$$

The total depletion width is

$$x_d = |x_n| + |x_p|, \tag{4.40}$$

and may be written in the most useful form by first squaring Eq. (4.40) to give

$$x_d^2 = |x_n|^2 + 2|x_n||x_p| + |x_p|^2. \tag{4.41}$$

From Eq. (4.26) for x_n^2 and Eq. (4.39) for V_{bi}^n, x_n^2 may be written as

$$x_n^2 = \frac{2\epsilon V_{bi}^n}{qN_d^+} = \frac{2\epsilon V_{bi}}{qN_d^+[1 + (N_d^+/N_a^-)]} = \frac{2\epsilon N_a^- V_{bi}}{qN_d^+(N_a^- + N_d^+)}. \tag{4.42}$$

With Eq. (4.21) for x_p^2 and Eq. (4.38) for V_{bi}^p, x_p^2 may be written as

$$x_p^2 = \frac{2\epsilon N_d^+ V_{bi}}{qN_a^-(N_a^- + N_d^+)}. \tag{4.43}$$

Substitution of Eqs. (4.42) and (4.43) into Eq. (4.41) gives

$$x_d^2 = \frac{2\epsilon N_a^- V_{bi}}{qN_d^+(N_a^- + N_d^+)} + 2\sqrt{\frac{2\epsilon N_a^- V_{bi}}{qN_d^+(N_a^- + N_d^+)}}\sqrt{\frac{2\epsilon N_d^+ V_{bi}}{qN_a^-(N_a^- + N_d^+)}} + \frac{2\epsilon N_d^+ V_{bi}}{qN_a^-(N_a^- + N_d^+)}, \tag{4.44}$$

then

$$x_d^2 = \frac{2\epsilon V_{bi}}{q(N_a^- + N_d^+)}\left[\frac{N_a^-}{N_d^+} + 2 + \frac{N_d^+}{N_a^-}\right] \tag{4.45}$$

or

$$x_d^2 = \frac{2\epsilon V_{bi}}{q(N_a^- + N_d^+)}\left[\frac{(N_a^-)^2 + 2N_a^- N_d^+ + (N_d^+)^2}{N_d^+ N_a^-}\right]. \tag{4.46}$$

Equation (4.46) may be written as

depletion width

$$\boxed{x_d = \sqrt{\frac{2\epsilon}{q}\left(\frac{N_a^- + N_d^+}{N_d^+ N_a^-}\right)V_{bi}}} \tag{4.47}$$

or

$$x_d = \sqrt{\frac{2\epsilon}{q}\left(\frac{1}{N_d^+} + \frac{1}{N_a^-}\right)V_{bi}}. \tag{4.48}$$

This expression for the depletion width will be used in Sec. 5.4.1 to find the depletion capacitance. It should be noted that for N_d^+ or N_a^- near a concentration of 1×10^{15} cm^{-3}, $x_d \approx 1\ \mu$m.

4.5 RELATION BETWEEN POTENTIAL AND CARRIER CONCENTRATION

One of the *most important relationships* in device physics is *the relationship between carrier concentration and potential.* This relationship between potential and carrier concentration is used for both the *p-n* junctions and the metal-oxide semiconductor (MOS) devices. The electrostatic potential, often called the potential, was given in Chapter 3 by Eq. (3.17) as

$$\psi(x) = -E_i(x)/q. \tag{4.49}$$

The reference for potential may be redefined by adding a constant potential. Addition of a constant potential such as the Fermi level E_f at thermal equilibrium does not affect potential differences, which are the important quantities. Addition of the Fermi level E_f to $-E_i$ gives

potential

$$\boxed{q\psi(x) = E_f - E_i(x)}. \tag{4.50}$$

With Eq. (4.50) for $q\psi(x)$, the previous expression for the electron concentration given in Eq. (2.109) may now be written as

$$n = n_i \exp[(E_f - E_i)/kT] = n_i \exp[q\psi(x)/kT]. \tag{4.51}$$

The expression for holes given in Eq. (2.110) becomes

$$p = n_i \exp[(E_i - E_f)/kT] = n_i \exp[-q\psi(x)/kT]. \tag{4.52}$$

The relationship between the variation of the potential $\psi(x)$ and the carrier concentrations $n(x)$ and $p(x)$ may be obtained by taking the ratio of the electron concentration at x_2 with $n(x_2) = n_2$ and $\psi(x_2) = \psi_2$ and at x_1 with $n(x_1) = n_1$ and $\psi(x_1) = \psi_1$:

$$\frac{n_2}{n_1} = \frac{n_i \exp(q\psi_2/kT)}{n_i \exp(q\psi_1/kT)} = \exp[q(\psi_2 - \psi_1)/kT]. \tag{4.53}$$

Then this relationship may be written as

n and ψ

$$\boxed{n_2 = n_1 \exp[q(\psi_2 - \psi_1)/kT]} \tag{4.54}$$

or

$$\boxed{\psi_2 - \psi_1 = (kT/q)\ln(n_2/n_1)}. \tag{4.55}$$

Note that for Eq. (4.54), if $n_1 = n_i$, then $E_{f_1} = E_{i_1}$ and $\psi_1 = 0$, and

$$n_2 = n_i \exp(q\psi_2/kT) = n_i \exp[(E_{f_2} - E_{i_2})/kT], \tag{4.56}$$

which is Eq. (4.51). In a similar manner for holes,

p and ψ

$$\boxed{p_2 = p_1 \exp[-q(\psi_2 - \psi_1)/kT]} \tag{4.57}$$

or

$$\boxed{\psi_2 - \psi_1 = -(kT/q)\ln(p_2/p_1)}. \tag{4.58}$$

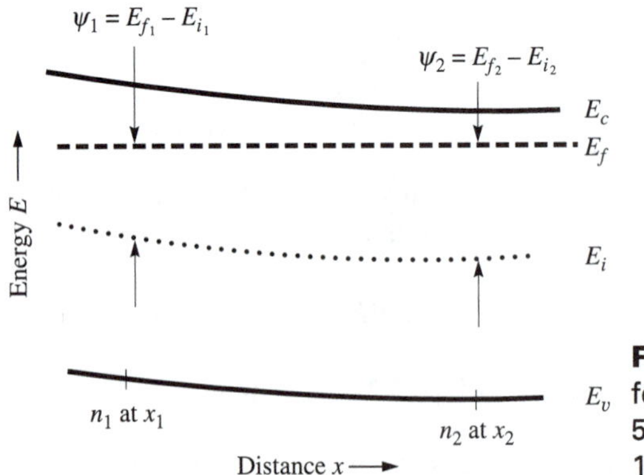

FIGURE 4.7 Energy-band diagram for nonuniformly doped Si with $n_1 = 5 \times 10^{15}$ cm^{-3} at x_1 and $n_2 = 7 \times 10^{16}$ cm^{-3} at x_2.

The relationship between potential and electron concentration is illustrated by the energy-band diagram in Fig. 4.7 for a nonuniformly doped n-type semiconductor. Because the Fermi level is closer to the conduction band at x_2 than x_1, $n_2 > n_1$ by Eq. (2.55).

EXAMPLE 4.2 In Fig. 4.7, let $n_1 = 5 \times 10^{15}$ cm^{-3} at $x = x_1$ and $n_2 = 7 \times 10^{16}$ cm^{-3} at $x = x_2$ for n-type Si with $n_i = 1.0 \times 10^{10}$ cm^{-3} at room temperature. Find ψ_2 and the potential difference $\psi_2 - \psi_1$.

Solution From Eq. (4.51):

$$\psi_2 = \frac{kT}{q} \ln \frac{n_2}{n_i}$$

$$\psi_2 = \frac{8.616 \times 10^{-5}\ \text{eV} - \text{K}^{-1} \times 300\ \text{K}}{1.6 \times 10^{-19}\ \text{coulombs}} \frac{1.6 \times 10^{-19}\ \text{joules}}{\text{eV}} \ln \frac{n_2}{n_I}$$

$$\psi_2 = 0.026 \frac{\text{joules}}{\text{coulombs}} \ln \frac{7 \times 10^{16}}{1.0 \times 10^{10}} = 0.410\ \text{V}.$$

From Eq. (4.55):

$$\psi_2 - \psi_1 = 0.026 \ln(7 \times 10^{16}/5 \times 10^{15}) = 0.069\ \text{V}.$$

■

The potential for a p-n junction was represented in Fig. 4.5 (c) and is relabeled in Fig. 4.8. A convenient expression may be obtained for the built-in potentialby application of the expressions relating carrier concentration to potential. In Fig. 4.8 at the edge of the depletion region on the p-side, $-x_p$, the carrier concentrations are designated n_2 and p_2 with potential ψ_2. At the edge of the depletion region on the n-side, x_n, the carrier concentrations are designated n_1 and p_1 with potential ψ_1. For holes at $-x_p$, $p_2 = N_a^-$, and $\psi_2 = 0$. At x_n, $p_1 = n_i^2/N_d^+$ and $\psi_1 = V_{bi}$. Substitution of those quantities in Eq. (4.58) gives

$$\psi_2 - \psi_1 = -(kT/q)\ln(p_2/p_1) = 0 - V_{bi} = -(kT/q)\ln\left(\frac{N_a^-}{n_i^2/N_d^+}\right) \tag{4.59}$$

or

built-in potential

$$\boxed{V_{bi} = (kT/q)\ln(N_a^- N_d^+/n_i^2)}. \tag{4.60}$$

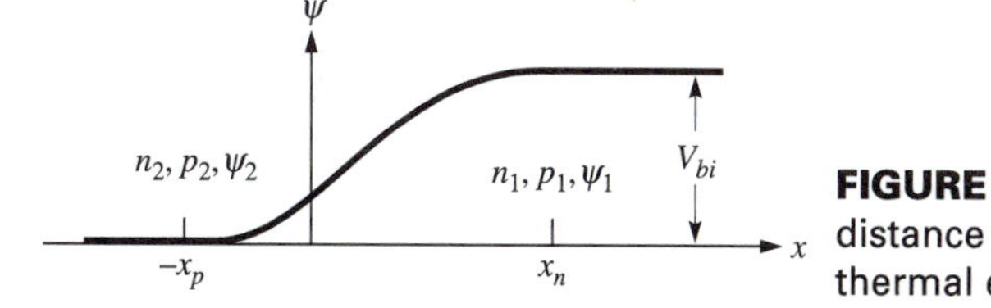

FIGURE 4.8 Potential variation with distance for an abrupt p-n junction at thermal equilibrium.

For electrons at $-x_p$, $n_2 = n_i^2/N_a^-$, and $\psi_2 = 0$, and at x_n, $n_1 = N_d^+$, and $\psi_1 = V_{bi}$. Substitution into Eq. (4.55) gives

$$-V_{bi} = (kT/q)\ln\left(\frac{n_i^2/N_a^-}{N_d^+}\right) \tag{4.61}$$

or

$$\boxed{V_{bi} = (kT/q)\ln(N_a^- N_d^+/n_i^2)}, \tag{4.62}$$

which is the same equation as obtained for holes. Equation (4.60) or (4.62) is much easier to evaluate for V_{bi} than Eq. (4.3).

4.6 *p-n* JUNCTION WITH BIAS

4.6.1 Quasi-Fermi Level

For an applied bias, it is assumed that *quasi-equilibrium* holds and is represented by *quasi-Fermi levels* for either carrier. Although the carrier concentrations no longer have their thermal equilibrium values, the carriers may still be represented by the thermal distribution given by the Fermi-Dirac distribution function, but with a Fermi level appropriate to the nonequilibrium concentrations, which is known as the quasi-Fermi level. The quasi-Fermi level, when substituted for the Fermi level in the expressions for the carrier concentrations, gives the concentration of that carrier under nonequilibrium conditions. The concentration of electrons for nonequilibrium conditions is given by Eq. (2.109) or (4.51) with E_f replaced by E_{f_n}:

$$n = n_i \exp[(E_{f_n} - E_i)/kT], \tag{4.63}$$

where E_{f_n} is the quasi-Fermi level for electrons. The concentration of holes under nonequilibrium is given by Eq. (2.110) as

$$p = n_i \exp[(E_i - E_{f_p})/kT], \tag{4.64}$$

where E_{f_p} is the quasi-Fermi level for holes.

Departures from thermal equilibrium carrier concentrations can be related to quasi-Fermi levels. With n_o for the thermal equilibrium electron concentration, Eq. (4.51) gives

$$n_o = n_i \exp[(E_f - E_i)/kT], \tag{4.65}$$

which gives

$$n_i \exp(-E_i/kT) = n_o \exp(-E_f/kT). \tag{4.66}$$

Then the nonthermal equilibrium electron concentration in Eq. (4.63) may be written as

$$n = \underbrace{n_i \exp(-E_i/kT)}_{n_o \exp(-E_f/kT)} \exp(E_{f_n}/kT) \tag{4.67}$$

or

$$\boxed{n = n_o \exp[(E_{f_n} - E_f)/kT]}. \tag{4.68}$$

Equation (4.68) demonstrates that the departure of n from the thermal equilibrium value n_o results in E_{f_n} different from the thermal equilibrium Fermi level E_f. Similar expressions may be written for the holes.

Quasi-Fermi levels are also useful in representing current flow in semiconductors. The electron current density was given in Eq. (3.37) as

$$J_n = q\mu_n n\mathcal{E} + qD_n \frac{dn}{dx}, \tag{4.69}$$

with the electric field $\mathcal{E}$ given by Eq. (3.15) as

$$\mathcal{E}(x) = \frac{1}{q}\frac{dE_i(x)}{dx}. \tag{4.70}$$

With Eq. (4.63) for n,

$$\frac{dn}{dx} = \underbrace{n_i \exp[(E_{f_n} - E_i)/kT]}_{n} \frac{1}{kT}\left(\frac{dE_{f_n}}{dx} - \frac{dE_i}{dx}\right) \tag{4.71}$$

or

$$\frac{dn}{dx} = \frac{n}{kT}\left(\frac{dE_{f_n}}{dx} - \frac{dE_i}{dx}\right). \tag{4.72}$$

Equation (4.69) may now be written as

$$J_n = q\mu_n n \frac{1}{q}\frac{dE_i}{dx} + qD_n \frac{n}{kT}\left(\frac{dE_{f_n}}{dx} - \frac{dE_i}{dx}\right), \tag{4.73}$$

and with the Einstein relation, $D_n = \mu_n(kT/q)$, Eq. (4.73) reduces to *electron current density*

$$\boxed{J_n = \mu_n n \frac{dE_{f_n}}{dx}}. \tag{4.74}$$

In a similar manner for holes,

hole current density

$$\boxed{J_p = \mu_p p \frac{dE_{f_p}}{dx}}. \tag{4.75}$$

Equations (4.74) and (4.75) apply for current flow by *both drift and diffusion*. Therefore, a quasi-Fermi-level gradient in an energy-band diagram indicates a current flow.

4.6.2 Energy-band Diagram at Nonequilibrium

When a bias voltage V_a is applied to a *p-n* junction, the potential across the *p-n* junction is altered by the polarity and magnitude of V_a. Voltage drops in the neutral bulk regions and at the contacts are taken as zero. The effects of voltage drops due to these effects are introduced in Sec. 4.7.5. The quasi-Fermi level for electrons in the neutral region on the *n*-side will be displaced from the quasi-Fermi level for holes in the neutral region on the *p*-side by V_a,

$$\boxed{E_{f_n} - E_{f_p} = qV_a}. \tag{4.76}$$

It is helpful to use the *schematic space-charge model* illustrated in Fig. 4.9 together with the energy-band diagram for thermal equilibrium. The circuit diagram symbol for the *p-n* junction diode is shown above the schematic space-charge model. Forward bias reduces the space charge: holes from the positive terminal flow into the *p*-side as illustrated in Fig. 4.10(*a*), with the positive charge neutralizing the negative space charge on the *p*-side. The energy-band diagram for forward bias is shown in Fig. 4.10(*b*). In this case the band bending is decreased from qV_{bi} at thermal equilibrium to give $q(V_{bi} - V_a)$. Reverse bias increases the space charge. For reverse bias, the negative terminal is con-

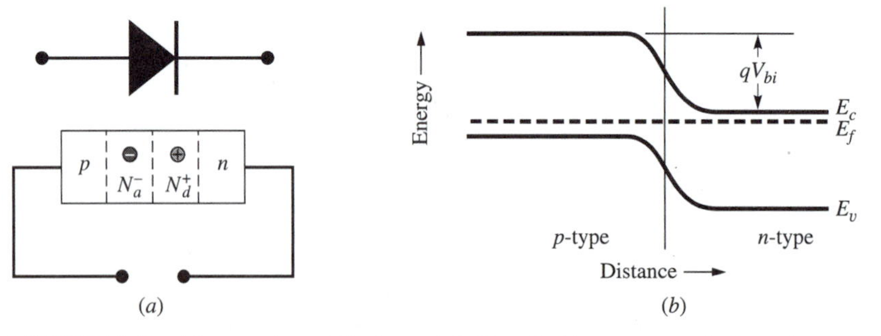

FIGURE 4.9 Thermal equilibrium. (*a*) Circuit diagram symbol and schematic space-charge model for *p-n* junction. (*b*) Energy-band diagram.

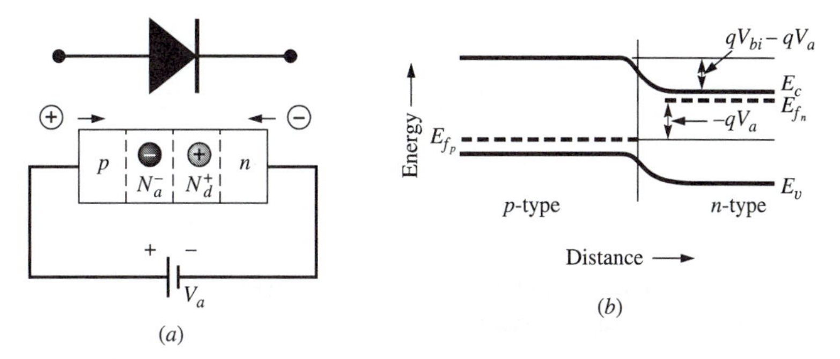

FIGURE 4.10 Forward bias. (*a*) Schematic space-charge model for *p-n* junction. (*b*) Energy-band diagram.

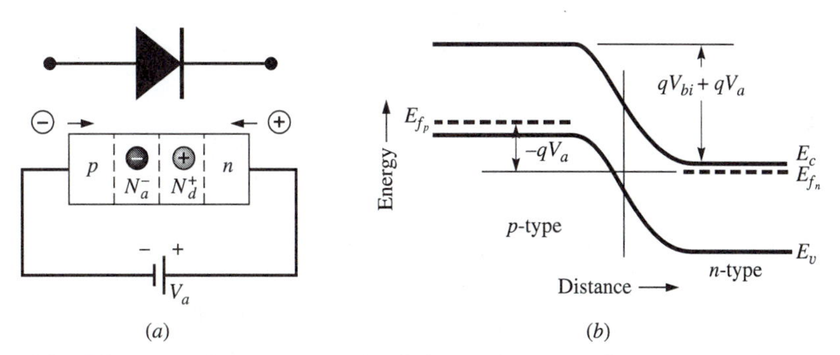

FIGURE 4.11 Reverse bias. (*a*) Schematic space-charge model for *p-n* junction. (*b*) Energy-band diagram.

nected to the *p*-side as illustrated in Fig. 4.11(*a*) with the negative charge increasing the negative space charge on the *p*-side and the positive charge increasing the positive space charge on the *n*-side. The energy-band diagram for reverse bias is shown in Fig. 4.11(*b*). In this case the band bending is increased from qV_{bi} at thermal equilibrium to give $qV_{bi} - (-qV_a) = q(V_{bi} + V_a)$.

The expressions for the potential and depletion widths at thermal equilibrium in Sec. 4.3 are modified when there is an applied voltage V_a. The potential at $x = 0$ now becomes $V_{bi}^p - V_a^p$. For example, Eq. (4.38) for the built-in potential on the *p*-side becomes

$$V_{bi}^p - V_a^p = (V_{bi} - V_a)/[1 + (N_a^-/N_d^+)], \tag{4.77}$$

and is used in Eq. (4.20) for the potential at $x = 0$. The built-in potential on the *n*-side becomes

$$V_{bi}^n - V_a^n = (V_{bi} - V_a)/[1 + (N_d^+/N_a^-)]. \tag{4.78}$$

Therefore, in the previous junction equations for equilibrium conditions, which contain V_{bi}^p, V_{bi}^n, and V_{bi}, these quantities are replaced by $V_{bi}^p - V_a^p$, $V_{bi}^n - V_a^n$, and $V_{bi} - V_a$.

The expression for the depletion width x_d in Eq. (4.47) becomes

depletion width voltage dependence

$$\boxed{x_d = \sqrt{\frac{2\epsilon}{q}\left(\frac{N_a^- + N_d^+}{N_d^+ N_a^-}\right)(V_{bi} - V_a)}}. \tag{4.79}$$

4.6.3 Minority Carrier Concentrations at the Depletion Boundaries

For thermal equilibrium, the potential variation with distance was illustrated in Fig. 4.8. With n_1, p_1, and ψ_1 at x_n and n_2, p_2, and ψ_2 at $-x_p$, Eq. (4.57) may be written as

$$\underbrace{p_2}_{N_a^-(-x_p)} = \underbrace{p_1}_{p_{no}(x_n)} \exp[-q(\underbrace{\psi_2}_{0} - \underbrace{\psi_1}_{V_{bi}})/kT] \tag{4.80}$$

or

$$p_{no}(x_n) = N_a^-(-x_p)\exp(-qV_{bi}/kT). \tag{4.81}$$

With forward bias V_a, the potential variation with distance shown in Fig. 4.8 is modified as shown in Fig. 4.12. For forward bias, the edges of the depletion region are decreased from $-x_p$ and x_n to $-x_p'$ and x_n', and V_{bi} is decreased by $-V_a$. For forward bias,

$$\underbrace{p_2}_{N_a^-(-x_p')} = \underbrace{p_1}_{p_n(x_n')} \exp[q\underbrace{(V_{bi} - V_a)}_{\psi_1}/kT] \tag{4.82}$$

or

$$p_n(x_n') = N_a^-(-x_p')\exp(-qV_{bi}/kT)\exp(qV_a/kT). \tag{4.83}$$

With Eq. (4.81) for $N_a^-(-x_p')\exp(-qV_{bi}/kT)$, then the hole concentration at x_n' may be written as

hole concentration at x_n' for forward bias

$$\boxed{p_n(x_n') = p_{no}(x_n')\exp(qV_a/kT)}. \tag{4.84}$$

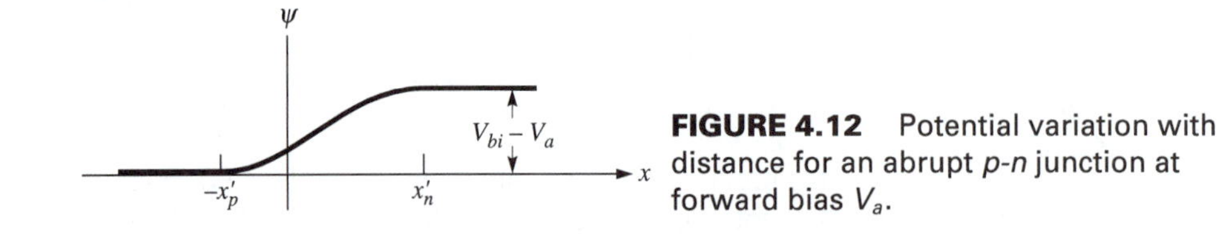

FIGURE 4.12 Potential variation with distance for an abrupt *p-n* junction at forward bias V_a.

Equation (4.84) demonstrates that the *minority carrier concentration* $p_n(x'_n)$ varies as $\exp(qV_a/kT)$, which is one of the most significant concepts in the behavior of semiconductors. An expression similar to Eq. (4.84) may be written for electrons at $-x'_p$ to give the minority carrier concentration as

electron concentration at $-x'_p$ for forward bias

$$\boxed{n_p(-x'_p) = n_{po}(-x'_p)\exp(qV_a/kT)}. \tag{4.85}$$

law of the junction

This exponential increase in the minority carrier concentrations at the edge of the depletion regions with applied voltage is known as the *law of the junction.* The np product at x'_n may be obtained by multiplying $p_n(x'_n)$ in Eq. (4.84) by n_{no} (note that outside the depletion region, the semiconductor is assumed to be neutral and $\Delta p = \Delta n \ll n_{no}$) to give

$$p_n(x'_n)n_{no}(x'_n) = p_{no}(x'_n)n_{no}(x'_n)\exp(qV_a/kT) = n_i^2\exp(qV_a/kT), \tag{4.86}$$

which demonstrates that the thermal equilibrium n_i^2 is increased by $\exp(qV_a/kT)$ at the edges of the depletion region.

4.7 *I–V* CHARACTERISTICS FOR *p-n* JUNCTION

4.7.1 Minority Carrier Concentration Variations

The derivation of the diffusion current for the *p-n* junction begins with the continuity equation to determine the variation of the minority carrier concentrations in the neutral semiconductor regions outside the depletion region. The continuity equation for holes on the *n*-side in the neutral region where the electric field $\mathscr{E}(x) = 0$ was given by Eq. (3.56) as

$$\frac{\partial p_n}{\partial t} = D_p\frac{\partial^2 p_n}{\partial x^2} + (G_p - R_p). \tag{4.87}$$

For steady state and no generation such as by absorption of light ($G_p = 0$), the continuity equation becomes

$$0 = D_p\frac{d^2 p_n}{dx^2} - R_p = D_p\frac{d^2 p_n}{dx^2} - \frac{p_n - p_{no}}{\tau_p} \tag{4.88}$$

or

$$\frac{d^2 p_n}{dx^2} - \frac{p_n}{D_p\tau_p} = -\frac{p_{no}}{D_p\tau_p}. \tag{4.89}$$

If the solution to Eq. (4.89) is assumed to be

$$p_n(x) = C_1 \exp(-x/L_p) + C_2 \exp(x/L_p) + p_{no}, \tag{4.90}$$

then substitution of Eq. (4.90) back into Eq. (4.89) shows that Eq. (4.90) is the general solution with $L_p = \sqrt{D_p \tau_p}$.

Boundary conditions at $x = \infty$ and at $x = x_n$ are used to evaluate C_1 and C_2. At $x = \infty$, which is far into the bulk region of the n-side, the minority carrier concentration becomes its thermal equilibrium value,

$$p_n(\infty) = p_{no}. \tag{4.91}$$

Application of this boundary condition to Eq. (4.90) gives

$$p_n(\infty) = C_1 \exp(-\infty) + C_2 \exp(\infty) + p_{no} = p_{no}. \tag{4.92}$$

The $\exp(-\infty)$ term goes to zero, and Eq. (4.92) can only be satisfied if $C_2 = 0$. At the edge of the depletion region on the n-side, $p_n(x_n)$ was given in Eq. (4.84) as $p_n(x_n) = p_{no} \exp(qV_a/kT)$, which goes to its thermal equilibrium value of $p_n = p_{no}$ when the applied voltage $V_a = 0$. This boundary condition gives for Eq. (4.90),

$$p_n(x_n) = p_{no} \exp(qV_a/kT) = C_1 \exp(-x_n/L_p) + p_{no}. \tag{4.93}$$

Solving for C_1 gives

$$C_1 = p_{no}[\exp(qV_a/kT) - 1] \exp(x_n/L_p), \tag{4.94}$$

which permits writing Eq. (4.90) as

$$\boxed{p_n(x) = p_{no}[\exp(qV_a/kT) - 1] \exp[-(x - x_n)/L_p] + p_{no}} \quad \text{for} \quad x \geq x_n. \tag{4.95}$$

An expression similar to Eq. (4.95) may be written for the electrons on the p-side in the neutral region outside the depletion region:

$$\boxed{n_p(x) = n_{po}[\exp(qV_a/kT) - 1] \exp[(x + x_p)/L_n] + n_{po}} \quad \text{for} \quad x \leq -x_p. \tag{4.96}$$

These minority carrier variations with V_a and x, which are given by Eqs. (4.95) and (4.96), are illustrated in Fig. 4.13.

The minority carrier concentrations through the depletion region may be found from Eq. (4.58) for the relationship between potential and hole concentrations:

$$\psi_2 - \psi_1 = -(kT/q) \ln(p_2/p_1) = (kT/q) \ln(p_1/p_2), \tag{4.97}$$

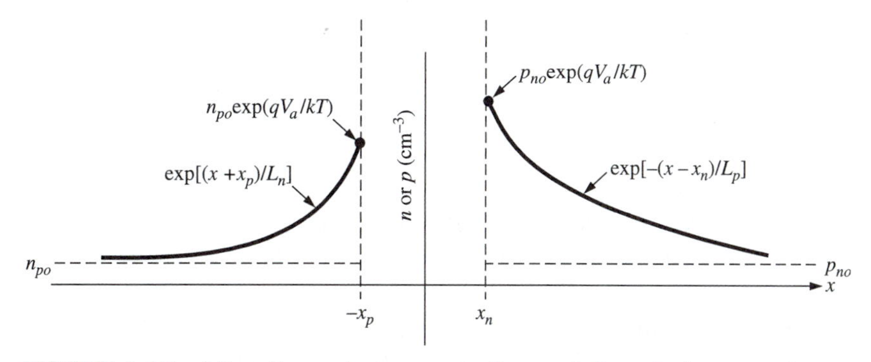

FIGURE 4.13 Minority carrier concentration variations in the neutral semiconductor as represented by Eq. (4.95) for $p_n(x)$ and by Eq. (4.96) for $n_p(x)$.

or from Eq. (4.55) for electrons:

$$\psi_2 - \psi_1 = (kT/q)\ln(n_2/n_1) = -(kT/q)\ln(n_1/n_2). \tag{4.98}$$

The potential on the p-side is given by Eq. (4.20) as

$$\psi_p(x) = (qN_a^-/2\epsilon)(x_p + x)^2 \qquad \text{for} \qquad -x_p \leq x < 0. \tag{4.99}$$

At $x = 0$, $\psi_p(x)$ becomes

$$\psi_p(0) = (qN_a^-/2\epsilon)x_p^2 = V_{bi}^p - V_a^p \tag{4.100}$$

and

$$x_p = \sqrt{\frac{2\epsilon(V_{bi}^p - V_a^p)}{qN_a^-}}. \tag{4.101}$$

The potential on the n-side is given by Eq. (4.23) with V_{bi} replaced by $V_{bi} - V_a$, which gives

$$\psi_n(x) = (V_{bi} - V_a) - (qN_d^+/2\epsilon)(x_n - x)^2 \qquad \text{for} \qquad 0 < x \leq x_n. \tag{4.102}$$

At $x = 0$, $\psi_n(x)$ becomes

$$\psi_n(0) = (V_{bi}^p - V_a^p) = \overbrace{(V_{bi}^p - V_a^p) + (V_{bi}^n - V_a^n)}^{(V_{bi}-V_a)} - (qN_d^+/2\epsilon)x_n^2 \tag{4.103}$$

or

$$(V_{bi}^n - V_a^n) = (qN_d^+/2\epsilon)x_n^2. \tag{4.104}$$

Then, the depletion width on the n-side is

$$x_n = \sqrt{\frac{2\epsilon(V_{bi}^n - V_a^n)}{qN_d^+}}. \tag{4.105}$$

An example can best illustrate the determination of the minority carrier concentrations in the depletion region for forward and reverse bias.

EXAMPLE 4.3 For a Si abrupt p-n junction with $p = N_a^- = 2 \times 10^{16}\ \text{cm}^{-3}$ and $n = N_d^+ = 1 \times 10^{16}\ \text{cm}^{-3}$, the determination of $p(x)$ and $n(x)$ between $-x_p$ and x_n begins with evaluation of the built-in potential from Eq. (4.60) as

$$\begin{aligned} V_{bi} &= (kT/q)\ln(N_a^- N_d^+/n_i^2) \\ &= 0.026\ln(2 \times 10^{16} \times 1 \times 10^{16})/(1.0 \times 10^{10})^2 \\ V_{bi} &= 0.736\ \text{V}. \end{aligned}$$

The built-in potential on the p-side from Eq. (4.38) is

$$V_{bi}^p = V_{bi}/[1 + (N_a^-/N_d^+)] = 0.736/(1 + 2) = 0.245\ \text{V},$$

and the built-in potential on the n-side from Eq. (4.39) is

$$V_{bi}^n = V_{bi}/[1 + (N_d^+/N_a^-)] = 0.736/(1 + 0.5) = 0.491\ \text{V}.$$

For an applied voltage $V_a = 0.5$ V, the applied voltage on the p-side is given by Eq. (4.77), with Eq. (4.38) for V_{bi}^p becoming

$$V_a^p = V_a/[1 + (N_a^-/N_d^+)] = 0.5/(1 + 2) = 0.167\ \text{V},$$

and with Eqs. (4.78) and (4.39), the applied voltage on the n-side is

$$V_a^n = V_a/[1 + (N_d^+/N_a^-)] = 0.5/(1 + 0.5) = 0.333\ \text{V}.$$

Therefore, the depletion region boundaries for forward bias become

$$x_p = \sqrt{\frac{2 \times 11.7 \times 8.85 \times 10^{-14}(0.245 - 0.167)}{1.6 \times 10^{-19} \times 2 \times 10^{16}}} = 7.11 \times 10^{-6}\ \text{cm},$$

and

$$x_n = \sqrt{\frac{2 \times 11.7 \times 8.85 \times 10^{-14}(0.491 - 0.333)}{1.6 \times 10^{-19} \times 1 \times 10^{16}}} = 1.43 \times 10^{-5}\ \text{cm}.$$

The potential on the p-side may be written with Eq. (4.99) as

$$\psi_p(x) = \frac{1.6 \times 10^{-19} \times 2 \times 10^{16}}{2 \times 11.7 \times 8.85 \times 10^{-14}}(x_p + x)^2 = 1.545 \times 10^9(x_p + x)^2,$$

and the potential on the n-side may be written with Eq. (4.102) as

$$\psi_n(x) = (V_{bi} - V_a) - \frac{1.6 \times 10^{-19} \times 1 \times 10^{16}}{2 \times 11.7 \times 8.85 \times 10^{-14}}(x_n - x)^2$$

$$\psi_n(x) = (0.736 - 0.5) - 7.726 \times 10^8 (x_n - x)^2,$$

With these expressions for the potential, Eqs. (4.97) and (4.98) may be used to obtain $p(x)$ and $n(x)$. The designations for p_2, n_2, ψ_2 and p_1, n_1, ψ_1 are illustrated in Fig. 4.14. The expression for $p(x)$ may now be written with Eq. (4.97) as

$$p(x) = p_2 \exp\{q[\psi_2 - \psi(x)]/kT\} = 2 \times 10^{16} \exp[-q\psi(x)/kT],$$

where $\psi_2 = 0$ at $-x_p$. The expression for $n(x)$ may now be written with Eq. (4.98) as

$$n(x) = n_2 \exp\{-q[\psi_2 - \psi(x)]/kT\} = \frac{n_i^2}{N_a^-} \exp(qV_a/kT) \exp[q\psi(x)/kT],$$

where Eq. (4.85) gave $n_2 = n_p(-x_p') \exp(qV_a/kT)$ with $n_p(-x_p') = n_i^2/N_a^-$ and then $(n_i^2/N_a^-) \exp(qV_a/kT) = 5.0 \times 10^3 \times 2.248 \times 10^8 = 1.124 \times 10^{12}$ cm^{-3}. *Note that with $p(x)$ and $n(x)$ given previously, the product $p(x)n(x) = n_i^2 \exp(qV_a/kT)$ throughout the depletion region.* The values for the potential and minority carrier concentrations between $-x_p$ and x_n are summarized in the table.

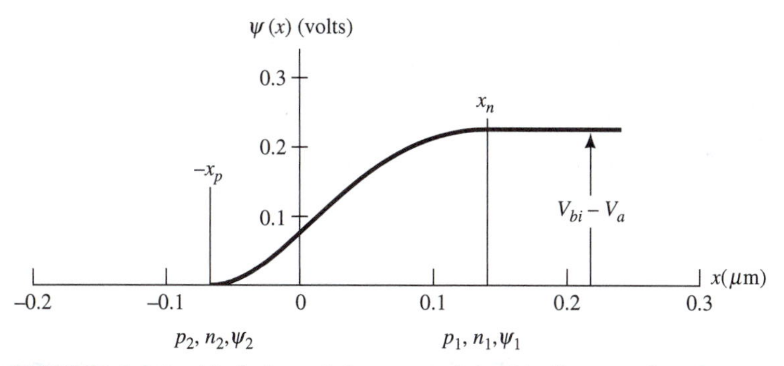

FIGURE 4.14 Variation of the potential with distance for a p-n junction with $p = N_a^- = 2 \times 10^{16}$ cm^{-3} and $n = N_d^+ = 1 \times 10^{16}$ cm^{-3} for a forward bias of 0.5 V.

x (μm)	Eq. (4.99) $\psi_p(x)$	Eq. (4.102) $\psi_n(x)$	Eq. (4.97) $p(x)$	Eq. (4.98) $n(x)$	$n(x)p(x) = n_i^2 \exp(qV_a/kT)$ (cm^{-6})
-7.11×10^{-2}	0		2.00×10^{16}	1.12×10^{12}	2.24×10^{28}
-3.55×10^{-2}	0.019		9.63×10^{15}	2.33×10^{12}	2.24×10^{28}
0	0.078	0.078	9.96×10^{14}	2.26×10^{13}	2.25×10^{28}
0.715×10^{-1}		0.197	1.02×10^{13}	2.20×10^{15}	2.25×10^{28}
1.43×10^{-1}		0.236	2.29×10^{12}	9.84×10^{15}	2.25×10^{28}
1.43×10^{-1}		0.23643	2.25×10^{12}	1.00×10^{16}	2.25×10^{28}

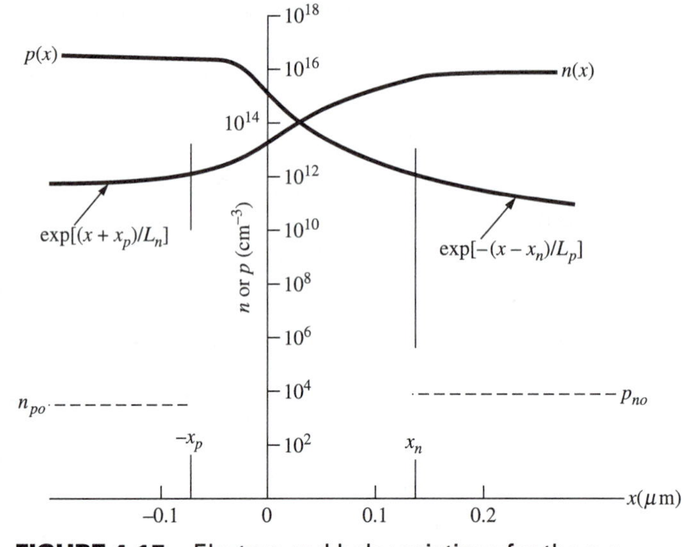

FIGURE 4.15 Electron and hole variations for the *p-n* junction with a forward bias of 0.5 V.

The carrier concentrations may now be plotted between $-x_p$ and x_n and are shown in Fig. 4.15. This example illustrates the numerical evaluation of the relationships between potential and carrier concentrations. It should be noted that at $x = 1.43 \times 10^{-1}$ μm, $\psi_n(x_n)$ has to be taken to five significant figures to give $n(x_n) = 1 \times 10^{16}$ cm^{-3}. *In the table, the last column emphasizes that* $n(x)p(x) = n_i^2 \exp(qV_a/kT)$, *not only at the edges of the depletion region but also throughout the depletion region.* ■

4.7.2 Diffusion Current

The hole diffusion current density on the *n*-side for $x \geq x_n$ may be found with Eq. (4.95) substituted for $p(x)$ in Eq. (3.35):

$$J_{p_n}(x) = -qD_p \frac{dp}{dx}, \tag{4.106}$$

which gives

$$J_{p_n}(x) = -qD_p p_{no}[\exp(qV_a/kT) - 1] \frac{d}{dx} \exp[-(x - x_n)/L_p] \tag{4.107}$$

or

$$J_{p_n}(x) = -qD_p(n_i^2/N_d^+)[\exp(qV_a/kT) - 1](-1/L_p)\exp[-(x - x_n)/L_p]. \tag{4.108}$$

Equation (4.108) gives the diffusion current density due to holes injected into the n-side at $x = x_n$ as

hole-diffusion current density

$$\boxed{J_{p_n}(x_n) = \frac{qD_p n_i^2}{L_p N_d^+}[\exp(qV_a/kT) - 1]}. \quad (4.109)$$

For $x > x_n$, $J_{p_n}(x) = -qD_p(dp/dx)$ gives

$$J_{p_n}(x) = \frac{qD_p n_i^2}{L_p N_d^+}[\exp(qV_a/kT) - 1]\exp[-(x - x_n)/L_p], \quad (4.110)$$

which shows that the diffusion current density decays as $\exp[-(x - x_n)/L_p]$ on the n-side.

In a similar manner, the electron diffusion current density due to electron injection into the p-side is

electron-diffusion current density

$$\boxed{J_{n_p}(-x_p) = \frac{qD_n n_i^2}{L_n N_a^-}[\exp(qV_a/kT) - 1]} \quad (4.111)$$

at the depletion edge $x = -x_p$. For $x < -x_p$ outside the depletion region,

$$J_{n_p}(x) = \frac{qD_n n_i^2}{L_n N_a^-}[\exp(qV_a/kT) - 1]\exp[(x + x_p)/L_n]. \quad (4.112)$$

The total diffusion current density J_t is the sum of Eqs. (4.109) and (4.111):

$$J_t = J_{p_n}(x_n) + J_{n_p}(-x_p), \quad (4.113)$$

which is

total diffusion current density

$$\boxed{J_t = q\left[\frac{D_p n_i^2}{L_p N_d^+} + \frac{D_n n_i^2}{L_n N_a^-}\right][\exp(qV_a/kT) - 1]}. \quad (4.114)$$

The prefactor is called the saturation diffusion current density and is

J_{s_d}

$$\boxed{J_{s_d} = q\left[\frac{D_p n_i^2}{L_p N_d^+} + \frac{D_n n_i^2}{L_n N_a^-}\right]}. \quad (4.115)$$

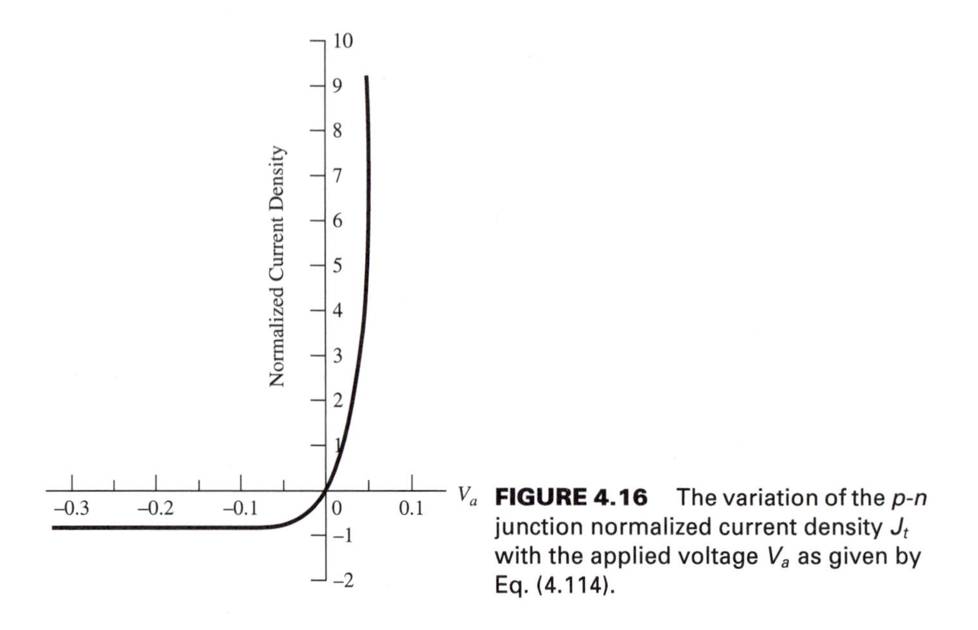

FIGURE 4.16 The variation of the *p-n* junction normalized current density J_t with the applied voltage V_a as given by Eq. (4.114).

The variation of the current density with V_a, as given by Eq. (4.114), is shown in Fig. 4.16. The diffusion current I_d is obtained as the product of the current density J_t and the cross-sectional area A:

ideal diode I–V behavior

$$\boxed{I_d = J_t A = J_{s_d} A[\exp(qV_a/kT) - 1] = I_{s_d}[\exp(qV_a/kT) - 1]}, \qquad (4.116)$$

where the saturation diffusion current $I_{s_d} = J_{s_d} A$. Equation (4.116) is the familiar ideal *I–V* characteristic for a *p-n* junction diode.

For the diffusion current, recombination is ignored in the depletion region and current due to recombination in the depletion region will be treated as a separate current component. Therefore, the hole diffusion current density will be the same at $x = -x_p$ as obtained at $x = x_n$. Also, the electron diffusion current density at $x = x_n$ will be the same as obtained at $x = -x_p$. The diffusion current density given by Eqs. (4.108) and (4.110) are shown in Fig. 4.17. Note that the diffusion current density goes to zero for distances much larger than L_p or L_n. Since the total current must be continuous, the continuity equation must be applied in the neutral regions outside the depletion region.

The total current density on the *n*-side for $x > x_n$ may be found from the continuity equation given in Eq. (3.50):

$$\frac{\partial n}{\partial t} = \frac{1}{q}\frac{\partial J_n}{\partial x} + (G_n - R_n). \qquad (4.117)$$

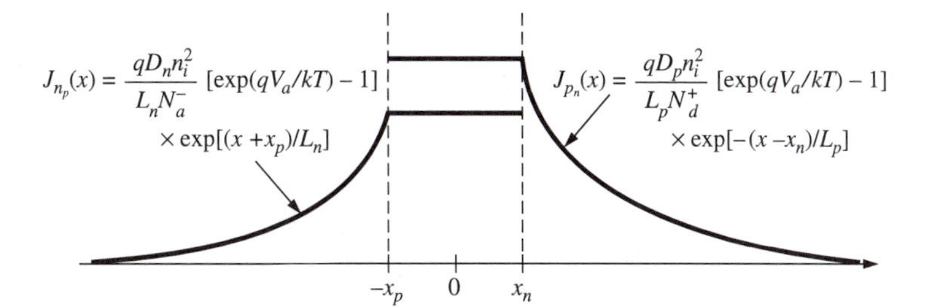

FIGURE 4.17 Variation of the hole and electron current densities in the neutral semiconductor outside the depletion region.

At steady state and for no external generation, $\partial n/\partial t = 0$ and $G_n = 0$. Then, the current density $J_n(x)$ represents the current density of electrons which recombine with the minority carrier holes plus the electron-diffusion current density, which is injected across the *p-n* junction to the *p*-side. The continuity equation in Eq. (4.117) becomes

$$\frac{dJ_n}{dx} = qR_n = qR_p, \tag{4.118}$$

because the electron and hole recombination rates are equal. Then,

$$\frac{dJ_n}{dx} = q\frac{\Delta p}{\tau_p} = q\frac{p_n(x) - p_{no}}{\tau_p}. \tag{4.119}$$

Integrating Eq. (4.119) with Eq. (4.95) for $p_n(x)$ gives

$$J_n(x) = \frac{q}{\tau_p}\int \{p_{no}[\exp(qV_a/kT) - 1]\exp[-(x - x_n)/L_p] + p_{no} - p_{no}\}dx, \tag{4.120}$$

which becomes

$$J_n(x) = \frac{qn_i^2}{\tau_p N_d^+}[\exp(qV_a/kT) - 1](-L_p)\exp[-(x - x_n)/L_p] + C. \tag{4.121}$$

By replacing the lifetime τ_p by the relationship $\tau_p = L_p^2/D_p$, the prefactor in Eq. (4.121) may be rewritten as $(qD_p n_i^2/L_p^2 N_d^+)$. The $(-L_p)$ in Eq. (4.121) is canceled by the L_p^2 term in the denominator of the prefactor to give

$$J_n(x) = -\frac{qD_p n_i^2}{L_p N_d^+}[\exp(qV_a/kT) - 1]\exp[-(x - x_n)/L_p] + C. \tag{4.122}$$

The constant of integration C may be evaluated from the boundary condition for $J_n(x_n) = J_{n_p}(-x_p)$, which was given in Eq. (4.111) for the electron current density at $x = -x_p$. Equation (4.122) may be written as

$$J_n(x_n) = J_{n_p}(-x_p) = \frac{qD_n n_i^2}{L_n N_a^-}[\exp(qV_a/kT) - 1]$$
$$= -\frac{qD_p n_i^2}{L_p N_d^+}[\exp(qV_a/kT) - 1] + C. \tag{4.123}$$

Solving for C gives

$$C = \left[\frac{qD_n n_i^2}{L_n N_a^-} + \frac{qD_p n_i^2}{L_p N_d^+}\right][\exp(qV_a/kT) - 1]. \tag{4.124}$$

With C in Eq. (4.123), the electron current density becomes

$$J_n(x) = \frac{qD_p n_i^2}{L_p N_d^+}[\exp(qV_a/kT) - 1]\{1 - \exp[-(x - x_n)/L_p]\}$$
$$+ \frac{qD_n n_i^2}{L_n N_a^-}[\exp(qV_a/kT) - 1]. \tag{4.125}$$

The first term in Eq. (4.125) represents the current density due to electron–hole recombination on the n-side for $x \geq x_n$ and goes to zero at $x = x_n$. The second term is the electron diffusion current density injected into the p-side and was given by Eq. (4.111).

A similar equation can be written for the hole current density on the p-side,

$$J_p(x) = \frac{qD_n n_i^2}{L_n N_a^-}[\exp(qV_a/kT) - 1]\{1 - \exp[(x + x_p)/L_n]\}$$
$$+ \frac{qD_p n_i^2}{L_p N_d^+}[\exp(qV_a/kT) - 1], \tag{4.126}$$

where the first term in Eq. (4.126) represents the hole current density due to electron–hole recombination on the p-side for $x \leq -x_p$ and goes to zero at $x = -x_p$. The second term is the hole current density injected into the n-side and was given by Eq. (4.109). The two current densities represented by Eqs. (4.125) and (4.126) are shown in Fig. 4.18. These current densities may be added to give the total current density at any x as $J_t = J_{p_n}(x) + J_n(x) = J_{n_p}(x) + J_p(x)$ and this sum is shown in Fig. 4.19 to demonstrate that the total current density is constant at any value of x. The quasi-Fermi levels E_{f_p} and E_{f_n} shown in Fig. 4.10(b) may be extended to the neutral semiconductor regions where the current density is related to the change in the quasi-Fermi levels as represented in Eqs. (4.74) and (4.75). The energy-band diagram illustrating the variation of the quasi-Fermi levels is shown in Fig. 4.20.

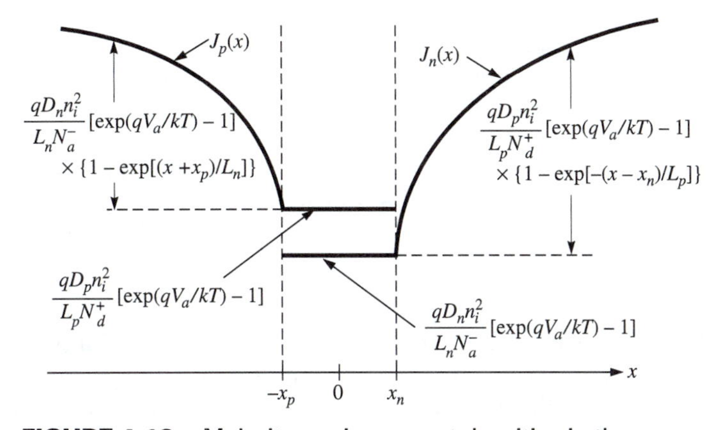

FIGURE 4.18 Majority carrier current densities in the neutral semiconductor outside the depletion region.

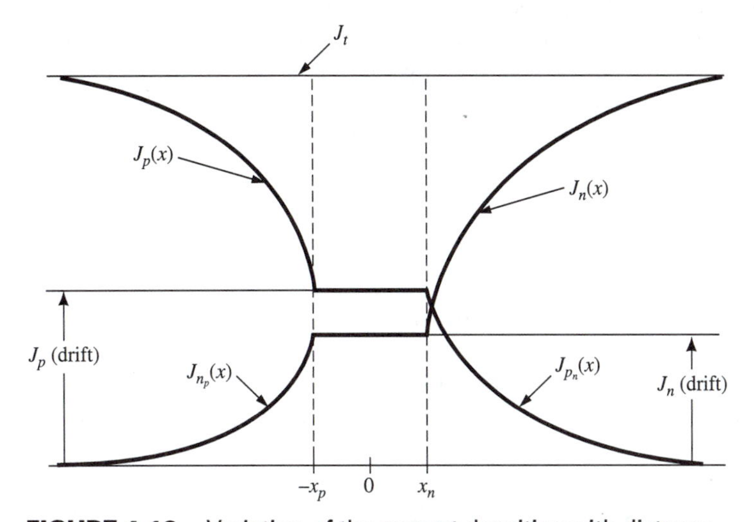

FIGURE 4.19 Variation of the current densities with distance for the *p-n* junction. On the *n*-side, the hole diffusion current density is $J_{p_n}(x)$ while the sum of the electron current density injected into the *p*-side and the electron current density due to recombination with holes is $J_n(x)$. On the *p* side, the electron diffusion current density is $J_{n_p}(x)$ while the sum of the hole current density injected into the *n*-side and the hole current density due to recombination with electrons is $J_p(x)$.

For the *I–V* behavior of *p-n* junctions, the representation given by Eqs. (4.115) and (4.116) is sufficient. The current density expressions for $x > x_n$ or $x < -x_p$ are required to understand the speed of response of *p-n* junctions and the behavior of bipolar transistors. Also, recombination in the depletion region has been neglected. In the Sec. 4.7.3, the space-charge recombination current density will be derived.

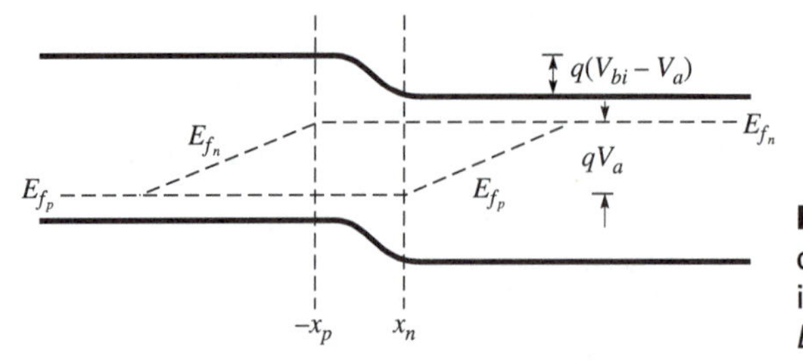

FIGURE 4.20 Energy-band diagram for forward bias to illustrate the quasi-Fermi levels E_{f_p} and E_{f_n}.

4.7.3 Diffusion Current for the Thin-layer Case

When the layer thickness (W_n on the n-side) is thin compared to the diffusion length (L_p on the n-side), the injected minority carrier distribution and the diffusion current density differ from the previous case for $L_p < W_n$. The thin-layer diode case of $L_p > W_n$ represents the minority carrier variation and the current density in the base of a bipolar transistor and is called the thin-base case.

For the p-n junction, the general solution of the continuity equation for minority carrier holes on the n-side was given by Eq. (4.90). If the depletion-layer thickness x_n is small compared to the n-side-layer thickness W_n, then x_n may be taken as zero ($x_n = 0$) and one boundary condition becomes

$$p_n(0) = p_{no} \exp(qV_a/kT) \qquad \text{at} \quad x = x_n = 0. \tag{4.127}$$

With this boundary condition, $p_n(x_n)$ becomes

$$p_n(x_n) = p_{no} \exp(qV_a/kT) = C_1 + C_2 + p_{no} \qquad \text{at} \quad x = x_n = 0. \tag{4.128}$$

For the other boundary condition taken as the case when the minority carrier concentration goes to zero at W_n, $p_n(W_n) = 0$ at $x = W_n$. This boundary condition gives

$$0 = C_1 \exp(-W_n/L_p) + C_2 \exp(W_n/L_p) + p_{no} \qquad \text{at} \quad x = W_n. \tag{4.129}$$

From this equation, C_1 may be obtained as

$$C_1 = -C_2 \exp(2W_n/L_p) - p_{no} \exp(W_n/L_p). \tag{4.130}$$

If it is assumed that $C_2 \exp(2W_n/L_p) \gg p_{no} \exp(W_n/L_p)$, the solution can be simplified and it will be found that C_2 varies as $p_{no} \exp(qV_a/kT)$, which is much larger than $p_{no} \exp(W_n/L_p)$ for $V_a > kT/q$ and $W_n/L_p < 0.1$. Therefore,

$$C_1 = -C_2 \exp(2W_n/L_p), \tag{4.131}$$

which can be substituted for C_1 in the equation for the boundary condition at $x = x_n = 0$:

$$p_{no} \exp(qV_a/kT) = -C_2 \exp(2W_n/L_p) + C_2 + p_{no}. \tag{4.132}$$

From this equation, C_2 may be obtained:

$$p_{no}\exp(qV_a/kT) = C_2[1 - \exp(2W_n/L_p)] + p_{no} \tag{4.133}$$

or

$$C_2 = -\frac{[p_{no}\exp(qV_a/kT) - p_{no}]\exp(-W_n/L_p)]}{[\exp(W_n/L_p) - \exp(-W_n/L_p)]}, \tag{4.134}$$

where $[1 - \exp(2W_n/L_p)]^{-1}$ has been written as $-\exp(-W_n/L_p)/[\exp(W_n/L_p) - \exp(-W_n/L_p)]$. Again, neglecting p_{no} for $V_a > kT/q$ in the expressions for C_2 and $p_n(x)$, $p_n(x)$ now becomes

$$p_n(x) = -C_2[\exp(2W_n/L_p)\exp(-x/L_p) - \exp(x/L_p)]. \tag{4.135}$$

With the expression for C_2,

$$p_n(x) = \left[\frac{p_{no}\exp(qV_a/kT)}{[\exp(W_n/L_p) - \exp(-W_n/L_p)]}\right] \times [\exp[(W_n - x)/L_p] - \exp[-(W_n - x)/L_p]. \tag{4.136}$$

For $x \ll L_p$ and $W_n \ll L_p$, the series expansion of the exponential terms such as $\exp(W_n/L_p)$ as $(1 + W_n/L_p)$ gives

$$p_n(x) = \frac{p_{no}\exp(qV_a/kT)}{[1 + W_n/L_p - (1 - W_n/L_p)]} \times \{1 + (W_n - x)/L_p - [1 - (W_n - x)/L_p]\} \tag{4.137}$$

or

$$\boxed{p_n(x) = p_{no}\exp(qV_a/kT)(1 - x/W_n)}. \tag{4.138}$$

The diffusion current density is obtained from $p_n(x)$ by $J_p = -qD_p\,dp/dx$ of Eq. (3.35) as

hole-diffusion current density for $L_p > W_n$ and $p_n(W_n) = 0$

$$\boxed{J_{p_n} = \frac{qD_p p_{no}}{W_n}\exp(qV_a/kT) = \frac{qD_p n_i^2}{W_n N_d^+}\exp(qV_a/kT)}, \tag{4.139}$$

which is the same as the diffusion current density given in Eq. (4.109) for $p_n(x) = p_{no}$ at $x \gg L_p$ and $V_a > kT/q$, but L_p in Eq. (4.109) is replaced by W_n and the -1 term is omitted. For p-n junction diodes in integrated circuits, layer thicknesses are very thin and can be expected to be less than the diffusion length.

At the contacts of the p-n junction diode, another boundary condition is to assume that the thermal equilibrium concentrations are maintained at all current densities: i.e., $p_n(W_n) = p_{no}$ and $n_p(-x_p) = n_{po}$. As shown by Eq. (4.140), a linear variation of the minority carrier concentration occurs when the diffusion length is larger than the layer thickness. Therefore, with $p_n(x)$ for $x_n = 0$ as $p_n(0) = p_{no}\exp(qV_a/kT)$ and $p_n(x) = p_{no}$ at $x = W_n$,

$$p_n(x) = p_n(0) - \left[\frac{p_n(0) - p_{no}}{W_n}\right]x. \tag{4.140}$$

The diffusion current density is obtained from $J_p = -qD_p\,dp/dx$ to give

hole-diffusion current density for $L_p > W_n$ and $p_n(W_n) = p_{no}$

$$\boxed{J_{p_n} = \frac{qD_p p_{no}}{W_n}[\exp(qV_a/kT) - 1] = \frac{qD_p n_i^2}{W_n N_a^-}[\exp(qV_a/kT) - 1]}, \tag{4.141}$$

which is similar to Eq. (4.139), but the boundary condition of $p_n(W_n) = p_{no}$ results in the -1 term as in Eq. (4.109). Similar expressions may be obtained for the diffusion current on the p-side, J_{n_p}.

4.7.4 Minority Carrier Mobility

Values for the minority carrier diffusivity are required for calculations and modeling of p-n junction diodes and bipolar transistors. Early measurements[7] suggested that the mobilities (diffusivity) of minority carriers at a given impurity concentration in n- and p-type Si at 300 K do not differ significantly from the values as majority carriers. Recent measurements[8] for minority carrier electrons in heavily doped p-type Si showed that at 296 K the mobility of electrons in p-type Si [μ_n(p-Si)] was approximately equal to the mobility of electrons in n-type Si [μ_n(n-Si)],

$$\mu_n(p\text{-Si}) = \mu_n(n\text{-Si}), \tag{4.142}$$

for $N_a^- \leq 5 \times 10^{18}$ cm^{-3}. For higher doping concentrations, the ratio of $\mu_n(p\text{-Si})/\mu_n(n\text{-Si}) \simeq 1$ to 1.27.

In GaAs for minority carrier holes in n-type GaAs,[9] μ_p(n-GaAs), was approximately equal to the hole mobility in p-type GaAs, μ_p(p-GaAs), up to electron concentrations of 1.8×10^{18} cm^{-3}. For concentrations greater than 1.8×10^{18}, $\mu_p(n\text{-GaAs})/\mu_p(p\text{-GaAs}) \simeq 1.3$.

These results for both Si and GaAs suggest that the best present choice is to use μ_n(minority carrier) $=$ μ_n(majority carrier) and μ_p(minority carrier) $=$ μ_p(majority carrier). At a given temperature, the majority or minority carrier mobility is determined by the total impurity concentration, $N_I = \Sigma N_a^- + \Sigma N_d^+$. The diffusivity is obtained from the Einstein relation.

EXAMPLE 4.4 The *n*-side of a *p-n* junction has an ionized donor concentration of $N_d^+ = 1.2 \times 10^{17}$ cm^{-3} and an ionized acceptor concentration of $N_a^- = 8.0 \times 10^{16}$ cm^{-3}. Find the minority carrier hole mobility and diffusivity.

Solution The total impurity concentration $N_I = 1.2 \times 10^{17} + 8.0 \times 10^{16} = 2.0 \times 10^{17}$ cm^{-3}. From Fig. 3.3, the hole mobility $\mu_p = 250$ cm^2/V-s at an impurity concentration of 2×10^{17} cm^{-3}, which gives a hole diffusivity $D_p = 0.026 \times 250 = 6.5$ cm^2/s. Note that the net electron concentration $n = 1.2 \times 10^{17} - 8.0 \times 10^{16} = 4.0 \times 10^{16}$ cm^{-3}. ■

4.7.5 Space-Charge Recombination Current

The derivation of the diffusion current in the previous part of this section did not consider recombination in the depletion region. This current is called space-charge recombination current for forward bias and its value can be calculated from the continuity equation. For determination of the space-charge recombination current (J_{scr}), the hole current density flowing from the *p*-side will be considered. This hole current flowing from the *p*-side will equal the electron current flowing from the *n*-side, because recombination requires equal numbers of recombining electrons and holes. The continuity equation for the hole current density $J_p(x)$ was given by Eq. (3.51) as

$$\frac{\partial p}{\partial t} = -\frac{1}{q}\frac{\partial J_p}{\partial x} + (G_p - R_p). \tag{4.143}$$

At steady state and no external generation, $\partial p/\partial t = 0$ and $G_p = 0$, and the continuity equation may be written as

$$-\frac{dJ_p}{dx} = qR_p. \tag{4.144}$$

Integration of Eq. (4.144) across the depletion region gives

$$-\int_{J_p(-x_p)}^{J_p(x_n)} dJ_p(x) = -[J_p(x_n) - J_p(-x_p)] = q\int_{-x_p}^{x_n} R_p\,dx, \tag{4.145}$$

where the hole recombination current density $J_p(x_n)$ is zero at x_n because $J_{scr} = J_p + J_n$ and $J_{scr}(x_n) = J_n$, and therefore $J_p(x_n) = 0$. At x_n, the space-charge recombination current density is due to electrons and is equal to the hole recombination current density on the *p*-side. With $J_p(-x_p) = J_{scr}$, the equation for the space-charge recombination current density becomes[10]

$$J_{scr} = q\int_{-x_p}^{x_n} R_p\,dx. \tag{4.146}$$

The recombination rate in the integral of Eq. (4.146) will vary with position in the depletion region, and therefore the integral is difficult to evaluate in closed-form. The usual assumption to obtain a closed-form solution of the integral is to use the maximum recombination rate. The recombination rate was given in Eq. (3.62) as

$$R_p = \frac{np - n_i^2}{\tau_p(p + n + n_1 + p_1)}. \tag{4.147}$$

The numerator of Eq. (4.147) can be rewritten with $np = n_i^2 \exp(qV_a/kT)$ within the depletion region. This relationship for $n(x)p(x)$ within the depletion region was given at the end of Example 4.3. Also, n in the denominator may be taken as $n_i^2 \exp(qV_a/kT)/p$, which gives

$$R_p = \frac{n_i^2[\exp(qV_a/kT) - 1]}{\tau_p[p + \dfrac{n_i^2 \exp(qV_a/kT)}{p} + n_1 + p_1]}. \tag{4.148}$$

To maximize R_p with respect to p, the derivative of R_p in Eq. (4.148) is taken with respect to p and set equal to zero:

$$\frac{dR_p}{dp} = 0 = R_p^2\left[1 - \frac{n_i^2 \exp(qV_a/kT)}{p^2}\right]. \tag{4.149}$$

Since $R_p \neq 0$, the term in the brackets must be zero:

$$1 - \frac{n_i^2 \exp(qV_a/kT)}{p^2} = 0, \tag{4.150}$$

which gives

$$p = n_i \exp(qV_a/2kT) \tag{4.151}$$

for maximum R_p. Also,

$$n = \frac{n_i^2 \exp(qV_a/kT)}{p} = \frac{n_i^2 \exp(qV_a/kT)}{n_i \exp(qV_a/2kT)} = n_i \exp(qV_a/2kT). \tag{4.152}$$

The maximum recombination rate with these values for n and p becomes

$$R_p(\max) = \frac{n_i^2[\exp(qV_a/kT) - 1]}{\tau_p[2n_i \exp(qV_a/2kT) + n_1 + p_1]}. \tag{4.153}$$

The -1 in the numerator and $n_1 + p_1$ in the denominator may be neglected when $V_a > 3kT/q$, and the maximum recombination rate becomes

$$R_p(\max) = \frac{n_i}{2\tau_p} \exp(qV_a/2kT). \tag{4.154}$$

Substitution of $R_p(\max)$ into Eq. (4.146) gives

$$J_{scr} = q\frac{n_i \exp(qV_a/2kT)}{2\tau_p} \int_{-x_p}^{x_n} dx, \tag{4.155}$$

which is

$$J_{scr} = \frac{qn_i \exp(qV_a/2kT)}{2\tau_p}[x_n - (-x_p)], \tag{4.156}$$

or with $x_d = x_n + x_p$, the space-charge recombination current density becomes

space-charge recombination current density

$$\boxed{J_{scr} = \frac{qn_i x_d}{2\tau_p} \exp(qV_a/2kT)}. \tag{4.157}$$

The prefactor is called the saturation space-charge current density and is

space-charge recombination saturation current density

$$\boxed{J_{S_{scr}} = \frac{qn_i x_d}{2\tau_p}}. \tag{4.158}$$

It should be noted that the diffusion current density given in Eq. (4.114) varies as $\exp(qV_a/kT)$ and the space-charge recombination current density given in Eq. (4.157) varies as $\exp(qV_a/2kT)$. The space-charge recombination current I_{scr} is obtained as the product of current density J_{scr} and the cross-sectional area A:

$$I_{scr} = J_{scr}A = I_{S_{scr}} \exp(qV_a/2kT), \tag{4.159}$$

where the saturation space-charge recombination current $I_{S_{scr}} = J_{S_{scr}}A$.

Examples of measured forward *I–V* characteristics are illustrated in Fig. 4.21. This figure helps to emphasize that the current for a *p-n* junction varies as $\exp(qV_a/\text{n}kT)$ with n designated as the *ideality factor*. For diffusion current, n = 1, and for space-charge recombination current, n = 2. In voltage ranges where n has values between n = 1 and n = 2, the current is a combination of diffusion and space-charge recombination current. The saturation current densities J_d and J_{scr} also influence which current mechanisms dominate; however, J_{scr} must be greater than J_d to observe I_{scr} with the n = 2 behavior. The n = 2 region for Si is due to space-charge recombination current, while the n = 2 region for GaAs is generally due to surface-recombination current, which is presented in the next part of this section.

ideality factor n

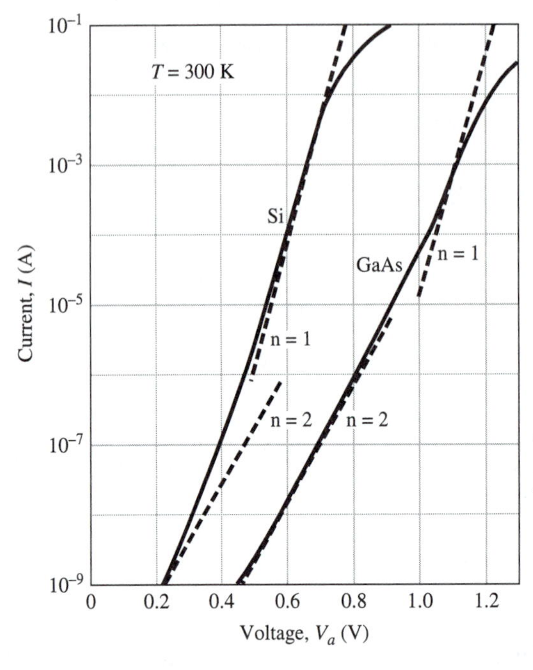

FIGURE 4.21 Comparison of the forward current-voltage characteristics of Si and GaAs *p-n* junction diodes at room temperature. The dashed lines indicate the regions where *I* varies as $\exp(qV_a/nkT)$ and diffusion current dominates for n = 1 and space-charge recombination current dominates for n = 2.[11]

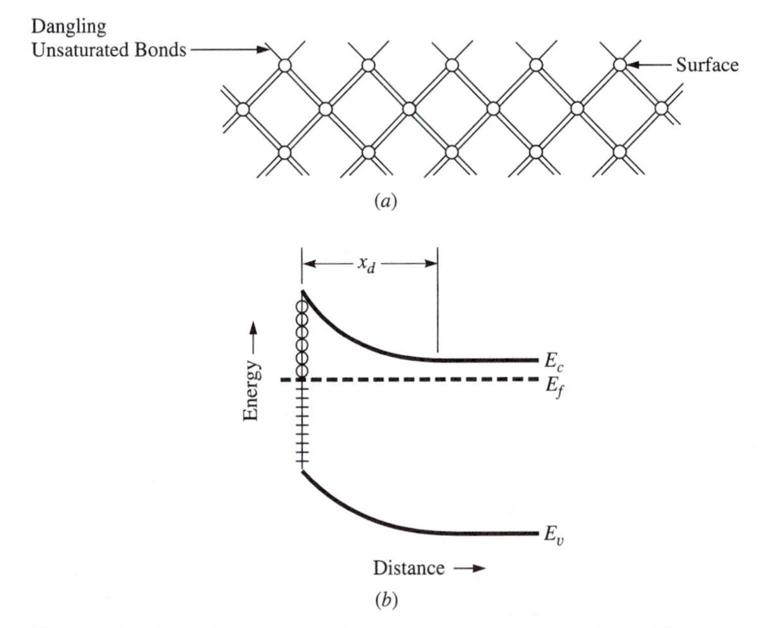

FIGURE 4.22 (*a*) Schematic bond representation for a Si surface. (*b*) Energy-band diagram for a Si surface illustrating neutral empty acceptor surface states above the Fermi level and negative occupied acceptor surface states below the surface Fermi level.

4.7.6 Surface-recombination Current

A clean [100] Si surface is represented in Fig. 4.22(a) to illustrate the surface dangling bonds.[12] Electrons can occupy the free orbitals so that surface atoms may become negatively charged. As for the impurities considered in Sec. 2.8.1, an impurity which can accept an extra electron and become negatively charged is termed an acceptor. These acceptor surface states are negative when occupied and neutral when empty. Therefore, these acceptor-like surface states will be negatively charged below the Fermi level (have an electron) and will be neutral (be empty) above the Fermi level (see Sec. 2.6). These surface states are allowed states in the energy gap at the surface, but are not present in the semiconductor bulk. The existence of these states was proposed by Tamm[13] and Shockley[14] and it was suggested that surface states should occur in the energy gap and their number should approximately equal the number of surface atoms [$N = (5 \times 10^{22}\ \text{cm}^{-3})^{2/3} \simeq 1 \times 10^{15}\ \text{cm}^{-2}$]. The energy-band diagram for a Si surface is shown in Fig. 4.22(b). In this representation it has been assumed that the charge in the depletion layer at the surface, $qN_d^+ x_d$, is equal and opposite to the charge per unit area on that surface. For device technology, an atomically clean Si surface would not be desirable because of surface states. It is more desirable to produce properties compatible with the device bulk properties.

For Si, the growth of a thermal silicon dioxide (SiO_2) surface layer greatly reduces the number of surface states, and surface-recombination current is generally not an issue for Si *p-n* junctions or bipolar transistors. However, for many III-V compound semiconductors such as GaAs, surface states lead to significant surface-recombination current for *p-n* junctions and bipolar transistors. (Surface recombination was introduced in Sec. 3.7.3.) A model for surface-recombination current has been developed.[15] In this model, a nearly constant ratio of n_s and p_s (where the subscript denotes the carrier concentration at the surface) exists. The approximately constant n_s/p_s ratio leads to a current variation with voltage as $\exp(qV_a/2kT)$, which is the same as for space-charge recombination current. That analysis led to an expression for the hole surface-recombination current I_{sr} on the n-side,

$$I_{sr} = q\pi d s_o L_{ps} n_i \exp(qV_a/2kT), \tag{4.160}$$

where s_o represents the surface recombination velocity defined in Eq. (3.91), πd represents the *p-n* junction perimeter for a circular *p-n* junction with diameter d, and L_{ps} represents the surface diffusion length for holes. This equation emphasizes that surface-recombination current varies with voltage as $\exp(qV_a/2kT)$, which is the same as the voltage dependence for space-charge recombination current. However, the space-charge recombination current density given in Eq. (4.157) is multiplied by the area of $\pi d^2/4$ for a circular *p-n* junction to obtain the current. Therefore, surface-recombination current varies linearly with d, while space-charge recombination current varies as d^2. This difference in the dependence on d may be used to determine if current which varies as $\exp(qV_a/2kT)$ is surface or bulk space-charge recombination current.

Growth of native oxides or dielectrics such as SiO_2 on GaAs has not resulted in reduced surface-recombination current. However, the growth of a high resistivity layer of $Al_xGa_{1-x}As$ on GaAs does reduce the surface-recombination current.[16] In Fig. 4.23(a),

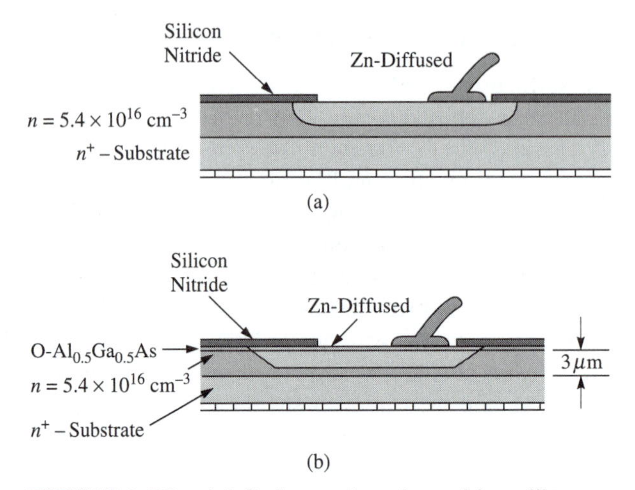

FIGURE 4.23 (*a*) GaAs *p-n* junction with a silicon nitride surface layer to serve as a diffusion mask to form the Zn diffused circular cross section. (*b*) GaAs *p-n* junction with a high resistivity $Al_xGa_{1-x}As$ layer on the GaAs to reduce the surface-recombination current.[16]

the cross section of a GaAs *p-n* junction formed by diffusion of the acceptor zinc through the circular opening in the silicon nitride layer is illustrated. A similar structure with a thin layer of $Al_xGa_{1-x}As$ between the GaAs and silicon nitride is illustrated in Fig. 4.23(*b*). The *I–V* behavior was measured for both structures for a series of devices with different diameters. It was found that for the silicon nitride surface case, the current varies linearly with the diameter as expected for surface-recombination current, while for the case of $Al_xGa_{1-x}As$ between the GaAs and silicon nitride, the current varies as the square of the diameter as expected for bulk space-charge recombination current. For semiconductors related to InP, the surface-recombination current is often not as severe a problem as for GaAs.

4.7.7 High Current Effects

At high current levels beyond the $\mathrm{n} = 1$ regions shown in Fig. 4.21, the current increases more slowly with forward voltage. This behavior is due to diode series resistance R_s and high-injection-level effects. The series resistance is due to the resistance of the bulk semiconductor outside the depletion regions and the resistance of the metal contact. This contact resistance is related to the formation of ohmic contacts and is introduced in Sec. 6.6. At current levels represented by the $\mathrm{n} = 2$ and $\mathrm{n} = 1$ regions in Fig. 4.21, the resistive voltage drop IR_s is usually small enough to be neglected. However, at 100 mA, an R_s of 0.5 ohm gives a voltage drop of 0.05 V. Then the voltage V across the junction is less than the external applied voltage V_a by IR_s, $V = V_a - IR_s$, and

$$I = J_{s_d} A \exp[q(V_a - IR_s)/kT] = I_s \exp(qV_a/kT)\exp(-qIR_s/kT), \quad (4.161)$$

where A is the cross-sectional area. Equation (4.161) illustrates the reduction of the ideal diffusion current by the series resistance.

At high current densities which are large enough for the injected *minority* carrier concentration to become comparable to the *majority* carrier concentration,

$$p_n(x_n) \simeq n_{no} \text{ or } n_p(-x_p) \simeq p_{po}. \tag{4.162}$$

At $x = x_n$, from Eq. (4.86),

$$n_n(x_n)p_n(x_n) = n_i^2 \exp(qV_a/kT) \simeq p_n^2(x_n) \qquad \text{for} \qquad p_n(x_n) \simeq n_n(x_n), \tag{4.163}$$

then,

$$p_n(x_n) \simeq n_i \exp(qV_a/2kT), \tag{4.164}$$

and the current begins to vary approximately as $\exp(qV_a/2kT)$.

4.7.8 Reverse Bias

Next, consider the reverse bias of $-V_a$. The hole concentration $p_n(x)$ is given by Eq. (4.95) and becomes

$$p_n(x) = p_{no}[\exp(-qV_a/kT) - 1]\exp[-(x - x_n)/L_p] + p_{no} \tag{4.165}$$

or

$$p_n(x) = p_{no}\{1 - \exp[-(x - x_n)/L_p]\}, \tag{4.166}$$

where the $\exp(-qV_a/kT)$ term goes to zero for the reverse bias in Eq. (4.166). From Eq. (4.96) for $n_p(x)$ at reverse bias $-V_a$,

for reverse bias $(-V_a)$, $p_n(x_n) = 0$ *and* $n_p(-x_p) = 0$

$$n_p(x) = n_{po}\{1 - \exp[(x + x_p)/L_n]\}. \tag{4.167}$$

Note that for $p_n(x)$ at $x = x_n$, $p_n(x_n) = 0$, and for $n_p(x)$ at $n_p(-x_p) = 0$. *These results demonstrate that for reverse bias* $(-\mathrm{V_a})$ *the minority carrier concentrations go to zero at the edges of the depletion region,* $-\mathrm{x_p}$ *and* $\mathrm{x_n}$. The negative electric field sweeps minority carriers across the depletion region. Equations (4.166) and (4.167) are sketched in Fig. 4.24.

The diffusion current on the n-side at $x = x_n$ is given by Eq. (4.106) with Eq. (4.165) for $p_n(x)$ as

$$\begin{aligned} J_{p_n}(x_n) &= -qD_p p_{no}\,[\exp(-qV_a/kT) - 1]\frac{d}{dx}\exp[-(x - x_n)/L_p]\Big|_{x=x_n} \\ &= -q\frac{D_p p_{no}}{L_p} = -q\frac{D_p n_i^2}{L_p N_d^+}. \end{aligned} \tag{4.168}$$

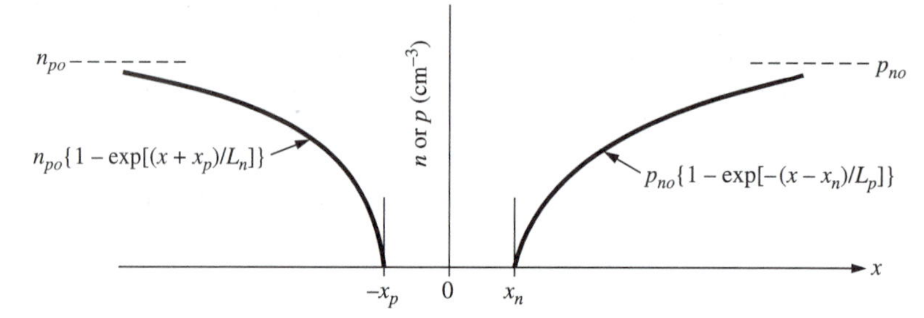

FIGURE 4.24 Variations of the minority carrier concentrations for reverse bias.

A similar expression may be obtained for $J_{n_p}(-x_p)$. At reverse bias for $-V_a$, the expression for diffusion current given in Eq. (4.114) when multiplied by the cross-sectional area A becomes

reverse-bias diffusion current

$$\boxed{I(-V_a) = -qA[D_p n_i^2/(L_p N_d^+) + D_n n_i^2/(L_n N_a^-)] = -I_{s_d}}, \qquad (4.169)$$

where the $\exp(-qV_a/kT)$ has been taken as zero. The expression given in Eq. (4.168), when multiplied by A, gives the hole current part of Eq. (4.169). These results emphasize that n_p and p_n go to zero at the edges of the depletion regions and Eq. (4.114) represents the diffusion current for both forward and reverse bias.

As mentioned earlier, reverse bias reduces the concentrations of carriers in the depletion region below the thermal equilibrium values because the applied reverse voltage sweeps the carriers out of the depletion region. The expression for the net recombination rate given in Eq. (3.62) for n and p less than n_i and with $n_1 = p_1 = n_i$, Eq. (3.62) becomes

$$U = -n_i/2\tau_o, \qquad (4.170)$$

where the lifetime τ_{nr} has been taken as the effective lifetime within the reverse-biased depletion region τ_o. The negative sign in Eq. (4.170) represents net *generation* rather than recombination. Equation (4.146) now becomes a generation current density

$$J_{scg} = -q\int_{-x_p}^{x_n} n_i/2\tau_o \, dx, \qquad (4.171)$$

which gives the space-charge generation current density for reverse bias as

reverse-bias space-charge generation current

$$\boxed{I_{scg} = -qn_i x_d A/2\tau_o}, \qquad (4.172)$$

where x_d is the voltage-dependent depletion width and A is the cross-sectional area.

4.7.9 Temperature Effects

Both the diffusion and recombination–generation currents have a strong temperature dependence. For forward bias, the ratio of the hole-diffusion current density from Eq. (4.109) to the recombination current density from Eq. (4.157) is

$$\begin{aligned}\frac{I_{p_n}}{I_{scr}} &= \frac{D_p n_i^2 / L_p N_d^+}{n_i x_d / 2\tau_p} \exp(qV_a/2kT) \\ &= \frac{2L_p n_i}{N_d^+ x_d} \exp(qV_a/2kT), \end{aligned} \tag{4.173}$$

with the -1 neglected in the $[\exp(qV_a/kT) - 1]$ term for the diffusion current, and $L_p^2 = D_p \tau_p$ used to replace $D_p \tau_p$. It should be noted that n_i varies as $\exp(-E_g/2kT)$ so that Eq. (4.173) may be written as

$$\frac{I_{p_n}}{I_{scr}} \sim \exp[-(E_g - qV_a)/2kT]. \tag{4.174}$$

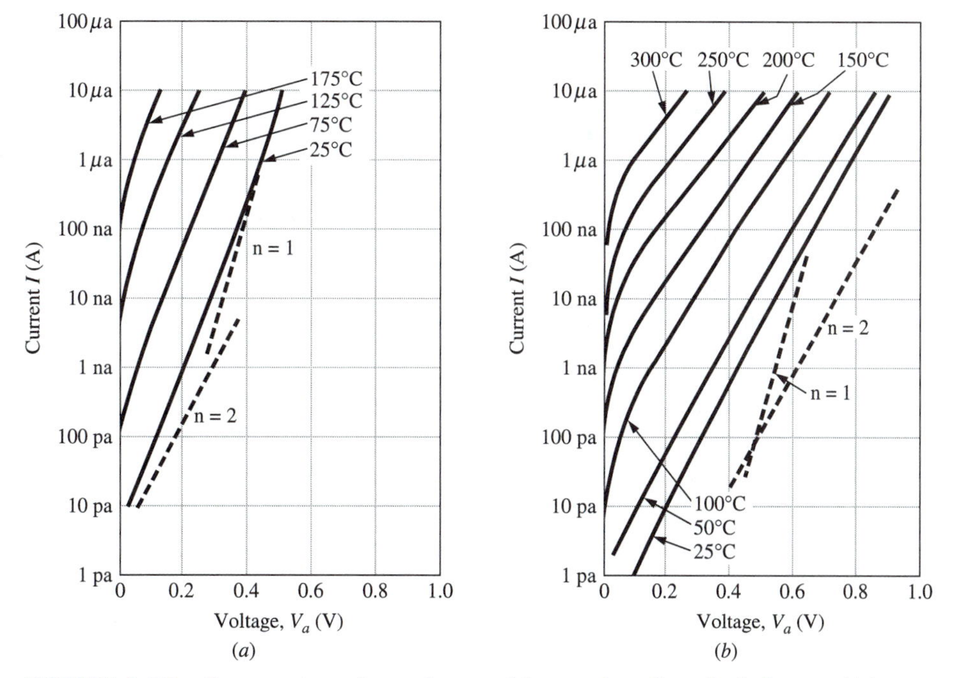

FIGURE 4.25 Temperature dependence of the *p-n* junction diode forward bias current. (*a*) Si. (*b*) GaAs.[17]

Equation (4.174) depends on both the temperature dependence of the energy gap and the exp$(1/kT)$ temperature dependence. As shown in Fig. 4.21 for room temperature, the recombination current dominates for small voltages, and then the diffusion current dominates for higher voltages. For a given forward voltage, the diffusion current will increase faster than the recombination current. The temperature dependence for forward bias is shown in Fig. 4.25(*a*) for Si and in Fig. 4.25(*b*) for GaAs.

For reverse bias, the voltage term is not present and the ratio becomes

$$\frac{I_{p_n}}{I_{scg}} = \frac{2L_p n_i}{N_d^+ x_d} \sim \exp(-E_g/2kT). \tag{4.175}$$

This ratio is proportional to n_i, and shows that the diffusion current should tend to dominate as the temperature is increased. Silicon with a smaller energy gap than GaAs will have a smaller ratio of diffusion current to space-charge generation current than GaAs. The reverse current for a Si *p-n* junction is shown in Fig. 4.26(*a*), and for a GaAs *p-n* junction diode is shown in Fig. 4.26(*b*). The rapid increase in current near 90 V for the Si diode and near 20 V for the GaAs diode is due to avalanche breakdown, which is covered in the next chapter.

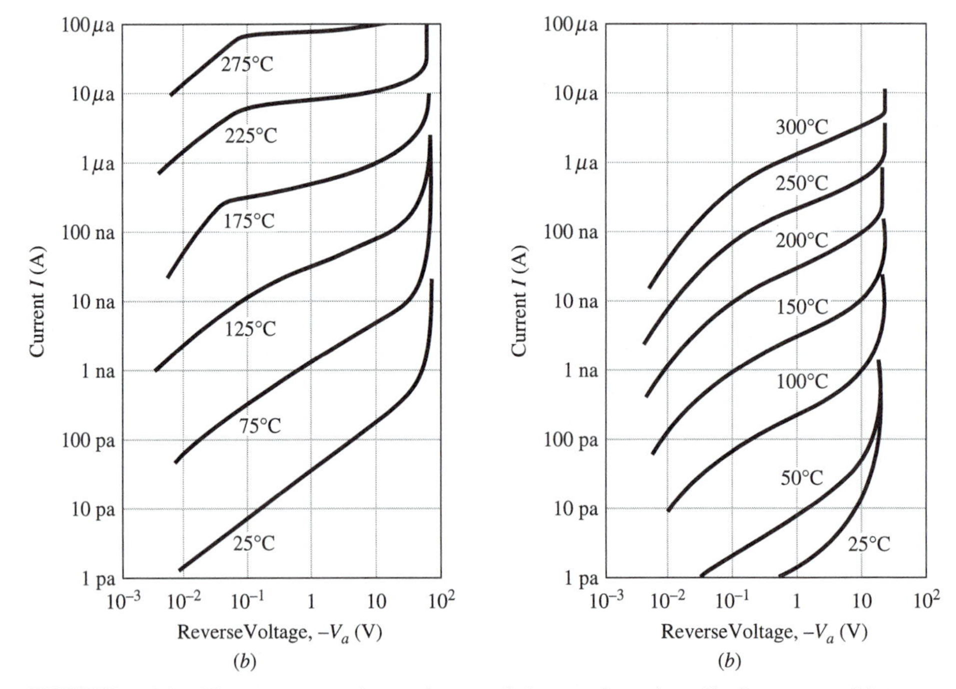

FIGURE 4.26 Temperature dependence of the *p-n* junction diode reverse bias current. (*a*) Si. (*b*) GaAs.[18]

4.8 SUMMARY AND USEFUL EXPRESSIONS

The following significant concepts were introduced in this chapter.

- Electron affinity qX is the difference in energy between the vacuum level (the energy of an electron at rest outside the semiconductor) and the conduction band.
- Work function $q\Phi$ is the energy difference between the vacuum level and the Fermi level.
- For the *p-n* junction, the space charge, electric field, and potential are related through Gauss' law.
- In the depletion approximation, it is assumed that all mobile carriers in the space-charge region may be neglected.
- The depletion-region width varies inversely with the square root of the impurity concentration and is approximately 1 μm at an impurity concentration of 1×10^{15} cm^{-3}.
- The potential given previously as $\psi(x) = -E_i(x)/q$ was redefined by the addition of a constant potential E_f/q to give $\psi(x) = (1/q)[E_f - E_i(x)]$.
- The relationship between potential and carrier concentrations shows that the minority carrier concentrations vary exponentially with potential.
- In what can be termed the law of the junction, the application of a forward bias V_a to a *p-n* junction increases the thermal equilibrium minority carrier concentrations n_{po} and p_{no} at the edges of the depletion regions as $n_{po}\exp(qV_a/kT)$ and $p_{no}\exp(qV_a/kT)$.
- The diffusion current I_d at the edges of the depletion region varies as $\exp(qV_a/kT)$: $I_d = I_s[\exp(qV_a/kT) - 1]$.
- Space-charge and surface-recombination current were found to vary as $\exp(qV_a/2kT)$.

The following parameters and expressions summarize the *p-n* junction *I–V* behavior.

Electron Affinity

$qX_{\text{Si}} = qX_{\text{GaAs}} = 4.05$ eV

Dielectric Constants

$\epsilon_{\text{Si}} = 11.7\epsilon_o$

$\epsilon_{\text{GaAs}} = 13.1\epsilon_o$

$\epsilon_o = 8.85 \times 10^{14}$ F/cm

Built-in Potential

$qV_{bi} = q\Phi_p - q\Phi_n$

$V_{bi} = (kT/q)\ln(N_a^- N_d^+ / n_i^2)$

$(V_{bi}^p - V_a^p) = (V_{bi} - V_a)/(1 + N_a^-/N_d^+)$

$(V_{bi}^n - V_a^n) = (V_{bi} - V_a)/(1 + N_d^+/N_a^-)$

Electrostatic Potential

$\mathscr{E} = -(d\psi/dx)$

$q\psi = E_f - E_i$

$\psi_2 - \psi_1 = (kT/q)\ln(n_2/n_1)$

$\psi_2 - \psi_1 = -(kT/q)\ln(p_2/p_1)$

Carrier Concentration

$n = n_i \exp(q\psi/kT)$

$p = n_i \exp(-q\psi/kT)$

Gauss's Law

$$\frac{d\mathscr{E}(x)}{dx} = \frac{\rho}{\epsilon}$$

Poisson's Equation

$$\frac{d^2\psi(x)}{dx} = \frac{q}{\epsilon}(N_a^- - N_d^+ + n - p)$$

Depletion Width

$$x_d = [(2\epsilon/q)(1/N_d^+ + 1/N_a^-)(V_{bi} - V_a)]^{1/2}$$
$$x_p = \sqrt{2\epsilon(V_{bi}^p - V_a^p)/qN_a^-}$$
$$x_n = \sqrt{2\epsilon(V_{bi}^n - V_a^n)/qN_d^+}$$

Nonequilibrium Minority Carrier Concentration

$$p_n(x_n) = p_{no}\exp(qV_a/kT) \qquad n_{no}(x_n)p_n(x_n) = n_i^2\exp(qV_a/kT)$$
$$p_n(x) = p_{no}[\exp(qV_a/kT) - 1] \times \exp[-(x - x_n)/L_p] + p_{no}$$
$$n_p(x) = n_{po}[\exp(qV_a/kT) - 1] \times \exp[(x + x_p)/L_n] + n_{po}$$

Diffusion Current Density

$$J_p(x_n) = (qD_p n_i^2/L_p N_d^+)[\exp(qV_a/kT) - 1]$$
$$J_n(-x_p) = (qD_n n_i^2/L_n N_a^-)[\exp(qV_a/kT) - 1]$$

Space-Charge Recombination Current Density

$$J_{scr} = (qn_i x_d/2\tau_p)\exp(qV_a/2kT)$$

REFERENCES

1. J. L. Moll, "The Evolution of the Theory for the Voltage-Current Characteristics of *P-N* Junctions," Proc. IRE **46,** 1076 (1958).
2. W. Shockley, "The Theory of *p-n* Junctions in Semiconductors and *p-n* Junction Transistors," Bell Syst. Tech. J. **28,** 435 (1949).
3. S. M. Sze, *Semiconductor Devices: Physics and Technology* (Wiley, New York, 1985), p. 71.
4. S. M. Sze, Ed., *VLSI Technology, 2nd ed.* (McGraw-Hill, New York, 1988), p. 272.
5. Sze, Ed., *VLSI Technology,* p. 327.
6. P. Bhattacharya, *Semiconductor Optoelectronic Devices, 2nd ed.* (Prentice-Hall, Upper Saddle River, N.J., 1997), p. 38.
7. J. Dziewior and D. Solber, "Minority-Carrier Diffusion Coefficients in Highly Doped Silicon," Appl. Phys. Lett. **35,** 170 (1979).
8. I.-Yun Leu and A. Neugroschel, "Minority-Carrier Transport Parameters in Heavily Doped *p*-type Silicon at 296 and 77 K," IEEE Trans. Electron Devices **ED-40,** 1872 (1993).
9. M. L. Lovejoy, M. R. Melloch, and M. S. Lundstrom, "Minority Hole Mobility in n^+ GaAs," Appl. Phys. Lett. **61,** 2683 (1992).
10. C. T. Sah, R. N. Noyce, and W. Shockley, "Carrier Generation and Recombination in P-N Junctions and P-N Junction Characteristics," Proc. IRE **45,** 1228 (1957).
11. A. S. Grove, *Physics and Technology of Semiconductor Devices* (Wiley, New York, 1967), p. 190.
12. M. M. Atalla, E. Tannenbaum, and E. J. Scheibner, "Stabilization of Silicon Surfaces by Thermally Grown Oxides," Bell Syst. Tech. J. **38,** 749 (1959).
13. I. Tamm, Phys. Z. Sowjetunion **1,** 733 (1932).
14. W. Shockley, "On the Surface States Associated with a Periodic Potential," Phys. Rev. **56,** 317 (1939).

15. C. H. Henry, "The Effect of Surface Recombination on Current in $Al_xGa_{1-x}As$ Heterojunctions," J. Appl. Phys. **49,** 3530 (1978).

16. H. C. Casey, Jr., A. Y. Cho, and P. W. Foy, "Reduction of Surface Recombination Current in GaAs *p-n* Junctions," Appl. Phys. Lett. **34,** 594 (1979).

17. Grove, *Physics and Technology,* p. 188.

18. Grove, *Physics and Technology,* p. 178.

PROBLEMS

All problems are for room temperature unless another temperature is specified.

4.1 **(a)** Find the work function for *p*-type Si at room temperature with $p = N_a^- = 1 \times 10^{17}$ cm^{-3}.
(b) Find the work function for *p*-type GaAs at room temperature with $p = N_a^- = 1\times10^{17}$ cm^{-3}.

4.2 **(a)** Find the work function for *n*-type Si at room temperature with $n = N_d^+ = 1 \times 10^{17}$ cm^{-3}.
(b) Find the work function for *n*-type GaAs at room temperature with $n = N_d^+ = 1\times10^{17}$ cm^{-3}.

4.3 For a Si *p-n* junction at room temperature, the work function on the *p*-side is 4.90 eV and on the *n*-side the work function is 4.20 eV.

(a) What is the built-in potential?

(b) What is $E_f - E_v$ on the *p*-side and $E_c - E_f$ on the *n*-side?

(c) What is the hole concentration in the neutral semiconductor on the *p*-side and the electron concentration in the neutral *n*-side?

4.4 At one location in a Si *p-n* junction at room temperature, the electron concentration is 2×10^{16} cm^{-3} and at another location the electron concentration is 4×10^7 cm^{-3}. What is the potential difference between these two locations?

4.5 For a Si *p-n* junction at 300 K, the properties on the *n*-side are $N_d^+ = 1 \times 10^{18}$ cm^{-3} and $L_p = 1.5 \times 10^{-4}$ cm, and on the *p*-side $N_a^- = 2 \times 10^{16}$ cm^{-3} and $L_n = 10 \times 10^{-4}$ cm.

(a) Find the total depletion width at thermal equilibrium.

(b) Find the total depletion width at a forward bias of $V_a = 0.4$ V.

(c) Find the ratio of the electron current density injected into the *p*-side to the hole current density injected into the *n*-side at a forward bias of $V_a = 0.4$ V.

4.6 For an abrupt Si *p-n* junction at room temperature and thermal equilibrium, at a point designated x_1 within the depletion region on the *n*-side, the hole concentration is 2.1×10^{12} cm^{-3} and the electric field is -3.0×10^3 V/cm.

(a) Find the hole concentration gradient dp/dx at x_1 and give the correct sign and units.

(b) Find the electron concentration and the electron concentration gradient at x_1 and give the correct sign and units.

4.7 For a *p-n* junction, the built-in potential on the *p*-side is 0.338 V and on the *n*-side is 0.471 V. What is the ratio of the ionized donor concentration to the ionized acceptor concentration?

4.8 At room temperature, consider Si doped so that at $x = x_2$ the electron concentration is 2.6×10^{17} cm^{-3}, and at $x = x_1$ the electron concentration is 6.3×10^{15} cm^{-3}. What is the potential difference between x_2 and x_1?

4.9 For a p-n junction with $p = N_a^- = 2\times10^{16}$ cm^{-3} and $n = N_d^+ = 1\times10^{16}$ cm^{-3}, $V_{bi} = 0.727$ V, $x_p = 0.125$ μm, and $x_n = 0.251$ μm at thermal equilibrium.

(a) Sketch $p(x)$ and $n(x)$ from $-2x_p < x < 2x_n$ with a linear scale for $p(x)$, $n(x)$, and x. (Use $x = -2x_p, -x_p, -x_p/2, 0, x_n/2, x_n$, and $2x_n$.)

(b) Repeat part (a), except use a log scale for $p(x)$ and $n(x)$.

(c) Does $n(x) = p(x) = n_i(x)$ on the p- or n-side of the junction interface at $x = 0$? [See your sketch in part (b).]

(d) Find $p_n(x_n)$ and $n_p(-x_p)$ for $V_a = +0.5$ V, and plot on the sketch in part (b).

(e) With $L_p = 4$ μm, sketch $p_n(x)$ for $V_a = +0.5$ V between $x_n < x < 10$ μm. Use a linear scale for $p(x)$ and x.

(f) Repeat part (e) with $V_a = -0.5$ V.

4.10 For the p-n junction of Problem 4.9, find the ratio of hole-diffusion current on the n-side to the electron-diffusion current on the p-side for $V_a = +0.5$ V and $L_n = L_p = 4.0$ μm. Assume the mobility of a carrier as a minority carrier is the same as when a majority carrier. (The mobility for a minority hole is obtained at the *donor* concentration.)

4.11 Consider a Si p^+-n junction at room temperature. The cross-sectional area is 1.0×10^{-4} cm^2, on the p^+-side the ionized acceptor concentration is $p = N_a^- = 1\times10^{20}$ cm^{-3}, and on the n-side the ionized donor concentration is $n = N_d^+ = 3\times10^{16}$ cm^{-3}. The measured I–V characteristic is shown in the figure.

(a) From the region dominated by space-charge recombination, find within a factor of 2 the minority carrier lifetime τ_p.

(b) From the diffusion current region, find within a factor of 2 the minority carrier diffusion length. Note that n in the expression, $I = I_o \exp(qV_a/nkT)$ may be found from

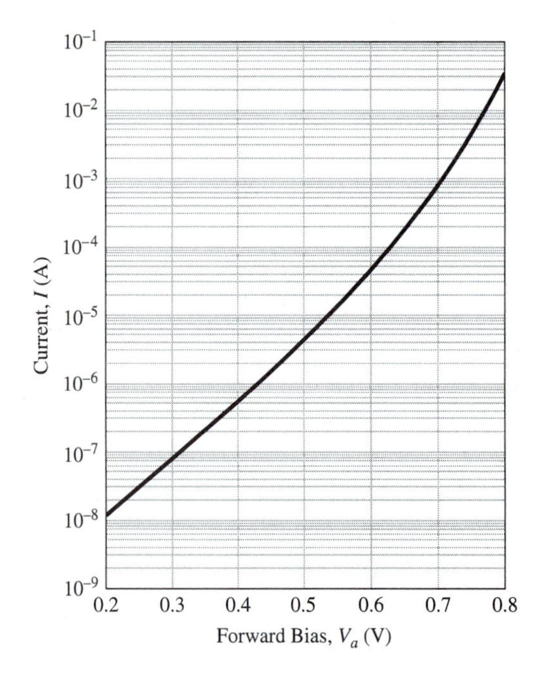

$$n = \frac{q}{kT}\frac{\Delta V}{\Delta \ln I}$$

in order to determine the recombination and diffusion current regions.

4.12 **(a)** Show that the maximum electric field for a p-n^+ junction, where $N_a^- \ll N_d^+$ so that $x_d = x_p$, is given by $\mathscr{E}_{max} = -[2qN_a^-(V_{bi} - V_a)/\epsilon]^{1/2}$.

(b) Evaluate $\mathscr{E}_{max}$ for Si at room temperature with $N_a^- = 2.5 \times 10^{16}$ cm^{-3} and $V_a = -40$ V.

4.13 Use the expression for the total hole current density J_p, which is zero at thermal equilibrium, to show that

$$\mathscr{E}(x) = \frac{1}{q}\frac{dE_v(x)}{dx}.$$

It is helpful to express the hole concentration as $p = N_v \exp[-(E_f - E_v)/kT]$. Note that the slope of the valence band dE_v/dx in an energy-band diagram gives the electric field.

4.14 Consider a Si p-n junction at room temperature. On the p-side, $p = N_a^- = 2.5 \times 10^{17}$ cm^{-3}, and on the n-side, $n = N_d^+ = 2.5 \times 10^{16}$ cm^{-3}. The electron affinity of Si is 4.05 eV.

(a) Find the work function on the p-side.

(b) Find the work function on the n-side.

(c) Find the built-in potential V_{bi}.

(d) Find the built-in potential on the p-side V_{bi}^p.

(e) Find the built-in potential on the n-side V_{bi}^n.

(f) Find the depletion width on the p-side x_p.

(g) Find the depletion width on the n-side x_n.

(h) Plot the potential $\psi(x)$ from $-2x_p < x < 2x_n$.
(Use $x = 0, -x_p, -x_p/2, x_n/2$, and x_n)

(i) Plot the electric field $\mathscr{E}(x)$ vs. distance.

(j) Plot the charge density $\rho(x)$ vs. distance.

4.15 Repeat Problem 4.14 except on the p-side, $p = N_a^- = 2 \times 10^{16}$ cm^{-3}, and on the n-side, $n = N_d^+ = 1 \times 10^{16}$ cm^{-3}.

4.16 A GaAs wafer at room temperature is doped so that the 0.8-μm-thick layer at the surface has $n = 5 \times 10^{18}$ cm^{-3} and the adjacent bulk region has $n = 2 \times 10^{16}$ cm^{-3}. What is the potential difference between these two regions?

4.17 A p-n junction has been designed to give the electric field vs. distance as shown below. Sketch the potential vs. distance.

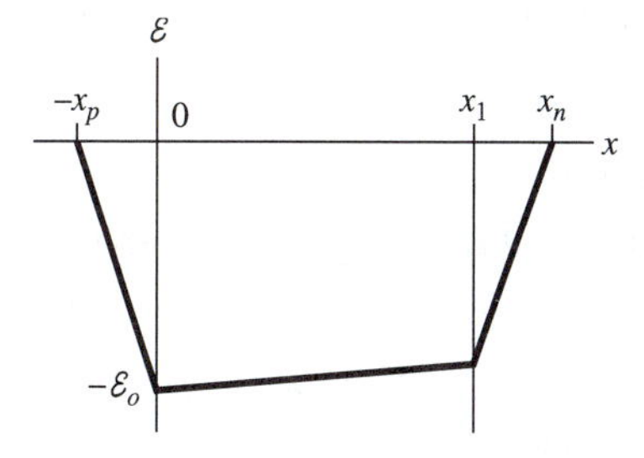

4.18 A Si *p-n* junction at room temperature has $p_{po} = N_a^- = 3 \times 10^{16}$ cm^{-3} and $n_{no} = N_d^+ = 4 \times 10^{16}$ cm^{-3}, $V_{bi} = 0.783$ V, $x_p = 0.14$ μm, and $x_n = 0.10$ μm at thermal equilibrium.

(a) Sketch $p(x)$ and $n(x)$ from $-2x_p < x < 2x_n$ with a log scale for $p(x)$ and $n(x)$. (Only use $-2x_p$, $-x_p$, $-x_p/2$, 0, $x_n/2$, x_n, and $2x_n$.)

(b) Repeat part (a) with a linear scale for $p(x)$ and $n(x)$.

(c) Find the position within the depletion region where $p(x) = n(x) = n_i$.

(d) Find $p_n(x_n)$ and $n_p(x_p)$ for $V_a = +0.5$ V and plot on the sketch in part (a).

(e) With $L_p = 3\mu$m, sketch $p_n(x)$ for $V_a = +0.5$ V between $x_n < x < 10$ μm.

(f) Repeat part (e) with $V_a = -0.5$ V.

4.19 For a Si *p-n* junction with $p = N_a^- = 2.5 \times 10^{18}$ cm^{-3} on the *p*-side and with $n = N_d^+ = 2.5 \times 10^{16}$ cm^{-3} on the *n*-side, $L_n = 2$ μm, and $L_p = 5$ μm. Assume that the mobility of a carrier as a minority carrier is the same as when a majority carrier. (The mobility for a minority carrier hole is obtained at the impurity concentration of the *n*-region so that μ_p is for an impurity concentration of 2.5×10^{16} cm^{-3}.) Find the ratio of the hole-diffusion current density on the *n*-side to the electron current density on the *p*-side.

4.20 For a Si *p-n* junction at 300 K, the properties on the *n*-side are $N_d^+ = 1 \times 10^{18}$ cm^{-3} and $L_p = 1.5 \times 10^{-4}$ cm, and on the *p*-side are $N_a^- = 2 \times 10^{16}$ cm^{-3} and $L_n = 10 \times 10^{-4}$ cm.

(a) Find the hole concentration at x_n on the *n*-side for $V_a = 0.45$ V.

(b) Find the hole concentration as in part (a), but for $x = L_p$.

4.21 The electron injection efficiency γ_n for a *p-n* junction when only the diffusion current is considered is the ratio of the injected electron-diffusion current to the total diffusion current. Derive an expression for γ_n in terms of N_a^-, N_d^+, L_n, L_p, D_n, and D_p.

4.22 The electron injection efficiency γ_n for a *p-n* junction is the ratio of the injected electron diffusion current I_{d_n} to the total junction current I_t due to I_{d_n}, the hole diffusion current I_{d_p}, the space-charge recombination current I_{scr}, and the surface-recombination current I_{sr}. Write an expression for γ_n in terms of these current components.

4.23 An abrupt step-junction diode has $N_a^- \ll N_d^+$ so that only the electron-diffusion current is significant. Let $N_a^- = 5 \times 10^{16}$ cm^{-3}.

(a) Compare the increase in current for a Si *p-n* junction diode and a GaAs *p-n* junction diode at a forward bias of $V_a = 0.5$ V when the temperature is raised from 300 K to 310 K.

(b) Which diode would be the more sensitive thermometer?

4.24 For a forward-biased *p-n* junction, show that the ratio of the electron-diffusion current to the hole-diffusion current can be controlled by varying N_a^- and N_d^+ on the two sides of the junction.

4.25 The doping profile for a p^+-ν-n^+ diode is shown on the next page, where p^+ represents a heavily doped *p*-type layer, ν represents a lightly doped *n*-type layer, and n^+ represents a heavily doped *n*-type layer.

(a) Sketch the potential vs. distance at thermal equilibrium.

(b) Sketch the energy-band diagram at thermal equilibrium.

4.26 How is the quasi-Fermi level separation related to the applied voltage?

4.27 Show that the depletion width for a p^+-*n* junction is near 1 μm when $n = N_d^+ = 1 \times 10^{15}$ cm^{-3}?

4.28 Consider an $Al_xGa_{1-x}As$ light-emitting diode (LED) at room temperature and assume that most of the light is generated by the injection of electrons into the *p*-side. If the nonradiative lifetime

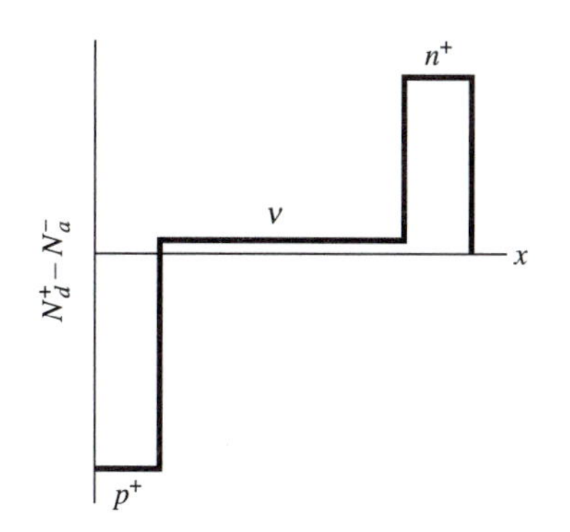

is assumed to remain constant as the acceptor concentration on the p-side is varied, should the p-side be lightly or heavily doped to increase the quantum efficiency?

4.29 For an abrupt Si p-n junction, why will the built-in potential become larger at low temperature? Assume that $p = N_a^-$ and $n = N_d^+$ remain constant with temperature.

4.30 Consider abrupt Si and GaAs p-n junctions at room temperature with equal cross-sectional areas, and identical ionized acceptor and donor concentrations on the p- and n-sides, respectively.

(a) At a given forward bias when diffusion current dominates, would the Si or GaAs p-n junction be expected to have the largest current?

(b) Why?

(c) Which p-n junction would have the largest built-in potential?

4.31 **(a)** Derive an expression for the built-in field in the quasi-neutral region on the n-side of a p-n junction where the donor concentration varies to give a majority carrier concentration variation $n(x)$ which may be represented by $N\exp(-x/\beta)$ for $x \geq x_n$, where x_n is the edge of the depletion region on the n-side.

(b) For $N = 5 \times 10^{18}$ cm^{-3} and $\beta = 0.1$ μm, find the magnitude of the electric field for $n(x)$ given in part (a).

(c) When the p-n junction is forward biased, show whether the field in the neutral region gives a drift current in the same or opposite direction as the injected minority carrier diffusion current.

4.32 A Si p-n junction diode at room temperature has a saturation space-charge recombination current density of 3×10^{-7} A/cm^2 and a saturation diffusion current density of 2×10^{-12} A/cm^2. At what forward bias will the space-charge recombination current and the diffusion current be equal?

4.33 For the p-n junction given in Problem 4.19, find the total minority carrier charge per unit area due to holes injected into the n-side for a forward bias of 0.75 V.

4.34 Show that Eq. (4.62) for V_{bi} reduces to Eq. (4.3).

4.35 For an abrupt Si p-n junction, the acceptor concentration on the p-side is measured to be 5×10^{16} cm^{-3} and the built-in potential V_{bi} is measured to be 0.871 V. Find the hole concentration on the n-side in the neutral semiconductor at thermal equilibrium.

4.36 The electron and hole concentrations as a function of distance in a Si p-n junction are shown on the next page. Assume $L_n = L_p$.

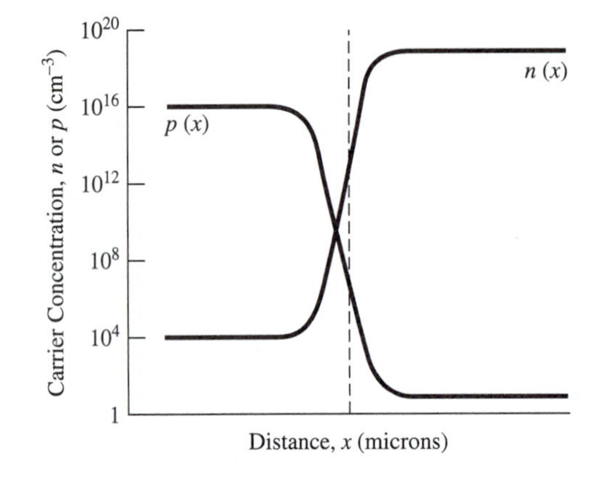

(a) At forward bias, will the injected electron diffusion current I_n or the hole diffusion current I_p be larger?

(b) Why?

4.37 For an abrupt Si *p-n* junction at room temperature with $p = N_a^- = 6 \times 10^{17}$ cm^{-3} and $n = N_d^+ = 2 \times 10^{16}$ cm^{-3}, find

(a) The built-in potential V_{bi}.

(b) The built-in potential on the *p*-side V_{bi}^p.

(c) The hole concentration $p(x = 0)$ and the electron concentration $n(x = 0)$ at the junction interface for thermal equilibrium.

4.38 How do you identify whether the *p-n* junction current for forward-bias is diffusion or space-charge recombination current?

4.39 Describe why surface recombination current is more of a problem for small-diameter *p-n* junctions.

4.40 What is meant by a quasi-Fermi level?

CHAPTER 5

p-n JUNCTIONS: REVERSE BREAKDOWN AND JUNCTION CAPACITANCE

5.1 INTRODUCTION

avalanche breakdown

tunneling

depletion capacitance

This chapter begins with an analysis of the abrupt increase in current for reverse voltages in excess of 5 to 10 V. This analysis in Sec. 5.2 has been simplified to permit illustration of the basic mechanisms and terminology associated with the avalanche breakdown process (also called impact ionization). The voltage for avalanche breakdown is found to depend on the energy gap of the semiconductor and the impurity concentration on the more lightly doped side of the *p-n* junction. If the current is limited to eliminate excessive thermal heating, the avalanche breakdown process is nondestructive. When the impurity concentration is sufficiently high that the depletion width is near 50 Å, the carriers can tunnel through the junction potential barrier. This mechanism is described in Sec. 5.3 and is called *tunneling*. Tunneling occurs for reverse voltages less than 5 to 10 V.

The junction capacitance has two components. The first component is called the *depletion capacitance* and is related to the variation of the depletion width with the applied voltage. The depletion capacitance for an abrupt junction is derived in Sec. 5.4.1. The abrupt junction is an adequate representation for shallow diffused or ion-implanted

Si junctions or compound semiconductor *p-n* junctions. A linearly graded approximation is used to represent a deeply diffused Si *p-n* junction. The depletion capacitance for a linearly graded junction is given in Sec. 5.4.2. The second capacitance component is called the *diffusion capacitance* and is due to the minority carriers injected into the neutral semiconductor for forward bias. The diffusion capacitance is derived in Sec. 5.4.3.

diffusion capacitance

The application of the simulation program with integrated-circuit emphasis (SPICE) to represent the *p-n* junction is given in Sec. 5.5. There are 12 SPICE2 or PSpice parameters which may be related to expressions derived for the current–voltage (*I–V*) behavior given in Chapter 4 and for the junction capacitance given in this chapter. Two additional SPICE parameters are used to represent noise. There are 11 additional model parameters used in PSpice which are also summarized in Sec. 5.5. Applications of *p-n* junctions are given in Sec. 5.6. These applications include microwave diodes, light-emitting diodes (LEDs), as well as heterojunctions which have different energy-gap semiconductors on each side of the *p-n* junction. A *p-n* junction is used for the gate electrode to control the current flow between the source and drain of the three terminal junction-field effect transistor (JFET). Because the JFET is no longer widely used in integrated-circuit applications, the sections on the voltage-current behavior of JFETs and on the application of PSpice to JFETs were deleted. These two sections may be found on the web site: www.ee.duke.edu/~hcc/DEVICE/JFET.html. The significant concepts introduced in this chapter are summarized in Sec. 5.7 together with the useful expressions derived in this chapter.

SPICE

PSpice

applications

JFET

5.2 AVALANCHE BREAKDOWN

As shown in Figs. 4.1 and 4.26 for large reverse voltages, the current increases abruptly due to a process called avalanche breakdown or impact ionization. At reverse voltages in excess of 5 to 10 V, avalanche breakdown occurs because free carriers in the space-charge region gain enough energy from acceleration by the high electric field to break covalent bonds upon collision with lattice electrons and create an electron–hole pair. These carriers can participate in a similar process, which leads to a multiplication of carriers in the space-charge region. This avalanche breakdown process results in an upper limit on the reverse voltage which may be applied on a *p-n* junction or on the base-collector junction of a bipolar transistor. Breakdown can occur due to tunneling at low voltages for heavily doped *p-n* junctions. As the impurity concentration in a *p-n* junction is increased, the depletion width decreases inversely as the square root of the impurity concentration as given in Eq. (4.79). For depletion widths near 50 Å, carriers can tunnel through the potential barrier. This lower voltage breakdown in heavily doped *p-n* junctions is called tunneling or Zener breakdown. Tunneling will be presented in Sec. 5.3, while avalanche breakdown will be given in this section.

The avalanche process will be illustrated by beginning with the consideration of an abrupt *p-n* junction which is asymmetrically doped with N_d^+ on the *n*-side much larger than the N_a^- on the *p*-side. The heavily doped *n*-side is represented by n^+ to give a p-n^+ junction. The minority carrier concentrations are represented at thermal

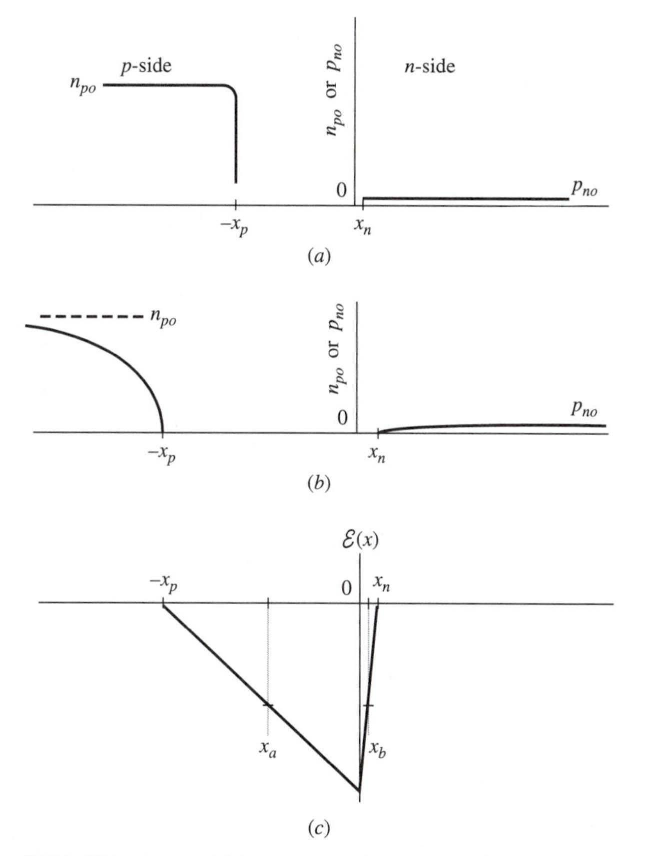

FIGURE 5.1 (*a*) Minority carrier concentrations at thermal equilibrium. (*b*) Minority carrier concentration at reverse bias. (*c*) Electric field for reverse bias with the boundaries of the high-field region for carrier multiplication designated by x_a and x_b.

equilibrium in Fig. 5.1 (*a*) and for a reverse bias in Fig. 5.1 (*b*). The minority carrier concentration variation with distance on the *p*-side, $n_p(x)$, is represented by Eq. (4.167). This equation illustrates that the minority carriers are depleted within a diffusion length of the edge of the depletion region $-x_p$. Because $n_{po} = n_i^2/N_a^- \gg p_{no} = n_i^2/N_d^+$, the minority-carrier holes entering the space-charge region from the *n*-side may be neglected. The electric field is represented in Fig. 5.1 (*c*).

The minority-carrier electrons entering the depletion region from the *p*-side are accelerated by the electric field. The collision process with the energetic electron and the lattice electron is illustrated in Fig. 5.2. A very approximate calculation of the energy necessary to break the bond of a lattice electron may be made by assuming that the initial electron and the resulting electron–hole pair all have the same mass. This process

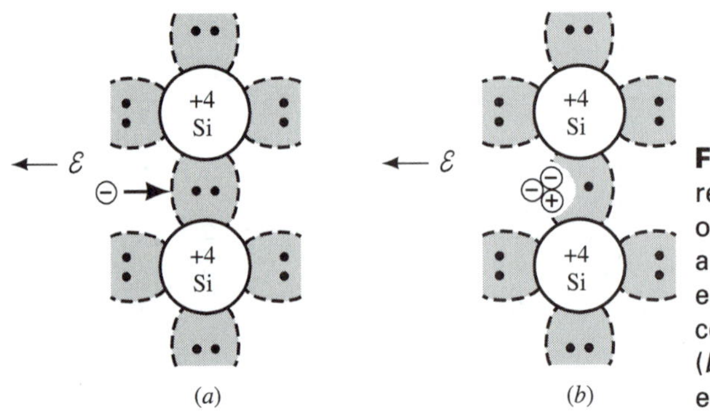

FIGURE 5.2 Schematic representation of collision of the incident electron with a lattice electron to create an electron-hole pair. (*a*) Before collision with lattice electron. (*b*) After collision with lattice electron.

requires conservation of energy and momentum for the initial electron shown in Fig. 5.2 (*a*) and the three particles shown in Fig. 5.2 (*b*) after the collision:

Example of approximate energy to break lattice electron bond

$$\frac{mv_1^2}{2} = E_g + \frac{3mv_2^2}{2} \qquad \text{and} \qquad mv_1 = 3mv_2 \qquad \text{or} \qquad v_1 = 3v_2.$$

Then

$$\frac{mv_1^2}{2} = E_g + \frac{3mv_1^2}{2 \times 9} \qquad \text{or} \qquad \frac{mv_1^2}{2} = 3E_g/2.$$

This approximate calculation shows that the larger the energy gap, the greater the energy required to create the hole–electron pair. Near the edges of the space-charge region, the electric field is not large enough for the carriers to gain enough energy to create an electron–hole pair. Therefore, the carrier multiplication region is confined to the central high-field region illustrated in Fig. 5.1 (*c*) with the boundaries of x_a and x_b.

The geometry used for consideration of the avalanche process is shown in Fig. 5.3. The space-charge region where the field is large enough to permit creation of electron–hole pairs is represented by the boundaries x_a and x_b. The electron concentration entering at x_a is n_o and the electron concentration leaving at x_b is n_f. The electron

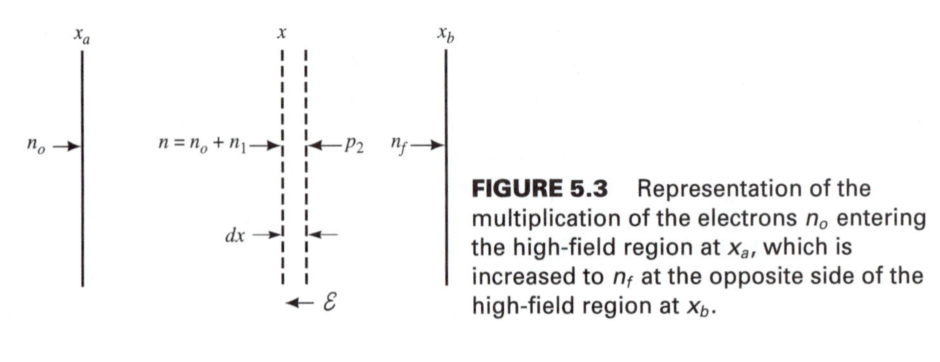

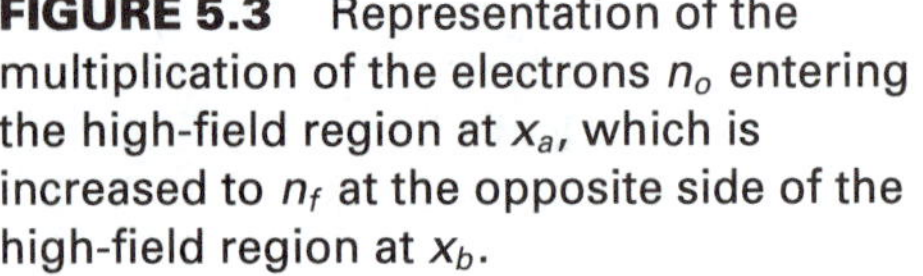

FIGURE 5.3 Representation of the multiplication of the electrons n_o entering the high-field region at x_a, which is increased to n_f at the opposite side of the high-field region at x_b.

concentration is considered at x in the incremental distance dx. The electron concentration entering the incremental distance dx at x is taken as $n = n_o + n_1$, where n_1 is the electron concentration gained between x_a and x. The hole concentration at x_b will be taken as zero, but the hole concentration will increase to p_2 at x due to electron–hole generation between x and x_b. The carrier multiplication M is defined as the ratio of n_f to n_o:

carrier multiplication

$$\boxed{M = n_f/n_o}. \tag{5.1}$$

Electrons entering the high-field region at x_a with the concentration n_o are increased by impact ionization to a concentration of $n_o + n_1$ at x as shown in Fig. 5.3. The probability of electron–hole pair creation in dx is given by the product of the distance dx and a proportionality factor called the *ionization coefficient for electrons* represented by α_n, which has units of number/cm. The hole ionization coefficient is represented by α_p. The added concentration of electrons (and holes) in dx due to n is dn_n (and dp_n) and is given by the product of the concentration of electrons n at x and the ionization probability:

ionization coefficient

$$dn_n = \alpha_n n\,dx = \alpha_n(n_o + n_1)\,dx, \tag{5.2}$$

where n_1 represents the increase in electron concentration from x_a to x.

Holes at x are due to holes generated as electron–hole pairs between x_b and x. The hole concentration at x_b is ignored because the n-side is heavily doped, which makes p_{no} very small. The added concentration of electrons (and holes) due to p_2 in dx is

$$dn_p = \alpha_p p_2\,dx. \tag{5.3}$$

The total increase in electron concentration dn in dx is the sum of dn_n and dn_p:

$$dn = \alpha_n(n_o + n_1)\,dx + \alpha_p p_2\,dx. \tag{5.4}$$

At x_b, the electron concentration is

$$n_f = n_o + n_1 + n_2 = n_o + n_1 + p_2, \tag{5.5}$$

where n_2 represents the electrons created between x and x_b. Because the concentration of electrons created between x and x_b must be the same as the concentration of holes created between x and x_b, n_2 in Eq. (5.5) may be represented by p_2. Solving for n_2 in Eq. (5.5) gives

$$n_2 = n_f - (n_o + n_1) = p_2. \tag{5.6}$$

Substitution of Eq. (5.6) for p_2 into Eq. (5.4) gives

$$dn = \alpha_n(n_o + n_1)\,dx + \alpha_p[n_f - (n_o + n_1)]\,dx \tag{5.7}$$

or

$$dn = (\alpha_n - \alpha_p)(n_o + n_1)\,dx + \alpha_p n_f\,dx. \tag{5.8}$$

In general, α_n and α_p are not equal. However, to simplify illustration of the avalanche breakdown process, α_n will be taken here to be equal to α_p. This simplification permits writing Eq. (5.8) as

$$dn = \alpha_p n_f\,dx = \alpha_n n_f\,dx. \tag{5.9}$$

Integration of Eq. (5.9) gives

$$\int_{n_o}^{n_f} dn = \int_{x_a}^{x_b} \alpha_n n_f\,dx, \tag{5.10}$$

which gives

$$n_f - n_o = n_f \int_{x_a}^{x_b} \alpha_n\,dx \tag{5.11}$$

or

$$n_o = n_f\left[1 - \int_{x_a}^{x_b} \alpha_n\,dx\right]. \tag{5.12}$$

The multiplication factor M defined in Eq. (5.1) may be obtained from Eq. (5.12) as

$$M = \frac{n_f}{n_o} = \frac{n_f}{n_f\left[1 - \int_{x_a}^{x_b} \alpha_n\,dx\right]}. \tag{5.13}$$

When the integral approaches unity, the ratio increases without bound and avalanche breakdown occurs when

condition for avalanche breakdown

$$\boxed{\int_{x_a}^{x_b} \alpha_n\,dx = 1}. \tag{5.14}$$

The ionization coefficients are strong functions of electric field and are determined experimentally. This field dependence can be represented in a simplified form as[1]

$$\alpha = A\exp\{[-B/\mathscr{E}(x)]^m\}, \tag{5.15}$$

where A, B, and m are empirical fitting parameters.

An approximate universal expression can be written for the breakdown voltage V_B in terms of the semiconductor energy gap E_g and the impurity concentration on the

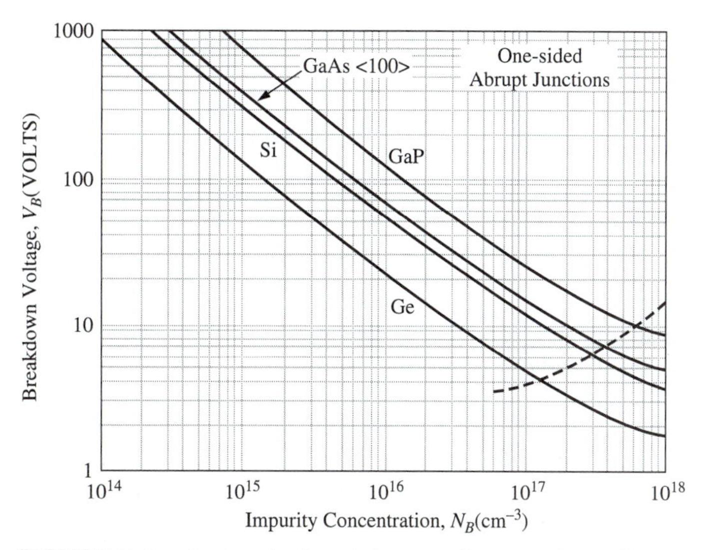

FIGURE 5.4 Avalanche breakdown voltage vs. impurity concentration for one-sided abrupt junctions in Ge, Si, GaAs, and GaP. The dashed line indicates the impurity concentration where tunneling begins to dominate the voltage breakdown.[2]

lightly doped side N_B of an asymmetrically doped (also called one-sided) *p-n* junction as[1]

breakdown voltage

$$\boxed{V_B \simeq 60(E_g/1.1)^{3/2}(10^{16}/N_B)^{3/4}\ \text{V}}. \tag{5.16}$$

Several useful plots have been given by Sze.[2] The avalanche breakdown voltage for one-sided abrupt junctions for several common semiconductors are shown in Fig. 5.4. The energy gaps at room temperature are $E_g(\text{Ge}) = 0.663$ eV, $E_g(\text{Si}) = 1.125$ eV, $E_g(\text{GaAs}) = 1.424$ eV, and $E_g(\text{GaP}) = 2.261$ eV. For voltages less than about $6E_g/q$, the breakdown voltage is mainly by tunneling[1] and is represented by the dashed line in Fig. 5.4.

The depletion width and maximum electric field at breakdown also may be obtained from previous expressions. Because the potential is related to the field by $V = -\int \mathscr{E}(x)\,dx$, the potential is the area of the $\mathscr{E}(x)$ plot in Fig. 5.1 (*c*), and the breakdown voltage V_B is related to the maximum electric field $\mathscr{E}_{max}$ as

$$V_B = \mathscr{E}_{max} x_p/2. \tag{5.17}$$

With $N_d^+ \gg N_a^-$, $x_p \simeq x_d$ and Eq. (4.79) for x_d becomes

$$x_d = \sqrt{\frac{2\epsilon}{qN_a^-}(V_{bi} + V_B)} \simeq \sqrt{\frac{2\epsilon}{qN_a^-}V_B} \tag{5.18}$$

for $V_B > V_{bi}$. Equation (5.17) with Eq. (5.18) gives

$$V_B^2 = \frac{\mathscr{E}_{\max}^2 2\epsilon V_B}{4qN_a^-} \tag{5.19}$$

or

$$\mathscr{E}_{\max} = \sqrt{2qN_a^- V_B/\epsilon}. \tag{5.20}$$

The depletion width W_m and maximum electric field at breakdown $\mathscr{E}_m$ as a function of impurity concentration on the lightly doped side of the one-sided junction are summarized in Fig. 5.5.

Investigations of avalanche breakdown have shown that there may also be very small localized areas of reduced breakdown voltage which are called *microplasmas.*[3] An example of microplasmas is shown in Fig. 5.6. Microplasmas have been shown to be

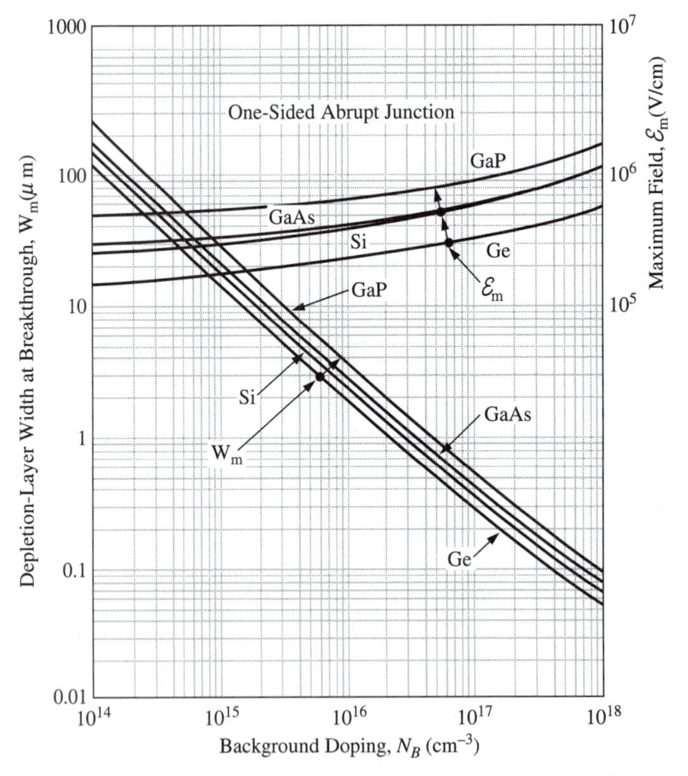

FIGURE 5.5 Depletion width and maximum electric field at breakdown for one-sided abrupt junctions in Ge, Si, GaAs, and GaP.[2]

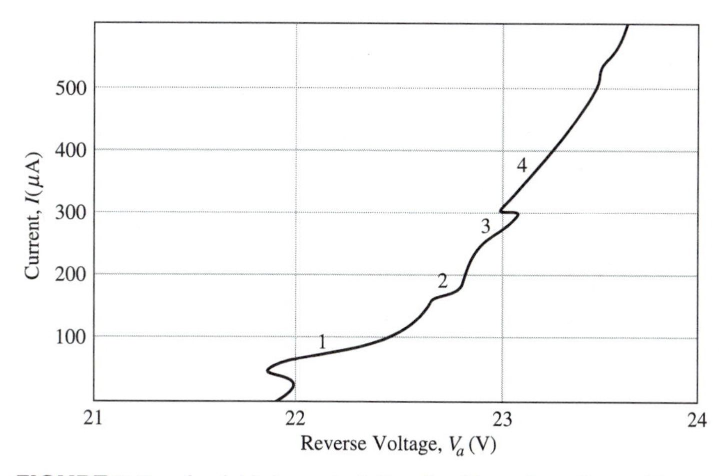

FIGURE 5.6 An I–V characteristic of a Si p-n junction with several microplasmas.[3]

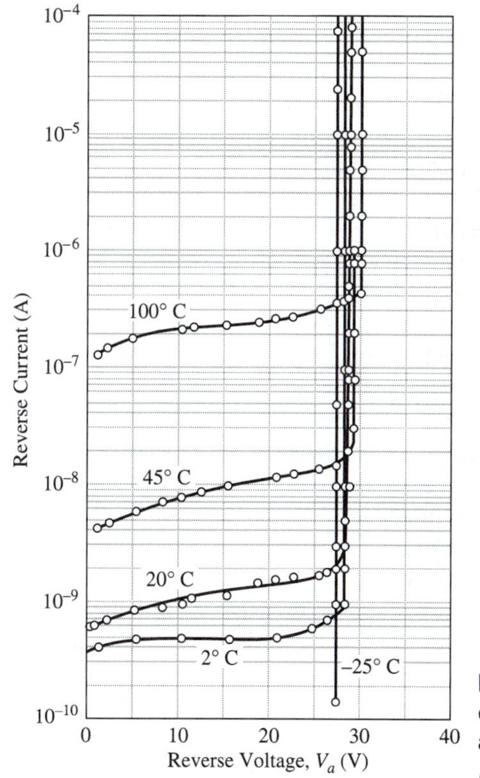

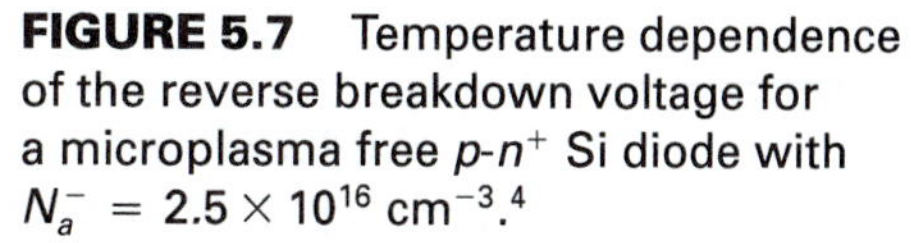

FIGURE 5.7 Temperature dependence of the reverse breakdown voltage for a microplasma free p-n^+ Si diode with $N_a^- = 2.5 \times 10^{16}$ cm^{-3}.[4]

caused by imperfections in the crystal lattice. This local defect results in an enhanced electrical field in this region and results in a breakdown voltage for the *p-n* junction which is lower than the bulk breakdown voltage. The quality of Si process technology has progressed to a status where microplasmas are an infrequent problem. However, in the more complex and less mature process technology for the III-V compound semiconductors, microplasmas can still be a problem.

The temperature dependence of the breakdown voltage is illustrated in Fig. 5.7 and demonstrates that the breakdown voltage becomes larger as the temperature increases. As was shown in Fig. 3.2, the mobility becomes smaller as the temperature increases due to energy losses due to phonon collisions. Therefore, carriers require higher voltage to obtain sufficient energy to generate an electron–hole pair.

5.3 TUNNELING

Because the depletion region width is related to $1/\sqrt{N_B}$, quantum-mechanical tunneling can become a significant current conduction mechanism at high-impurity concentrations. This reverse-voltage conduction mechanism is also called Zener breakdown. The transmission probability for quantum-mechanical tunneling generally depends exponentially on the barrier thickness, and at depletion-region thicknesses less than 50 Å, tunneling can become a significant conduction mechanism. Although a detailed analysis is beyond the scope of this text, a qualitative graphic description can be useful. The energy-band diagram for a *p-n* junction heavily doped on both the *p*- and *n*-sides is shown in Fig. 5.8 (*a*) for thermal equilibrium. The doping is sufficiently high that the Fermi level is within both the valence and conduction bands. In this case, $p > 0.1N_v$ and $n > 0.1N_c$, which requires the use of the Nilsson expression given in Eq. (2.58) for the Fermi level rather than the expressions based on the Boltzmann approximation. At forward bias shown in Fig. 5.8 (*b*), electrons in the conduction band can tunnel into the empty hole states in the valence band. In reverse bias shown in Fig. 5.8 (*c*), filled electron states in the valence band can tunnel into unoccupied states in the conduction band.

As seen in Fig. 5.9, the tunneling current at a given voltage increases with temperature. At higher temperatures, the carrier velocities increase. The increased velocities increase the rate at which the carriers collide with the junction potential barrier, and hence the tunneling current increases. The *I–V* behavior for tunneling depends on the impurity concentrations on the *n*- and *p*-sides. As the impurity concentration becomes larger than for the *I–V* behavior shown in Fig. 5.9, the reverse current at a given reverse voltage exceeds the forward current at the same value of forward voltage. This diode is known as a *backward diode.* Further increases in impurity concentrations can result in a negative differential resistance for forward bias. This diode is called a *tunnel diode.*

backward diode

tunnel diode

For tunneling breakdown, the reverse voltage for a fixed current becomes smaller as the temperature is increased, which is opposite temperature behavior as compared to avalanche breakdown. By the proper selection of the impurity concentrations, it is

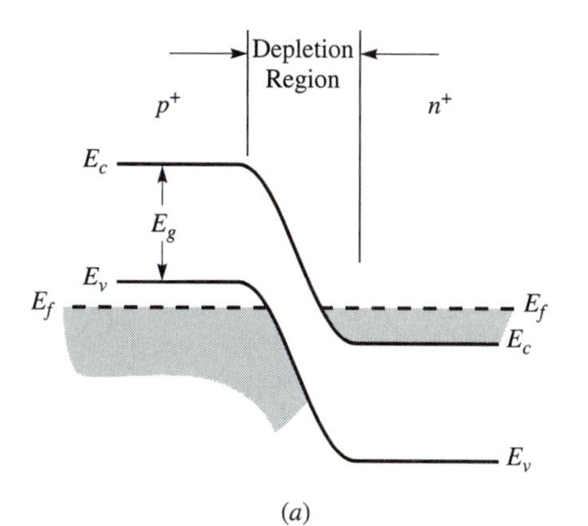

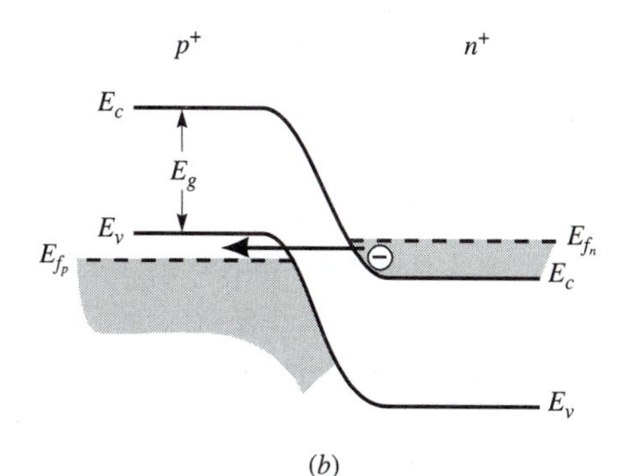

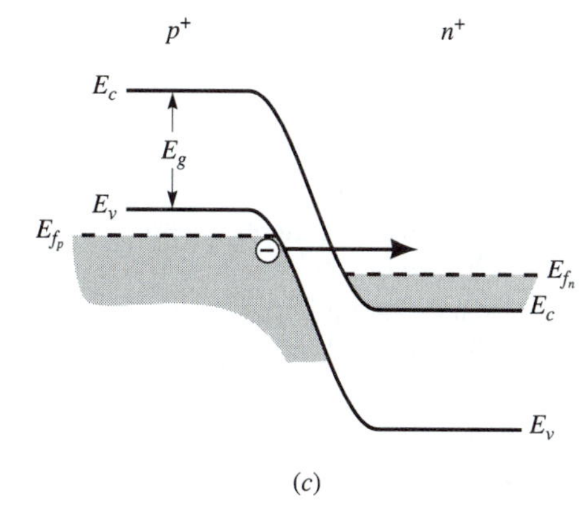

FIGURE 5.8 Energy-band diagram for a heavily doped *p-n* junction. (*a*) Thermal equilibrium. (*b*) Electron tunneling from the conduction band to the valence band for forward bias. (*c*) Electron tunneling from filled states in the valence band to empty states above the Fermi level in the conduction band for reverse bias.

possible to design a *p-n* junction which conducts by both avalanche breakdown and tunneling for reverse bias. Because the temperature coefficients for these two mechanisms are opposite, the temperature sensitivity of the voltage at a fixed current can be very small. Devices with this reduced temperature sensitivity are used as voltage reference diodes. A more complete analysis has been given by Sze.[6]

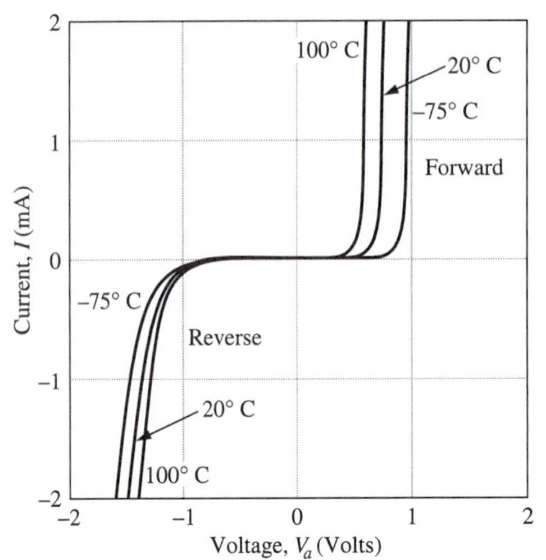

FIGURE 5.9 Current-voltage characteristics for the tunneling conduction mechanism in a heavily doped *p-n* junction.[5]

5.4 JUNCTION CAPACITANCE

5.4.1 Abrupt-Junction Depletion Capacitance

Not only is the *I–V* behavior an important property of the *p-n* junction, but the capacitance–voltage (*C–V*) behavior is also an important property. The static capacitance is defined as $C_{static} \equiv Q/V$, where Q is the total charge on the capacitor and V is the applied voltage. The *differential* capacitance is

differential capacitance

$$\boxed{C_{dif} \equiv \left|\frac{dQ}{dV}\right|}. \tag{5.21}$$

The static and differential capacitances for the *p-n* junction differ because the charge on the *p-n* junction varies nonlinearly with the applied voltage. For the parallel-plate capacitor, there is a linear relationship between the charge and voltage, and the static and differential capacitance will be the same. In this part of Sec. 5.4, the capacitance for an abrupt junction due to the space-charge depletion region will be derived. The abrupt junction represents shallow diffused and ion-implanted Si junctions and epitaxially grown compound semiconductor junctions.

To find the depletion capacitance, it is necessary to obtain an expression for the charge in terms of the voltage. The space charge per unit volume was shown in Fig. 4.5 (*a*) and the total positive and negative space charge Q_{sc} must be equal, which gives

$$|Q_{sc}| = |-qN_a^- x_p A| = |qN_d^+ x_n A|, \tag{5.22}$$

where the depletion width on the *p*-side was given in Eq. (4.32) as

$$x_p = \sqrt{2\epsilon V_{bi}^p / qN_a^-}, \tag{5.23}$$

and the depletion width on the n-side was given in Eq. (4.33) as

$$x_n = \sqrt{2\epsilon V_{bi}^n / qN_d^+}. \tag{5.24}$$

For an applied bias (see Sec. 4.6.2), V_{bi}^p is replaced by $V_{bi}^p - V_a^p$, and x_p becomes

$$x_p = \sqrt{2\epsilon(V_{bi}^p - V_a^p)/qN_a^-}, \tag{5.25}$$

where $V_{bi}^p - V_a^p$ was given by Eq. (4.77) as

$$V_{bi}^p - V_a^p = (V_{bi} - V_a)/[1 + (N_a^-/N_d^+)]. \tag{5.26}$$

Substitution of Eqs. (5.25) and (5.26) into Eq. (5.22) for x_p gives

$$|Q_{sc}| = qAN_a^- \sqrt{\frac{2\epsilon N_d^+(V_{bi} - V_a)}{qN_a^-(N_d^+ + N_a^-)}}, \tag{5.27}$$

or

$$|Q_{sc}| = A\sqrt{\frac{2q\epsilon N_a^- N_d^+(V_{bi} - V_a)}{N_d^+ + N_a^-}}. \tag{5.28}$$

The depletion capacitance C_j may now be found from Eq. (5.21) with Eq. (5.28) for $|Q_{sc}|$:

$$C_j \equiv \left|\frac{dQ_{sc}}{dV_a}\right| = A\sqrt{\frac{2q\epsilon N_a^- N_d^+}{N_d^+ + N_a^-}}\frac{1}{2}(V_{bi} - V_a)^{-1/2} \tag{5.29}$$

or

abrupt-junction depletion capacitance C_j

$$\boxed{C_j = A\sqrt{\frac{q\epsilon N_a^- N_d^+}{2(V_{bi} - V_a)(N_d^+ + N_a^-)}}}. \tag{5.30}$$

The change in space charge dQ_{sc} with voltage dV is illustrated in Fig. 5.10.

Several useful results may be obtained by squaring Eq. (5.30). Consider an abrupt one-sided junction with $N_d^+ \gg N_a^-$ so that

$$C_j^2 = A^2 q\epsilon N_a^- / 2(V_{bi} - V_a) \tag{5.31}$$

or

$$1/C_j^2 = 2(V_{bi} - V_a)/A^2 q\epsilon N_a^-. \tag{5.32}$$

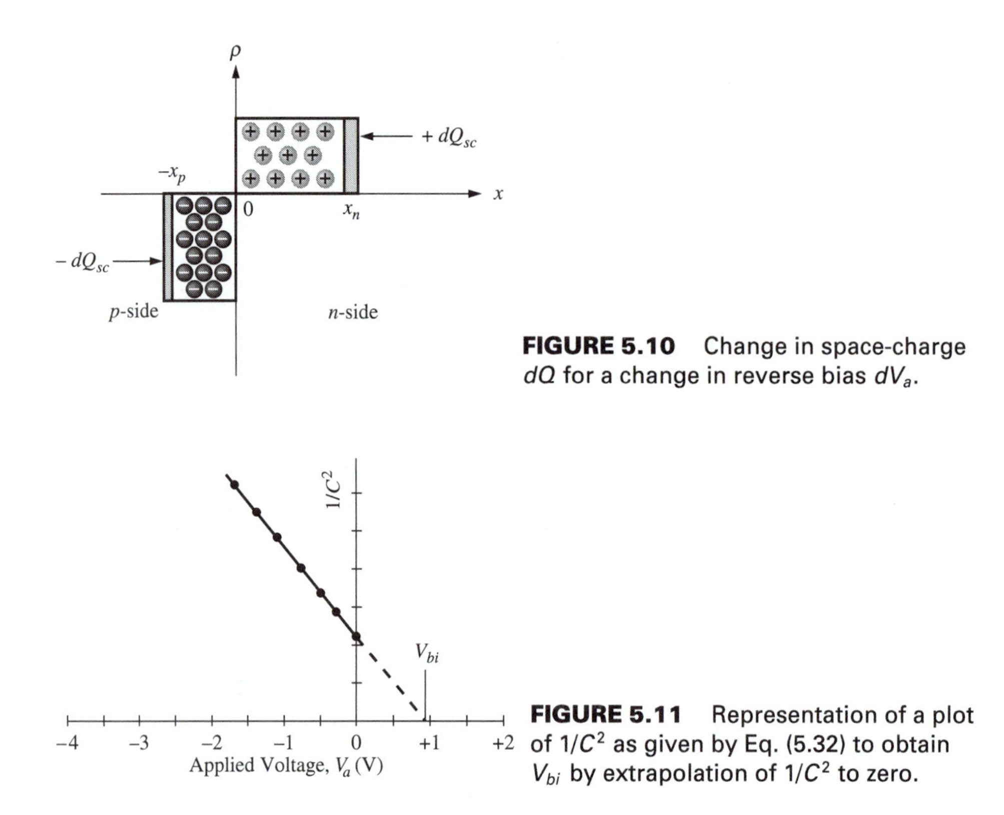

FIGURE 5.10 Change in space-charge dQ for a change in reverse bias dV_a.

FIGURE 5.11 Representation of a plot of $1/C^2$ as given by Eq. (5.32) to obtain V_{bi} by extrapolation of $1/C^2$ to zero.

A plot of the measured $1/C^2$ for an abrupt one-sided junction as a function of V_a is illustrated in Fig. 5.11. Extrapolation of $1/C^2$ to zero occurs at $V_a = V_{bi}$ to give an experimentally determined value for V_{bi}. If the cross-sectional area A is known, N_a^- may be obtained from Eq. (5.31) with V_{bi} obtained from the $1/C^2$ plot.

Equation (5.30) may be written in a form used in SPICE as

$$C_j = A\sqrt{\frac{q\epsilon N_a^- N_d^+}{2V_{bi}(N_d^+ + N_a^-)}}\sqrt{\frac{V_{bi}}{V_{bi} - V_a}} \tag{5.33}$$

or

$$C_j = \mathrm{CJO}\sqrt{\frac{V_{bi}}{V_{bi} - V_a}} = \mathrm{CJO}\left[\frac{V_{bi}}{V_{bi} - V_a}\right]^{0.5}, \tag{5.34}$$

where CJO (CJ uppercase ‘OH’) or lower case cjo represents the *zero* bias depletion capacitance, and the abrupt-junction depletion capacitance has a $\underline{-1/2}$ voltage dependence. In reverse bias, the capacitance is depletion capacitance. In forward bias, capacitance due to the minority carriers injected into the neutral semiconductor must also be considered and is given in Sec. 5.4.3.

5.4.2 Linearly Graded Junction Depletion Capacitance

For deeply diffused junctions, the linearly graded approximation is generally used. The linearly graded space charge is shown in Fig. 5.12 (a), and may be represented by

$$q(N_d^+ - N_a^-) = qax, \tag{5.35}$$

where a is the impurity gradient in cm^{-4}. Gauss's law was given in Eq. (4.4), and with the depletion approximation becomes

$$d\mathscr{E}(x)/dx = (qax)/\epsilon \qquad \text{for} \qquad -W/2 \leq x \leq W/2. \tag{5.36}$$

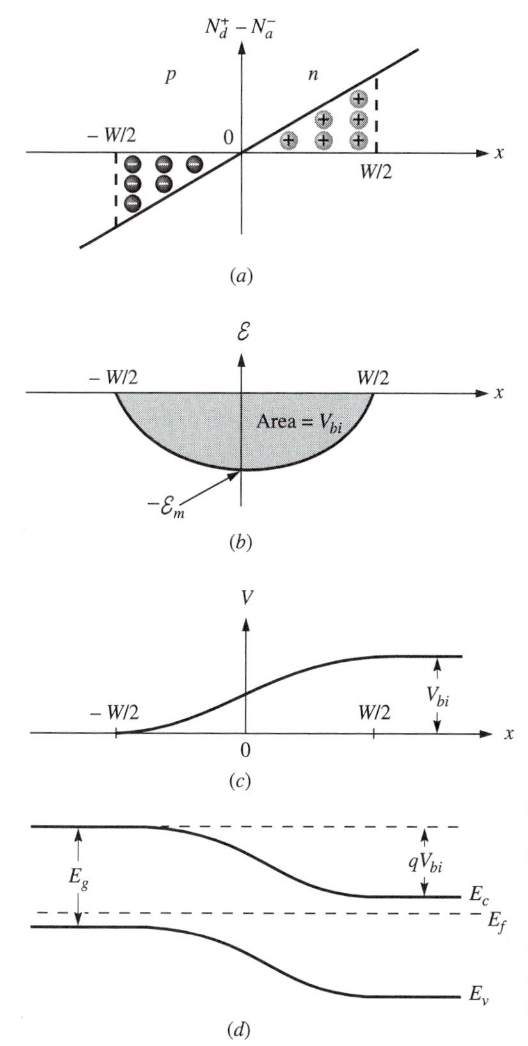

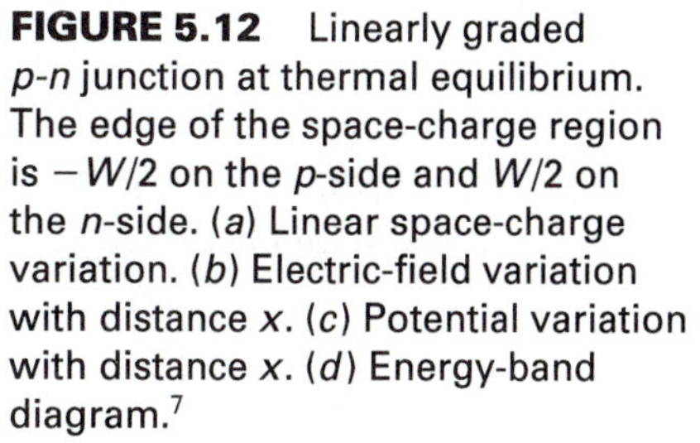

FIGURE 5.12 Linearly graded p-n junction at thermal equilibrium. The edge of the space-charge region is $-W/2$ on the p-side and $W/2$ on the n-side. (a) Linear space-charge variation. (b) Electric-field variation with distance x. (c) Potential variation with distance x. (d) Energy-band diagram.[7]

Integration of Eq. (5.36) gives

$$\mathscr{E}(x) = qax^2/2\epsilon + A_1, \tag{5.37}$$

where A_1 is an arbitrary integration constant to be determined by the boundary condition. At $x = -W/2$, the electric field is zero, which gives

$$A_1 = -qa(W/2)^2/2\epsilon, \tag{5.38}$$

and $\mathscr{E}(x)$ becomes

$$\mathscr{E}(x) = -qa[(W/2)^2 - x^2]/2\epsilon. \tag{5.39}$$

The potential $\psi(x) = -\int \mathscr{E}(x)\,dx$ and is obtained by integrating Eq. (5.39) to give

$$\psi(x) = qa[(W/2)^2 x - x^3/3]/2\epsilon + B_1. \tag{5.40}$$

Let $\psi(x) = 0$ at $x = -W/2$, and then

$$B_1 = qaW^3/24\epsilon, \tag{5.41}$$

and the potential is given by

$$\psi(x) = \frac{qa}{2\epsilon}\left(\frac{W^3}{12} + \frac{W^2}{4}x - \frac{x^3}{3}\right). \tag{5.42}$$

At $x = W/2$, $\psi(W/2)$ is the built-in potential:

$$V_{bi} = \frac{qaW^3}{12\epsilon}. \tag{5.43}$$

The variation of the electric field with distance is shown in Fig. 5.12 (*b*) and the variation of the potential with distance is shown in Fig. 5.12 (*c*).

The ionized impurity concentration at $x = W/2$ is $qaW/2$ and the space-charge per unit area is the triangular area in Fig. 5.12 (*a*), and is $(1/2)(qaW/2)(W/2)$. The space-charge for the linearly graded junction becomes

$$Q_{sc} = AqaW^2/8. \tag{5.44}$$

With V_{bi} in Eq. (5.43) replaced by $(V_{bi} - V_a)$ and used to replace W in Eq. (5.44), the space-charge may be written as

$$Q_{sc} = A\frac{qa}{8}\left[\frac{12\epsilon(V_{bi} - V_a)}{qa}\right]^{2/3}. \tag{5.45}$$

The depletion capacitance C_j may now be found from Eq. (5.21) with Eq. (5.45) for $|Q_{sc}|$:

$$C_j \equiv \left|\frac{dQ_{sc}}{dV_a}\right| = \frac{Aqa}{8}\left[\frac{12\epsilon}{qa}\right]^{2/3}\frac{2}{3}(V_{bi} - V_a)^{-1/3}, \tag{5.46}$$

or

linearly graded junction depletion capacitance C_j

$$\boxed{C_j = A\left[\frac{qa\epsilon^2}{12(V_{bi} - V_a)}\right]^{1/3}}. \tag{5.47}$$

Note that for a linearly graded junction the capacitance has a cube-root dependence on voltage, while for the abrupt junction the capacitance [see Eq. (5.30)] varies as the square root of the voltage.

5.4.3 Diffusion Capacitance

The minority carrier concentration variations in the neutral semiconductor were illustrated in Fig. 4.13 and the excess minority carriers for a p^+-n junction are indicated by the shaded area in Fig. 5.13. These minority carriers are due to the injected diffusion current and are called the diffusion capacitance or the storage capacitance C_s. The storage capacitance due to holes on the n-side is given by

$$C_{s_p} = \frac{|dQ_p|}{|dV_a|}, \tag{5.48}$$

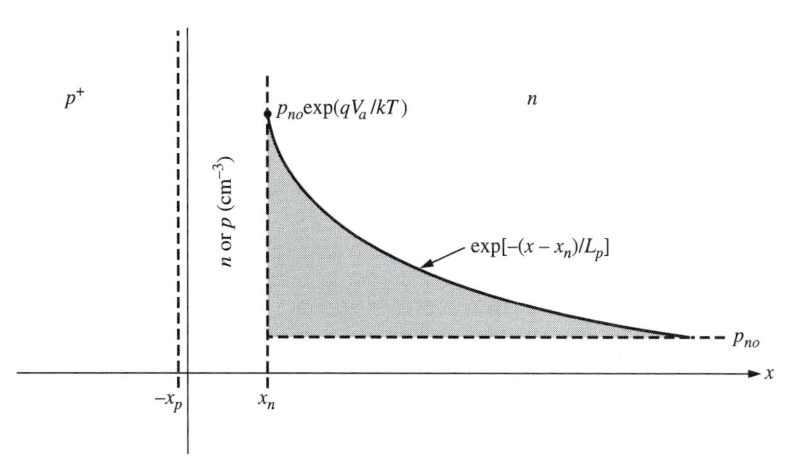

FIGURE 5.13 Excess minority carrier concentrations $p_n(x)$ in the quasi-neutral semiconductor for an abrupt p^+-n junction due to the diffusion current as indicated by the shaded area.

where the charge Q_p due to excess holes is given by

$$Q_p = qA\int_{x_n}^{\infty} [p_n(x) - p_{no}]\,dx. \quad (5.49)$$

With Eq. (4.95) for $p_n(x)$ in Eq. (5.49), Q_p may be written as

$$Q_p = qA\int_{x_n}^{\infty} \{p_{no}[\exp(qV_a/kT) - 1]\exp[-(x - x_n)/L_p] + p_{no} - p_{no}\}\,dx. \quad (5.50)$$

Integration of Eq. (5.50) gives

$$Q_p = qAp_{no}[\exp(qV_a/kT) - 1](-L_p)\exp[-(x - x_n)/L_p]|_{x_n}^{\infty}. \quad (5.51)$$

Evaluation of Eq. (5.51) at the integration limits gives

$$Q_p = -qA\{p_{no}L_p[\exp(qV_a/kT) - 1]\}\{[\exp(-\infty) - \exp[-(x_n - x_n)/L_p]\}, \quad (5.52)$$

or

$$Q_p = qAp_{no}L_p[\exp(qV_a/kT) - 1]. \quad (5.53)$$

By replacing p_{no} by n_i^2/N_d^+ and multiplying by L_p/L_p with $L_p^2 = D_p\tau_p$, Eq. (5.53) may be written as

$$Q_p = \frac{qAD_pn_i^2}{L_pN_d^+}\tau_p[\exp(qV_a/kT) - 1]. \quad (5.54)$$

From Eq. (4.109) for $J_{p_n}(x_n)$, Q_p may be written as

$$\boxed{Q_p = J_{p_n}(x_n)A\tau_p = I_d\tau_p}, \quad (5.55)$$

where the diffusion current $I_d = J_{p_n}(x_n)A$ and demonstrates that the stored charge is the product of the diffusion current and the minority-carrier lifetime. The diffusion or storage capacitance from Eq. (5.48) with Eq. (5.54) becomes

$$C_{s_p} = \frac{|dQ_p|}{|dV_a|} = \frac{qAD_pn_i^2}{L_pN_d^+}\tau_p\exp(qV_a/kT)\frac{q}{kT} \quad (5.56)$$

diffusion or storage capacitance C_s

or

$$\boxed{C_{s_p} = \frac{q}{kT}\tau_pI_s\exp(qV_a/kT)}, \quad (5.57)$$

where the saturation current $I_s = qAD_p n_i^2/L_p N_d^+$ for the minority carrier concentration length L_p smaller than the layer thickness W_n: $W_n > L_p$. For $V_a > 3kT/q$, the -1 in the $[\exp(qV_a/kT) - 1]$ expression for the diffusion current may be neglected and $I_s \exp(qV_a/kT)$ represents the diffusion current I_d.

In a similar manner for the minority-carrier electrons on the p-side,

$$Q_n = qAn_{po}L_n[\exp(qV_a/kT) - 1], \tag{5.58}$$

and the diffusion capacitance on the n-side C_{s_n} is given by $|dQ_n/dV_a|$. With $N_a^- \gg N_d^+$, $Q_n \ll Q_p$. When N_a^- and N_d^+ are not greatly different, contributions to the diffusion capacitance must be considered from both the p- and n-sides. Since the total diffusion current density is the sum of the hole and electron current densities: $J_d = J_{p_n} + J_{n_p}$, which is $Q_d/t = Q_p/t + Q_n/t$. Then with $Q = CV$, the total diffusion capacitance C_s times the applied voltage V_a becomes $C_s V_a = C_{s_p} V_a + C_{s_n} V_a$. Therefore, the total diffusion capacitance is the sum of the contributions from both the p- and n-sides.

For the thin-diode case when the layer thickness W_n is less than the hole-diffusion length L_p on the n-side, $W_n < L_p$, the charge due to excess holes is

$$Q_p = qA\int_0^{W_n} p_n(x)\,dx = qA\int_0^{W_n} p_{no}\exp(qV_a/kT)(1 - x/W_n)\,dx, \tag{5.59}$$

where $p_n(x)$ was given by Eq. (4.138). Integration of Eq. (5.59) gives

$$Q_p = qAp_{no}\exp\left(\frac{qV_a}{kT}\right)\frac{W_n}{2} = \frac{qAn_i^2}{N_d^+}\exp\left(\frac{qV_a}{kT}\right)\frac{W_n}{2}. \tag{5.60}$$

To convert Q_p to a form which will give a diffusion capacitance with an expression similar to Eq. (5.57), the transit time for the minority-carrier holes to cross the thin n-region of thickness W_n will be derived. The transit time is given by

$$\tau_D = \int_0^{W_n} \frac{dx}{v_p(x)}, \tag{5.61}$$

where $v_p(x)$ is the hole velocity, which may be found from the hole current I_p. The hole velocity is related to the current through the hole flux F_p, which is

$$F_p = v_p(x)p_n(x) = v_p(x)p_{no}\exp(qV_a/kT)(1 - x/W_n). \tag{5.62}$$

The flux is multiplied by A and q to give $I_p = qAv_p(x)p_{no}\exp(qV_a/kT)(1 - x/W_n)$, which is also given by the expression for the diffusion current [Eq. (4.106)] $I_p = -qD_pA(dp/dx) = -qAD_p\frac{d}{dx}[p_{no}\exp(qV_a/kT)(1-x/W_n)]$. By equating these two expressions for I_p,

$$I_p = qAv_p(x)p_{no}\exp(qV_a/kT)(1 - x/W_n) = qAD_p p_{no}\exp(qV_a/kT)(1/W_n), \tag{5.63}$$

the hole velocity becomes

$$v_p(x) = \frac{D_p}{W_n(1 - x/W_n)} = \frac{D_p}{(W_n - x)}. \tag{5.64}$$

Equation (5.61) gives

transit time τ_D

$$\boxed{\tau_D = \int_0^{W_n} \frac{W_n - x}{D_p}\, dx = \frac{W_n^2}{2D_p}}. \tag{5.65}$$

If W_n is taken as 1 μm and D_p is taken as 1 cm^2/s, τ_D becomes 5×10^{-9} s.

From Eq. (5.65) for the transit time $\tau_D = W_n^2/2D_p$, multiply Eq. (5.60) for Q_p by $W_n/2 = D_p\tau_D/W_n$ to give

$$Q_p = \frac{qAD_pn_i^2}{W_nN_d^+}\tau_D \exp\left(\frac{qV_a}{kT}\right) = \tau_D I_d. \tag{5.66}$$

The diffusion capacitance from Eq. (5.48) with Eq. (5.66) for Q_p becomes

thin-layer diode diffusion capacitance C_s

$$C_s = \frac{|dQ_p|}{|dV_a|} = \frac{qAD_pn_i^2}{W_nN_d^+}\tau_D \exp\left(\frac{qV_a}{kT}\right)\frac{q}{kT} \tag{5.67}$$

or

$$\boxed{C_s = \frac{q}{kT}\tau_D I_p}, \tag{5.68}$$

where I_p/A is J_{p_n} in Eq. (4.141). Note that the expression for diffusion capacitance given in Eq. (5.67) for the thin-diode case is the same as the expression given in Eq. (5.57) with the transit time τ_D replacing the minority-carrier lifetime τ_p and the layer thickness W_n replacing the diffusion length L_p. The diffusion capacitance for the thin-diode case will be used in the bipolar junction transistor in Chapter 9.

5.5 SPICE MODEL FOR *p-n* JUNCTIONS

5.5.1 Element and Model Lines

As mentioned in Sec. 3.2.4, PSpice is now intended to be run with the schematics front end, Schematics. The line command version of PSpice illustrated here is helpful to learn before using Schematics. An introduction to PSpice for the *p-n* junction diode has been

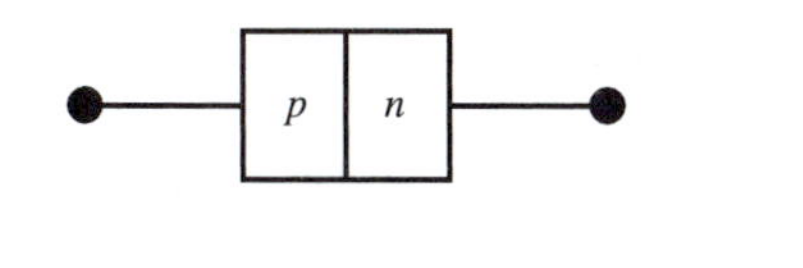

Anode Cathode

FIGURE 5.14 Circuit symbol for the *p-n* junction diode.

given by Banzhaf,[8] where the *element line,* also called the General Form, for a *p-n* junction diode is

p-n junction diode element line

```
DXXXXXX N+ N- MODNAME <AREA> <OFF> <IC=VD>
```

where `DXXXXXX` is the name of the *p-n* junction diode, `N+` is the anode node, `N-` is cathode node, and `MODNAME` is the model name which is used in an associated `.MODEL` control line. These items are required in the *p-n* junction diode element line. The optional parameters are the quantities in the $< \cdots >$, and an element line may be continued by entering a `+` sign at the start of the next line. The circuit symbol for the *p-n* junction diode with the anode and cathode identified are shown in Fig. 5.14. The meanings of the optional parameters are:

`AREA` The area parameter specifies how many of the *p-n* junction diode model `MODNAME` are connected in parallel to make one `DXXXXXX`. This parameter also affects the model parameters

TABLE 5.1 SPICE2 and PSpice *p-n* Junction Diode Model Parameters.

No.	Text Symbol	SPICE Keyword	Parameter Name	Default Value	Units
1	I_s	IS	saturation current	1.0E-14	A
2	n	N	ideality factor (ranges from 1.0 to 2.0)	1.0	—
3	R_s	RS	ohmic resistance	0	Ohm
4	V_{bi}	VJ	built-in voltage	1	V
5	E_g	EG	energy gap	1.11	eV
6	—	XTI	saturation current temperature exponent	3.0	—
7	V_B	BV	reverse breakdown voltage	infinite	V
8	—	IBV	current at breakdown voltage	1.0E-3	A
9	τ_D or τ_o	TT	transit time or minority carrier lifetime	0	s
10	$C_j(0)$	CJO	zero-bias depletion capacitance	0	F
11	m	M	grading coefficient	0.5	—
12	—	FC	coefficient for forward-bias depletion capacitance expression	0.5	—
13	k_f	KF	flicker noise coefficient	0	—
14	a_f	AF	flicker noise exponent	1	—

`IS`, `RS`, `CJO`, and `IBV` which are listed in Table 5.1. The default value is 1.

`OFF`	The initial condition of `DXXXXXX` for DC analysis.
`IC-VD`	SPICE will use `VD` as the initial condition for the *p-n* junction diode voltage instead of the quiescent operating point voltage when a transient analysis is made.

The *model line*, also called the Model Form, for the *p-n* junction diode is

p-n junction diode `.MODEL` *line*

```
.MODEL MODNAME D<(PAR1=PVAL1 PAR2=PVAL2 ...)>,
```

where `MODNAME` is the model name given to a *p-n* junction in the element line, and `D` denotes that the device is a *p-n* junction diode. `PAR` is the parameter name of one of the optional parameters listed in Table 5.1 for SPICE2 and PSpice. Additional optional parameters have been added in PSpice.[9] The additional PSpice parameters are given in Table 5.2 in Sec. 5.5.3. `PVAL` is the value of the designated parameter. Care must be taken to assign the correct units, which are also designated in the tables.

5.5.2 SPICE2 and PSpice Model Parameters

`IS` `N` `RS` `VJ` `EG` `XTI`

The optional parameters, as described by Massobrio and Antognetti,[10] which are used to represent the *p-n* junction diode are given in Table 5.1. Parameter no. 1 is the saturation current `IS` and would be the value for diffusion current [see Eqs. (4.115) and (4.116)] with the ideality factor in parameter no. 2 taken as `N` = 1. If `IS` represents space-charge recombination, the `IS` is given by Eqs. (4.157) and (4.158) with parameter no. 2 taken as `N` = 2. Surface recombination current, which was given in Eq. (4.159), would also be used with `N` = 2. Often, `IS` is determined by measurement of a test structure. It should be noted that a small conductance is added by default in SPICE with every *p-n* junction to aid convergence. This conductance value is 10^{-12} mho. The third parameter is the series resistance `RS`, which was represented in Eq. (4.161) in Sec. 4.7.7. The built-in potential, which was given in Eq. (4.62), is represented by `VJ`, which is parameter no. 4. The energy gap `EG` given by parameter no. 5 will depend on the semiconductor being used and the temperature. The saturation current temperature exponent given by parameter no. 6 as `XTI` refers to temperature-dependent terms in the saturation current. For example, the diffusion current saturation current [see Eq. (4.115)] is proportional to n_i^2, which has the product $N_c N_v$, which varies as T^3. Therefore, `XTI` is 3.0 for diffusion current. Space-charge recombination current [see Eq. (4.157)] varies as n_i, and therefore, `XTI` would be 1.5.

`BV` `IBV`

The reverse breakdown *I-V* characteristic is shown in Fig. 5.15. The reverse breakdown voltage is represented by `BV`, which is parameter no. 7. The current at the breakdown voltage is specified by parameter no. 8, which is `IBV`. These first eight parameters in Table 5.1 describe the dc behavior of the *p-n* junction. Both `BV` and `IBV` are specified as positive numbers.

The next four parameters describe the *C–V* behavior. The total capacitance C_D for the *p-n* junction is the sum of the depletion capacitance C_j given in Eq. (5.34) or

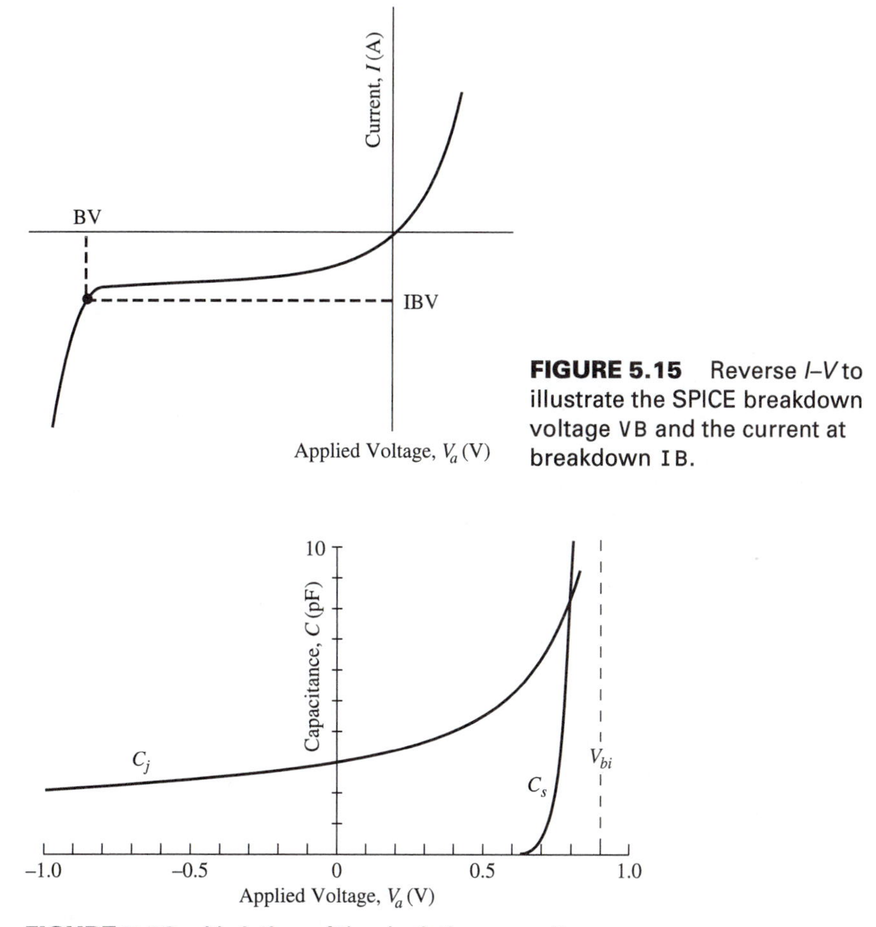

FIGURE 5.15 Reverse *I–V* to illustrate the SPICE breakdown voltage VB and the current at breakdown IB.

FIGURE 5.16 Variation of the depletion capacitance C_j and diffusion capacitance C_s with applied voltage. Note that C_s depends on I_s and τ_p.

Eq. (5.47) and the diffusion capacitance C_s given in Eq. (5.58) or Eq. (5.68). These capacitances are illustrated in Fig. 5.16 and emphasizes that for reverse bias only C_j applies, but for forward bias both capacitances are important. The details of the capacitance modeling are given by Massobrio and Antognetti.[11] The equivalent circuit is shown in Fig. 5.17 with a current source I_D for the *p-n* junction current, the series resistance R_S, and a parallel capacitance C_D.

Parameter no. 9 is related to C_s. The charge injected into the neutral semiconductor for forward bias was given in Eq. (5.55) when $W_n > L_p$ as $Q_p = \tau_p I_d$, where τ_p is the minority-carrier lifetime and I_d is the diffusion current. When $W_n < L_p$, the charge in the quasi-neutral semiconductor for forward bias was given in Eq. (5.66) as $Q_p = \tau_D I_d$, where τ_D is the minority-carrier transit time and I_d is the diffusion current.

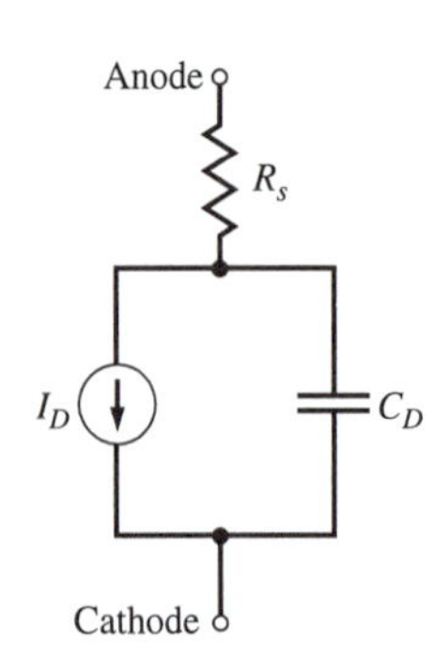

FIGURE 5.17 SPICE equivalent circuit for *p-n* junction diode.

CJO

M

FC

Parameter no. 10 was given in Eq. (5.34) and is the zero bias depletion capacitance represented by `CJO`. The depletion capacitance given in Eq. (5.34) varies as $V_a^{-0.5}$ for the abrupt step junction. When the impurity concentration profile varies linearly, the capacitance was shown in Eq. (5.47) to vary as $V_a^{-1/3}$. This capacitance dependence on voltage is given by parameter no. 10, which is called the grading coefficient `M` and has a value of 0.5 or 0.333. SPICE uses several different expressions for the capacitance.[11] One expression is used from $V_a = 0$ to a voltage which is a fraction of the built-in potential. This fraction of the built-in potential is represented by `FC` and is generally 0.5, which is a voltage of $0.5V_{bi}$.

KF & AF

The noise parameters used in SPICE are listed as parameter nos. 13 and 14 and are used to represent the flicker noise. A value of noise at a given time t is unpredictable so that noise is characterized as the mean squared value $\overline{i^2}$. Usual measurement involves a narrow bandwidth Δf. Flicker noise ($1/f$ noise) is found in all types of active devices and is associated with the flow of direct current and displays a spectral density of the form

$$\overline{i_{D_F}^2} = k_f \frac{I_D^{a_f}}{f} \Delta f, \tag{5.69}$$

where Δf is a small bandwidth at the frequency f, I_D is the DC drain current, k_f is the flicker-noise coefficient (values range from 0.5 to 2), parameter `KF`, and a_f is the flicker-noise exponent, parameter `AF`. The flicker noise is due mainly to random fluctuations of carriers.

5.5.3 Additional PSpice *p-n* Junction Diode Model Parameters

ISR

NR

The additional optional parameters given in PSpice are listed in Table 5.2.[12] Parameter no. 15 is `ISR` and would be the value for the space-charge recombination saturation current given in Eq. (4.158) or the surface-recombination saturation current given in Eq. (4.160). Parameter no. 16, which is `NR`, represents the ideality factor for `ISR` and would be taken as `N` $= 2$ in the $\exp(qV_a/nkT)$ term for the space-charge or surface-recombination current. These two parameters with parameters no. 1 and 2 permit

TABLE 5.2 Additional PSpice *p-n* Junction Diode Model Parameters.

No.	Text Symbol	SPICE Keyword	Parameter Name	Default Value	Units
15	$I_{s_{scr}}$ or $I_{s_{sr}}$	ISR	space-charge recombination or surface-recombination saturation current	0	A
16	n	NR	ideality factor	2.0	—
17	—	IKF	high-injection "knee" current	infinite	A
18	n	NBV	reverse breakdown ideality factor	1 or 2	—
19	—	IBVL	low-level reverse breakdown "knee" current	0	A
20	n	NBVL	low-level reverse breakdown ideality factor	1 or 2	—
21	—	TIKF	IKF temperature coefficient (linear)	0	$°C^{-1}$
22	—	TBV1	BV temperature coefficient (linear)	0	$°C^{-1}$
23	—	TBV2	BV temperature coefficient (quadratic)	0	$°C^{-2}$
24	—	TRS1	RS temperature coefficient (linear)	0	$°C^{-1}$
25	—	TRS2	RS temperature coefficient (quadratic)	0	$°C^{-2}$

including *both* the diffusion current and the space-charge or surface-recombination currents in the SPICE representation for the *p-n* junction.

As presented in Sec. 4.7.7, when the injected minority carrier concentration becomes comparable with the majority carrier concentration, the current begins to vary as $\exp(qV_a/2kT)$. This behavior is represented by IKF, which is parameter no. 17. In this case, the forward-bias diffusion current I_d is represented by[13]

IKF

$$I_d = \left[\frac{1}{1 + \texttt{IS}[\exp(qV_a/kT) - 1]/\texttt{IKF}}\right]^{0.5} \texttt{IS}[\exp(qV_a/kT) - 1], \qquad (5.70)$$

where IS was parameter no. 1 in Table 5.1 and is shown in Fig. 5.18. Curve a represents Eq. (5.70), curve b represents the diffusion current IS$\exp(qV_a/kT)$, while curve c gives IKF = 0.8 A.

PSpice also modifies the space-charge recombination current to take into account the variation of the depletion width with forward bias. The space-charge recombination saturation current density J_{scr} was given in Eq. (4.157) and the depletion width x_d was given in Eq. (4.79). With these two equations for a one-sided p^+-n step junction, the space-charge recombination saturation current density may be written as

$$J_{s_{scr}} = \frac{qn_i x_d}{2\tau_p} = \frac{qn_i}{2\tau_p}\sqrt{\frac{2\epsilon}{qN_d^+}V_{bi}(1 - V_a/V_{bi})^{0.5}}, \qquad (5.71)$$

with $N_a^- \gg N_d^+$. In PSpice, the voltage term is written as[13]

$$(1 - V_a/V_{bi})^{0.5} \Rightarrow [(1 - V_a/V_{bi})^2 + 0.005]^{\texttt{M}/2}, \qquad (5.72)$$

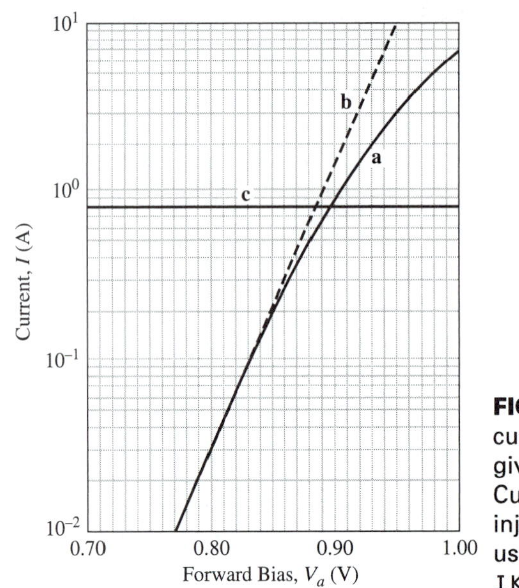

FIGURE 5.18 Illustration of the current behavior at high injection as given by the PSpice parameter IKF. Curve a is the current given at high injection by Eq. (5.70), curve b is the usual diffusion current, and curve c is IKF = 0.8 A.

where M is parameter no. 11, the grading coefficient, and 0.005 is an empirical constant. The space-charge recombination current I_{scr} becomes

$$I_{scr} = \mathrm{ISR}[(1 - \underbrace{V_a/V_{bi}}_{V_a/\mathrm{VJ}})^2 + 0.005]^{\mathrm{M}/2}[\exp(qV_a/2kT) - 1]. \tag{5.73}$$

The total forward current I_{fwd} becomes the sum of Eqs. (5.70) and (5.73) to give

$$I_{fwd} = I_d + I_{scr}$$

$$I_{fwd} = \left[\frac{1}{1 + \mathrm{IS}[\exp(qV_a/kT) - 1]/\mathrm{IKF}}\right]^{0.5} \mathrm{IS}[\exp(qV_a/kT) - 1] + [(1 - \underbrace{V_a/V_{bi}}_{V_a/\mathrm{VJ}})^2 + 0.005]^{\mathrm{M}/2}[\exp(qV_a/2kT) - 1]. \tag{5.74}$$

NBV

The reverse breakdown voltage was specified in Table 5.1 and Fig. 5.15 as VB while the breakdown current was IBV. Parameter no. 18 is NBV and is the ideality factor for the saturation current IBV. In PSpice, a high reverse current $I_{rev_{high}}$ is specified as[13]

$$I_{rev_{high}} = \mathrm{IBV} \exp[q(\mathrm{BV} - V_a)/\mathrm{NBV}kT], \tag{5.75}$$

IBVL NBVL

where BV is a negative number so that when $|\mathrm{BV}| = |V_a|$, $I_{rev_{high}} = \mathrm{IBV}$ and becomes large when $|\mathrm{BV}| < |V_a|$. An additional reverse breakdown current is given by parameters 19 and 20, IBVL and NBVL. This current is represented in PSpice as $I_{rev_{low}}$ as[13]

$$I_{rev_{low}} = \mathrm{IBVL} \exp[q(\mathrm{BV} - V_a)/\mathrm{NBV}kT]. \tag{5.76}$$

The total reverse breakdown current I_{rev} is the sum[13]

$$I_{rev} = I_{rev_{high}} + I_{rev_{low}}$$
$$= \texttt{IBV} \exp[q(\texttt{BV} - V_a)/\texttt{NBV}kT] + \texttt{IBVL} \exp[q(\texttt{BV} - V_a)/\texttt{NBVL}kT]. \qquad (5.77)$$

The first temperature effect is for the saturation diffusion current [Eq. (4.115)] IS, which is proportional to n_i^2:

$$\texttt{IS} \propto n_i^2 \propto N_c N_v \exp(-E_g/kT), \qquad (5.78)$$

where N_c and N_v are given by Eqs. (2.66) and (2.67), respectively, and each varies as $T^{3/2}$. Therefore, the temperature dependence of IS when $\texttt{N} = 1$ is given by[13]

$$\texttt{IS}(T) = \texttt{IS}(T/300)^{\texttt{XTI}} \exp[-\texttt{EG}(1 - T/300)/kT], \qquad (5.79)$$

with a linear representation for $E_g(T)$ near room temperature. For the space-charge recombination current [Eq. (4.157)] or the surface-recombination current [Eq. (4.160)], ISR is proportional to n_i:

$$\texttt{ISR} \propto n_i \propto (N_c N_v)^{1/2} \exp(-E_g/2kT). \qquad (5.80)$$

The temperature dependence of ISR when $\texttt{NR} = 2$ is given in a similar manner by[13]

$$\texttt{ISR}(T) = \texttt{ISR}(T/300)^{\texttt{XTI}/2} \exp[-\texttt{EG}(1 - T/300)/2kT]. \qquad (5.81)$$

The temperature dependence of the high current knee (parameter no. 17) is given as[13]

$$\texttt{IKF}(T) = \texttt{IKF}[1 + \texttt{TIKF}(T - 300)], \qquad (5.82)$$

TIKF

where TIKF is the IKF temperature coefficient, which is given as parameter no. 21.

The next two parameters represent the temperature dependence of the reverse breakdown voltage BV as

$$\texttt{BV}(T) = \texttt{BV}[1 + \texttt{TBV1}(T - 300) + \texttt{TRS2}(T - 300)^2], \qquad (5.83)$$

TBV1
TRV2

where TBV1 is the linear BV temperature coefficient (parameter no. 22) and TRV2 is the quadratic BV temperature coefficient (parameter no. 23). Parameters no. 24 and 25 represent the temperature dependence of the diode series resistance RS as[13]

$$\texttt{RS}(T) = \texttt{RS}[1 + \texttt{TRS1}(T - 300) + \texttt{TRS2}(T - 300)^2], \qquad (5.84)$$

TRS1
TRS2

where TRS1 is the linear RS temperature coefficient and TRS2 is the quadratic RS temperature coefficient.

The temperature dependence of the built-in potential V_{bi} (VJ) is included in PSpice. From Eq. (4.62) for V_{bi} with Eq. (2.71) for n_i^2, the built-in potential may be written as

$$V_{bi} = \frac{kT}{q} \ln \frac{N_a^- N_d^+}{n_i^2} = \frac{kT}{q} \ln \frac{N_a^- N_d^+}{N_c N_v \exp(-E_g/kT)}. \tag{5.85}$$

At $T = 300$ K,

$$V_{bi}(300) = \frac{300k}{q} \left\{ \ln \frac{N_a^- N_d^+}{N_c(300) N_v(300) \exp[-E_g(300)/300k]} \right\}, \tag{5.86}$$

and at temperature T,

$$V_{bi} = \frac{kT}{q} [\ln(N_a^- N_d^+) - \ln N_c(300) N_v(300)(T/300)^3 + E_g/kT]. \tag{5.87}$$

Rewriting Eq. (5.87) and adding and subtracting $E_g(300)/300k$ gives

$$V_{bi}(T) = \frac{kT}{q} \left[\ln \frac{N_a^- N_d^+}{N_c(300) N_v(300) \exp[-E_g(300)/300k]} - \frac{E_g(300)}{300k} + \frac{E_g(T)}{kT} \right] \tag{5.88}$$

or

$$V_{bi}(T) = \underbrace{\frac{300}{300} \frac{kT}{q} \ln \frac{N_a^- N_d^+}{n_i^2(300)}}_{V_{bi}(300)(\frac{T}{300})} - \frac{E_g(300)}{q} \left(\frac{T}{300} \right) - \frac{E_g(T)}{q}. \tag{5.89}$$

Equation (5.89) when written with the PSpice parameters becomes[13]

$$\text{VJ}(T) = \text{VJ}(T/300) - [E_g(300)/q](T/300) + E_g(T)/q, \tag{5.90}$$

where $E_g(T)$ is given in PSpice as $E_g(T) = 1.16 - 7.02 \times 10^{-4} T^2/(T + 1108)$, which is somewhat different than $E_g(T)$ given in Eq. (2.72) for Si.

The temperature dependence of V_{bi} is used in the temperature dependence of the zero-bias depletion capacitance given in Eq. (5.33) as

$$\text{CJO} = A\sqrt{q\epsilon N_a^- N_d^+ / 2(N_a^- + N_d^+) V_{bi}}.$$

The temperature dependence is given as[13]

$$\text{CJO}(T) = \text{CJO}\{1 + \text{M}[4.0 \times 10^{-4}(T - 300) + (1 - \text{VJ}(T)/\text{VJ})]\}, \tag{5.91}$$

which appears to be an empirical representation with the previous grading coefficient M (parameter no. 11).

In these temperature-dependent expressions, the normal (nominal) room temperature of 300 K is referred to in PSpice as Tnom. This Tnom may be changed with the `.OPTIONS` control line such as

```
.OPTIONS TNOM=297
```

for the case of a reference temperature of 297 K rather than 300 K. There are four additional PSpice temperature parameters[12] which may be used in the `.MODEL` control line to specify temperatures for the *p-n* junction diode as well as for other semiconductor devices. These parameters are used to specify temperatures such as the temperature for which the parameters in the `.MODEL` control line were measured.[14] These parameters will not be presented here.

5.5.4 Transient Analysis

For switching applications, the time required for the transition from forward-bias current flow to reverse bias where the current flow is very small is generally required to be very fast. The transient behavior of the *p-n* junction was given by Kingston,[15] and several SPICE examples have been given by Pulfrey and Tarr.[16] The circuit for the SPICE transient response analysis is shown in Fig. 5.19.

The voltage source for the transient analysis requires a piecewise-linear voltage, which is illustrated in detail by Banzhaf[17] and is given in Appendix I at the end of this text. The element line for the piecewise voltage source is given by

piecewise-linear voltage source element line

```
VXXXXXX N+ N- PWL(T1 V1 <T2 V2 T3 V3 ...>),
```

where `N+` is the node of the positive terminal of the voltage source and `N-` is the negative terminal of the voltage source. The voltage at a given time is given by `T2 V2 T3 V3 ...` for the piecewise voltage source designated by `PWL`. The `.TRAN` control line is used to plot the results of a transient analysis,[18]

`.TRAN` *control line*

```
.TRAN TSTEP TSTOP <TSTART <TMAX>> <UIC>.
```

`TSTEP` is the time increment used for plotting the results of the transient analysis, not the computing time step. The time of the last transient analysis is `TSTOP`. The default value of `TSTART` is zero and designation of a value may be omitted. `TMAX` is the largest computing time step SPICE will use and if `TMAX` is not specified, the computing time

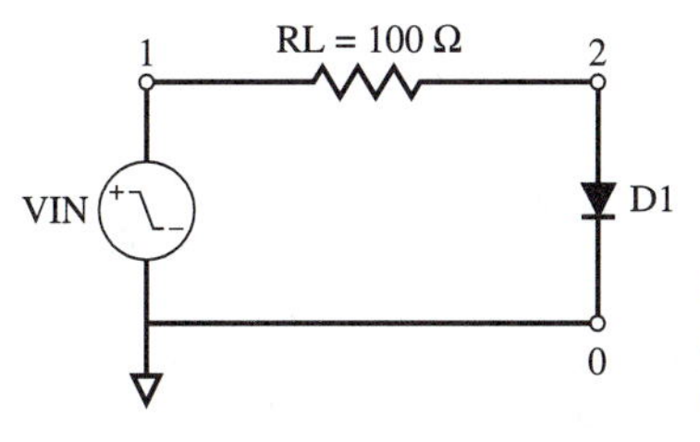

FIGURE 5.19 Circuit diagram for the SPICE analysis of the *p-n* junction diode transient response.

step will be the smaller of `TSTEP` or `(TSTOP-TSTART)/50`. The parameter `UIC` means use initial conditions. For further details, Banzhaf[18] should be consulted. The use of the transient analysis for the recovery time of a *p-n* junction diode may best be illustrated by an example.

EXAMPLE 5.1 Use PSpice to evaluate the current response of a *p-n* junction diode as the bias is changed from forward to reverse bias. The diode has the following SPICE parameters: `IS` $= 1.1 \times 10^{-15}$ A, `TT` $= \tau_p = 1.0 \times 10^{-8}$ s, the built-in voltage is `VJ` $= 0.986$ V, and the zero-bias capacitance `CJO` $= 5.0 \times 10^{-12}$ F. The cross-sectional area is 1×10^{-4} cm^2, $N_a^- = 1 \times 10^{20}$ cm^{-3}, and $N_d^+ = 3 \times 10^{16}$ cm^{-3}. Only consider diffusion current. A 100-Ω resistor is in series with the diode specified earlier. The time variation of the applied voltage is a piecewise step that is +5.0 V at $t = 0$ and at $t = 10$ ns. Then, at $t = 10.2$ ns the voltage switches to −5.0 V. The circuit diagram is shown in Fig. 5.19.

Plot the time variation of the current through the diode from 0 to 50 ns. Let the plotting interval `TSTEP` in the transient analysis be 0.25 ns. However, PSpice would use a computing time interval of (`TSTOP−TSTART`)/50 = 1 ns unless `TMAX` is also specified as `TMAX` = 0.25 `NS`. Note that the diode current reverses for reverse bias, but the diode current continues to flow, then rapidly decays to zero.

Solution The PSpice input file which is named `diode.cir` is:

```
diode turn-off TT=1.0E-8 diode.cir
VIN 1 0 PWL(0NS 5 10NS 5  small +10.2NS -5)
RL 1 2 100
```

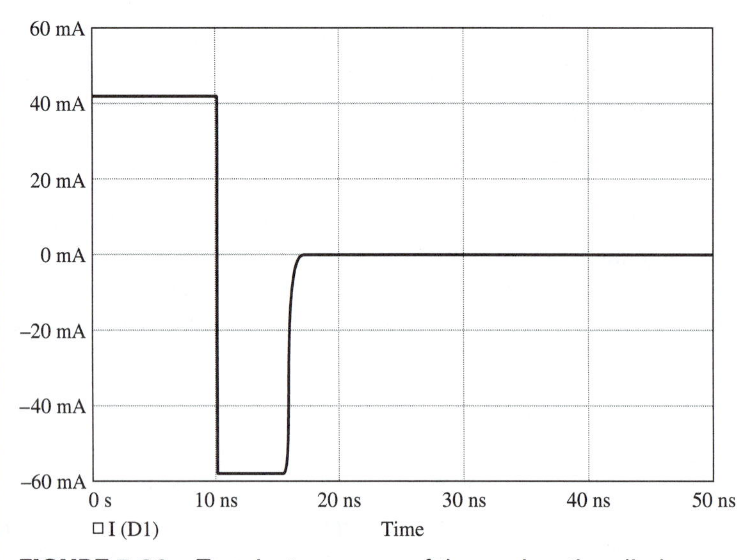

FIGURE 5.20 Transient response of the *p-n* junction diode obtained from the PSpice analysis and plotted with probe.

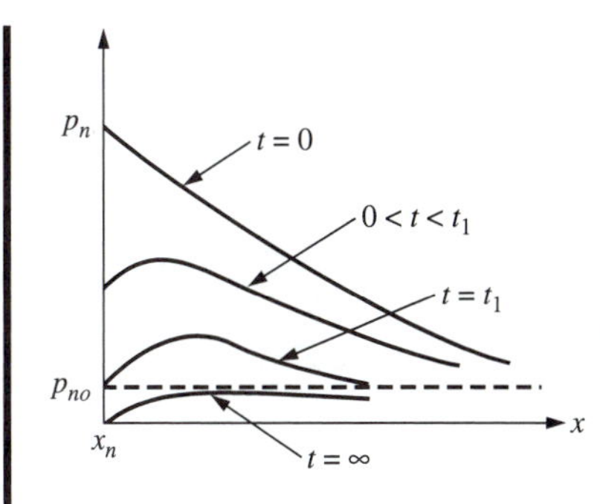

FIGURE 5.21 Hole minority carrier concentration variation with time on the *n*-side of a *p-n* junction diode after the bias is switched from forward to reverse bias at $t = 0$.[15]

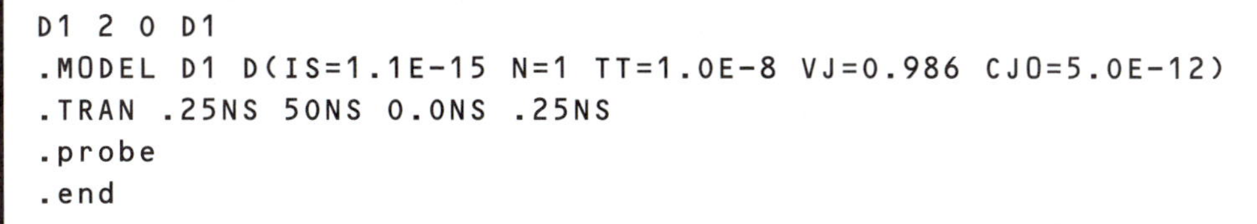

```
D1 2 0 D1
.MODEL D1 D(IS=1.1E-15 N=1 TT=1.0E-8 VJ=0.986 CJO=5.0E-12)
.TRAN .25NS 50NS 0.0NS .25NS
.probe
.end
```

The transient response is shown in Fig. 5.20. The reason for the reverse current when the bias is switched from forward to reverse bias is illustrated in Fig. 5.21 for the variation of the minority hole concentration in the neutral region on the *n*-side. At forward bias, the hole concentration p_n is shown at $t = 0$. As the bias is reversed, the current continues to flow in the reverse direction until all the excess holes are removed from the *n*-side by the recombination of the holes. Therefore, the shorter the lifetime, the more rapidly the current goes to zero. ■

5.6 APPLICATIONS

5.6.1 Planar *p-n* Junction Diode

A planar *p-n* junction diode for use as a circuit element in an integrated circuit is illustrated in Fig. 5.22. In part (*a*), an *n*-type epitaxial layer[19] is grown with a uniform impurity concentration on the *p*-type bulk substrate which has a less closely controlled impurity concentration. This *p-n* junction serves to electrically isolate the devices formed in the epitaxial layer. A layer of silicon dioxide (SiO_2) serves as the mask for formation of the diffused[20] or ion-implanted[21] *p*-type regions shown in Fig. 5.22 (*a*). These *p*-type regions provide electrical isolation between adjacent devices. In part (*b*), an n^+-region permits contact to the *n*-type region, while a diffused or ion-implanted *p*-type region is formed to give the *p-n* junction diode with Al contacts.

A discrete *p-n* junction designed to rectify alternating current is a common application with very low resistance to current in one direction and high resistance to the other direction. Rectifiers generally are large area and their slow switching speed remains faster than the application requires. Detailed descriptions of rectifier applications may be found in most introductory electronics texts.

The material in this section is not necessary for continuity in the remainder of this book and may be omitted.

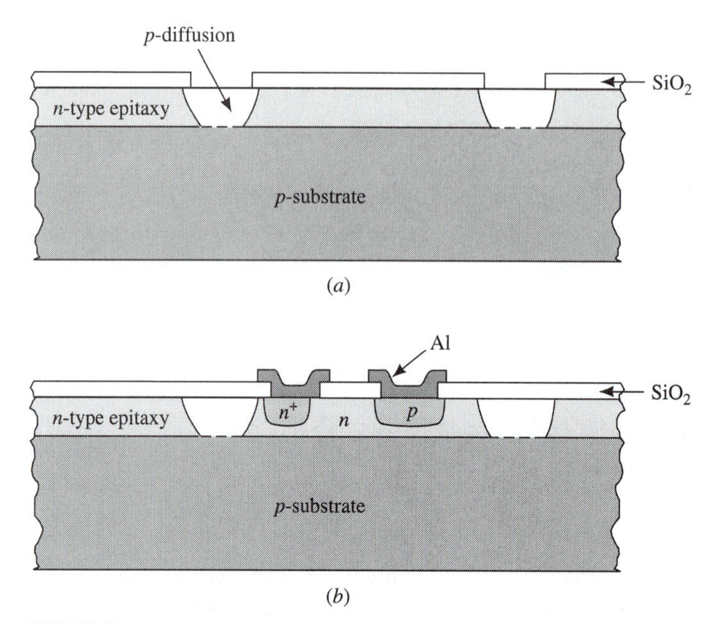

FIGURE 5.22 (*a*) Application of *p-n* junctions to provide electrical isolation between *n*-type regions. (*b*) Planar *p-n* junction for an integrated circuit.

A *p-n* junction can be operated in the reverse breakdown mode to serve as a voltage regulator. The voltage is limited to the breakdown voltage. For voltages larger than $6E_g/q$, the breakdown mechanism is avalanche breakdown, which was introduced in Sec. 5.2. At smaller breakdown voltages, the reverse conduction mechanism is tunneling, which was presented in Sec. 5.3. Most voltage regulators are made with Si because the advanced Si technology permits easier preparation of microplasma-free devices. The temperature coefficient for avalanche breakdown shown in Fig. 5.7 is opposite to the temperature coefficient for tunneling shown in Fig. 5.9. Connection of a positive-temperature coefficient diode in series with a negative-temperature coefficient diode can provide a temperature-insensitive voltage reference.

5.6.2 Microwave Diodes

There are several microwave applications of *p-n* junction diodes.[22] One of the most common is the varactor diode which is used as a variable reactance circuit element.[23] Both Si and GaAs are used. The *p-n* junction is biased in the forward direction to provide a voltage-variable capacitance. Applications include parametric amplification, harmonic generation, mixing, detection, and voltage-variable tuning.[22]

IMPATT diode

The avalanche and transit properties of *p-n* junctions may be used to produce negative resistance at microwave frequencies. One example is the IMPATT diode (Impact avalanche and transit time).[24,25] The negative resistance results from two delays which cause the current to lag behind the voltage. One delay is the avalanche delay caused by the time required for the build up of the avalanche current. The other delay is the transit time delay due to the finite time for the carriers to cross the high-field depletion region. When these two delays equal a half-cycle time, the diode electronic resistance is negative at this frequency. To obtain microwave generation, the *p-n* junction diode is biased into reverse avalanche breakdown and mounted in a microwave cavity.

snap-back diode

Another interesting *p-n* junction diode, which will be classified here as a microwave diode because it can serve as a harmonic generator, is called the charge-storage diode[26] or snap-back diode. This diode is capable of transition from a large reverse voltage to zero voltage in time less than a nanosecond. Applications include pulse generation, wave shaping, as well as harmonic generation. These diodes are characterized by a very abrupt interruption of reverse current in the turn-off transient. In the conventional *p-n* junction diode illustrated by the reverse transient shown in Fig. 5.20, the constant reverse current phase is called the *storage phase*. The storage phase is related to the carrier lifetime and lasts until the injected minority carrier concentration near the edge of the depletion region has reached zero. The *decay phase* represents the current due to the remaining injected minority carriers, and this current decays in a time comparable to the carrier lifetime. These time-dependent carrier concentrations were illustrated in Fig. 5.21. In the snap-back diode, the storage phase lasts for a time comparable to the lifetime, but the time for the decay phase may be several orders of magnitude less than the carrier lifetime. This abrupt transition is obtained by designing the diode so that when the storage phase ends essentially all the minority carriers have been removed. Therefore, there no longer is a decay phase. The *p-n* junctions are designed with majority-carrier profiles which result in retarding fields for the minority carriers outside the depletion region, so that the injected minority carriers are constrained to the region adjacent to the depletion region. This structure can provide 100-V steps in less than ns times.

5.6.3 Optoelectronic Diodes

Optoelectronic diodes are devices in which the photon is affected. Such devices include photodetectors that convert photons into electrons and light-emitting diodes that convert the current due to a forward bias into photons. The generation of carriers by the absorption of photons is governed by Lambert's law, which was given in Eq. (3.76), and the absorption coefficients for Si and GaAs, which were given in Fig. 3.18. A *p-n* junction photodetector absorbs photons which generate electron–hole pairs near the *p-n* junction. These electron–hole pairs are collected by the *p-n* junction and result in a current through the external circuit. A schematic representation is shown in Fig. 5.23. Silicon is used for photodetectors in the 0.8- to 0.9-μm wavelength region, while III-V compound semiconductors are used for the 1.0- to 1.6-μm region. Photodetectors are generally intended to operate only over a narrow wavelength region of a particular optical source. Photodetectors may be integrated with amplifying transistors to give an integrated optical amplifier. These integrated circuits are called OEICs (optoelectronic integrated circuits).

photodetector

light-emitting diode

The light-emitting diode (LED) is a *p-n* junction in which the recombination of the injected minority carriers is largely radiative recombination which was presented

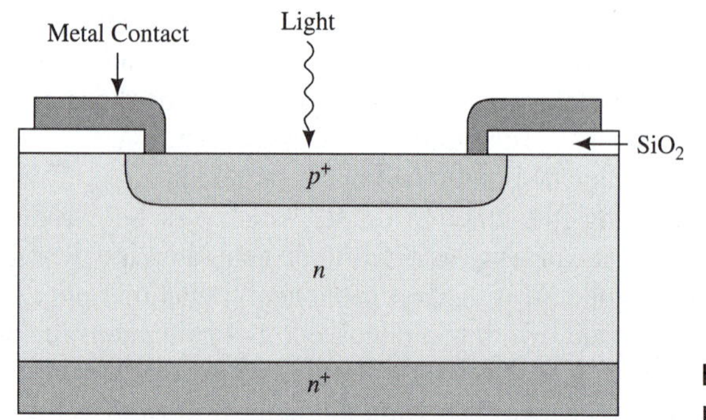

FIGURE 5.23 A *p-n* junction photodiode.[27]

in Sec. 3.5.3. The quantum efficiency was given in Eq. (3.72) and depends on both the nonradiative and radiative lifetimes. The emission is generally in the visible part of the spectrum with wavelengths from 0.4 μm to 0.7 μm or in the near infrared with wavelengths between 2.0 μm and 0.7 μm. Visible LEDs are used for numeric displays or as indicator lamps. Blue LEDs are made with gallium nitride (GaN). Infrared LEDs are used in optoisolators, in television remote controls, and as sources in communication systems. A typical LED is a chip which is $250 \times 250 \times 250$ μm in size and is in a reflective cavity on a reflective lead frame as illustrated in Fig. 5.24 (*a*). The diode is encapsulated in epoxy, both to hold the structure together and to provide refractive-index matching to increase the amount of light which can escape the semiconductor chip. A red LED made with $Al_xGa_{1-x}As$ (see Fig. 2.30) is illustrated in Fig. 5.24 (*b*). The $Al_xGa_{1-x}As$ upper and lower confining layers serve to confine the recombining injected carriers to the $Al_xGa_{1-x}As$ active layer. These confining layers have a larger Al composition and larger energy gap than the active layer where the light is generated. These

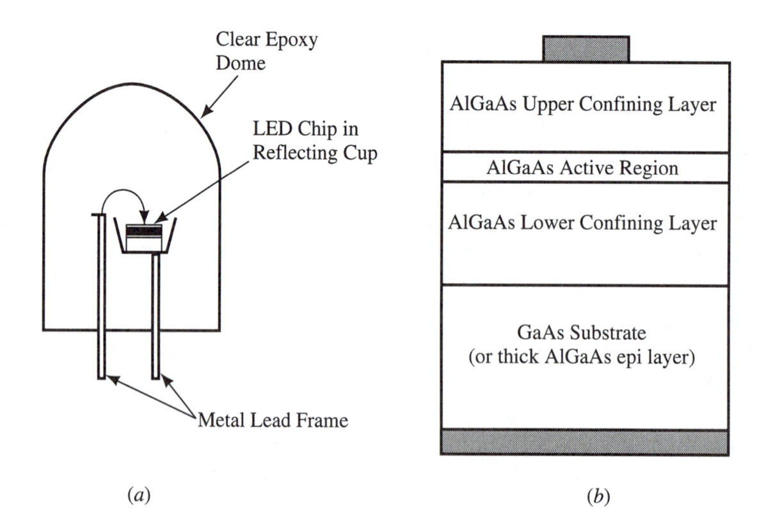

FIGURE 5.24 Typical LED device and chip configuration. (*a*) Cross-sectional view of an LED lamp. (*b*) The LED chip.[28]

confining layers are also transparent to the emitted light and help to reduce the light absorbed in the LED chip. These junctions between semiconductors with different energy gaps are called *heterojunctions.*

5.6.4 Heterojunction Diodes

A heterojunction is a junction in a single crystal between two dissimilar semiconductors. The distinguishing difference is generally the band structure, and the most important feature of the band structure is the energy gap. It is convenient to represent the narrow energy-gap semiconductor by a lower case n or p for its conductivity type, and to represent the larger energy-gap semiconductor by an uppercase N or P.

The heterojunction energy-band diagrams are obtained by assuming that the bulk properties are retained up to the heterojunction interface as shown in Fig. 5.25. In this figure, the energy-band profile for separated p-GaAs and N-$Al_xGa_{1-x}As$ is shown in a diagram similar to Fig. 4.3 (*a*). It may be readily seen that the discontinuity in the conduction band is simply the difference in the electron affinities:

$$\Delta E_c = q(X_1 - X_2) \tag{5.92}$$

and

$$E_{g_2} = \Delta E_c + E_{g_1} + \Delta E_v. \tag{5.93}$$

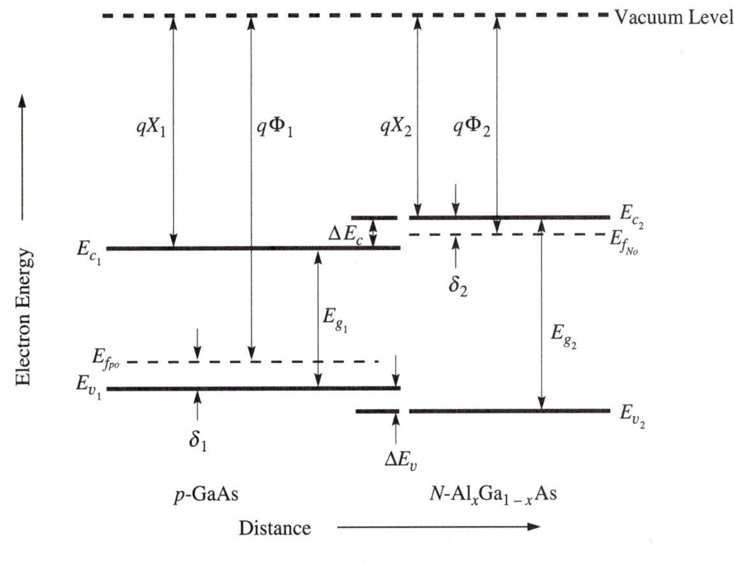

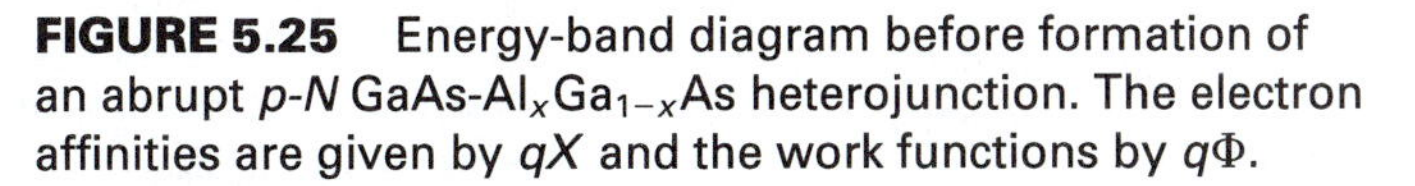
FIGURE 5.25 Energy-band diagram before formation of an abrupt p-N GaAs-$Al_xGa_{1-x}As$ heterojunction. The electron affinities are given by qX and the work functions by $q\Phi$.

Equation (5.93) gives the discontinuity in the valence band as

$$\Delta E_v = (E_{g_2} - E_{g_1}) - \Delta E_c, \tag{5.94}$$

and emphasizes that

$$\Delta E_c + \Delta E_v = E_{g_2} - E_{g_1}. \tag{5.95}$$

Although Eq. (5.93) is an obvious result, it is very important because ΔE_c and ΔE_v are accurately known for GaAs and $Al_xGa_{1-x}As$, while the difference between the electron affinities is not well established. These conduction- and valence-band differences are given by

$$\Delta E_c/(E_{g_2} - E_{g_1}) = 0.35 \tag{5.96}$$

and

$$\Delta E_v/(E_{g_2} - E_{g_1}) = 0.65. \tag{5.97}$$

Detailed expressions for the energy-band diagram for heterojunctions may be obtained from Gauss' law by the analysis given for the *p-n* junction in Sec. 4.3. For the heterojunction, the dielectric constant is different for each side of the heterojunction.[29] A convenient way to draw heterojunction energy-band diagrams is to break the expressions for the variation of the conduction and valence bands with distance x into two

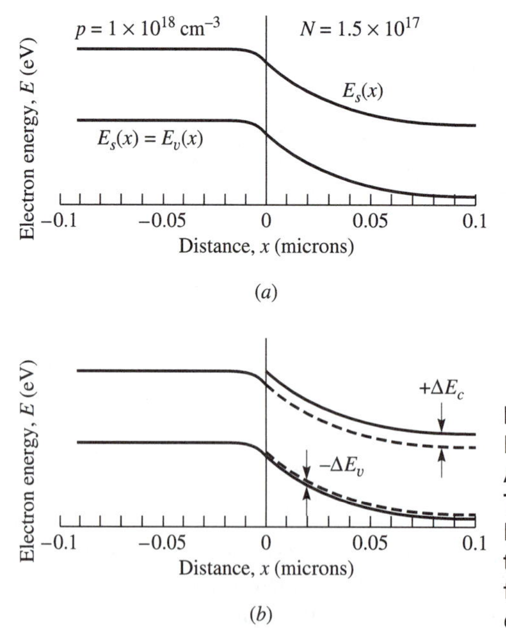

FIGURE 5.26 Energy-band diagram by superposition for the GaAs and $Al_xGa_{1-x}As$ *p-N* heterojunction. (*a*) The *p-n* junction portion of the energy-band diagram. (*b*) Addition of ΔE_c to $E_c(x)$ and the subtraction of ΔE_v from $E_v(x)$ to obtain the heterojunction energy-band diagram.

parts. The energy-band diagram is then obtained by adding the two parts together. One part, represented as E_s, is related to the difference between the built-in potential and the applied bias in the same manner as for the *p-n* junction in Sec. 4.3. The other part is due to the difference in energy gaps and is represented by ΔE_c and ΔE_v.

An example of a *p-N* abrupt heterojunction is shown in Fig. 5.26. In Fig. 5.26 (*a*), $E_s(x)$ is shown, which is the same as for the abrupt *p-n* junction. In Fig. 5.26 (*b*), the heterojunction energy-band diagram is obtained by adding ΔE_c to $E_c(x)$ for $x > 0$ and subtracting ΔE_v from $E_v(x)$ for $x > 0$. This technique is useful when sketching energy-band diagrams to obtain approximate representations to determine the behavior of the electrons and holes.

The derivation of the diffusion current for the *p-N* heterojunction follows the same procedure as given in Sec. 4.7 for the *p-n* junction. The diffusion current density due to holes J_P injected into the *N*-side at $x = x_N$ was given in Eq. (4.109) and may be written for the *p-N* heterojunction as

$$J_P = J_{P_N}(x_N) = \frac{qD_P N_i^2}{L_P N_d^+}[\exp(qV_a/kT) - 1], \tag{5.98}$$

where uppercase P and N have been used to indicate that the holes are in the wide-energy-gap *N*-side. Also, N_i^2 represents the intrinsic carrier concentration in the wide-energy-gap *N*-side. The electron current density J_n due to electron injection into the *p*-side at $x = -x_p$ was given by Eq. (4.111) as

$$J_n = J_{n_p}(-x_p) = \frac{qD_n n_i^2}{L_n N_a^-}[\exp(qV_a/kT) - 1]. \tag{5.99}$$

The influence of the heterojunction on the diffusion current can be illustrated by taking the ratio $|J_n/J_P|$ with Eq. (2.71) for n_i^2 and Eqs. (2.54) and (2.63) for N_c and N_v:

$$\begin{aligned}\left|\frac{J_n}{J_P}\right| &= \frac{D_n L_P N_d^+ n_i^2}{D_P L_n N_a^- N_i^2} = \frac{D_n L_P N_d^+ N_{v_1} N_{c_1}\exp(-E_{g_1}/kT)}{D_P L_n N_a^- N_{v_2} N_{c_2}\exp(-E_{g_2}/kT)}\\ &= \frac{D_n L_P N_d^+ (m_n^* m_p^*)^{3/2}}{D_P L_n N_a^- (m_N^* m_P^*)^{3/2}}\exp[(E_{g_2} - E_{g_1})/kT],\end{aligned} \tag{5.100}$$

where E_{g_2} is the energy gap on the wide-energy-gap *N*-side and E_{g_1} is the energy gap on the narrower energy-gap *p*-side, so that $E_{g_2} > E_{g_1}$, $m_n^* m_p^*$ are the effective masses on the *p*-side, and $m_N^* m_P^*$ are the effective masses on the *N*-side. For a *p-n* junction with $E_{g_2} = E_{g_1}$,

$$\left|\frac{J_n}{J_P}\right| = \frac{D_n L_p N_d^+}{D_p L_n N_a^-}. \tag{5.101}$$

Let $N_a^- = 100 N_d^+$, and neglect the small differences in $D_n L_p$ and $D_p L_n$, then $J_p = 100 J_n$. For the *p-n* junction the more heavily doped *p*-side is responsible for the diffusion current which is injected from the p-side into the n-side. For a p-N heterojunction with $E_{g_2} - E_{g_1} = 0.3$ eV, the $\exp(0.3/kT)$ term at room temperature is 1.0×10^5 so that

heterojunction emitter

$$J_n = \frac{1.0 \times 10^5}{100} J_P = 1.1 \times 10^3 J_P. \tag{5.102}$$

For the p-N heterojunction, the diffusion current is determined by the injection of electrons from the wider energy-gap N-side rather than from the more heavily doped p-side, as in the case of the p-n junction.

5.7 SUMMARY AND USEFUL EXPRESSIONS

This chapter extended the basic concepts of p-n junctions to reverse breakdown and junction capacitance.

- Avalanche breakdown occurs when the carriers which are swept across the space-charge region gain enough energy from acceleration by the high electric field to break covalent bonds upon collision with lattice electrons and create electron–hole pairs.
- The more lightly doped side N_B of an asymmetrically doped p-n junction determines the avalanche breakdown voltage V_B, which varies as $(1/N_B)^{3/4}$.
- For a given impurity concentration, V_B increases with the energy gap as $E_g^{3/2}$.
- For impurity concentrations on both sides of the p-n junction of approximately 10^{19} cm^{-3}, the depletion region becomes thin enough for carriers to tunnel through the junction potential barrier, and tunneling occurs for voltages less than 5 to 10 V.
- The depletion capacitance is due to the variation of the depletion width with applied voltage.
- The diffusion or storage capacitance is due to the minority carriers injected into the neutral semiconductor for forward bias.
- The minority carrier transit time, $\tau_D = W_n^2/2D_p$, in the neutral semiconductor may apply to the diffusion capacitance rather than the minority carrier lifetime τ_o.

The following expressions for p-n junctions were obtained in this chapter.

Avalanche Breakdown Voltage

$V_B \simeq 60(E_g/1.1)^{3/2}(10^{16}/N_{(d \text{ or } a)})^{3/4}$ V

Abrupt Junction Capacitance

$C_j = A[q\epsilon N_a^- N_d^+/2(N_a^- + N_d^+)(V_{bi} - V_a)]^{1/2}$

For $N_a^- \gg N_d^+$, $N_d^+ = (C_j/A)^2 2(V_{bi} - V_a)/q\epsilon$

Diffusion Capacitance

$C_s = (q/kT)\tau_o I_s \exp(qV_a/kT)$

Linearly Graded Junction Capacitance
$C_j = A[qa\epsilon^2/12(V_{bi} - V_a)]^{1/3}$

Minority Carrier Transit Time
$\tau_D = W^2_{p \text{ or } n}/2D_{n \text{ or } p}$

Heterojunction Diffusion Current

$$\left|\frac{J_n}{J_P}\right| = \frac{D_n L_P N_d^+ (m_n^* m_p^*)^{3/2}}{D_P L_n N_a^- (m_N^* m_P^*)^{3/2}} \exp[(E_{g_2} - E_{g_1}/kT]$$

REFERENCES

1. S. M. Sze and G. Gibbons, "Avalanche Breakdown Voltages of Abrupt and Linearly Graded *p-n* Junctions in Ge, Si, GaAs, and GaP," Appl. Phys. Lett. **8**, 111 (1966).
2. S. M. Sze, *Physics of Semiconductor Devices,* 2nd ed. (Wiley, New York, 1981), pp. 101–103.
3. R. H. Haitz, A. Goetzberger, R. M. Scarlett, and W. Shockley, "Avalanche Effects in Silicon *p-n* Junctions. I. Localized Photomultiplication Studies on Microplasmas," J. Appl. Phys. **34**, 1581 (1963).
4. A. Goetzberger, B. McDonald, R. H. Haitz, and R. M. Scarlett, "Avalanche Effects in Silicon *p-n* Junctions. II. Structurally Perfect Junctions," J. Appl. Phys. **34** 1591 (1963).
5. M. O. J. Strutt, *Semiconductor Devices,* Vol. 1, *Semiconductor and Semiconductor Diodes* (Academic Press, New York, 1966), Chap. 2.
6. S. M. Sze, *Physics of Semiconductor Devices,* pp. 513–565.
7. Ibid., p. 82.
8. W. Banzhaf, *Computer-Aided Circuit Analysis using PSpice,* 2nd ed. (Regents/Prentice Hall, Englewood Cliffs, N.J., 1992), p. 109.
9. MicroSim Corporation, 20 Fairbanks, Irvine, Calif. 92718, now Orcad, 9300 S.W. Nimbus Ave., Beaverton, OR 97008.
10. G. Massobrio and P. Antognetti, Eds., *Semiconductor Device Modeling with SPICE,* 2nd ed. (McGraw-Hill, New York, 1993), pp. 11–43.
11. Ibid., p. 22.
12. *The Design Center Circuit Analysis Reference Manual* (MicroSim Corp., Irvine, Calif., 1994), p. 119.
13. Ibid., p. 122.
14. Ibid., p. 60.
15. R. H. Kingston, "Switching Time in Junction Diodes and Junction Transistors," Proc. IRE **42**, 829 (1954).
16. D. L. Pulfrey and N. G. Tarr, *Introduction to Microelectronic Devices* (Prentice Hall, Englewood Cliffs, N.J., 1989), pp. 167–170.
17. Banzhaf, *Computer-Aided Circuit Analysis,* p. 26.
18. Ibid., p. 64.
19. S. M. Sze, Ed., *VLSI Technology,* 2nd. ed. (McGraw-Hill, New York, 1988), p. 55.
20. Ibid., p. 272.
21. Ibid., p. 327.

22. H. A. Watson, Ed., *Microwave Semiconductor Devices and Their Circuit Applications* (McGraw-Hill, New York, 1969), p. 194.
23. Ibid., p. 149.
24. R. L. Johnson, B. C. DeLoach, Jr., and B. G. Cohen, "A Silicon Diode Microwave Oscillator," Bell Syst. Tech. J. **44**, 369 (1965).
25. Sze, *Physics of Semiconductor Devices,* p. 566.
26. J. L. Moll, S. Krakauer, and R. Shen, " *P-N* Junction Charge-Storage Diodes," Proc. IRE **50**, 43 (1962).
27. Sze, *Physics of Semiconductor Devices,* p. 759.
28. M. G. Craford, "LEDs Challenge the Incandescents," IEEE Circuits Devices **CD-8**, 24 (1992).
29. H. C. Casey, Jr., and M. B. Panish, *Heterostructure Lasers, Part A: Fundamental Principles* (Academic Press, New York, 1978) pp. 207–244.

PROBLEMS

All problems are for room temperature unless another temperature is specified.

5.1 Design an abrupt Si p^+-n avalanche breakdown diode.

(a) What net donor concentration of what impurity will give a breakdown voltage $V_B = -60$ V?

(b) For a cross-sectional area of 5×10^{-4} cm^2, plot the capacitance from $V_a = 0$ to V_B.

(c) Plot $1/C^2$ versus V_a from -5.0 to 0 V and indicate V_{bi}.

(d) What is the depletion width at V_B?

5.2 Design an abrupt Si p^+-n avalanche breakdown diode.

(a) What net donor concentration of what impurity will give a breakdown voltage of $V_B = -20$ V?

(b) For a cross-sectional area of 4×10^{-5} cm^2, plot the capacitance from $V_a = 0$ to -10 V.

(c) Plot $1/C^2$ versus V_a from $V_a = 0$ to -10 V.

(d) From the plot in part (c), estimate V_{bi}.

(e) What is the depletion width at V_B?

5.3 Repeat Problem 5.2 for GaAs.

5.4 Repeat Problem 5.1 for a p-n^+ GaN diode with $E_g = 3.40$ eV, $\epsilon = 9.5\epsilon_0$, $m_n^* = 0.20m_0$, and $m_p^* = 0.80m_0$.

5.5 In Problem 4.11 of Chapter 4, a measured *I–V* characteristic for a Si p^+-n junction was given. The cross-sectional area was 1×10^{-4} cm^2. The ionized acceptor concentration on the p^+-side is $p = N_a^- = 1 \times 10^{20}$ cm^{-3}, and on the n-side the ionized donor concentration is $n = N_d^+ = 3 \times 10^{16}$ cm^{-3}. From the measured *I–V* characteristic in the region where $I \propto \exp(qV_a/2kT)$, the lifetime for the space-charge recombination current was 5.0×10^{-9} s, and in the region where $I \propto \exp(qV_a/kT)$, the diffusion length for holes on the n-side was $L_p = 4.6 \times 10^{-4}$ cm.

(a) Give the following SPICE parameters for the diffusion current: `IS`, `N`, `TT`, `VJ`, and `CJO`.

(b) Repeat part (a) for the space-charge recombination current. Use the zero-bias depletion width and ignore its voltage dependence.

5.6 The Si p^+-n junction in Problem 5.1 at room temperature has a breakdown voltage of -60 V. From Fig. 5.4, a breakdown voltage of 60 V requires that $n = N_d^+ = 1 \times 10^{16}$ cm^{-3}. Let the acceptor concentration of the p^+-region be 1×10^{19} cm^{-3}, and the minority carrier lifetime be 1×10^{-7} s (100 ns). The cross-sectional area is 5×10^{-4} cm^2. Give the SPICE parameters for the diffusion current: `IS`, `N`, `TT`, `VJ`, and `CJO`.

5.7 This problem uses PSpice to plot the I–V characteristics for the diode with the SPICE parameters in Problem 5.5. Use two different diodes in parallel to represent the total diode current. One diode represents the diffusion current and the other diode represents the space-charge recombination current.

(a) Sketch and label a circuit diagram to measure the forward I–V characteristic.

(b) Print your PSpice input file.

(c) Plot the $\log I$ vs. V for V_a varied from 0 to 0.8 V. Use a current scale from 1×10^{-8} A to 1×10^{-1} A.

(d) Repeat part (c) for a different space-charge recombination current when the lifetime is changed to 5.0×10^{-8} s.

(e) Repeat part (c) for the space-charge recombination current with the original lifetime of 5.0×10^{-9} s, but with a different diffusion current for $L_p = 1.0 \times 10^{-4}$ cm.

(f) Describe the effect of τ_p and L_p on the intersection of the $I \propto \exp(qV_a/2kT)$ and $I \propto \exp(qV_a/kT)$ currents.

5.8 **(a)** Show that the maximum electric field for a p^+-n junction where $N_a^- \gg N_d^+$, so that $x_d = x_n$ is given by $\mathscr{E}_{max} = [2qN_d^+(V_{bi} - V_a)/\epsilon]^{1/2}$.

(b) Evaluate $\mathscr{E}_{max}$ for Si at room temperature and $V_a = V_B = -35$ V with $N_d^+ = 2 \times 10^{16}$ cm^{-3}.

5.9 The capacitance of an abrupt GaAs p^+-n junction diode as a function of reverse bias was measured with a capacitance meter with a 1-MHz frequency and is given below:

Reverse bias, $-V_a$ (V)	0	-0.5	-1.0	-1.5	-2.0	-2.5	-3.0
Capacitance, C (pF)	4.41	3.68	3.23	2.91	2.67	2.48	2.32

With a cross-sectional area of 7.85×10^{-5} cm^2, find the built-in potential and the ionized donor concentration N_d^+.

5.10 Start with Eq. (5.30) for C_j and show that C_j may be written as $C_j = \epsilon A/x_d$.

5.11 When the reverse bias of an abrupt Si p^+-n junction diode is increased to -2.55 V, the capacitance becomes one-half the value at zero bias. What is the built-in potential?

5.12 Use PSpice to illustrate the effect of series resistance on a p^+-n junction diode with $I_s = 7.3 \times 10^{-13}$ A and an ideality factor of $n = 1$. Plot the current I from 1×10^{-6} A to 1×10^{-1} A for the series resistance R_s of 0.1 Ω, 1.0 Ω, and 10.0 Ω.

5.13 This problem uses PSpice to evaluate the current response as the bias is changed from forward to reverse bias for the diode with the following SPICE parameters: `IS` $= 1.1 \times 10^{-15}$ A, `TT` $= \tau_p = 5.0 \times 10^{-9}$ s, the built-in voltage is `VJ` $= 0.986$ V, and the zero-bias capacitance `CJO` $= 5.0 \times 10^{-12}$ F. The cross-sectional area is 1×10^{-4} cm^2, $N_a^- = 1 \times 10^{20}$ cm^{-3}, and $N_d^+ = 3 \times 10^{16}$

cm.$^{-3}$ Only consider diffusion current. A 100-Ω resistor is in series with the diode specified earlier. The time variation of the applied voltage is a piecewise step that is +5.0 V at $t = 0$ and at $t = 10$ ns. Then, at $t = 10.2$ ns the voltage switches to −5.0 V.

(a) Sketch and label your circuit diagram.

(b) Plot the time variation of the current through the diode from 0 to 50 ns. Let the time interval in the transient analysis be 0.25 ns. Note that the diode current reverses for reverse bias, but the diode current continues to flow, then rapidly decays to zero.

(c) Repeat part (b) with TT $= 1 \times 10^{-8}$ s.

(d) Repeat part (b) with TT $= 2 \times 10^{-8}$ s.

(e) Repeat part (b) with TT $= 4 \times 10^{-8}$ s.

(f) Is there an approximate relationship to the minority-carrier lifetime and the time delay of the decay of the current?

(g) Why does the current continue to flow when the bias is initially reversed?

5.14 **(a)** For Problem 5.13 (b) with $\tau_p = 5.0 \times 10^{-9}$ s, include a second plot to the diode current plot with both the input voltage and the voltage across the diode.

(b) Why does the voltage across the diode correspond to the time response of the diode current rather than the input pulse?

5.15 Repeat Problem 5.13, but let the applied voltage be a piecewise step that remains at −5.0 V from $t = 0$ until $t = 10$ ns. Then, at $t = 10.2$ ns the voltage switches to +5.0 V.

(a) Repeat part (b) of Problem 5.13.

(b) In this case, the diode is initially reverse biased and then forward biased. Explain why the time response is so different from the result in Problem 5.13 when the diode is initially forward biased and then reverse biased.

5.16 Use PSpice to compare the effect of $R_s = 10$ Ω and for $R_s = 0$, but IKF $= 1 \times 10^{-3}$ A. Let $I_s = 7.3 \times 10^{-13}$ A and the ideality factor n = 1. Plot the current I from 1×10^{-6} to 0.1 A.

5.17 Repeat parts (a) and (b) of Problem 5.13 for the D1N4148 switching diode with parameters selected from the PSpice parts library: IS = 2.682n, N = 1.836, RS = 0.5664, CJO = 4p, M = 0.333, VJ = 0.5, Fc = 0.5, TT = 11.54n.

5.18 An abrupt Si p^+-n junction at room temperature is to be designed to have a −35-V breakdown voltage and is gold diffused to have a lifetime of 1.5×10^{-7} s.

(a) What gold concentration is required to give $\tau_p = 1.5 \times 10^{-7}$s?

(b) What donor concentration is required to give the −35-V breakdown voltage?

(c) What is the depletion width at breakdown?

(d) What is the saturation current density for the diffusion current?

(e) What is the saturation current density for the space-charge recombination current?

(f) Let the cross-sectional area be 2.5×10^{-4} cm^2. Use PSpice to plot the current with a semi-logarithmic scale for a forward bias from 0 to 0.8 V.

5.19 Show the effect of the series resistance of a light-emitting diode (LED) by plotting with PSpice probe the log I vs. the applied voltage for one LED with $R_s = 0$ and the other LED with $R_s = 25$ Ω. Let $I_s = 4.9 \times 10^{-19}$ A and n = 1.8. Use two diodes in parallel so that both plots are on the same sheet and use a current scale from 10^{-7} A to 10^{-2} A for a voltage variation from 0 to 2.0 V.

5.20 A simple rectifier circuit is shown below:

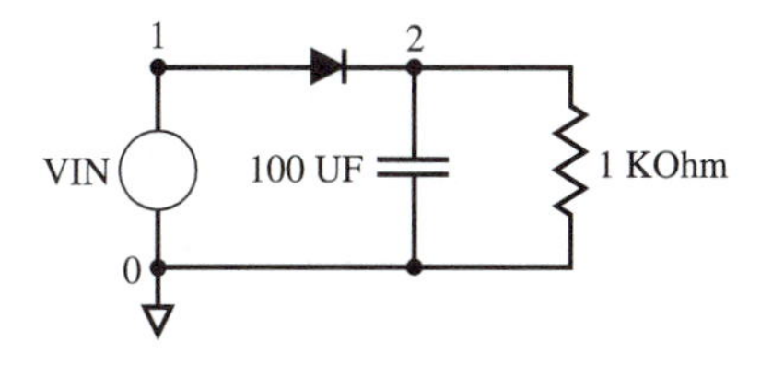

Let `VIN` be a 60-Hz voltage source with a 10-V amplitude. With PSpice, plot `V(1)` and `V(2)` from $t = 0$ s to $t = 6 \times 10^{-2}$ s with increments of 3×10^{-4} s. The diode SPICE parameters are `IS` $= 5.0 \times 10^{-10}$ A, the ideality factor is 1.0, `TT` $= 1 \times 10^{-7}$ s, the built-in potential is 0.90 V, and `CJO` $= 5.0 \times 10^{-10}$ F.

5.21 For the diode circuit in Problem 5.20, let the resistance–capacitance (*RC*) product of the capacitor and resistor increase and decrease by a factor of 10. The diode SPICE parameters are `IS` $=$ 5.0×10^{-10} A, the ideality factor is 1.0, `TT` $= 1 \times 10^{-7}$ s, the built-in potential is 0.90 V, and `CJO` $= 5.0 \times 10^{-10}$ F. Let `VIN` be a 60-Hz voltage source with a 10-V amplitude.

(a) Plot `V(1)` and `V(2)` from $t = 0$ s to $t = 6 \times 10^{-2}$ s with increments of 3×10^{-4} s for both 0.1 *RC* and 10 *RC*.

(b) Give an explanation of how the circuit works and how the *RC* time constant affects the voltage across the parallel resistor and capacitor.

5.22 A diode circuit called a voltage doubler is shown below:

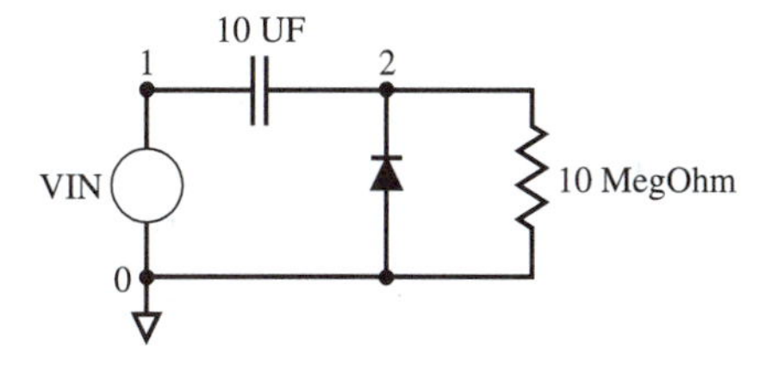

Use the same SPICE diode parameters and plotting conditions as for Problem 5.20.

(a) Print the input and output voltages for the same time scale as in Problem 5.20.

(b) Give an explanation of how the circuit works.

5.23 This problem represents an amplitude detector for an amplitude-modulated signal such as the signal used in commercial AM radio. The amplitude modulation of a higher frequency signal with an audio-frequency signal is represented by

$$\tilde{v}(t) = V_c[1 + m\cos\omega_s t]\cos\omega_c t,$$

where the low-frequency audio-frequency modulation is $\tilde{v}(t) = V_m \cos\omega_s t$, the high-frequency carrier is $\tilde{v}(t) = V_c \cos\omega_c t$, and the ratio $V_m/V_c = m$, the modulation index.

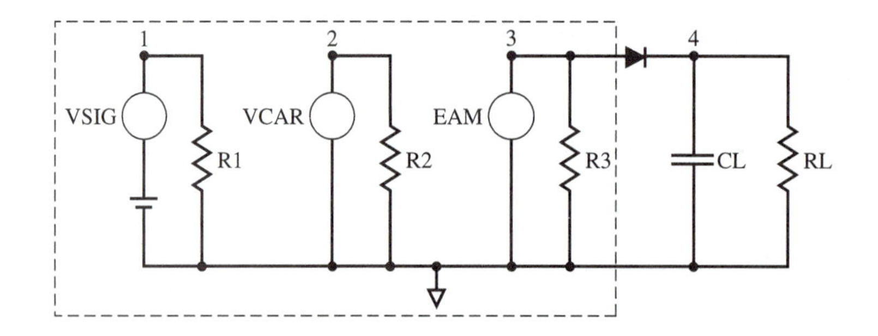

SPICE does not have an amplitude-modulation generator, but one can be created by using a polynomial nonlinear voltage-controlled voltage source (VCVS). The amplitude generator is represented in the figure by the elements in the dashed box. The product of the two signals may be represented by the VCVS given by

```
vsig 1 0 sin(1 1.0 0.8k)
vcar 2 0 sin(0 2.0 50k)
eam 3 0 poly(2) 1,0 2,0 0 0 0 0 1
```

where vsig is the audio signal with a 1-V offset and 800-Hz frequency, vcar is the high-frequency carrier signal of 50 kHz, and eam is the amplitude-modulated signal given by the poly source where the four zeros represent zero coefficients while the last one represents multiplication of vsig and vcar. Let R1 = R2 = R3 = 1 Ω. The model line for the diode is

```
.model gediode D IS=1.77e-6 n=1.7
```

Design the amplitude detector circuit shown earlier to recover the audio signal from the modulated signal by the selection of resistance (*RL*) and capacitance (*CL*) to minimize the ripple due to carrier while the recovered audio signal goes to zero at its minimum.
Give your values of *RL* and *CL* and plot `V(1)`, `V(3)`, and `V(4)` on separate plots on the same page to demonstrate how well your demodulator works. Note that the simple *RC* filter needs a small impedance to the carrier signal and a high impedance to the audio signal, and the *RC* circuit must be able to discharge fast enough to follow the audio signal. More complex filters are used to improve the filtering beyond what this simple *RC* filter can achieve.

5.24 For an $Al_{0.3}Ga_{0.7}As$-GaAs *N-p* heterojunction diode, find the ratio of the injected electron current to the injected hole current if $N = 3 \times 10^{16}$ cm^{-3} and $p = 1 \times 10^{19}$ cm^{-3}. As a first-order approximation, take the effective masses, diffusivities, and diffusion lengths in $Al_{0.3}Ga_{0.7}As$ and GaAs to be equal. Use the energy gap for $Al_{0.3}Ga_{0.7}As$ as given in Sec. 2.7.

5.25 Show that the effective density of states for electrons in $Al_xGa_{1-x}As$ when all three conduction bands are considered is given by

$$N_c = 2.5 \times 10^{19}[(m_n^\Gamma/m_o)^{3/2} + (m_n^L/m_o)^{3/2} \exp(-\Delta E^{L-\Gamma}/kT) + (m_n^X/m_o)^{3/2} \exp(-\Delta E^{X-\Gamma}/kT)](T/300)^{3/2},$$

where $\Delta E^{L-\Gamma} = E_g^L - E_g^\Gamma$ and $\Delta E_g^{X-\Gamma} = E_g^X - E_g^\Gamma$.

5.26 A GaAs light-emitting diode (LED) has a p-type layer doped with Zn to give $p = 1.2 \times 10^{18}$ cm^{-3} and an n-type layer doped with Sn to give $n = 1 \times 10^{17}$ cm^{-3}. Light emission predominately results from the injection of electrons into the p-GaAs layer.

(a) If the radiative constant B is 3.2×10^{-10} cm^3/s, what is the radiative lifetime in the p-type GaAs layer?

(b) If the overall lifetime τ is 1.5×10^{-10} s, what is the nonradiative lifetime?

(c) What is the internal quantum efficiency?

5.27 Consider a Si p^+-n junction doped to give $p = N_a^- = 1 \times 10^{19}$ cm^{-3} and $n = N_d^+ = 2.0 \times 10^{16}$ cm^{-3}. The junction area is 3.0×10^{-4} cm^2, and the minority-carrier lifetime is 1.0×10^{-8} s. Plot the depletion capacitance C_j and the storage capacitance C_s as a function of the applied voltage up to $C_j = C_s$.

5.28 Compare the avalanche breakdown voltage for Si, SiC, and GaN p^+-n junctions with $n = N_d^+ = 3 \times 10^{16}$ cm^{-3}. The energy gaps are given in Table 2.2.

5.29 For the p^+-n junction in Prob. 5.28, find the maximum electric fields at avalanche breakdown. ($\epsilon_{SiC} = 10.0\epsilon_0$, $\epsilon_{GaN} = 9.5\epsilon_0$)

5.30 Show that the total diode capacitance is the sum of the depletion and diffusion capacitance.

CHAPTER 6

SCHOTTKY-BARRIER DEVICES

6.1 INTRODUCTION

unipolar devices

Semiconductor devices in which only one type of carrier predominantly participates in the conduction process may be termed *unipolar devices*. Two unipolar devices are considered in this chapter: the metal-semiconductor diode or Schottky-barrier diode and the metal-semiconductor field-effect transistor (MESFET). The next unipolar device, which is presented in Chapter 7, is the metal-oxide semiconductor (MOS) capacitor, while the metal-oxide semiconductor field-effect transistor (MOSFET) is the subject of Chapter 8.

early history

The first practical semiconductor device was the metal-semiconductor contact in the form of a point-contact rectifier, which was a metallic whisker pressed against a semiconductor surface.* Guglielmo Marconi, considered the father of wireless telegraphy, used Ferdinand Braun's crystal detector, which comprised a spring with a sharpened needle tip on the bluish gray galena crystal pyrite, galena and other sulfur-containing crystals. They shared the 1906 Nobel Prize for these contributions. These

*A very interesting account of the evolution of semiconductor devices from the early rectifier of Ferdinand Braun in 1876 has been given by Hans Queisser in *The Conquest of the Microchip* (Harvard University Press, Cambridge, 1988).

point-contact rectifiers required a special *forming process*,† which often was unreliable. Some sort of localized melting and recrystallization probably occurs during the forming process. In 1938, Walter Schottky[1] suggested that the rectification between a metal and a semiconductor could arise as a result of stable space charge in the semiconductor, and this structure has become known as the Schottky-barrier diode. Schottky-barrier diodes are still used for detectors in microwave receivers as well as rectifiers in high current applications. Metal contacts on heavily doped semiconductors exhibit ohmic behavior with very low resistance and are used as *ohmic contacts* to provide connections to semiconductor devices. The metal-semiconductor diode may also be used as the gate electrode to control the current flow between the source and drain of the three-terminal MESFET.

Schottky-barrier diodes

The analysis of the metal-semiconductor contact begins with the energy-band diagram in Sec. 6.2 which illustrates the built-in potential of Schottky barriers with n- and p-type semiconductors. In Sec. 6.3, Gauss's law is used to obtain the depletion width in the semiconductor at the metal-semiconductor interface. Then the depletion capacitance is obtained in Sec. 6.4. The current–voltage (I–V) characteristic for the Schottky-barrier diode is derived in Sec. 6.5. At high impurity concentrations, the depletion width becomes narrow and carriers can tunnel through the potential barrier. The current–voltage behavior of the metal-semiconductor contact under these conditions becomes ohmic and is described in Sec. 6.6.

MESFET

SPICE

The three-terminal MESFET is introduced in Sec. 6.7. MESFETs made with n-type GaAs are useful in monolithlic microwave integrated circuits (MMICs), low-noise amplifiers, and high-speed logic circuits. Derivation of the source-drain current I_D versus the source-drain voltage V_{DS} as a function of gate voltage V_{GS} is also given in Sec. 6.7. Then, the PSpice MESFET model is given in Sec. 6.8. A brief discussion of MESFET integrated-circuit applications is given in Sec. 6.9. The concepts introduced in this chapter are summarized in Sec. 6.10 and the expressions useful for representation of Schottky-barrier diodes and MESFETs are listed.

6.2 ENERGY-BAND DIAGRAM

The characteristics of point-contact rectifiers were unstable and not reproducible. They have largely been replaced by Schottky-barrier diodes fabricated by the evaporation of a metal layer on the semiconductor surface. A schematic representation of such a device is shown in Fig. 6.1 (a). To fabricate the device, a window is opened in the oxide layer, and a metal layer is deposited in a vacuum system. The metal layer covering the window is subsequently defined by a lithographic step. Analysis of the Schottky-barrier diode will consider a one-dimensional structure as represented in Fig. 6.1 (b) and corresponds to the central section of Fig. 6.1 (a), which is between the dashed lines. The circuit diagram symbol is shown in Fig. 6.1 (c) for metal on an n-type semiconductor.

†The forming process requires overloads with forward or reverse currents and is still an empirical art and is not well understood.

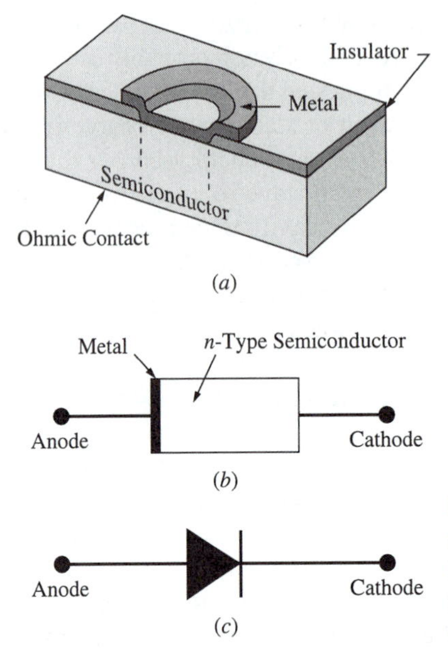

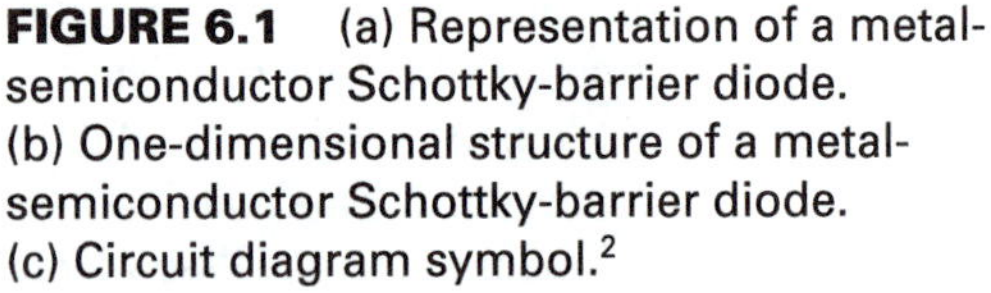

FIGURE 6.1 (a) Representation of a metal-semiconductor Schottky-barrier diode. (b) One-dimensional structure of a metal-semiconductor Schottky-barrier diode. (c) Circuit diagram symbol.[2]

work function $q\Phi$

vacuum level E_0

electron affinity $q\text{X}$

Figure 6.2 (a) shows the energy-band diagram of an isolated metal adjacent to an isolated n-type semiconductor and represents the allowed energies E in the bulk of the material versus position, which is given by the distance x. The metal work function $q\Phi_m$ is the difference between the vacuum or free-electron energy E_0 and the Fermi level E_f and is generally different from the semiconductor work function. For the semiconductor, the difference between E_0 and E_f depends on the impurity concentration in the semiconductor. The difference between E_0 and the conduction band edge E_c is a constant for a particular semiconductor. This difference between E_0 and the E_c is called the *electron affinity* and is represented by $q\text{X}$. The semiconductor work function $q\Phi_s$ is now $q\text{X} + (E_c - E_f)$. The energy-band diagram at initial contact before equilibrium can be established is shown in Fig. 6.2 (b). It is clear that $q\Phi_m > q\Phi_s$, and the electrons in the semiconductor have a higher potential energy than the electrons in the metal. Therefore, electrons in the semiconductor will instantaneously transfer from the semiconductor to the metal to establish thermal equilibrium. At thermal equilibrium, the Fermi levels in the metal and semiconductor must be equal and the vacuum level must be continuous. These two requirements determine a unique energy-band diagram for the ideal metal-semiconductor Schottky-barrier diode, as shown in Fig. 6.2 (c).

For this ideal case, the built-in potential for a Schottky-barrier diode with an n-type semiconductor V_{bi}^n is the difference in the work functions:

built-in potential n-type

$$\boxed{V_{bi}^n = \Phi_m - \Phi_s = \Phi_m - [\text{X} + (E_c - E_f)/q]}. \tag{6.1}$$

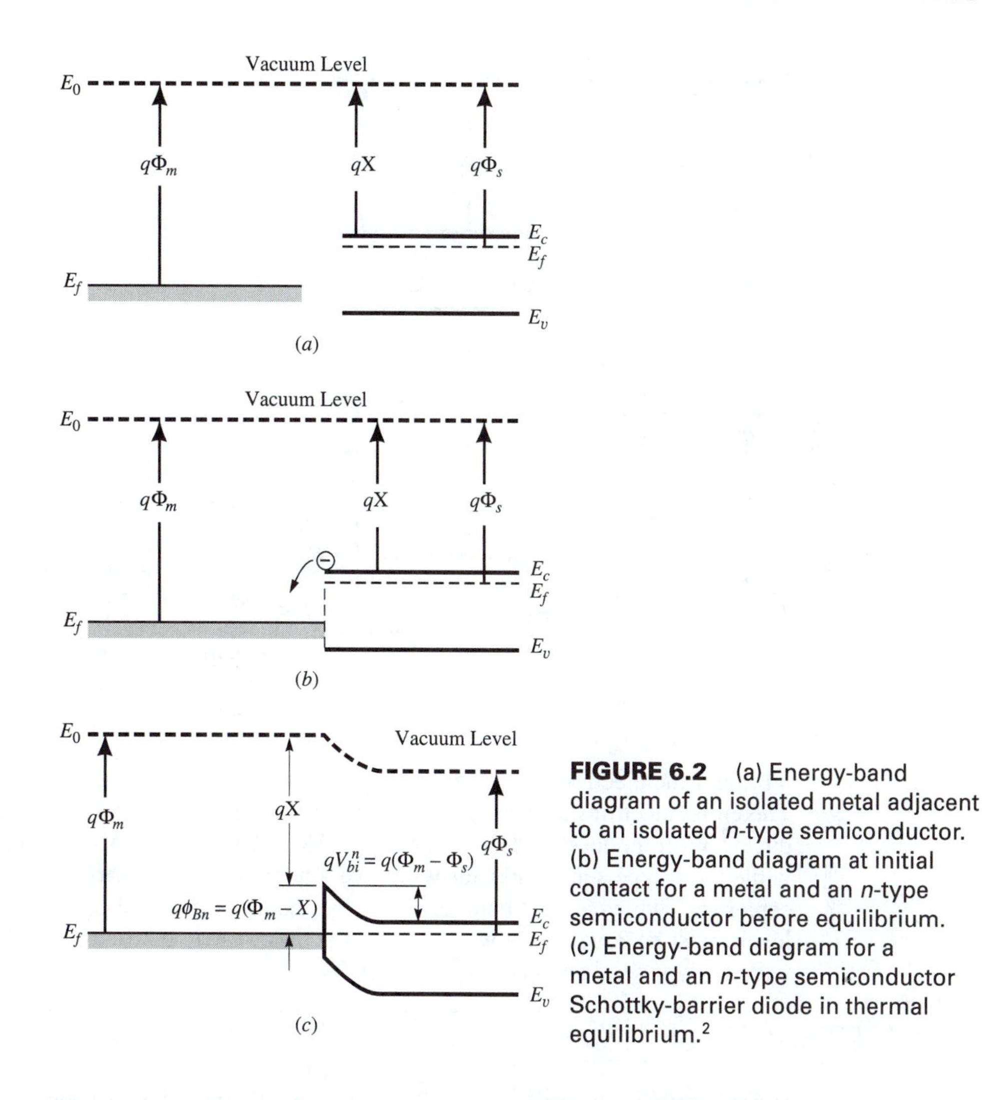

FIGURE 6.2 (a) Energy-band diagram of an isolated metal adjacent to an isolated *n*-type semiconductor. (b) Energy-band diagram at initial contact for a metal and an *n*-type semiconductor before equilibrium. (c) Energy-band diagram for a metal and an *n*-type semiconductor Schottky-barrier diode in thermal equilibrium.[2]

The *barrier height* $q\phi_{Bn}$ for an *n*-type semiconductor is the difference between the metal work function and the semiconductor electron affinity:

barrier height n-type

$$\boxed{q\phi_{Bn} = q(\Phi_m - X)}. \tag{6.2}$$

Both $q\phi_{Bn}$ in eV and ϕ_{Bn} in V are referred to as the barrier height. Substitution of Eq. (6.2) into Eq. (6.1) gives V_{bi}^n as

$$V_{bi}^n = \phi_{Bn} - (E_c - E_f)/q. \tag{6.3}$$

Figure 6.3 (a) shows the energy-band diagram for a metal and *p*-type semiconductor at initial contact before equilibrium can be established. In this case, $q\Phi_m <$

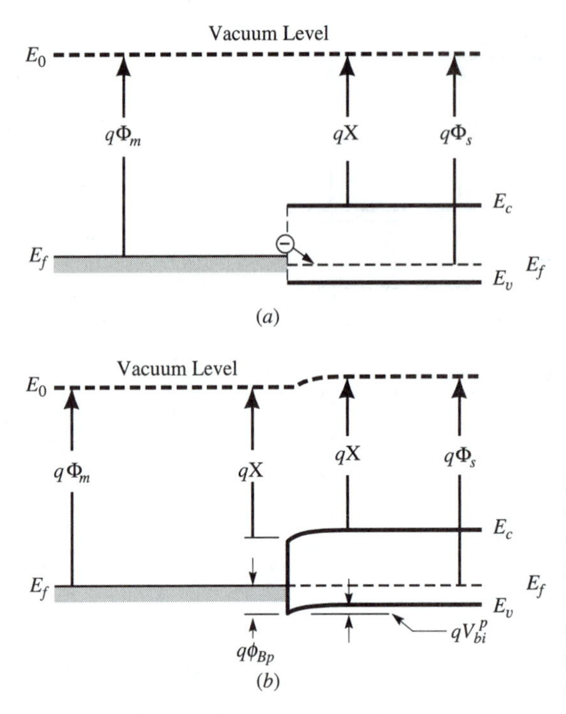

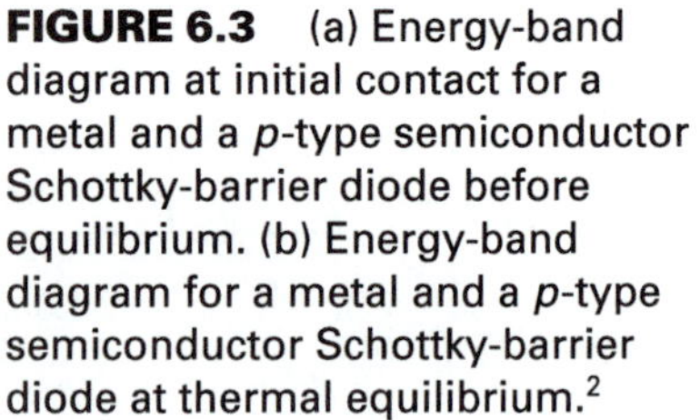
FIGURE 6.3 (a) Energy-band diagram at initial contact for a metal and a *p*-type semiconductor Schottky-barrier diode before equilibrium. (b) Energy-band diagram for a metal and a *p*-type semiconductor Schottky-barrier diode at thermal equilibrium.[2]

$q\Phi_s$, and the electrons in the metal have a higher energy than in the semiconductor. Therefore, electrons at the metal surface will instantaneously transfer to the semiconductor. With the loss of electrons, the metal will have a thin layer of positive charge, while the *p*-type semiconductor will have a negative space charge due to ionized acceptors N_a^- which are not neutralized by free holes. At thermal equilibrium, the Fermi levels must be equal as shown in Fig. 6.3 (b). From Fig. 6.3 (b), the built-in potential with a *p*-type semiconductor V_{bi}^p is given by

built-in potential p-type

$$\boxed{V_{bi}^p = \Phi_m - \Phi_s = \Phi_m - [X + E_g/q - (E_f - E_v)/q]}, \tag{6.4}$$

and is negative because $\Phi_m < \Phi_s$. The barrier height $q\phi_{Bp}$ for the *p*-type semiconductor is also obtained from Fig. 6.3 (b) as

barrier height p-type

$$\boxed{q\phi_{Bp} = qX + E_g - q\Phi_m}. \tag{6.5}$$

By adding Eqs. (6.2) and (6.5),

$$\boxed{q\phi_{Bn} + q\phi_{Bp} = (q\Phi_m - qX) + (qX + E_g - q\Phi_m) = E_g}, \tag{6.6}$$

which demonstrates that the sum of the barrier heights is equal to the energy gap.

To evaluate the built-in potential, numerical values are needed for the metal-work functions and the semiconductor electron affinities. Unfortunately, well-established experimental values are not known and the built-in potential is generally determined experimentally from Schottky-barrier diode capacitance measurements as described in Sec. 6.4. To permit numerical illustrations, the work function of Au is taken[3] as $q\Phi_{Au} = 4.75$ eV and the work function of Al is taken as $q\Phi_{Al} = 4.1$ eV.[4] A more generally accepted value[4] of $q\Phi_{Au}$ is 5.1 eV; however, the value of 4.75 eV has been selected here to assure that V_{bi}^{p} is negative for Au on p-type Si. In actual devices, surface states control the position of the Fermi level at the surface, and experimentally measured barrier heights are the more useful quantity. Use of the work function is helpful in introducing the terminology for Schottky barriers, but does not, in general, give results which numerically agree with experimental measurements. The electron affinities of Si and GaAs are taken as equal, $q\mathrm{X}_{Si} = q\mathrm{X}_{GaAs} = 4.05$ eV.

Au and Al work functions

Si and GaAs electron affinities

6.3 DEPLETION WIDTH

The first electrical property of Schottky-barrier diodes to be derived will be the depletion width x_d. The application of Gauss's law to obtain x_d is much simpler than for the p-n junction in Chapter 4. The energy-band diagram for a Schottky-barrier diode with Au on n-type Si is illustrated in Fig. 6.4 (a) for $n = N_d^+ = 1 \times 10^{15}$ cm^{-3}. The work function for Au, $q\Phi_{Au} = 4.75$ eV, and the electron affinity for Si, $q\mathrm{X} = 4.05$ eV, are also shown in Fig. 6.4 (a). As was shown in Fig. 6.2 (b), electrons are transferred from Si to the metal. These electrons exist in a plane at the metal surface as negative charge, while the n-type Si is depleted of free electrons to give a positive charge N_d^+ over a distance x_d, as illustrated in Fig. 6.4 (b).

The electric field $\mathscr{E}(x)$ is related to the space charge $\rho(x)$ by Gauss's law as

$$\frac{d\mathscr{E}(x)}{dx} = \frac{\rho}{\epsilon_{Si}} = \frac{qN_d^+}{\epsilon_{Si}}. \tag{6.7}$$

Integration of Eq. (6.7) gives

$$\mathscr{E}(x) = (qN_d^+/\epsilon_{Si})x + A_1, \tag{6.8}$$

where A_1 is an arbitrary integration constant to be determined by the boundary conditions. At the edge of the space-charge region at $x = x_d$, the field must be zero. This condition gives

$$A_1 = -(qN_d^+/\epsilon_{Si})x_d \tag{6.9}$$

and

$$\mathscr{E}(x) = -(qN_d^+/\epsilon_{Si})(x_d - x). \tag{6.10}$$

This electric field is plotted in Fig. 6.4 (c).

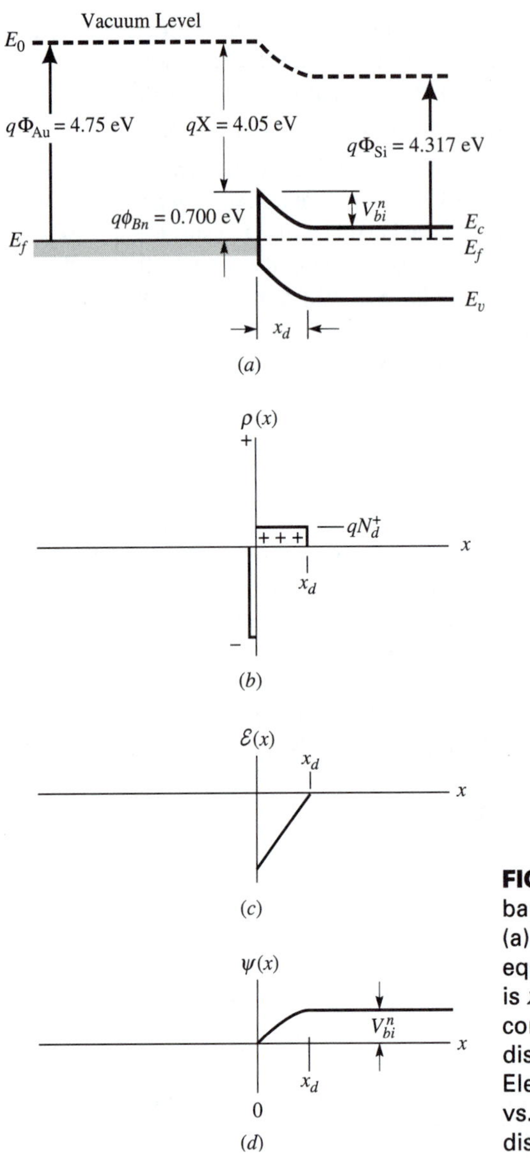

FIGURE 6.4 Gold/n-Si Schottky-barrier diode at thermal equilibrium. (a) Energy-band diagram at thermal equilibrium. The space-charge width is x_d and the ionized donor concentration is N_d^+. (b) Charge distribution $\rho(x)$ vs. distance x. (c) Electric field $\mathscr{E}(x)$ vs. distance x. (d) Potential $\psi(x)$ vs distance x.[2]

For a static electric field, the potential $\psi(x)$ is related to the field by

$$\mathscr{E}(x) = -d\psi(x)/dx, \tag{6.11}$$

and with Eq. (6.10) for $\mathscr{E}(x)$,

$$d\psi(x)/dx = (qN_d^+/\epsilon_{\text{Si}})(x_d - x) \qquad \text{for } 0 < x \leq x_d. \tag{6.12}$$

Integration of Eq. (6.12) gives

$$\psi(x) = (qN_d^+/\epsilon_{\rm Si})(x_d x - x^2/2) + B_1 \qquad \text{for } 0 < x \leq x_d. \tag{6.13}$$

For the boundary condition at $x = x_d$ with $\psi(x_d) = V_{bi}^n$,

$$B_1 = V_{bi}^n - (qN_d^+/\epsilon_{\rm Si})(x_d^2 - x_d^2/2) = V_{bi}^n - (qN_d^+/\epsilon_{\rm Si})x_d^2/2, \tag{6.14}$$

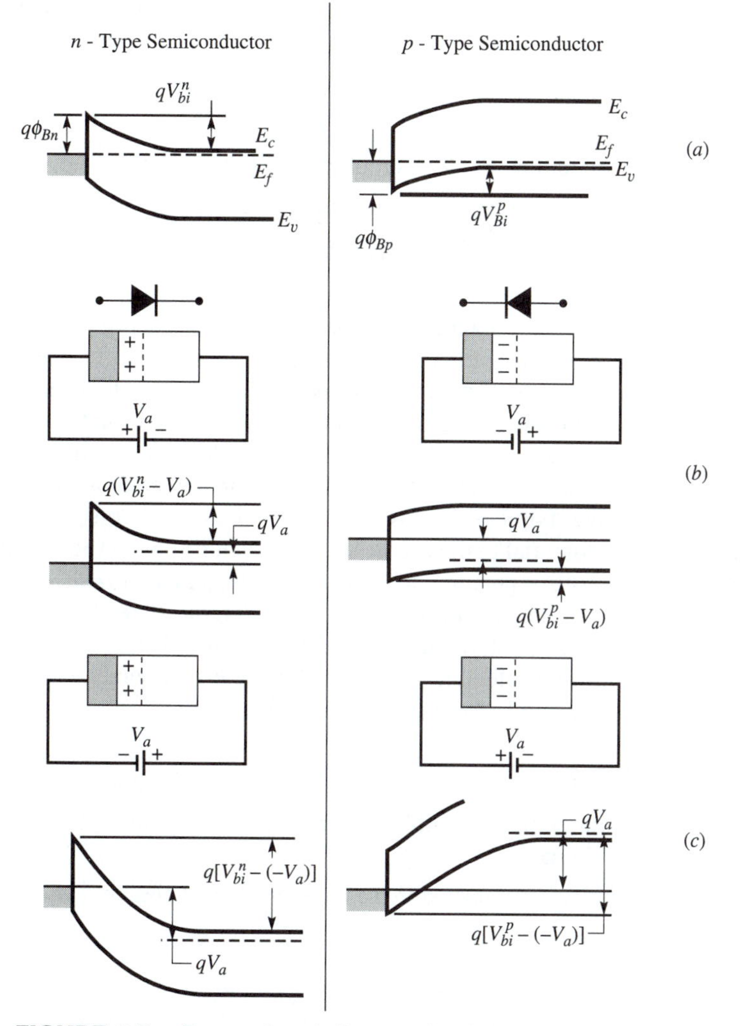

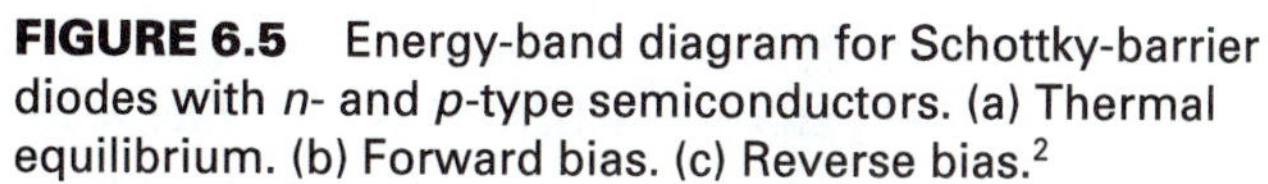

FIGURE 6.5 Energy-band diagram for Schottky-barrier diodes with *n*- and *p*-type semiconductors. (a) Thermal equilibrium. (b) Forward bias. (c) Reverse bias.[2]

which gives for Eq. (6.13),

$$\psi(x) = (qN_d^+/\epsilon_{\text{Si}})(x_d x - x^2/2) + V_{bi}^n - (qN_d^+/\epsilon_{\text{Si}})x_d^2/2, \tag{6.15}$$

or

$$\begin{aligned}\psi(x) &= V_{bi}^n - (qN_d^+/\epsilon_{\text{Si}})(x_d^2/2 - x_d x + x^2/2) \\ &= V_{bi}^n - (qN_d^+/2\epsilon_{\text{Si}})(x_d - x)^2.\end{aligned} \tag{6.16}$$

The potential $\psi(x)$ is plotted in Fig. 6.4 (d). The depletion width is given by Eq. (6.16) with $x = 0$ as

depletion width

$$x_d = \sqrt{2\epsilon_{\text{Si}} V_{bi}^n / qN_d^+}. \tag{6.17}$$

With applied bias V_a, V_{bi}^n is replaced by $(V_{bi}^n - V_a)$ and Eq. (6.17) becomes

$$\boxed{x_d = \sqrt{2\epsilon_{\text{Si}}(V_{bi}^n - V_a)/qN_d^+}}. \tag{6.18}$$

The energy-band diagrams for metals on both n- and p-type semiconductors are shown in Fig. 6.5. Figure 6.5 (a) is for thermal equilibrium. As was done for p-n junctions, it is helpful to use the *schematic space charge model* to determine the polarity of the bias connected to the semiconductor (and metal) for forward or reverse bias. The top part of Fig. 6.5 (b) shows the Schottky-barrier diode circuit symbol and the bias polarities for forward bias. For the n-type semiconductor, the semiconductor space charge due to N_d^+ is positive. Forward bias is for a reduced space-charge width x_d, and connecting a negative potential to the n-type semiconductor *reduces* the positive space charge. Connecting a positive potential to the p-type semiconductor *reduces* the negative space charge due to N_a^-. The resulting energy-band diagrams are shown in the bottom parts of Fig. 6.5 (b). In Fig. 6.5 (c), the reverse-bias polarity is for an increased space-charge width x_d. A positive potential is connected to the n-type semiconductor, and a negative potential is connected to the p-type semiconductor.

EXAMPLE 6.1

(a) What is the depletion width x_d for a Au/n-Si Schottky-barrier diode with $n = N_d^+ = 1 \times 10^{15}\ \text{cm}^{-3}$?

(b) What polarity is connected to the n-Si for reverse bias, and what is x_d for a reverse bias of 10 V?

Solution

(a) $V_{bi}^n = \Phi_m - [\mathrm{X} + (E_c - E_f)/q]$

$$V_{bi}^n = 4.75 - [4.05 + (-0.026 \ln 1 \times 10^{15}/2.84 \times 10^{19})]$$

$$V_{bi}^n = 4.75 - (4.05 + 0.267) = 0.433\ \text{V}$$

$$x_d = \sqrt{\frac{2 \times 11.7 \times 8.85 \times 10^{-14} \times 0.433}{1.6 \times 10^{-19} \times 1 \times 10^{15}}} = 0.749\ \mu\text{m}.$$

(b) To increase the positive space charge for reverse bias, the positive potential is connected to the n-Si. At a reverse bias of 10 V,

$$x_d = \sqrt{\frac{2 \times 11.7 \times 8.85 \times 10^{-14} \times (0.433 + 10.0)}{1.6 \times 10^{-19} \times 1 \times 10^{15}}} = 3.675\ \mu\text{m}.$$

It should be noted that in Example 6.1 an impurity concentration of 1×10^{15} cm^{-3} gives a depletion width of $x_d \approx 1\ \mu$m. Thus, x_d can be estimated by $x_d \approx 1\ \mu\text{m}/\sqrt{N_d^+/1 \times 10^{15}}$ so that $x_d \approx 0.1\ \mu$m for $N_d^+ = 1 \times 10^{17}$ cm^{-3}. ■

6.4 DEPLETION CAPACITANCE

The differential depletion capacitance C_d can be found in a similar manner to the p-n junction capacitance C_j given in Chapter 5. The space charge Q_{sc} with Eq. (6.18) for x_d and a cross-sectional area A is given by

$$Q_{sc} = qN_d^+ x_d A = A\sqrt{2q\epsilon_{\text{Si}} N_d^+ (V_{bi} - V_a)}. \tag{6.19}$$

The differential depletion capacitance C_d becomes

$$C_d = \left|\frac{dQ_{sc}}{dV}\right| = A(2q\epsilon_{\text{Si}} N_d^+)^{1/2}(1/2)(V_{bi}^n - V_a)^{-1/2}, \tag{6.20}$$

or

depletion capacitance

$$\boxed{C_d = A\sqrt{q\epsilon_{\text{Si}} N_d^+/2(V_{bi}^n - V_a)} = \epsilon_{\text{Si}} A/x_d}. \tag{6.21}$$

Several additional useful expressions may be obtained by squaring Eq. (6.21). The ionized donor concentration is given as

$$N_d^+ = (C_d/A)^2 2(V_{bi}^n - V_a)/q\epsilon_{\text{Si}}, \tag{6.22}$$

and

$$1/C_d^2 = 2(V_{bi}^n - V_a)/A^2 q\epsilon_{\text{Si}} N_d^+. \tag{6.23}$$

Because C/A occurs frequently in Schottky-barrier and MOS devices, it has become common to use capacitance per unit area C' as

$$C' = C_d/A, \tag{6.24}$$

and Eq. (6.23) becomes

$$1/C'^2 = 2(V^n_{bi} - V_a)/q\epsilon_{\text{Si}}N^+_d. \tag{6.25}$$

Another useful form of Eq. (6.25) is obtained by taking the derivative to give

$$\frac{-d(1/C'^2)}{dV} = \frac{2}{q\epsilon_{\text{Si}}N^+_d}, \tag{6.26}$$

and the ionized donor concentration N^+_d is given by

$$N^+_d = \frac{2}{q\epsilon_{\text{Si}}}\left[\frac{-1}{d(1/C'^2)/dV}\right]. \tag{6.27}$$

Similar equations may be written for Schottky-barrier diodes with p-type semiconductors.

Measurement of the Schottky-barrier diode capacitance as a function of voltage can provide the built-in potential, the ionized impurity concentration, and the barrier height. To determine these quantities, it is necessary to know the Schottky-barrier

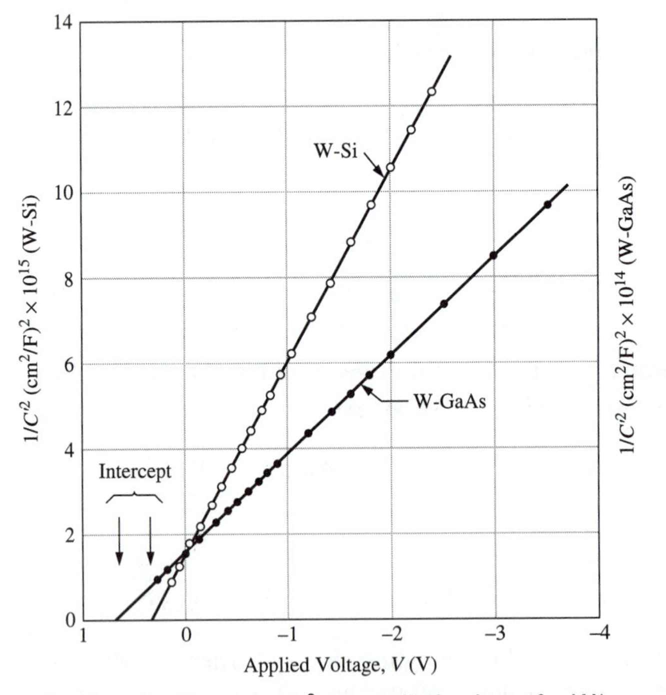

FIGURE 6.6 Plots of $1/C'^2$ vs. applied voltage for W/n-Si and W/n-GaAs Schottky-barrier diodes.[5]

cross-sectional area. If N_d^+ is constant throughout the depletion region, a straight line is obtained by plotting $1/C^2$ or $1/C'^2$ versus the voltage. Because there is no minority-carrier storage (or diffusion capacitance) in the neutral semiconductor for a forward-biased Schottky-barrier diode, Eq. (6.21) gives the total capacitance for forward or reverse bias. Figure 6.6 is a plot of the measured capacitance per unit area versus voltage for tungsten/n-Si and tungsten/n-GaAs Schottky-barrier diodes.[5] The measured capacitance has been divided by the cross-sectional area in cm^2 for the plots in Fig. 6.6. From Eq. (6.25), it may be seen that $1/C'^2$ goes to zero when $V_a = V_{bi}^n$ so that the intercept at $1/C'^2 = 0$ in Fig. 6.6 gives V_{bi}^n. Once V_{bi}^n is obtained, the ionized donor concentration $N_d^+ = n$ can be obtained from Eq. (6.22) or Eq. (6.27). With the Fermi level given by $(E_c - E_f) = -kT \ln n/N_c$ and V_{bi}^n obtained from the $1/C'^2$ intercept, the barrier height may be calculated from Eq. (6.3) as

$$\phi_{Bn} = V_{bi}^n + (E_c - E_f)/q. \tag{6.28}$$

Schottky-barrier diode capacitance–voltage measurements have been one of the most useful techniques for measuring the impurity concentration of epitaxial layers of III-V compound semiconductors such as GaAs.

Figure 6.7 shows the measured barrier heights for n-type Si and n-type GaAs.[6] Note that the barrier height $q\phi_{Bn}$ increases with increasing metal work function $q\Phi_m$.

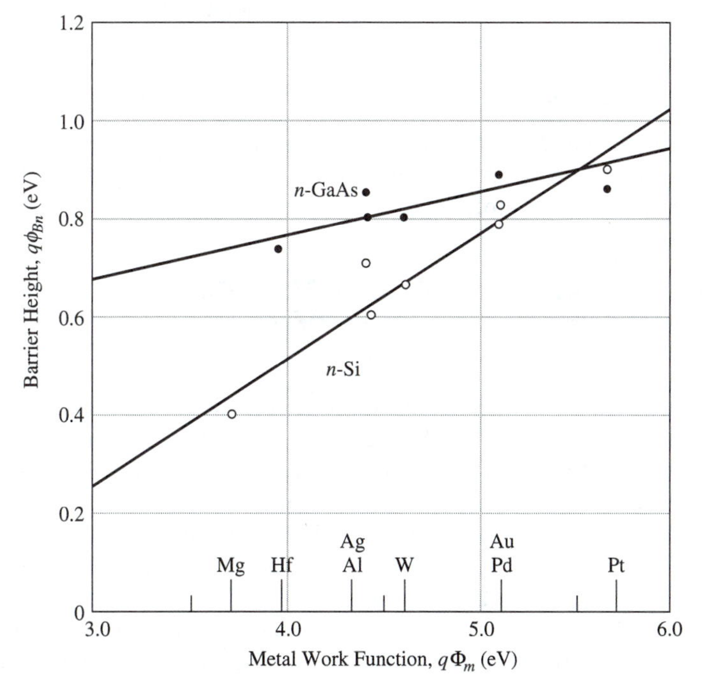

FIGURE 6.7 Measured barrier height for metal-Si and metal-GaAs Schottky-barrier diodes.[6]

However, the dependence is not as strong as predicted by Eq. (6.2). This difference occurs in real Schottky-barrier diodes because there is Schottky-barrier lowering due to an applied electric field,[7] and the disruption of the crystal lattice at the semiconductor surface produces a large number of surface states located in the energy gap. These surface states can act as donors or acceptors which influence the final determination of the barrier height. The barrier height for the ideal Au/n-Si Schottky-barrier diode was given in Fig. 6.4 as 0.700 eV, while the measured value is given in Fig. 6.7 as 0.80 eV.

EXAMPLE 6.2 Find the donor concentration and the barrier height for the tungsten/n-Si Schottky-barrier diode from the data in Fig. (6.6).

Solution The plot of $1/C'^2$ versus V is a straight line, which implies that the donor concentration is constant throughout the depletion region. The $1/C'^2$ intercept gives

$$V_{bi}^n = 0.42 \text{ V}.$$

The ionized donor concentration may be obtained from Eq. (6.22) with

$$1/C'^2 = 1/(C/A)^2 = 1.8 \times 10^{15} \text{ (cm}^2\text{/F)}^2 \text{ at } V = 0.$$

$$N_d^+ = (C/A)^2 2V_{bi}^n/q\epsilon_{\text{Si}}$$

$$N_d^+ = \frac{1}{1.8 \times 10^{15}} \frac{2 \times 0.42}{1.6 \times 10^{-19} \times 11.7 \times 8.85 \times 10^{-14}} = 2.8 \times 10^{15} \text{ cm}^{-3}.$$

Then,

$$\phi_{Bn} = V_{bi}^n + (E_c - E_f)/q = 0.42 - 0.026 \ln 2.8 \times 10^{15}/2.84 \times 10^{19} = 0.66 \text{ V}.$$

For any particular case, it is necessary to experimentally determine the built-in potential. The built-in potential is dependent on the surface preparation before evaporation and the metallization process. ■

6.5 CURRENT–VOLTAGE CHARACTERISTICS

6.5.1 Theory

The current transport in metal-semiconductor diodes is due mainly to majority carriers, in contrast to p-n junctions, where current transport is due mainly to minority carriers. The transport of majority carriers can be adequately described by the thermionic emission theory for the transport of electrons over the potential barrier. The thermionic emission theory by Bethe[8] assumes that (1) the barrier height $q\phi_{Bn}$ or $q\phi_{Bp}$ is much larger than kT, (2) thermal equilibrium is established at the plane that determines emission, and (3) the existence of a net current flow does not affect this equilibrium. Therefore, the current flux from the metal to the semiconductor and the current flux from the semiconductor to the metal can be superimposed. Because of these assumptions, the shape of the barrier profile is immaterial and the current flow depends solely on the barrier height.

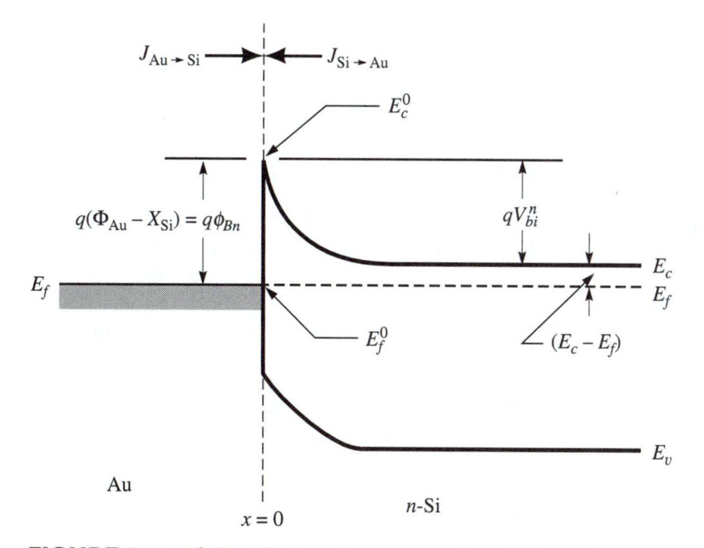

FIGURE 6.8 Schottky-barrier energy-band diagram at thermal equilibrium to illustrate the balance of the current flows between the metal and the semiconductor.

For the Au/n-Si Schottky barrier illustrated[9] in Fig. 6.8 at thermal equilibrium, the current density is balanced by two equal and opposite flows of carriers, thus there is zero net current. Electrons in the semiconductor tend to be emitted into the metal from the semiconductor and are represented by the current density $J_{\mathrm{Si}\rightarrow\mathrm{Au}}$. The opposing balanced flow of electrons from the metal into the semiconductor is represented by the current density $J_{\mathrm{Au}\rightarrow\mathrm{Si}}$. These current components are proportional to the electron concentration at the metal-semiconductor interface at $x = 0$, n_s. At $x = 0$ in Fig. 6.8, the conduction-band edge is given by E_c^0, and the Fermi level is given by E_f^0. By Eq. (2.55),

$$\begin{aligned} n(x = 0) = n_s &= N_c \exp[-(E_c^0 - E_f^0)/kT] \\ &= N_c \exp(-q\phi_{Bn}/kT). \end{aligned} \tag{6.29}$$

It may also be seen in Fig. 6.8 that

$$q\phi_{Bn} = qV_{bi}^n + (E_c - E_f), \tag{6.30}$$

and Eq. (6.29) may be written as

$$n_s = N_c \exp\{-[qV_{bi}^n + (E_c - E_f)]/kT\}. \tag{6.31}$$

At thermal equilibrium,

$$|J_{\mathrm{Au}\rightarrow\mathrm{Si}}| = |J_{\mathrm{Si}\rightarrow\mathrm{Au}}| \propto n_s, \tag{6.32}$$

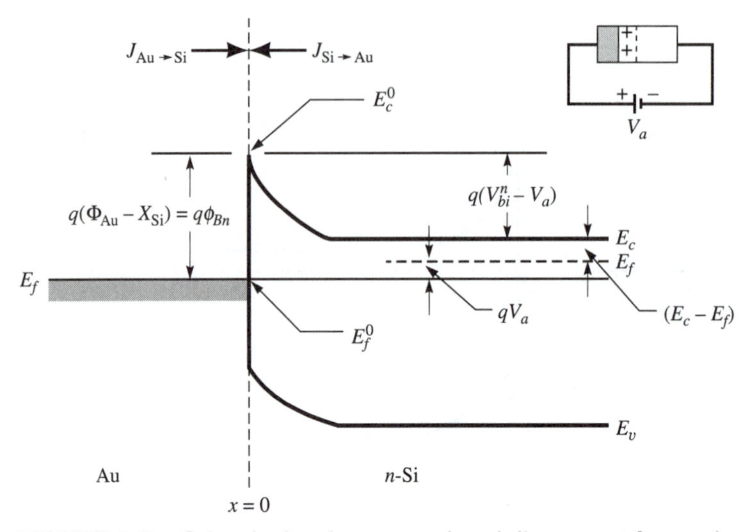

FIGURE 6.9 Schottky-barrier energy-band diagram at forward bias to illustrate the imbalance in current flow.

or,

$$|J_{Au \to Si}| = |J_{Si \to Au}| = C_1 N_c \exp\{-\underbrace{[qV_{bi}^n + (E_c - E_f)]}_{q\phi_{Bn}}/kT\}, \tag{6.33}$$

where C_1 is a proportionality constant.

The energy-band diagram is shown in Fig. 6.9 for forward bias V_a. The built-in potential V_{bi}^n is replaced by $V_{bi}^n - V_a$, and the electron concentration at the interface is increased. Equation (6.31) now becomes

$$n_s = N_c \exp\{-[q(V_{bi}^n - V_a) + (E_c - E_f)]/kT\}, \tag{6.34}$$

but Eq. (6.30) permits replacement of $qV_{bi}^n + (E_c - E_f)$ by $q\phi_{Bn}$. The electron concentration at the interface for forward bias can be written as

$$n_s = N_c \exp[-q(\phi_{Bn} - V_a)/kT]. \tag{6.35}$$

The current density $J_{Si \to Au}$ that results from the electron flow out of the semiconductor is therefore altered by the same factor, $\exp(qV_a/kT)$. The current density from the metal to the semiconductor, however, remains the same because the barrier $q\phi_{Bn}$ remains at its equilibrium value. The net current density under forward bias is then

$$J = J_{Si \to Au} - J_{Au \to Si} = C_1 N_c \exp[-q(\phi_{Bn} - V_a)/kT] - C_1 N_c \exp(-q\phi_{Bn}/kT), \tag{6.36}$$

which gives

$$J = C_1 N_c \exp(-q\phi_{Bn}/kT)[\exp(qV_a/kT) - 1]. \tag{6.37}$$

For reverse bias, the expression for the net current density is identical to Eq. (6.37) except that V_a is replaced by $-V_a$.

Richardson constants

The more rigorous theory which considers the electron velocities[10] permits designation of the coefficient $C_1 N_c$ as $A^* T^2$, where A^* is called the *effective Richardson constant* with units of A/K^2-cm^2. The values of A^* depend on the effective mass. As shown in Table 2.3, the hole effective mass in Si varies with temperature and at room temperature would give $A^* \simeq 130$ A/K^2-cm^2 for *p*-type Si and $A^* \simeq 250$ A/K^2-cm^2 for *n*-type Si. For GaAs, $A^* \simeq 64$ A/K^2-cm^2 for *p*-type and $A^* \simeq 8$ A/K^2-cm^2 for *n*-type. The current–voltage (*I–V*) characteristic of a Schottky-barrier diode for thermionic emission is therefore given by the current density as

$$\boxed{J = J_s[\exp(qV_a/kT) - 1]} \tag{6.38}$$

and

$$\boxed{J_s \equiv A^* T^2 \exp(-q\phi_{Bn}/kT)}, \tag{6.39}$$

where J_s is the saturation current density and the applied voltage V_a is positive for forward bias and negative for reverse bias. Experimental forward *I–V* characteristics of two Schottky-barrier diodes are shown in Fig. 6.10. By extrapolating the forward *I–V* curve to $V = 0$, J_s may be obtained. Then with J_s in Eq. (6.39), the barrier height may be evaluated.

The experimental *I–V* characteristics for Schottky-barrier diodes in forward bias may be represented by

I–V characteristic

$$\boxed{I = I_s[\exp(qV_a/\mathrm{n}kT) - 1]}, \tag{6.40}$$

where n is termed the *ideality factor* and generally has a value between 1.0 and 1.1. The ideality factor may be obtained from the experimental *I–V* characteristic by plotting as log I versus V_a and is given by

$$\mathrm{n} = \frac{q}{kT} \frac{\Delta V_a}{\Delta \log I}. \tag{6.41}$$

In reverse bias, Schottky-barrier diodes also exhibit avalanche breakdown as described for *p-n* junctions.

SPICE

The previous *p-n* junction SPICE model applies to the Schottky-barrier diode, but I_s is given by the current density in Eq. (6.39) times the cross-sectional area A. The transit time parameter `TT` is set to zero because of the absence of minority-carrier injection.

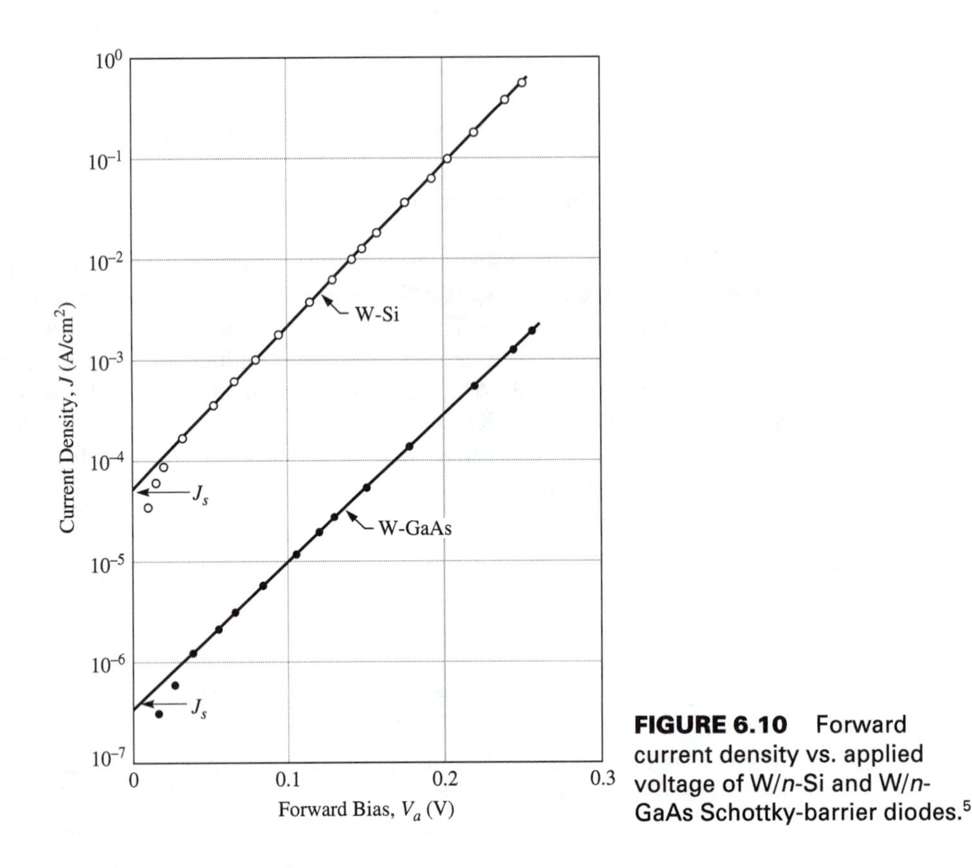

FIGURE 6.10 Forward current density vs. applied voltage of W/*n*-Si and W/*n*-GaAs Schottky-barrier diodes.[5]

Also, the saturation current temperature-dependence exponent parameter `XTI` is set to 2.0 because I_s varies as A^*T^2.

6.5.2 Applications

The *I–V* relationship given in Eq. (6.40) is similar to the expression for *p-n* junctions given in Chapter 4, especially the one-sided p^+-n and n^+-p junctions. There are two significant differences which make Schottky-barrier diodes useful in certain applications. The Schottky-barrier diode operates as a majority-carrier device, which means that there is no minority-carrier storage. The absence of minority-carrier storage results in a faster response than for a *p-n* junction. This difference makes the Schottky-barrier diode useful in high-speed switching applications and as a detector at microwave frequencies. Also, the saturation current I_s for a Schottky-barrier diode is several orders of magnitude larger than for the *p-n* junction. This difference means that at a fixed forward current, the voltage across the Schottky-barrier diode is about one-half the voltage across a *p-n* junction. In power rectifier applications, less power will be dissipated by a Schottky-barrier diode than a *p-n* junction. Conversely, at a given voltage, the

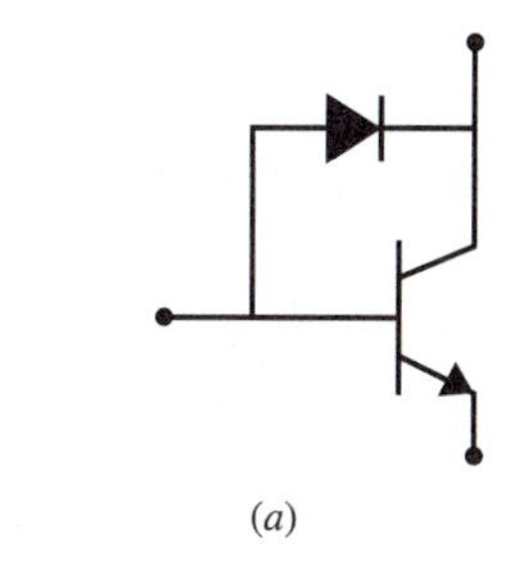

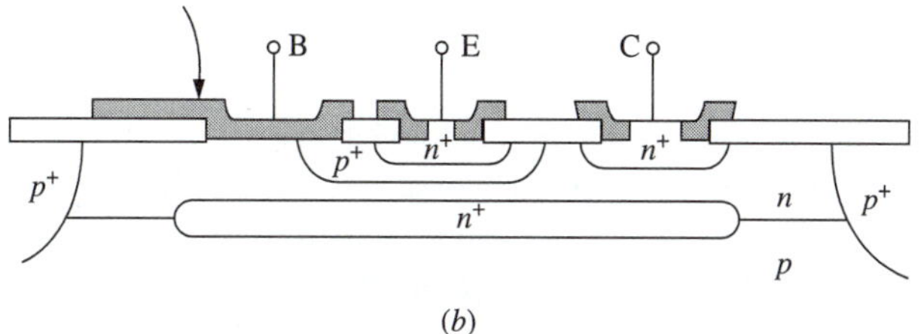

FIGURE 6.11 (*a*) Schematic representation of a Schottky-barrier diode connected between the base and collector of an *npn* transistor. (*b*) Schematic cross section of the circuit in part (*a*).[11]

current through a Schottky-barrier diode is much greater than through a *p*-*n* junction. This current difference will be illustrated in an application with bipolar transistors.

In bipolar transistor logic circuits such as those used in the transistor–transistor logic (TTL), the minority-carrier storage in the base can be avoided by connecting a Schottky-barrier diode between the base and collector of an *npn* transistor (see Problem 9.33) as shown in Fig. 6.11 (*a*). The schematic cross section for the combined structure is shown in Fig. 6.11 (*b*).[11] The Al on the p^+-base region provides an ohmic contact, and the Al on the lightly doped collector layer provides a rectifying contact. When a large signal pulse is placed on the base, the Schottky-barrier diode has a much larger current than the base-collector junction and diverts most of the excess base current through the Schottky-barrier diode. By preventing the transistor from saturating, the charge storage in the transistor is very small and it can switch very rapidly.

6.6 OHMIC CONTACTS

An ohmic contact is defined as a metal-semiconductor contact that has a negligible contact resistance relative to the bulk or series resistance of the semiconductor. A satisfactory ohmic contact should not significantly degrade device performance, and it can pass the required current with a voltage drop that is small compared with the drop across the active region of the device.

A figure of merit for ohmic contacts is the *specific contact resistance* defined as

specific contact resistance

$$R_c \equiv \left[\frac{\partial J}{\partial V}\right]^{-1}_{V=0} \quad \Omega\text{-cm}^2. \tag{6.42}$$

For metal-semiconductor contacts with low-impurity concentrations, the thermionic-emission current dominates the current transport as represented by Eqs. (6.38) and (6.39). The specific contact resistance is obtained from Eq. (6.42) as

$$R_c = (k/qA^*T)\exp(q\phi_{Bn}/kT), \tag{6.43}$$

and does not result in a negligible contact resistance.

To obtain ohmic contacts, it is necessary to contact heavily doped layers. Therefore, the semiconductor is generally heavily doped in the contact region to give an n^+-region on an n-type semiconductor or a p^+-region on a p-type semiconductor. When the impurity concentration is about 3×10^{19} cm^{-3}, the depletion width given by Eq. (6.18) nears 50 Å, and the electrons in the conduction band can tunnel through the potential barrier at the metal-semiconductor interface, as illustrated in Fig. 6.12. The tunneling current for n-type Si may be written as[12]

$$I \sim \exp\left[\frac{-2\sqrt{\epsilon_{\text{Si}} m_n^*}}{\hbar}\left(\frac{\phi_{Bn} - V_a}{\sqrt{N_d}}\right)\right]. \tag{6.44}$$

From Eq. (6.42), the contact resistance when tunneling current dominates is

$$R_c \sim \exp\left[\frac{2\sqrt{\epsilon_{\text{Si}} m_n^*}}{\hbar}\left(\frac{\phi_{Bn}}{\sqrt{N_d}}\right)\right]. \tag{6.45}$$

Equation (6.45) shows that in the tunneling region the specific contact resistance depends strongly on the impurity concentration and varies exponentially with the factor $\phi_{Bn}/\sqrt{N_d}$. Similar expressions may be written for other semiconductors and p-type material.

The calculated values of R_c are plotted in Fig. 6.12 as a function of $1/\sqrt{N_d}$. For $N_d \geq 10^{19}$ cm^{-3}, R_c is dominated by the tunneling process and decreases rapidly with increased impurity concentration. For $N_d \leq 10^{17}$ cm^{-3}, the current is due to thermionic emission, and R_c is independent of doping. Also shown in Fig. 6.12 are experimental data for platinum silicide-Si (PtSi-Si) and aluminum-Si (Al-Si) diodes. They are in close agreement with the calculated values. It is common practice with Si to evaporate a metal such as platinum, which will react with Si to form the silicide PtSi when heated to 500°C. Other widely used silicides are Pd_2Si, $NiSi_2$, and $TiSi_2$. These silicides have resistivities of metals. The process of silicide formation consumes a thin surface layer

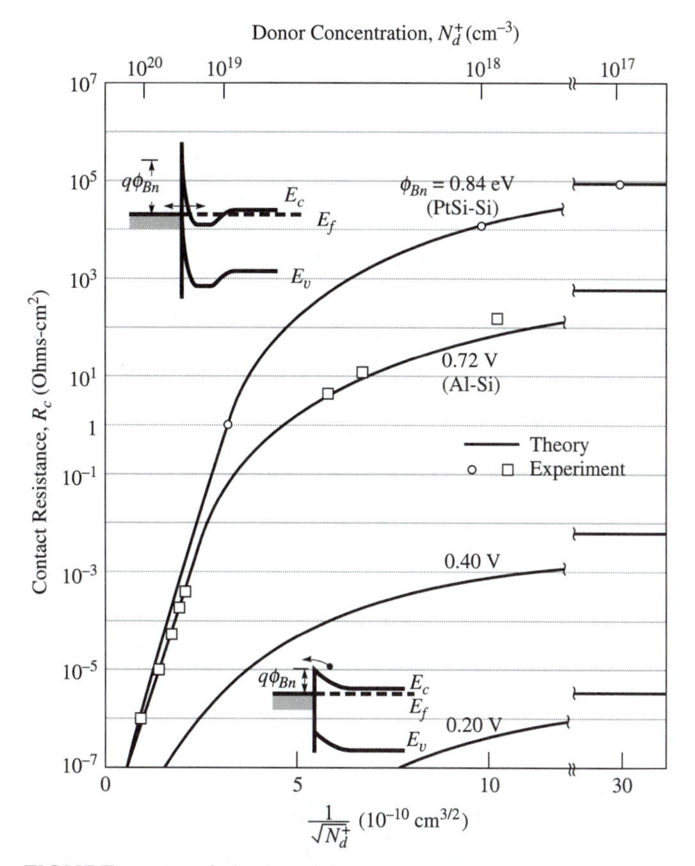

FIGURE 6.12 Calculated and measured values of specific contact resistance. Upper insert shows the tunneling process. Lower insert shows thermionic emission over the barrier.[12]

of Si and makes the Schottky-barrier insensitive to the cleaning process before metal evaporation. Figure 6.12 shows that high-impurity concentration, low barrier height, or both must be used to obtain low values of R_c. Therefore, metal/p^+-p and metal/n^+-n regions make low-resistance ohmic contacts.

Another technique to make ohmic contacts is to alloy an element such as Al on p-type Si. The Al is a highly soluble acceptor impurity in Si and would give a p^+-region on Si. In general, alloyed contacts are very complex to understand because often intermetallic compounds form upon recrystallization after melting in the region where the metal contacts the semiconductor. The melting temperature for Al with Si or Au with Si is very low compared to the melting temperature of Si. Often an element such as Au with a small percentage of a donor or acceptor impurity will be used for an alloyed ohmic contact. The Au determines the alloy temperature, while the impurity makes the alloyed region n^+ or p^+. For the III-V compound semiconductors, ohmic contacts is a very complex subject. Ohmic contact to n-type GaAs is often made by evaporation of layers of Au/Ge/Ni and subsequent alloying near 400°C.[13]

6.7 THE MESFET

6.7.1 Introductory Description

The metal-semiconductor field-effect transistor (MESFET) is basically a voltage-controlled resistor. In the MESFET, the conduction process involves predominately one kind of carrier, and therefore it is called a *unipolar* transistor to distinguish it from the bipolar junction transistor. The MESFET was proposed by Mead[14] in 1966 and subsequently was fabricated by Hooper and Lehrer[15] using an n-type GaAs epitaxial layer on a semi-insulating GaAs substrate. A schematic representation of the MESFET is shown in Fig. 6.13 and consists of a conductive channel with two ohmic contacts, one acting as the carrier source and the other as the drain. The gate metal forms a Schottky-barrier diode which gives a depletion region between the source and drain. The gate depletion region and the semi-insulating substrate form the boundaries of

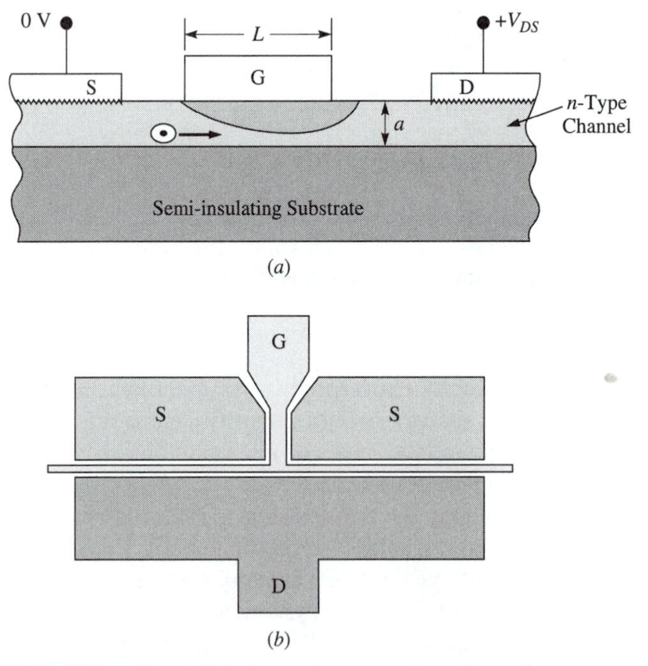

FIGURE 6.13 (a) Schematic representation of a MESFET with an n-type GaAs conducting layer of thickness a. (b) Top view of a T-gate geometry MESFET.[16]

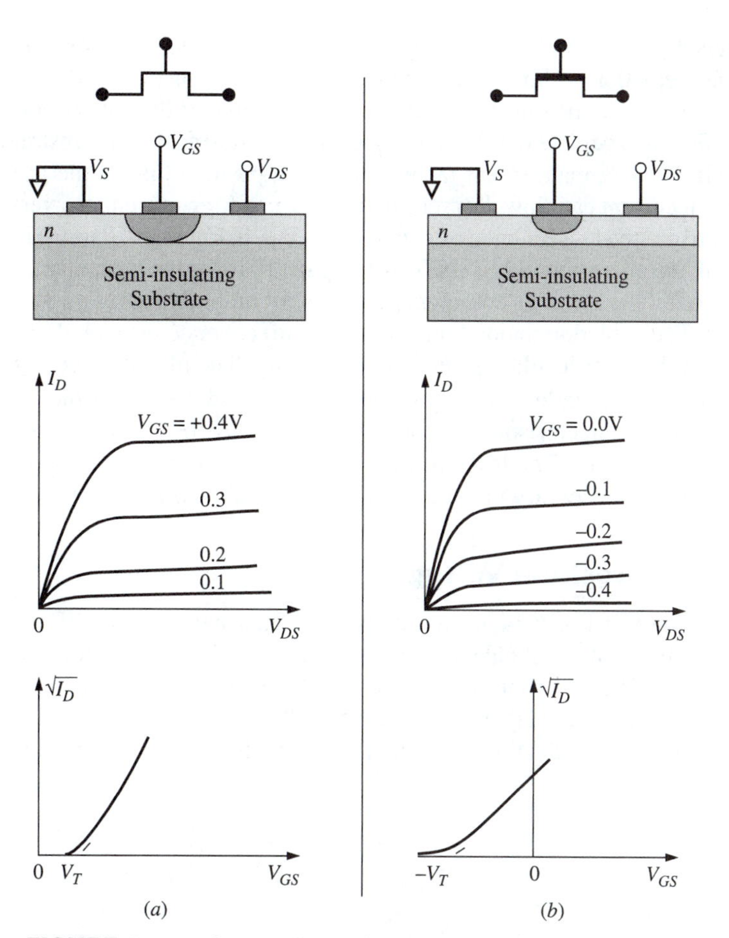

FIGURE 6.14 Comparison of I_D-V_{DS} characteristics. (a) Normally off enhancement-mode MESFET. (b) Normally on depletion-mode MESFET.[17]

the conducting channel. The gate voltage changes the width of the depletion region, which varies the width of the conducting channel. This gate voltage variation of the conducting channel provides a modulation of the source-to-drain current.

Comparison of the electron drift velocity v_d versus the electric field $\mathscr{E}$ for Si in Fig. 3.24 and for GaAs in Fig. 3.25 shows that the maximum drift velocity, which is called the *saturation velocity* v_{sat}, for GaAs is twice the value for Si. This drift velocity difference and the availability of semi-insulating substrates has made GaAs and other III-V compound semiconductors with high values of v_{sat} the preferred material for MESFETs. The high saturation velocities result in higher cutoff frequencies.

There are two basic types of MESFETs. In one type, the normally on (or depletion mode) device, the MESFET has a conductive channel with the Schottky-barrier gate voltage $V_{GS} = 0$, and a negative gate bias must be applied to increase the gate Schottky-barrier depletion width to reach the semi-insulating substrate and cut off the source-to-drain current. In the other type, the built-in potential of the gate Schottky-barrier diode is sufficient to deplete the channel region to the semi-insulating substrate. This normally off MESFET (enhancement mode) requires a positive bias to the gate before source-to-drain current can flow. Representative source–drain currents I_D versus drain voltage V_{DS} are plotted as a function of gate voltage V_{GS} in Fig. 6.14. Part (a) is for the normally off enhancement-mode MESFET, while part (b) is for the normally on depletion-mode MESFET. The symbols commonly used in circuit diagrams are given at the top of the figure. The depletion region for the normally off enhancement-mode MESFET is shown to reach the semi-insulating substrate in part (a). The plot of I_D versus V_{DS} for various values of V_{GS} is called the output characteristic and the plot of the $\sqrt{I_D}$ versus V_{GS} is the transfer characteristic. For the normally off MESFET, the positive V_{GS} necessary to turn the MESFET on is the *threshold voltage* V_T. For the normally on MESFET, the negative V_{GS} necessary to cut the MESFET off is the *threshold voltage* V_T.

6.7.2 Threshold Voltage†

As illustrated in Fig. 6.14, the normally off and normally on MESFETs are distinguished by their different threshold voltages V_T. One is positive and the other is negative. For the schematic MESFET representation in Fig. 6.13(a), with a uniform donor concentration N_d^+ in the conducting layer, V_T is the gate voltage V_{GS} necessary for the depletion width x_d to equal the conducting layer thickness a with $V_{DS} = 0$. From Eq. (6.18),

$$\begin{aligned} x_d &= a \\ &= \sqrt{2\epsilon_{\text{GaAs}}(V_{bi}^n - V_{GS})/qN_d^+}. \end{aligned} \tag{6.46}$$

Then,

pinch-off voltage V_p

$$\boxed{V_{bi}^n - V_{GS} = \frac{qN_d^+ a^2}{2\epsilon_{\text{GaAs}}} = V_p}, \tag{6.47}$$

where V_p is the potential difference across the depletion region at $V_{GS} = V_T$, and V_p is called the *pinch-off voltage*.

For the normally off enhancement-mode MESFET, the Schottky-barrier gate electrode must be forward biased to decrease the depletion width and permit conduction. Therefore, $V_p < V_{bi}^n$ and V_T is given by

$$V_{GS} = V_T = V_{bi}^n - V_p. \tag{6.48}$$

†The material in this part of Section 6.7 is not necessary for continuity in the remainder of this book and may be omitted.

With Eq. (6.1) for V_{bi}^n and Eq. (6.47) for V_p, then

normally off V_T

$$\boxed{V_T = \{\Phi_m - [X_{\text{GaAs}} - (E_c - E_f)/q]\} - qN_d^+ a^2/2\epsilon_{\text{GaAs}}}. \tag{6.49}$$

For the normally on depletion-mode MESFET, the Schottky-barrier gate electrode must be reverse biased to increase the depletion width and reach threshold at cutoff. Therefore, $V_p > V_{bi}^n$, and at threshold, V_{GS} is negative:

$$V_{bi}^n - (-V_{GS}) = V_{bi}^n + V_{GS} = \frac{qN_d^+ a^2}{2\epsilon_{\text{GaAs}}} = V_p, \tag{6.50}$$

Then, the threshold voltage becomes

$$V_T = -V_{GS} = V_{bi}^n - V_p, \tag{6.51}$$

and again substituting Eq. (6.1) for V_{bi}^n and Eq. (6.47) for V_p gives,

normally on V_T

$$\boxed{V_T = \{\Phi_m - [X_{\text{GaAs}} - (E_c - E_f)/q]\} - qN_d^+ a^2/2\epsilon_{\text{GaAs}}}. \tag{6.52}$$

6.7.3 I_D-V_{DS} DC Characteristic for Small V_{DS}

To obtain the I_D-V_{DS} characteristic, first consider a normally on depletion-mode MESFET with V_{DS} small so that the thickness of the conducting channel x_c may be considered to be constant from $y = 0$ to $y = L$. The coordinate system is given in Fig. 6.15 with the gate length L and gate width W. The thickness of the conducting channel is given by

$$x_c = a - x_d. \tag{6.53}$$

The resistance of the channel may be written as

$$R = \frac{\rho L}{x_c W}, \tag{6.54}$$

where ρ is the resistivity of the conducting layer with $N_d^+ = n$. The drain current for this case of V_{DS} small is

$$I_D = \frac{V_{DS}}{R} = q\mu_n n \frac{W}{L} x_c V_{DS}. \tag{6.55}$$

The gate voltage dependence is given by x_c in Eq. (6.53) to give

$$I_D = q\mu_n n \frac{W}{L}(a - x_d)V_{DS} = q\mu_n n \frac{W}{L} a(1 - x_d/a)V_{DS}. \tag{6.56}$$

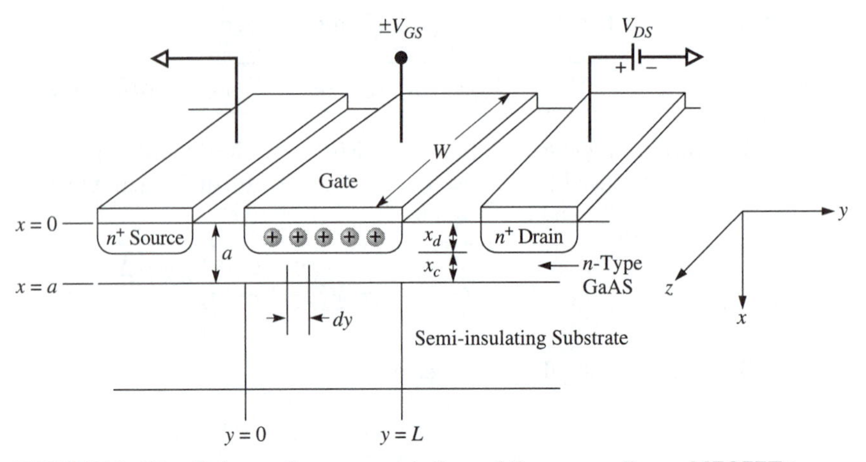

FIGURE 6.15 Schematic representation of the normally on MESFET to illustrate the coordinate system, the depletion width x_d, the conducting layer thickness a, and the thickness of the conducting channel x_c. The drain voltage V_{DS} is small.

With Eq. (6.18) for x_d,

I_D-V_{DS} for small V_{DS}

$$I_D = q\mu_n n \frac{W}{L} a \left(1 - \sqrt{\frac{2\epsilon_{\text{GaAs}}(V_{bi}^n - V_{GS})}{qN_d^+ a^2}} \right) V_{DS}, \tag{6.57}$$

where

$$q\mu_n n \frac{W}{L} a \equiv G_0 = \frac{1}{R_0} \tag{6.58}$$

represents the conductance of the epitaxial layer between the source and drain with no depletion region. Note that with the constraint of small V_{DS}, I_D has a linear dependence on V_{DS}. Current flowing into the drain is a positive current.

6.7.4 I_D-V_{DS} DC Characteristics for Arbitrary V_{DS}

For the more general behavior, the restriction of small V_{DS} is removed. This analysis assumes that the channel and the depletion-region widths vary slowly from the source to the drain. This model is called the *gradual channel approximation.* This condition means that $\mathscr{E}_y << \mathscr{E}_x$ and a one-dimensional analysis may be used for the variation of the depletion region along the channel. For small gate lengths, $L \leq 1$ μm, two-dimensional analysis is required.[18]

For the gradual channel approximation with Eq. (6.55), the voltage $dV(y)$ across an incremental channel length dy at y, as illustrated in Fig. 6.15, becomes

$$dV(y) = I_D dR = \frac{I_D dy}{Wq\mu_n n(a - x_d)}, \tag{6.59}$$

with $x_c = (a - x_d)$, $L \rightarrow dy$, $R \rightarrow dR$, and $V_{DS} \rightarrow dV(y)$. The depletion depth x_d is now controlled by the potential $[V_{bi}^n - V_{GS} + V(y)]$,

$$x_d = \sqrt{2\epsilon_{\text{GaAs}}[V_{bi}^n - V_{GS} + V(y)]/qN_d^+}. \tag{6.60}$$

The I_D-V_{DS} characteristic may be obtained by substituting Eq. (6.60) into Eq. (6.59) and integrating from $y = 0$ to $y = L$, which gives

$$I_D \int_0^L dy = Wq\mu_n n \int_0^{V_{DS}} \left[a - \sqrt{\frac{2\epsilon_{\text{GaAs}}[V_{bi}^n - V_{GS} + V(y)]}{qN_d^+}} \right] dV(y). \tag{6.61}$$

Integration gives

$$I_D = q\mu_n n \frac{W}{L} a \left\{ V(y)\Big|_0^{V_{DS}} - \frac{2}{3}\sqrt{\frac{2\epsilon_{\text{GaAs}}}{qN_d^+ a^2}} [V_{bi}^n - V_{GS} + V(y)]^{3/2} \Bigg|_0^{V_{DS}} \right\}. \tag{6.62}$$

Evaluation with the limits gives

I_D vs. V_{DS} for a given V_{GS}

$$\boxed{I_D = q\mu_n n \frac{W}{L} a \left\{ V_{DS} - \frac{2}{3}\sqrt{\frac{2\epsilon_{\text{GaAs}}}{qN_d^+ a^2}} [(V_{bi}^n - V_{GS} + V_{DS})^{3/2} - (V_{bi} - V_{GS})^{3/2}] \right\}}. \tag{6.63}$$

Note that the factor

$$\sqrt{\frac{2\epsilon_{\text{GaAs}}}{qN_d^+ a^2}} = \frac{1}{V_p^{1/2}} \tag{6.64}$$

was given in Eq. (6.47). Equation (6.63), reduces to the previous Eq. (6.57) for small V_{DS}.

For the drain current I_D given by Eq. (6.63), I_D for a given V_{GS} increases with V_{DS} until a maximum is reached which is the limit of the validity for this analysis. This maximum in I_D is called $I_{D_{sat}}$, and the voltage is called $V_{DS_{sat}}$. The current for $V_{DS} > V_{DS_{sat}}$ is taken as constant. The saturation voltage $V_{DS_{sat}}$ is taken as the voltage where the conducting channel is completely depleted near the drain. This voltage is given by Eq. (6.47) with $(V_{bi}^n - V_{GS})$ replaced by $(V_{bi}^n - V_{GS} + V_{DS_{sat}})$ as

$$V_{bi}^n - V_{GS} + V_{DS_{sat}} = \frac{qN_d^+ a^2}{2\epsilon_{\mathrm{GaAs}}}, \tag{6.65}$$

or

$$V_{DS_{sat}} = \frac{qN_d^+ a^2}{2\epsilon_{\mathrm{GaAs}}} - (V_{bi}^n - V_{GS}). \tag{6.66}$$

With V_p given by Eq. (6.47) and V_T given by Eq. (6.48),

$V_{DS_{sat}}$

$$\boxed{V_{DS_{sat}} = V_p - V_{bi}^n + V_{GS} = V_{GS} - V_T}. \tag{6.67}$$

The drain saturation current $I_{D_{sat}}$ at V_{DS} is given by Eq. (6.63) with V_{DS} replaced by $V_{DS_{sat}}$ of Eq. (6.67):

$$\begin{aligned} I_{D_{sat}} = q\mu_n n \frac{W}{L} a \Bigg\{ & \frac{qN_d^+ a^2}{2\epsilon_{\mathrm{GaAs}}} - (V_{bi}^n - V_{GS}) \\ & - \frac{2}{3}\sqrt{\frac{2\epsilon_{\mathrm{GaAs}}}{qN_d^+ a^2}} \left[(V_{bi}^n - V_{GS}) + \frac{qN_d^+ a^2}{2\epsilon_{\mathrm{GaAs}}} - (V_{bi}^n - V_{GS})\right]^{3/2} \\ & + \frac{2}{3}\sqrt{\frac{2\epsilon_{\mathrm{GaAs}}}{qN_d^+ a^2}} (V_{bi}^n - V_{GS})^{3/2} \Bigg\}. \end{aligned} \tag{6.68}$$

This equation reduces to

$$\begin{aligned} I_{D_{sat}} = q\mu_n n \frac{W}{L} a \Bigg\{ & \frac{qN_d^+ a^2}{6\epsilon_{\mathrm{GaAs}}} \\ & - (V_{bi}^n - V_{GS}) \left[1 - \frac{2}{3}\sqrt{\frac{2\epsilon_{\mathrm{GaAs}}(V_{bi}^n - V_{GS})}{qN_d^+ a^2}} \right] \Bigg\}. \end{aligned} \tag{6.69}$$

A simpler form for $I_{D_{sat}}$ may be written by replacing $[q\mu_n n(W/L)a]$ by G_0, as given in Eq. (6.58), and by replacing $(qN_d^+ a^2/2\epsilon_{\mathrm{GaAs}})$ by V_p:

$I_{D_{sat}}$

$$\boxed{I_{D_{sat}} = \frac{G_0 V_p}{3}\left[1 - \frac{3(V_{bi}^n - V_{GS})}{V_p} + 2\left(\frac{V_{bi}^n - V_{GS}}{V_p}\right)^{3/2}\right]}. \tag{6.70}$$

The I_D-V_{DS} characteristics were illustrated in Fig. 6.14.

6.7.5 Small-signal AC Model[†]

The effectiveness of the gate voltage in control of the drain current is given by the *transconductance, g_m*. The transconductance (g_m) is the change in the output current (I_D) with respect to the change in the input voltage (V_{GS}) and is given by

transconductance, g_m

$$\boxed{g_m \equiv \left.\frac{\partial I_D}{\partial V_{GS}}\right|_{V_{DS}}}. \tag{6.71}$$

With Eq. (6.70) for I_D in the saturation region, the transconductance is

$$g_{m_{sat}} = G_0\left[1 - \sqrt{\frac{V_{bi}^n - V_{GS}}{V_p}}\right]. \tag{6.72}$$

To compare transconductance for various devices, it has become standard practice to normalize g_m by dividing through by the gatewidth W in mm. Common values of g_m range from 10 to 100 millisiemens per millimeter (mS/mm). The output conductance is the slope of the I-V output characteristic in the saturation region and is given by

$$g_d \equiv \left.\frac{\partial I_D}{\partial V_{DS}}\right|_{V_{GS}}, \tag{6.73}$$

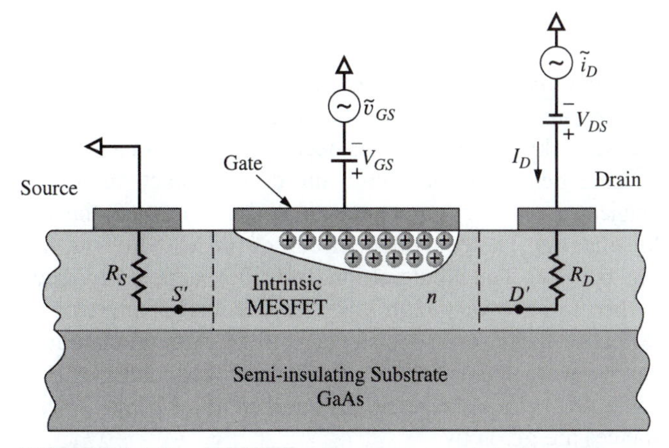

FIGURE 6.16 MESFET with source and drain resistances due to finite conductivity in the semiconductor regions between the end of the active channel and the source and drain terminals.

[†]The material in this part of Section 6.7 is not necessary for continuity in the remainder of this book and may be omitted.

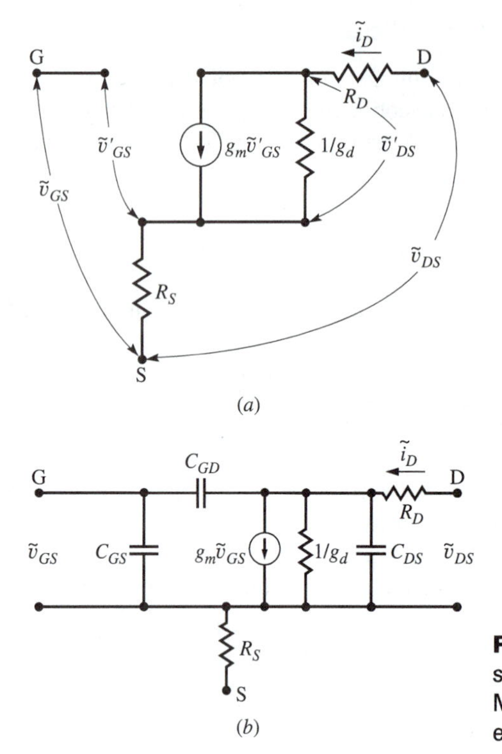

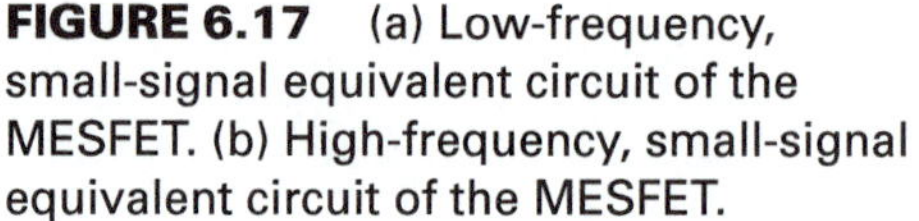

FIGURE 6.17 (a) Low-frequency, small-signal equivalent circuit of the MESFET. (b) High-frequency, small-signal equivalent circuit of the MESFET.

where the *output* or *channel conductance* has units of reciprocal ohms, or mhos.

In the previous discussion, only the resistance in the channel which can be modulated by the Schottky-barrier gate has been considered. In a practical device, there are series resistances near both the source and drain contacts, as illustrated in Fig. 6.16. These resistances give current-resistance (IR) drops between the source and drain contacts and the channel. The equivalent circuit of the MESFET with these resistances is shown in Fig. 6.17 (a). The input portion (left side) is open-circuited because the MESFET is assumed to have very high input impedance for the reversed-biased Schottky gate. The effect of these series resistances on the channel conductance and transconductance can be evaluated as follows. The center section of Fig. 6.16 is the "intrinsic" MESFET. The AC voltages across the terminals of a MESFET with these source and drain resistances are given by

$$\tilde{v}_{DS} = \tilde{v}'_{DS} + (R_S + R_D)\tilde{i}_D, \tag{6.74}$$

and

$$\tilde{v}_{GS} = \tilde{v}'_{GS} + R_S\tilde{i}_D, \tag{6.75}$$

where $\tilde{v}'_{DS}$ and $\tilde{v}'_{GS}$ are the intrinsic MESFET drain and gate voltages, respectively. The drain current $\tilde{i}_D$ is given by

$$\tilde{i}_D = g_d \tilde{v}'_{DS} + g_m \tilde{v}'_{GS}. \tag{6.76}$$

Substitution of Eqs. (6.74) and (6.75) into Eq. (6.76) gives

$$\tilde{i}_D = g_d[\tilde{v}_{DS} - (R_S + R_D)\tilde{i}_D] + g_m(\tilde{v}_{GS} - R_S \tilde{i}_D). \tag{6.77}$$

Solving for $\tilde{i}_D$ gives

$$\begin{aligned} \tilde{i}_D = & \left[\frac{g_m}{1 + R_S g_m + (R_S + R_D) g_d}\right] \tilde{v}_{GS} \\ & + \left[\frac{g_D}{1 + R_S g_m + (R_S + R_D) g_d}\right] \tilde{v}_{DS}. \end{aligned} \tag{6.78}$$

Thus, the terminal transconductance and channel conductance for the MESFET with source and drain resistances are given by

$$g'_m = \frac{g_m}{1 + R_S g_m + (R_S + R_D) g_d}, \tag{6.79}$$

and

$$g'_d = \frac{g_d}{1 + R_S g_m + (R_S + R_D) g_d}, \tag{6.80}$$

which shows that the measured g'_m and g'_d will be reduced by the source and drain resistances.

The high-frequency equivalent circuit shown in Fig. 6.17 (b) includes the capacitance between the gate and source C_{GS}, the capacitance between the drain and source C_{DS}, and the capacitance between the gate and drain C_{GD}. This equivalent circuit is used for AC circuits. A common figure of merit for high frequency applications is the cutoff frequency f_T, which is defined as the frequency at which the output current equals the input current with the output short-circuited. The equivalent circuit of Fig.

cutoff frequency f_T

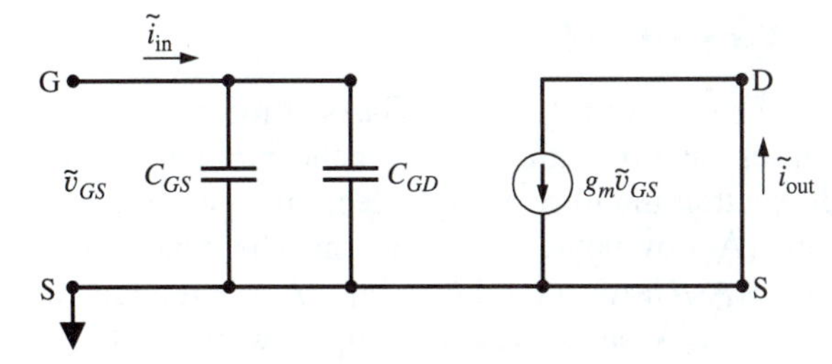

FIGURE 6.18 Equivalent circuit with the output short-circuited for the determination of f_T as the frequency where $\tilde{i}_{out}/\tilde{i}_{in} = 1$.

6.17 (b) with the output short-circuited and the source and drain resistances neglected is shown in Fig. 6.18. The unity gain condition for $\tilde{i}_{\text{out}}/\tilde{i}_{\text{in}} = 1$ is reached when the current through the input capacitance is equal to the output drain current. The input current is

$$\tilde{i}_{\text{in}} = 2\pi f_T(C_{GS} + C_{GD})\tilde{v}_G = 2\pi f_T C_G \tilde{v}_G, \tag{6.81}$$

where $C_G = C_{GS} + C_{GD}$. The output current is

$$\tilde{i}_{\text{out}} = g_m \tilde{v}_g. \tag{6.82}$$

Equating Eqs. (6.81) and (6.82) gives

$$f_T = \frac{g_m}{2\pi C_G}. \tag{6.83}$$

For g_m, $g_m = G_o$ when the square-root term in Eq. (6.72) is neglected. Equation (6.58) gave G_0 as $q\mu_n N_d^+(W/L)a$ with $n = N_d^+$. The capacitances, C_{GS} and C_{GD}, have been derived by Pulfrey and Tarr.[19] As an approximation, the Schottky-barrier capacitance given by Eq. (6.21) will be used for C_G as $\epsilon_{\text{GaAs}} WL/a$ with $x_d \approx a$. Then,

$$f_T \approx \frac{q\mu_n N_d^+(W/L)a}{2\pi\epsilon_{\text{GaAs}} WL/a} \tag{6.84}$$

or

$$f_T \approx \frac{q\mu_n a^2 N_d^+}{2\pi\epsilon_{\text{GaAs}} L^2}. \tag{6.85}$$

Equation (6.85) illustrates that to improve high-frequency performance, the MESFET should have a high carrier mobility and short channel length. As illustrated in Figs. 3.3 and 3.4, the electron mobility is higher than the hole mobility and the electron mobility is higher for GaAs than for Si. The f_T given here is an approximate representation of a very complex problem. In general, the representation as $f_T = g_m/2\pi C$ is useful, but it is necessary to depend on experimental measurements of devices to determine an actual f_T. Such issues as velocity saturation at high fields and stray capacitance add to the difficulty of representing calculations which require two-dimensional modeling.[18]

6.7.6 MODFET or HEMT

The electron mobility shown in Fig. 3.4 for GaAs at room temperature illustrates that the mobility decreases rapidly with impurity concentration due to ionized impurity scattering. A high electron mobility at high electron concentrations may be obtained in GaAs (or $Ga_xIn_{1-x}As$) by physically separating the donors from the free carriers by using an $Al_xGa_{1-x}As$/GaAs (or InP/$Ga_xIn_{1-x}As$) heterojunction,[20] as shown in Fig. 6.19. The n-$Al_xGa_{1-x}As$ layer is heavily doped, while the thin $Al_xGa_{1-x}As$ layer

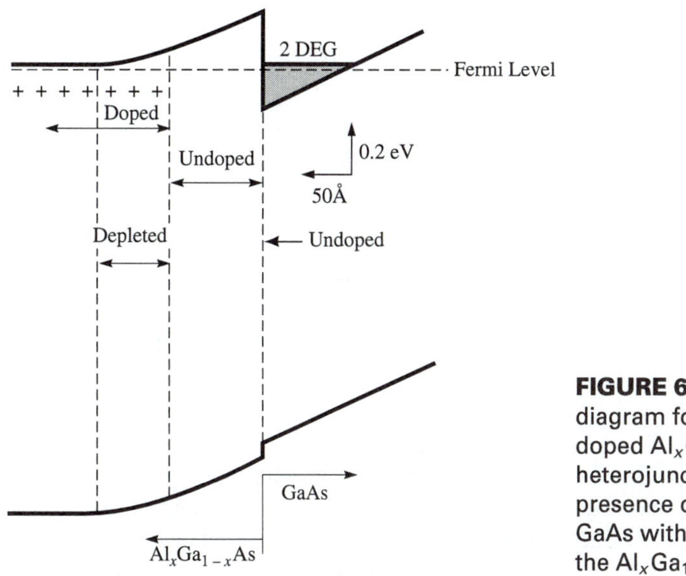

FIGURE 6.19 Energy-band diagram for a modulation-doped $Al_xGa_{1-x}As/GaAs$ heterojunction to illustrate the presence of free electrons in the GaAs with the ionized donors in the $Al_xGa_{1-x}As$.

and the GaAs layer are not intentionally doped. The thin layer of $Al_xGa_{1-x}As$ at the interface is left undoped to reduce the Coulombic interactions between ionized donors in the $Al_xGa_{1-x}As$ and the free electrons in the GaAs. Ionized impurities in the $Al_xGa_{1-x}As$ near the interface would degrade the electron mobility in the GaAs layer. This technique of spatially separating the donor impurities from the free electrons is called *modulation doping*.

modulation doping

The band bending observed in Fig. 6.19 results in a potential well in the GaAs layer at the interface where the carriers are confined. The thickness of the well is of the order of 100 Å, which causes quantization of the density of states in one direction. The

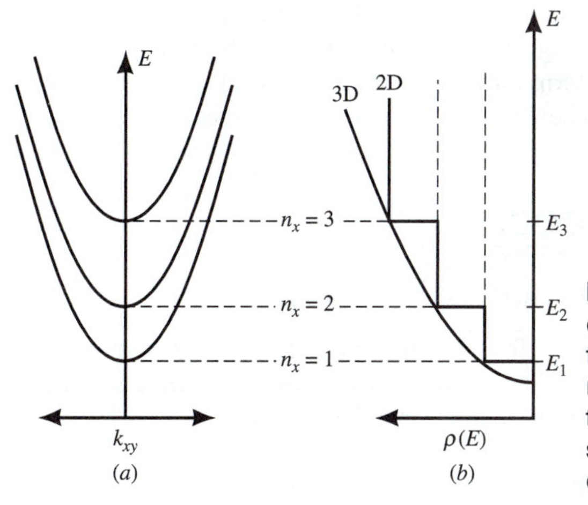

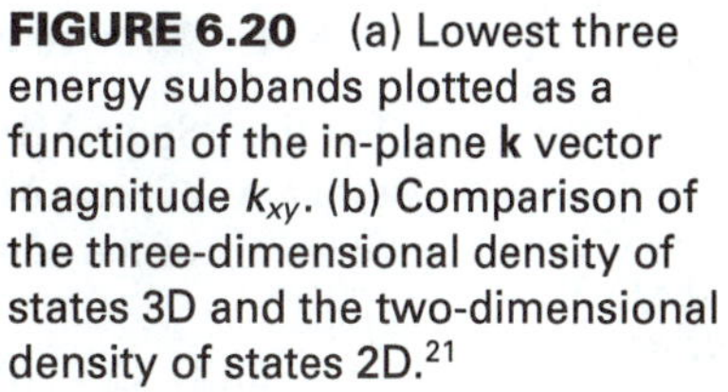

FIGURE 6.20 (a) Lowest three energy subbands plotted as a function of the in-plane **k** vector magnitude k_{xy}. (b) Comparison of the three-dimensional density of states 3D and the two-dimensional density of states 2D.[21]

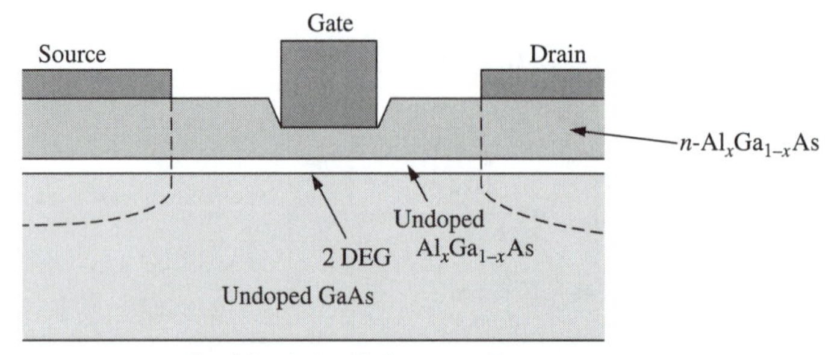

FIGURE 6.21 Modulation-doped Schottky-barrier gate FET.

quantization results from the width of the GaAs potential well at the interface, which is less than the electron de Broglie wavelength of approximately 500 Å. The electrons exist in the GaAs potential well as a 2-dimensional electron gas (2 DEG) as illustrated. In Fig. 6.20 (a), the lowest three energy subbands of a quantum well are plotted as a function of the in-plane **k** vector magnitude k_{xy}. The density of states is found from a circle in **k**-space rather than the sphere, as was done in Sec. 2.5, and becomes

density of states 2 DEG

$$\rho(E) = \frac{\iota m_n^*}{\pi\hbar^2}, \tag{6.86}$$

where $\iota = 1, 2, 3, \ldots$ for the bands of integer ι. This quantized density of states, labeled 2D, is shown in Fig. 6.20 (b) and is compared to the usual three-dimensional parabolic density of states, which is labeled 3D.

The Schottky-barrier gate FET with the modulation-doped heterojunction is illustrated in Fig. 6.21. Various names have been applied to this device. These names include the modulation-doped FET (MODFET), the two-dimensional electron gas FET (TEGFET), and the high electron mobility transistor (HEMT). The HEMT has become the more preferred term for this structure. The I_D-V_{DS} characteristics are similar to the MESFET. Detailed device behavior has been summarized in several books.[22–24]

6.8 PSPICE MESFET MODEL

6.8.1 Element and Model Lines

For the PSpice MESFET model, the equivalent circuit is shown in Fig. 6.22.[25] In addition to PSpice, MESFET models are available in HSPICE and SPICE3. The general form[26,27] of the *element line* for the MESFET is

PSpice MESFET element line

```
BXXXXXX ND NG NS MODNAME <AREA>
```

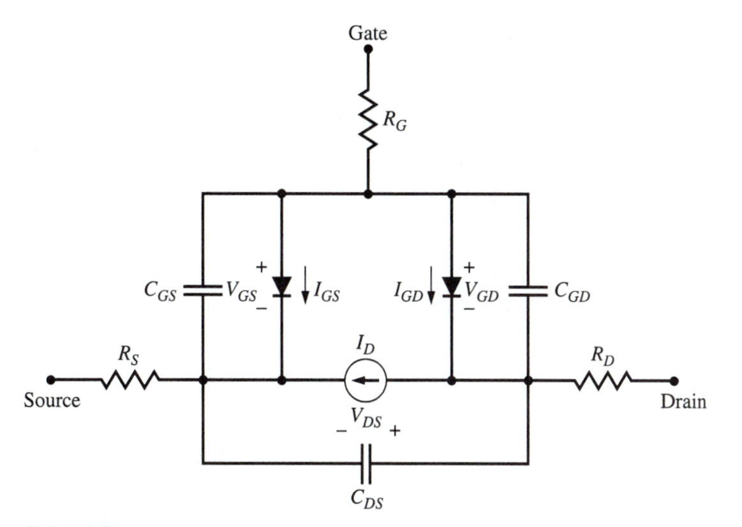

FIGURE 6.22 PSpice equivalent circuit for the *n*-channel GaAs MESFET.[25]

where `BXXXXXX` is the name of the transistor with seven or less alphanumeric characters. The drain, gate, and source nodes are represented by `ND`, `NG`, and `NS`, respectively. The model name is given by `MODNAME` and is used again in the `.MODEL` control line. The `AREA` factor gives how many of the GaAs MESFETs are put in parallel to give one `BXXXXXX`.

The model form or the *model line* for the MESFET is:

MESFET model line

```
.MODEL MODNAME GASFET<(PAR1=PVAL1 PAR2=PVAL2 . . .)>
```

The `MODNAME` is the model name given to the MESFET in the element line. Only the *n*-channel MESFET is used in PSpice. The minimum specification of a MESFET is `GASFET`.

6.8.2 PSpice Model Parameters

The PSpice dc model[25] uses a further simplification of the expression for $I_{D_{sat}}$ given in Eq. (6.70). First, $V_T = V_{bi}^n - V_p$ from Eq. (6.48) is used to eliminate V_{bi}^n in Eq. (6.70):

$$I_{D_{sat}} = \frac{G_0 V_P}{3}\left[1 - \frac{3(V_p + V_T - V_{GS})}{V_P} + 2\left(\frac{V_p + V_T - V_{GS}}{V_P}\right)^{3/2}\right]. \tag{6.87}$$

The $2(\ldots)^{3/2}$ term may be written as $2[1 - (V_{GS} - V_T)/V_p]^{3/2}$ and expanded in a three term binomial series to give

$$2\left[1 - \left(\frac{V_{GS} - V_T}{V_p}\right)\right]^{3/2} = 2\left[1 - \frac{3}{2}\left(\frac{V_{GS} - V_T}{V_p}\right) + \frac{3}{2}\frac{1}{2}\frac{1}{2}\left(\frac{V_{GS} - V_T}{V_p}\right)^2\right]. \tag{6.88}$$

Equation (6.87) now becomes

$$I_{D_{sat}} = \frac{G_0 V_p}{3}\left[1 - 3 + 3\left(\frac{V_{GS} - V_T}{V_p}\right) + 2 - 3\left(\frac{V_{GS} - V_T}{V_p}\right) + \frac{3}{4}\left(\frac{V_{GS} - V_T}{V_p}\right)^2\right]. \tag{6.89}$$

Cancellation of common terms gives

$$I_{D_{sat}} = \frac{G_0}{4V_p}(V_{GS} - V_T)^2. \tag{6.90}$$

Previously G_0 was given in Eq. (6.58) as $q\mu_n n(W/L)a$, and V_p was given in Eq. (6.50) as $qna^2/2\epsilon_{\mathrm{GaAs}}$. Then,

$$\frac{G_0}{4V_p} = \frac{q\mu_n n}{4}\frac{W}{L}a\frac{2\epsilon_{\mathrm{GaAs}}}{qna^2} = \frac{\mu_n}{2a}\frac{W}{L}\epsilon_{\mathrm{GaAs}}, \tag{6.91}$$

or

MESFET β for PSpice

$$\boxed{\beta \equiv \frac{\mu_n}{2a}\frac{W}{L}\epsilon_{\mathrm{GaAs}}}. \tag{6.92}$$

Equations (6.90) and (6.92) may be combined to give

$$\boxed{I_{D_{sat}} = \beta(V_{GS} - V_T)^2}. \tag{6.93}$$

Equation (6.93) is a commonly used approximation for the MESFET.

In the saturation region, the depletion region near the drain will increase as V_{DS} exceeds $V_{D_{sat}}$. This maximum in the depletion depth will move toward the source and effectively shortens the channel length L much like the Early voltage effects in bipolar junction transistors. This effect is called *channel length modulation* and is represented by the parameter λ. The PSpice representation becomes

channel length modulation

$$I_{D_{sat}} = \beta(V_{GS} - V_T)^2(1 + \lambda V_{DS}), \tag{6.94}$$

where λ is related to the output conductance given by

$$g_d \equiv \left.\frac{\partial I_D}{\partial V_{DS}}\right|_{V_{GS}} = \beta(V_{GS} - V_T)^2\lambda. \tag{6.95}$$

The PSpice parameters are given in Table 6.1.[25–27] Several PSpice MESFET models are used and are designated as Level 1, 2, and 3 in parameter no. 1. The differences in the models are largely the representation of the drain current and the parasitic

Level = 1, 2 or 3

TABLE 6.1 PSpice MESFET Model Parameters.

No.	Text Symbol	SPICE Keyword	Parameter Name	Default Value	Units
1	—	LEVEL	model index (1, 2, or 3)	1	—
2	V_p	VTO	pinch-off voltage	−2.5	V
3	β	BETA	transconductance parameter	0.1	A/V^2
4	λ	LAMBDA	channel-length modulation	0	V^{-1}
5	R_G	RG	gate ohmic resistance	0	Ω
6	R_D	RD	drain ohmic resistance	0	Ω
7	R_S	RS	source ohmic resistance	0	Ω
8	C_{GS}	CGS	zero-bias gate-source capacitance	0	F
9	C_{GD}	CGD	zero-bias gate-drain capacitance	0	F
10	C_{DS}	CDS	zero-bias drain-source capacitance	0	F
11	V_{bi}	VBI	gate built-in potential	1	
12	FC	FC	coefficient for forward-bias depletion capacitance	0.5	
13	m	M	grading coefficient	0.5	—
14	α	ALPHA	saturation voltage parameter	2.0	1/V
15	—	VDELTA	capacitance transition voltage (Level 2 & 3)	0.2	V
16	—	VMAX	capacitance limiting voltage (Level 2 & 3)	0.5	V
17	I_s	IS	gate saturation current	10^{-14}	A
18	n	N	gate current ideality factor	1.0	—
19	τ	TAU	drain current delay time	0	sec
20	B	B	doping tail parameter (Level 2)	0.3	V^{-1}
21	γ	GAMMA	static feedback parameter (Level 3)	0	—
22	δ	DELTA	output feedback parameter (Level 3)	0	$A\cdot V^{-1}$
23	Q	Q	power law parameter (Level 3)	2	—
24	E_g	EG	energy gap (barrier height)	1.11	eV
25	—	XTI	IS temperature exponent	0	—
26	—	VTOTC	VTO temperature coefficient	0	V/°C
27	—	BETATCE	BETA exponential temperature coefficient	0	%/°C
28	—	TRG1	RG temperature coefficient (linear)	0	°C^{-1}
29	—	TRD1	RD temperature coefficient (linear)	0	°C^{-1}
30	—	TRS1	RS temperature coefficient (linear)	0	°C^{-1}
31	k_f	KF	flicker-noise coefficient	0	
32	α_f	AF	flicker-noise exponent	1	

VTO

BETA

LAMBDA

capacitances. Parameter no. 2 is the pinch-off voltage. It should be noted that V_p, given by Eq. (6.47), is related to the threshold voltage V_T, and V_p depends on the carrier concentration and thickness of the conducting layer. The next parameter is BETA which was given in Eq. (6.92) and contains the gate width W and length L. LAMBDA, the channel length modulation parameter, was introduced in Eq. (6.94).

RG, RD and RS

The parasitic gate resistance R_G is parameter no. 5. The drain resistance, R_D, and the source resistance, R_S, are parameters nos. 6 and 7. The gate-to-source and

CGS CGD CDS gate-to-drain capacitances are represented by C_{GS} (parameter no. 8) and C_{GD} (parameter no. 9), respectively. The drain-to-source capacitance, C_{DS}, is parameter no. 10.
VBI Parameter no. 11 is the Schottky gate built-in potential V_{bi}^n as given in Eq. (6.1) or
FC (6.3) or a measured value. Parameter no. 12 is given by `FC` and is used to represent the voltage dependences of the parasitic capacitances in a similar manner as for the *p-n* junction. The capacitances C_{GS} and C_{GD} are represented in the Level 1 model by[26]

$$C_{GS} = C_{GS}(0)(1 - V_{GS}/V_{bi}^n)^{-m} \text{ for } V_{GS} \leq FC \times V_{bi}^n \tag{6.96}$$

and

$$C_{GS} = C_{GS}(0)(1 - FC)^{-(1+m)}[1 - FC(1+m) + mV_{GS}/V_{bi}^n] \text{ for } V_{GS} > FC \times V_{bi}^n. \tag{6.97}$$

The expressions for C_{GD} is the same as for C_{GS} except that V_{GS} is replaced by V_{GD}. No voltage dependence is taken with C_{DS}. For the Schottky-barrier gate, the capacitance
M grading parameter $m = 0.5$.

More complex expressions are used for the Level 2 and Level 3 models of C_{GS} and C_{GD}:[25]

$$\begin{aligned} C_{GS,GD} &= \frac{1}{2}\frac{C_{GS}(0)}{\sqrt{1 - V_n/V_{bi}}}\left[1 + \frac{V_e - V_T}{\sqrt{(V_e - V_T)^2 + \text{VDELTA}^2}}\right] \\ &\quad \times \frac{1}{2}\left[1 \pm \frac{V_{GS} - V_{GD}}{\sqrt{(V_{GS} - V_{GD})^2 + (1/\alpha)^2}}\right] \\ &\quad + \frac{1}{2}C_{GD}(0)\left[1 \mp \frac{V_{GS} - V_{GD}}{\sqrt{(V_{GS} - V_{GD})^2 + (1/\alpha)^2}}\right]. \end{aligned} \tag{6.98}$$

The upper sign of $\pm$ and $\mp$ in the square brackets applies to C_{GS} and the lower sign applies to C_{GD}. In Eq. (6.98), V_e and V_n are given by

$$V_e = \frac{1}{2}\left[V_{GS} + V_{GD} + \sqrt{(V_{GS} - V_{GD})^2 + (1/\text{ALPHA})^2}\right] \tag{6.99}$$

$$V_n = \frac{1}{2}[V_e + V_T + \sqrt{(V_e - V_T)^2 + \text{VDELTA}^2}] \quad \text{for} \quad V_n \leq \text{VMAX} \tag{6.100}$$

$$V_n = \text{VMAX} \qquad \text{for} \quad V_n > \text{VMAX} \tag{6.101}$$

α In these equations, α is parameter no. 14 and determines drain-source voltage
VDELTA V_{DS} for the onset of the saturation of I_D. Parameter no. 15, which is given by `VDELTA`, represents the voltage range (such as 0.2 V) for the transition of C_{GS} and C_{GD} through the discontinuity at $V_{DS} = 0$. The voltage V_e gives a smooth interpolation of C_{GS} and C_{GD} at $V_{DS} = 0$, with V_n given by V_e before pinch-off and by V_T after pinch-off (V_n
VMAX is the smaller of $-V_T$ and $-V_e$). The `VMAX` (parameter no. 16) term limits the value of V_n to avoid imaginary numbers or dividing by zero.

The gate leakage currents I_{GS} and I_{GD}, shown in the equivalent circuit of Fig. 6.22, are given by

$$I_{GS} = I_S[\exp(qV_{GS}/\mathrm{n}kT) - 1] \tag{6.102}$$

and

$$I_{GD} = I_S[\exp(qV_{GD}/\mathrm{n}kT) - 1], \tag{6.103}$$

IS N

where I_S (parameter no. 17) is the saturation current and the ideality factor n is parameter no. 18.

Level 1 model

The Curtice model[28] is the Level 1 model. An additional effect is also included in the Level 1 Curtice model which takes into account that a change in $V_{GS}(t)$ does not cause an instantaneous change in I_D. This delay is the time needed for the gate depletion region to respond to the change in gate voltage. The current source in the equivalent circuit of Fig. 6.22 is taken as

$$I_D(V, t) = I_D(V) - \tau \frac{dI_D(t)}{dt}, \tag{6.104}$$

TAU

where τ (parameter no. 19) is the source-to-drain transit time given by $\tau = L/v_{\text{drift}}$.

Level 2

A further improvement in the I_D-V_{DS} characteristic was made by Statz et al.[29] and is the Level 2 model.[25] An empirical expression was chosen as

$$I_{D_{52+}} = \frac{\beta(V_{GS} - V_T)^2}{1 + B(V_{GS} - V_T)} \tag{6.105}$$

B

to represent the gradual transition in doping from the channel into the substrate which is caused by diffusion and/or implant doping tails. Parameter no. 20 is B in this expression. These current expressions are the Level 2 model and are given as

$$\begin{aligned} I_D &= \frac{\beta(V_{GS} - V_T)^2}{1 + B(V_{GS} - V_T)}(1 + \lambda V_{DS}) \\ &\quad \times \left[1 - \left(1 - \frac{\alpha V_{DS}}{3}\right)^3\right] \qquad \text{for} \quad 0 < V_{DS} < 3/\alpha \end{aligned} \tag{6.106}$$

and

$$I_D = \frac{\beta(V_{GS} - V_T)^2}{1 + B(V_{GS} - V_T)}(1 + \lambda V_{DS}) \qquad \text{for} \quad V_{DS} \geq 3/\alpha. \tag{6.107}$$

This model also gave the capacitance expressions given in Eqs. (6.98)–(6.101).

The GaAs MESFET model given by McCamant et al.[30] has several additional parameters for I_D. This model is also referred to as the TriQuint model and is the Level 3 model for I_D. This model is intended to be more accurate at low currents where V_{GS} is near cutoff and the drain conductance is not a good fit to measured characteristics. In the TriQuint model, V_T is made to be a function of the drain voltage to give a new threshold voltage V'_T which is represented as

Level 3

$$V'_T = V_T - \gamma V_{DS}, \tag{6.108}$$

γ

where γ is parameter no. 21. The next change is to represent the drain current as

$$I_D = \frac{I_{D_o}}{1 + \delta V_{DS} I_{D_o}}, \tag{6.109}$$

δ

where δ is parameter no. 22. In Eq. (6.94),

$$I_{D_o} = \beta(V_{GS} - V_T)^Q \tanh(\alpha V_{DS}), \tag{6.110}$$

Q

where Q (parameter no. 23) is used to model the non-square-law dependence of I_D. The new parameters γ and δ replace λ and B in the expressions given in Eq. (6.106).

The remaining parameters are the energy gap E_g, which for the Schottky-barrier gate should become the barrier height and parameter temperature dependences. The temperature exponent for `IS` should be `XTI`=2 because the Richardson constant varies as T^2. Parameters nos. 26–30 represent temperature dependences:

SPICE Keywords are always upper or lower case 'OH', not zero

$$V_T(T) = V_T + \texttt{VTOTC}(T - 300), \tag{6.111}$$
$$\beta(T) = \beta \times 1.01^{\texttt{BETATCE}}(T - 300), \tag{6.112}$$
$$R_G(T) = R_G[1 + \texttt{TRG1}(T - 300)], \tag{6.113}$$
$$R_D(T) = R_D[1 + \texttt{TRD1}(T - 300)], \tag{6.114}$$
$$R_S(T) = R_S[1 + \texttt{TRS1}(T - 300)]. \tag{6.115}$$

The noise parameters (nos. 31 and 32) were presented with Eq. (5.69).

6.9 APPLICATIONS

Selective commercial applications have been made of both MESFETs and HEMTs in integrated circuits. These circuits do not approach the complexity of Si integrated circuits, but generally have high-speed performance. Several books are available which summarize GaAs integrated circuits. The book by Kanopoulos[31] is devoted to GaAs integrated-circuit design. Gallium arsenide microprocessor design is given in the book by Milutinović.[32] The design of GaAs digital logic circuits is different from the Si-based technologies. Because GaAs MESFETs and HEMTs have potential for higher speed operation than Si circuits, their applications are for circuits whose performance is limited

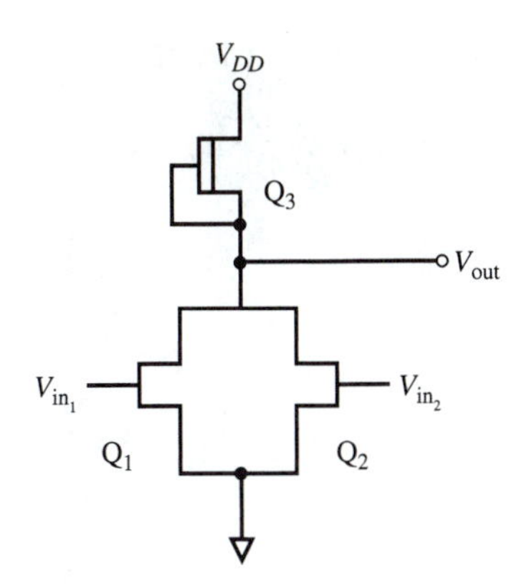

FIGURE 6.23 Two-input NOR gate implemented with direct-coupled FET logic.

by transistor speed. Since MESFETs do not have insulated gates like Si metal-oxide semiconductor (MOS) circuits, the gate forward bias is limited to about 0.7 V, which limits the logic swing. Lack of threshold uniformity limits the yield of large numbers of transistors on a chip.

Logic families with enhancement-mode and depletion-mode MESFETs are designed with direct-coupled FET logic (DCFL) and have low power dissipation. An example of a DCFL NOR gate is shown in Fig. 6.23. The depletion mode device Q_3 serves as the load for the two normally off enhancement mode MESFETs Q_1 and Q_2. The supply voltage V_{DD} is approximately 1 V. The output V_{out} remains at approximately V_{DD} unless V_{in_1} or V_{in_2} are large enough to turn Q_1 or Q_2 on. Then, the output drops to approximately 0 V. A large number of logic technologies and circuits are described by Kanopoulos[31] and Milutinović.[32] Advances in Si technologies continues to diminish the advantages of many GaAs digital applications.

Microwave applications of GaAs and other III-V compound semiconductor MESFETs and HEMTs as discrete transistors as well as in monolithic microwave integrated circuits (MMICs) continue to be superior to the performance obtainable with Si. Maximum frequency of oscillation f_{max} for GaAs MESFETs is expected to be approximately 700 GHz, while the gain–bandwidth product should exceed 200 GHz.[33] Innovative MMIC circuit functions have been developed for commercial and military applications. Circuits are used in low-noise amplifiers, intermediate and high-power amplifiers, limiters, mixers, fixed-tuned and semiconductor device capacitor tuned oscillators, attenuators, switches, filters, multiplexers, frequency multipliers, phase shifters, modulators, active power spliters, phase detectors, and gain equalizers.[34] An example of a GaAs monolithic, 6 GHz, two-stage, low-noise amplifier with a measured 1.7 dB noise figure and 21 dB gain, is shown in Fig. 6.24.[35] A photomicrograph of the chip is shown in Fig. 6.24 (a), and part (b) illustrates the circuit diagram. The circuit diagram shows that the chip contains two MESFETs, resistors, capacitors, and inductors. The inductors are the looped interconnects which look like the number "9." Many microwave applications are being reported in the published research literature and at conferences.

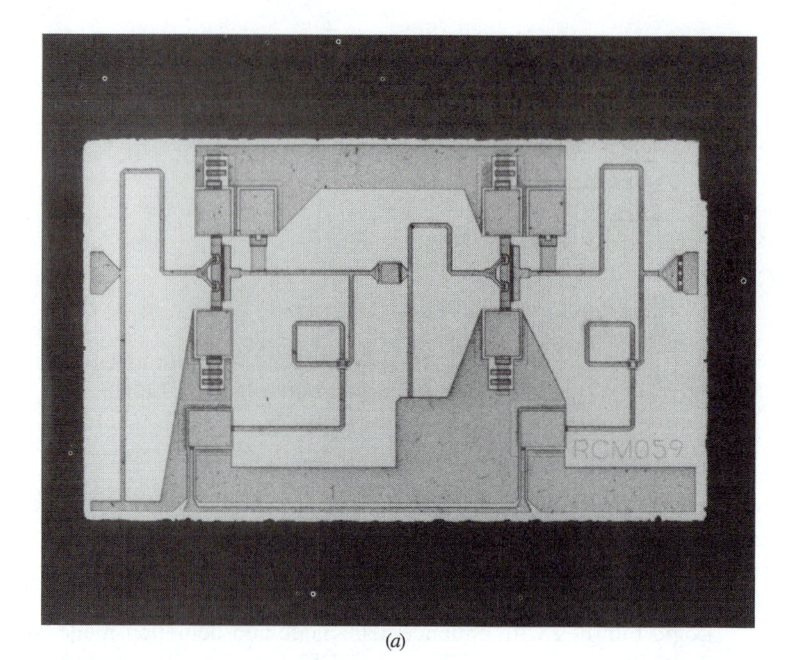

(a)

(b)

FIGURE 6.24 (*a*) Photomicrograph of a 6 GHz low-noise amplifier. (*b*) Equivalent circuit of the amplifier in part (*a*).[35]

6.10 SUMMARY AND USEFUL EXPRESSIONS

This chapter introduced metal-semiconductor two-terminal Schottky-barrier diodes and the three-terminal MESFETs.

- Current flow across the metal-semiconductor interface is by majority carriers.
- The Schottky-barrier current–voltage behavior is given by the same expression as for a *p-n* junction, $I = I_s[\exp(qV_a/nkT) - 1]$ with the ideality factor n between 1.0 and 1.1.
- I_s is given by $area \times A^*T^2 \exp(-q\phi_{Bn}/kT)$, where A^* is the effective Richardson constant and ϕ_{Bn} is the barrier height for an *n*-type semiconductor.
- The previous *p-n* junction SPICE model applies to the Schottky-barrier diode with the lifetime TT set to zero because there is no minority carrier storage.
- A Schottky barrier on a heavily doped semiconductor gives an ohmic contact.
- A Schottky-barrier diode as a gate electrode between the source and drain contacts gives the three terminal MESFET with the current carried by majority carriers.
- High electron mobilities at high electron concentrations are obtained in GaAs (or $Ga_xIn_{1-x}As$) by physically separating the donors from the free carriers by using an $Al_xGa_{1-x}As$/GaAs (or InP/$Ga_xIn_{1-x}As$) heterojunction.
- MESFETs and HEMTs are used in low-noise amplifiers as well as in monolithic microwave integrated circuits (MMICs) which integrate MESFETs or HEMTs with resistors, capacitors, and inductors on the semiconductor chip.

The following expressions for Schottky-barrier diodes and MESFETs were obtained in this chapter.

Schottky-barrier diodes

$$q X_{Si} = q X_{GaAs} = 4.05 \text{ eV}$$
$$q\Phi_{Au} = 4.75 \text{ eV}$$
$$q\Phi_{Al} = 4.1 \text{ eV}$$
$$\phi_{Bn} = \Phi_{metal} - X_{semiconductor}$$
$$\phi_{Bp} = X_{semiconductor} + E_g/q - \Phi_{metal} = E_g/q - \phi_{Bn}$$
$$V_{bi}^n = \Phi_{metal} - \Phi_{semiconductor}$$
$$x_d = \sqrt{2\epsilon(V_{bi}^n - V_a)/qN_d^+}$$
$$x_d \approx 1\,\mu\text{m}/\sqrt{N_d^+/1 \times 10^{15}}$$
$$C_d = A\sqrt{q\epsilon N_d^+/2(V_{bi}^n - V_a)} = \epsilon A/x_d$$
$$J_s = A^*T^2 \exp(-q\phi_{Bn}/kT)$$
$$I = I_s[\exp(qV_a/nkT) - 1]$$

MESFET

$$V_p = qN_d^+ a^2/2\epsilon$$
$$V_T = V_{bi}^n - V_p \text{ normally off MESFET}$$
$$G_o = q\mu_n n(W/L)a$$
$$V_{DS_{sat}} = V_{GS} - V_T$$
$$\beta = \mu_n \epsilon (W/L)/2a$$
$$I_{D_{sat}} = \beta(V_{GS} - V_T)^2(1 + \lambda V_{DS})$$
$$g_m = \partial I_D/\partial V_{GS}|_{V_{DS}}$$
$$f_T = g_m/2\pi C_G$$

PSpice MESFET Element Line

```
BXXXXXX ND NG NS MODNAME <AREA>
```

PSpice MESFET Model Line

```
.MODEL MODNAME GASFET<(PAR1=PVAL1 PAR2=PVAL2 . . .)>
```

REFERENCES

1. W. Schottky, "Halbleitertheorie der Sperrschicht," Naturwiss **26**, 843 (1938).
2. S. M. Sze, *Semiconductor Devices: Physics and Technology* (Wiley, New York, 1986) pp. 161–165.
3. R. S. Muller and T. I. Kamins, *Device Electronics for Integrated Circuits*, 2nd ed. (Wiley, New York, 1986) p. 157.
4. H. B. Michaelson, "The Work Function of the Elements and Its Periodicity," J. Appl. Phys. **48**, 4729 (1977).
5. C. R. Crowell, J. C. Sarace, and S. M. Sze, "Tungsten-Semiconductor Schottky-Barrier Diodes," Trans. Met. Soc. AIME **233**, 478 (1965).
6. A. M. Cowley and S. M. Sze, "Surface States and Barrier Height of Metal-Semiconductor Systems," J. Appl. Phys. **36**, 3213 (1965).
7. S. M. Sze, *Physics of Semiconductor Devices*, 2nd ed. (Wiley, New York, 1981) p. 250.
8. H. A. Bethe, "Theory of the Boundary Layer of Crystal Rectifiers," MIT Radiation Laboratory Report 43 (1942).
9. V. L. Rideout, "Theory of the Boundary Layer of Crystal Rectifiers," Thin Solid Films **48**, 261 (1978).
10. Sze, *Phisics of Semiconductor Devices,* p. 254.
11. R. N. Noyce, R. E. Bohn, and H. T. Chua, "Schottky Diodes Make IC Sense," Electronics **42**, 74 (July 21, 1969).
12. Sze, *Semiconductor Devices,* p. 170.
13. A. Callegari, E. T.-S. Pau, and M. Murakami, "Uniform and Thermally Stable AuGeNi Ohmic Contacts to GaAs," Appl. Phys. Lett. **46**, 1141 (1985).
14. C. A. Mead, "Schottky Barrier Gate Field-Effect Transistor," Proc. IRE **40**, 1365 (1966).
15. W. W. Hooper and W. I. Lehrer, "An Epitaxial GaAs Field-Effect Transistor," Proc. IEEE **55**, 1237 (1967).
16. M. A. Hollis and R. A. Murphy, *High-Speed Semiconductor Devices* S. M. Sze, Ed. (Wiley, New York, 1990), p. 217.
17. Sze, *Semiconductor Devices,* p. 184.
18. M. Ino and T. Takada, "GaAs LSI Circuit Design," *Very High Speed Integrated Circuits: Gallium Arsenide LSI, Semiconductors and Semimetals,* Vol. 29, T. Ikoma, Ed. (Academic Press, Boston, 1990), p. 159.
19. D. L. Pulfrey and N. G. Tarr, *Introduction to Microelectronic Devices* (Prentice Hall, Englewood Cliffs, N.J., 1989), p. 306.
20. R. Dingle, H. L. Störmer, A. C. Gossard, and W. Wiegmann, "Electron Mobilities in Modulation-Doped Semiconductor Heterojunction Superlattices," J. Appl. Phys. **33**, 665 (1978).
21. S. W. Corzine, R.-H. Yan, and L. A. Coldren, *Quantum Well Lasers,* P. S. Zory, Ed. (Academic Press, San Diego, 1993), p. 25.
22. R. Dingle, Ed., *Semiconductors and Semimetals,* Vol. 24, *Applications of Multiquantum Wells, Selective Doping, and Superlattices* (Academic Press, Boston, 1987).
23. Hollis and Murphy, *High-Speed Semiconductor Devices,* p. 221.
24. S. Tiwari, *Compound Semiconductor Device Physics,* (Academic Press, San Diego, 1992) pp. 299–401.

25. G. Massobrio and P. Antognetti, Eds., *Semiconductor Device Modeling with SPICE,* 2nd ed. (McGraw-Hill, New York, 1993), pp. 375–390.

26. *The Design Center Circuit Analysis Reference Manual* (MicroSim Corp., Irvine, Calif., 1994) pp. 4-4–4-10.

27. W. Banzhaf, *Computer-Aided Circuit Analysis using PSpice,* 2nd ed. (Regents/Prentice Hall, Englewood Cliffs, N.J., 1992) pp. 122–124.

28. W. R. Curtice, "A MESFET Model for Use in the Design of GaAs Integrated Circuits," IEEE Trans. Microwave Theory Tech. **MMT-28**, 448 (1980).

29. H. Statz, P. Newman, I. W. Smith, R. A. Pucel, and H. A. Haus, "GaAs FET Device and Circuit Simulation in SPICE," IEEE Trans. Electron. Devices **ED-34**, 160 (1987).

30. A. J. McCamant, G. D. McCormack, and D. H. Smith, "An Improved GaAs MESFET Model for SPICE," IEEE Trans. Microwave Theory Tech. **MTT-38**, 822 (1990).

31. N. Kanopoulos, *Gallium Arsenide Digital Integrated Circuits: A Systems Perspective* (Prentice Hall, Englewood Cliffs, N.J. 1989).

32. V. Milutinović, Ed., *Microprocessor Design for GaAs Technology* (Prentice Hall, Englewood Cliffs, N.J., 1990).

33. J. M. Golio and R. J. Golio, "Projected Frequency Limits of GaAs MESFET's," IEEE Trans. Microwave Theory Tech. **MTT-39**, 142 (1991).

34. Special Issue on MMIC's and Their System Applications, IEEE Trans. on Microwave Theory Tech. **MTT-38** (Sept. 1990).

35. R. C. Mott, "A GaAs 6 GHz Low-Noise Amplifier for Satellite Receivers," IEEE Trans. Microwave Theory Tech. **MTT-37**, 565 (1989).

PROBLEMS

All problems are for room temperature unless another temperature is specified.

6.1 Consider an ideal Au/*n*-Si Schottky-barrier diode with $n = N_d^+ = 4 \times 10^{16}$ cm.

(a) Sketch the energy-band diagram at thermal equilibrium and label the barrier height and built-in potential.

(b) Calculate the depletion distance in cm.

(c) Calculate the barrier height.

(d) For a forward bias of 0.2 V, plot the charge density ρ versus distance x. What is x_d in cm?

(e) Plot the electric field versus the distance for thermal equilibrium and a forward bias of 0.2 V.

(f) Sketch the energy-band diagram for a positive bias of 0.2 V.

(g) Sketch the energy-band diagram for a negative bias of -1.0 V.

(h) Calculate the capacitance for an area of 5×10^{-4} cm^2 at $V_a = 0$.

6.2 Repeat Problem 6.1 for $p = N_a^- = 6 \times 10^{15}$ cm^{-3}, but for a positive bias of 0.1 V in parts (d), (e), and (f).

6.3 Repeat Problem 6.1 for *n*-type GaAs.

6.4 Consider an ideal Au/*n*-Si Schottky-barrier diode with $n = N_d^+ - N_a^- = 1 \times 10^{16}$ cm^{-3}.

(a) What is the built-in potential?

(b) What is the depletion distance?

(c) Sketch the energy-band diagram.

(d) Sketch the charge density ρ versus distance x.

(e) Sketch the electric field versus distance and give $\mathscr{E}_{max}$ in V/cm.

(f) Sketch the potential versus distance.

(g) What is the zero-bias capacitance if the area of the Au dot is 2×10^{-4} cm^2?

(h) Sketch the energy-band diagram for a positive bias of +0.35 V and show which polarity is connected to the Au contact.

(i) Sketch the energy-band diagram for a negative bias of −2.0 V and show which polarity is connected to the Au contact.

6.5 Repeat Problem 6.4 for $p = N_a^- = 2 \times 10^{16}$ cm^{-3}, but for a positive bias of 0.1 V in part (h).

6.6 Repeat Problem 6.4 for n-type GaAs.

6.7 For a Schottky-barrier diode on n-type Si, the metal work function is 4.6 eV and $(E_c - E_f)$ is 0.310 eV in the neutral semiconductor bulk.

(a) What is the built-in potential?

(b) Sketch the energy-band diagram at thermal equilibrium. Indicate the barrier height, built-in potential, and depletion width.

(c) For a reverse bias of −1 V, sketch the energy-band diagram and show the polarity connected to the metal.

6.8 Repeat Problem 6.7 for n-type GaAs.

6.9 A Si Schottky-barrier diode was determined by C–V measurements to have a built-in potential of 0.310 V and a carrier concentration $N_d^+ = n = 5.0 \times 10^{16}$ cm^{-3}.

(a) Sketch the energy-band diagram. Label the built-in potential and barrier height.

(b) Find the barrier height.

(c) Find the electron concentration in the Si at the metal/Si interface.

6.10 **(a)** What is the zero-bias depletion width for an Au/n-Si Schottky-barrier diode with $n = N_d^+ = 5 \times 10^{15}$ cm^{-3}.

(b) Repeat part (a) for a negative bias of −3.0 V.

6.11 **(a)** Describe how you would use a Schottky-barrier diode to determine the net donor or acceptor concentration in a GaAs wafer.

(b) How would you use a Schottky-barrier diode to determine if a semiconductor is n- or p-type?

6.12 The capacitance of a Au/n-Si Schottky-barrier diode was measured as a function of applied voltage. The cross-sectional area is 2×10^{-4} cm^2. For the data given below, find the built-in potential and the ionized donor concentration.

Bias, V_a (V)	0.0	+0.1	+0.2	+0.3	-0.5	-1.0	-2.0
Capacitance, C (pF)	20.6	22.3	24.4	27.3	15.8	13.2	10.5

6.13 A current–voltage characteristic is given by

$$I = 2 \times 10^{-13}[\exp(qV_a/1.04kT) - 1].$$

(a) Plot I versus V on a semilog scale from 1×10^{-6} A to 1×10^{-2} A.

(b) Plot I versus V on a linear scale for I between 0 and 10 mA.

6.14 What net donor concentration is required to obtain an Au/n-Si ohmic contact with $x_d < 40$ Å?

6.15 Consider an Au/n-GaAs Schottky-barrier diode with E_f pinned at the surface at $2E_g/3$ above E_v.

(a) What is the built-in potential with $N_d^+ = 1 \times 10^{16}$ cm^{-3}.

(b) What is the electron concentration at the metal-semiconductor interface (x=0)?

6.16 Design an Au/n-Si Schottky-barrier diode to give a capacitance of 10 pF at a reverse bias of -4.0 V. The net donor concentration is 2×10^{16} cm^{-3}.

6.17 Find the depletion width for an Au/n-Si Schottky-barrier diode with $N_d^+ = 3 \times 10^{16}$ cm^{-3} at avalanche breakdown.

6.18 A Au/n-Si Schottky-barrier diode can be represented by $I = 8.4 \times 10^{-8}[\exp(qV_a/1.06kT) - 1]$. The cross-sectional area is 4.6×10^{-6} cm^2, the free-electron concentration in the bulk semiconductor is 1×10^{16} cm^{-3}, and the series resistance is 0.1 ohm.

(a) Find the capacitance at thermal equilibrium.

(b) Give the values for the SPICE parameters `IS`, `RS`, `N`, and `VJ`.

(c) Use PSpice to obtain a linear plot of the I–V characteristic for the voltage range from 0.0 to 0.48 V in steps of 0.01 V.

6.19 A PtSi Schottky-barrier diode on n-type Si with $n = N_d^+ = 1 \times 10^{17}$ cm^{-3} has a barrier height of 0.89 eV and a cross-sectional area of 1×10^{-4} cm^2. The I–V behavior is given by $I = 2.7 \times 10^{-12}[\exp(qV/1.05kT) - 1]$.

(a) Give the following parameters: `IS`, `N`, `TT`, `VJ`, and `CJO`.

(b) Sketch and label your circuit diagram to use PSpice to evaluate the current response as the bias is changed from forward to reverse bias for the Schottky-barrier diode in series with a 100-Ω resistor.

(c) Plot the variation of the current through the diode from 0 to 50 ns. Let the $TSTEP$ in the transit analysis be 0.25 ns.

(d) Why is the response time much faster for the Schottky diode than for a p^+n diode?

6.20 Capacitance–voltage measurements for a Au-Schottky-barrier diode on n-type GaAs gave a zero-bias capacitance of 11.2 pF and a built-in voltage of 0.635 V. The cross-sectional area is 1.57×10^{-4} cm^2.

(a) Find the donor concentration of the n-type GaAs.

(b) Find the avalanche breakdown voltage.

6.21 The schematic representation of the MESFET was shown in Fig. 6.15. The n-channel region is 0.2 μm in thickness and has an impurity concentration of 1×10^{16} cm^{-3}. This channel region overlies a semi-insulating GaAs substrate, which can be considered as an insulator for purposes of this problem. The barrier height of the Al gate is 0.8 eV.

(a) Show that the channel region is completely depleted when there is no bias between the gate and the source.

(b) Determine the forward bias that must be applied to the gate to reduce the depletion width of the Schottky-barrier gate electrode to 0.15 μm. What is the Schottky-barrier gate current density at this bias level?

6.22 Use PSpice to evaluate the current response of the MBD101 Schottky-barrier diode as the bias is changed from forward to reverse bias. The diode has the following SPICE parameters: `is` = 192.1p, `rs` = 0.1, `n` = 1, `ikf` = 0, `xti` = 3, `eg` = 1.1, `cjo` = 893.8f, `m` = 98.2m, `vj` = 0.75, `fc` = 0.5, `isr` = 16.91n, `nr` = 2, `bv` = 5, and `ibv` = 10u. A 100-Ω resistor is in series with the diode specified above. The time variation of the applied voltage is a piecewise step that is +5.0 V at $t = 0$ and at $t = 10$ ns. Then, at $t = 10.2$ ns the voltage switches to −5.0 V.

(a) Plot the time variation of the current through the diode from 0 to 50 ns.

(b) Why does the time response not exhibit the time delay shown for the *p-n* junction diode in Example 5.1?

6.23 Use PSpice to obtain the drain current versus the drain voltage for a MESFET with `VTO` = -1.0 V, `VBI` = 0.75, `ALPHA` = 2.0, `BETA` = 6.0×10^{-3}, `LAMBDA` = 0.03, and `IS` = 3×10^{-14}. Let the gate voltage vary from −0.5 V to 0.5 V in 0.25-V steps and 0.05-V steps for the drain voltage from 0 to 5 V.

6.24 Depletion-mode (normally on with zero input voltage) MESFETs have been used for implementing GaAs integrated circuits. A simplified inverter is illustrated in the circuit where `FET2` serves as the load to the input `FET1` as shown in the figure. The input voltage is selected to turn `FET1` on and off. The input voltage `vin` is a pulse which has a 0 V value at $t = 0$ and a pulsed value of −3.0 V with no time delay. The pulse rise and fall times are 0.5 ns, the pulse width is 10 ns, and the pulse period is 50 ns. Let the MESFET parameters be `VTO` = −2.4 V, `BETA` = 30×10^{-6}, `ALPHA` = 1.6, `VBI` = 0.6, `CGS` = 2.5×10^{-15}, and `CGD` = 0.5×10^{-15}. Find the time response of the output voltage.

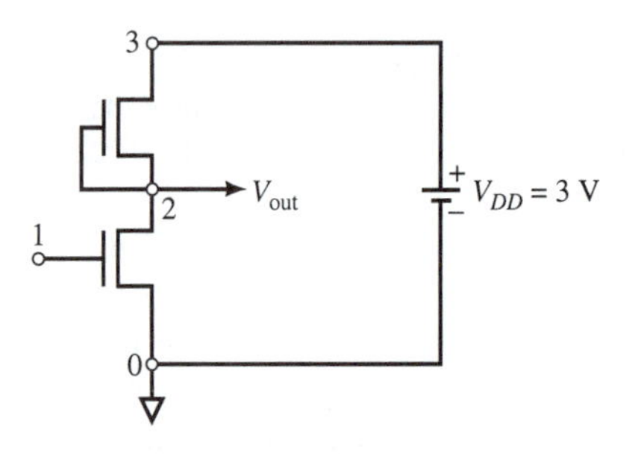

6.25 Calculate the density of states for a two-dimensional electron gas. Take the electron degeneracy as two.

CHAPTER 7

MOS CAPACITORS

7.1 INTRODUCTION

MOSFET or IGFET

The second type of unipolar devices in which only one type of carrier predominately participates in the conduction process are the metal-oxide semiconductor (MOS) devices. In this chapter, the two-terminal MOS capacitor is considered, while the three-terminal metal-oxide semiconductor field-effect transistor (MOSFET) is presented in the next chapter. The MOSFET is also called the insulated-gate FET (IGFET). Silicon is the only semiconductor considered here, but these concepts apply to any metal-insulator semiconductor system. Not only is the MOS capacitor used to study the electrical properties of the MOS system and monitor integrated-circuit fabrication processes, but it is also useful as a storage capacitor in integrated circuits such as switched-capacitor filters, semiconductor memory, and image sensors. This chapter is devoted to understanding the basic concepts needed to apply MOS capacitors to integrated circuits.

initial MOS studies

The development of the MOS technology took a rather meandering path to the development of commercial devices. In 1957, Frosch and Derick[1] demonstrated that SiO_2 on Si provided protection to the Si surface at 1000°C, and in addition to surface passivation, the SiO_2 surface layer also could serve as a selective mask to provide patterned areas of donor and acceptor diffusion. The initial studies by Atalla et al.[2] in 1959

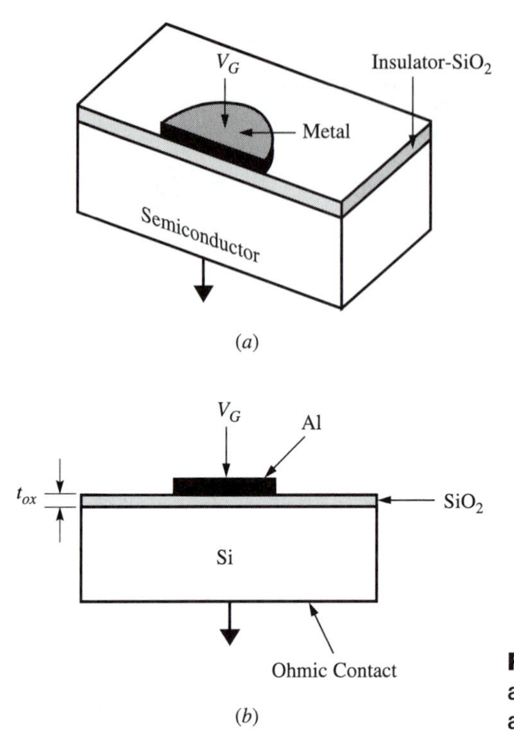

FIGURE 7.1 (*a*) Representation of an MOS capacitor. (*b*) Cross section of an Al/SiO_2/Si MOS capacitor.[10]

of SiO_2 grown on Si were devoted to the reduction of surface leakage to passivate *p-n* junctions and bipolar transistors. The MOS capacitor structure was first proposed by Moll[3] in 1959. Studies of SiO_2 grown on Si by Ligenza and Spitzer[4] during this period were devoted to the mechanism of Si oxidation and the effects of crystal orientation on the oxidation rates of Si.[5] In 1960, Kahng and Atalla[6] used a thermally oxidized Si structure to form a surface field-effect transistor which is now the MOSFET. The MOS capacitor was used by Terman[7] to study thermally oxidized Si surfaces. More detailed studies were undertaken by Grove et al.[8] A comprehensive treatment of the analysis and measurement of the MOS capacitor was given by Nicollian and Brews.[9]

first MOSFET

A schematic representation of an MOS capacitor is shown in Fig. 7.1 (*a*), and as represented in Fig. 7.1 (*b*), t_{ox} is the thickness of the SiO_2 layer, and V_G is the voltage on the Al gate electrode. Although the MOS capacitor shown in Fig. 7.1 is two dimensional, a one-dimensional analysis will be used here which neglects the two-dimensional edge effects. The bias voltage V_G is positive when the Al gate is connected to the positive terminal of the bias voltage, and V_G is negative when the Al gate is connected to the negative terminal of the bias voltage.

The basic behavior of the MOS capacitor can be introduced by considering the similarities and differences with the Schottky-barrier diode which was presented in the previous chapter. The schematic space-charge model for Al on *p*-type Si is shown in

Fig. 7.2 (*a*) with the depletion space-charge region of $-qN_a^-$. The space-charge $\rho(x)$ vs. distance is shown in part (*b*) and the electric field obtained from $\mathscr{E}(x) = (1/\epsilon)\int \rho(x)\,dx$ is shown in part (*c*). The potential $\psi(x)$ obtained from $\psi(x) = -\int \mathscr{E}(x)\,dx$ is given in part (*d*). The solution of these equations with appropriate boundary conditions for *n*-type Si were given in Sec. 6.3. The next step is to insert an ideal (no charge) SiO_2 layer between the Al and *p*-type Si. The schematic space-charge model shown in Fig. 7.2 (*e*) is similar to the Schottky barrier, but the positive charge on the Al and the negative space-charge $-qN_a^-$ are now separated by the SiO_2 layer. The space charge $\rho(x)$ is shown in part (*f*), and the electric field obtained from $\mathscr{E}(x) = (1/\epsilon)\int \rho(x)\,dx$ is shown in part (*g*). The potential $\psi(x) = -\int \mathscr{E}(x)\,dx$ is given in part (*h*). As will be shown in Sec. 7.3.4, the electric field at the SiO_2/Si interface, $\mathscr{E}_{\mathrm{Si}} = 0.33\mathscr{E}_{ox}$, where $\mathscr{E}_{ox}$ is the electric field in the SiO_2. The 0.33 term is the result of the difference in dielectric constants for SiO_2 ($3.9\epsilon_0$) and Si ($11.7\epsilon_0$).

The built-in potential V_{bi}^p for the Schottky barrier was given in Eq. (6.4) for *p*-type Si as the difference in the metal and Si work functions, $V_{bi}^p = \Phi_{\mathrm{Al}} - \Phi_{\mathrm{Si}}$. For positive applied voltages approaching V_{bi}^p, the current increases rapidly and large currents cause failure due to excessive heating. As shown in Sec. 7.2 for the MOS capacitor, the difference in the metal (or heavily doped polycrystalline Si) electrode work function and the semiconductor work function is termed the *flat-band voltage*, $V_{FB}^{\circ} = \Phi_{\mathrm{Al}} - \Phi_{\mathrm{Si}}$, which is the same as the built-in potential for the Schottky-barrier diode. In the MOS capacitor, current flow is prevented by the insulator.

flat-band voltage

The detailed analysis of the MOS capacitor begins with the thermal-equilibrium energy-band diagram in Sec. 7.2. When the flat-band voltage is applied to the MOS capacitor, the difference in work functions is exactly compensated and the energy bands do not vary with distance. Bias voltages different from the flat-band voltage are considered next. Bias conditions which give a majority carrier concentration at the SiO_2–Si interface greater than the majority carrier concentration in the neutral bulk are called *accumulation* or *enhancement*. Bias conditions which give a minority carrier concentration at the SiO_2–Si interface greater than the majority carrier concentration in the neutral bulk are called *inversion*.

accumulation

inversion

The bias at which the electron concentration at the SiO_2/Si interface is equal to the hole concentration in the neutral bulk of a *p*-type semiconductor is termed the *threshold voltage* and is introduced in Sec. 7.3. Gauss's law is used to relate the charge in the semiconductor to the applied bias to obtain an expression for the threshold voltage. The charges associated with the nonideal oxide are also introduced in Sec. 7.3. Section 7.4 begins with the ideal MOS capacitance–voltage (*C–V*) behavior and also considers the *C–V* behavior of the nonideal oxide with fixed oxide charge and interface trapped charge. Analysis of the behavior of the small-signal differential capacitance provides further understanding of the electrical behavior of the MOS system. The application of the MOS capacitor to charge-coupled devices (CCDs) for imaging is described in Sec. 7.5. The significant concepts introduced in this chapter are summarized in Sec. 7.6 and the expressions used to represent the behavior of MOS capacitors are listed. The inversion layer thickness which is taken as the centroid of the inversion layer charge is derived in Appendix C.1. The solution of Poisson's equation for the differential capacitance is given as Appendix C.2.

threshold voltage

C–V behavior

charge-coupled devices

inversion layer thickness

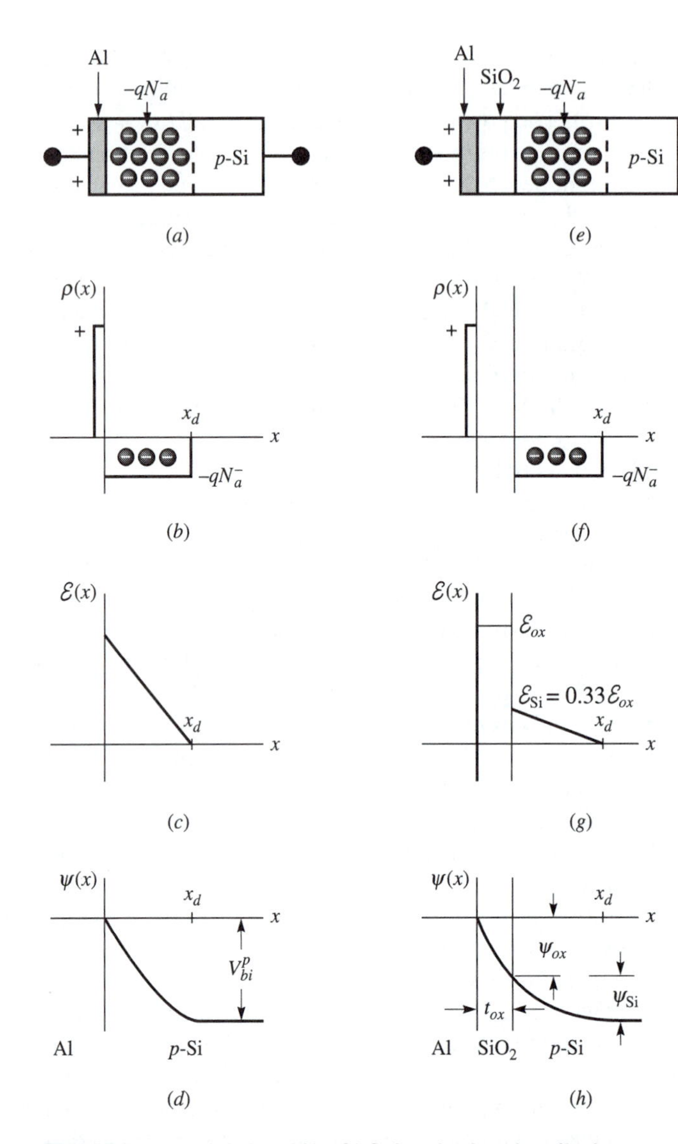

FIGURE 7.2 (*a*)–(*d*) Al/*p*-Si Schottky-barrier diode. (*a*) Schematic space-charge model. (*b*) Charge density $\rho(x)$ vs. distance x with depletion width x_d. (*c*) Electric field $\mathcal{E}(x)$ vs. distance x. (*d*) Potential $\psi(x)$ vs. distance x where V_{bi}^p is the built-in potential. (*e*)–(*h*) Al/SiO_2/*p*-Si MOS capacitor. (*e*) Schematic space-charge model. (*f*) Charge density $\rho(x)$ vs. distance x. (*g*) Electric field $\mathcal{E}(x)$ vs. distance x where $\mathcal{E}_{ox}$ is the electric field in the SiO_2 and $\mathcal{E}_{Si}$ is the electric field at the SiO_2/Si interface. (*h*) Potential $\psi(x)$ vs. distance x where t_{ox} is the oxide thickness, ψ_{ox} is the potential across the oxide, and ψ_{Si} is the potential across the *p*-Si.

7.2 IDEAL MOS CAPACITOR

7.2.1 Thermal-equilibrium Energy-band Diagram

Al work function

electron affinity for SiO_2 and Si

Figure 7.3 shows the energy-band diagram of an isolated metal (Al) adjacent to an isolated insulator (SiO_2) which is adjacent to an isolated semiconductor (*p*-type Si) and represents the allowed energies E in the bulk of the material versus position, which is given by the distance x. The Al work function $q\Phi_{Al}$ is the difference between the vacuum level or free electron energy E_0 and the Fermi level E_{f_m} and for Al is $\boxed{E_0 - E_{f_m} = q\Phi_{Al} = 4.1 \text{ eV}}$. The difference between the vacuum level E_0 and the conduction-band edge E_c is the electron affinity qX. For the SiO_2, $\boxed{qX_{ox} = 0.95 \text{ eV}}$ and for Si, $\boxed{qX_{Si} = 4.05 \text{ eV}}$. It should be noted that the work function differences depend on the Si orientation[12] as well as whether the Al is single crystal or polycrystalline. These effects will be neglected here, but are necessary for a careful comparison between theory and experiment.

The Si work function is shown in Fig. 7.3 as

$$q\Phi_{Si} = qX_{Si} + (E_c - E_f), \tag{7.1}$$

but can be put in a more convenient form by adding and subtracting E_v to give

$$q\Phi_{Si} = qX_{Si} + \underbrace{E_c - E_v}_{E_g} + E_v - E_f$$

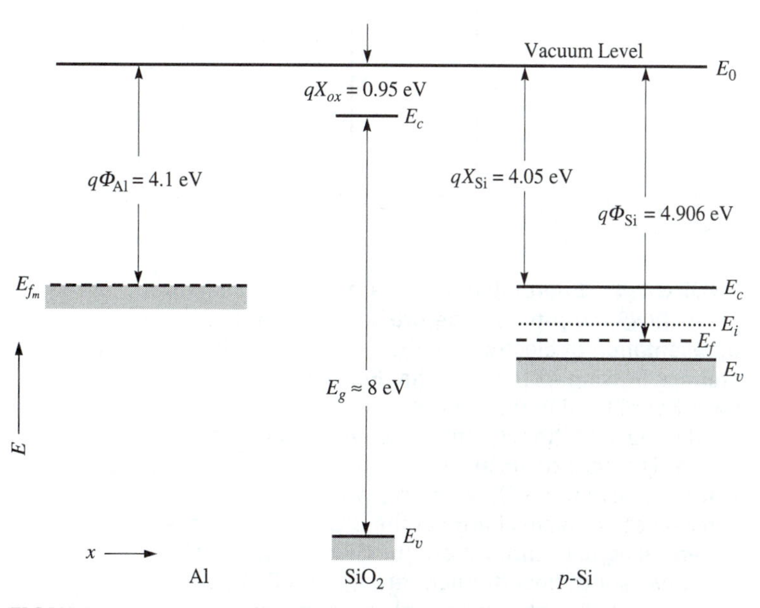

FIGURE 7.3 Energy levels in the separated metal (Al), insulator (SiO_2), and semiconductor (*p*-type Si) with $p = 1 \times 10^{15}$ cm^{-3}.[11]

or

work function for p-type Si

$$q\Phi_{Si} = qX_{Si} + E_g - (E_f - E_v). \tag{7.2}$$

For $p = 1 \times 10^{15}$ cm^{-3} at 300 K, Eq. (2.65) gives $(E_f - E_v) = -0.026 \ln(1 \times 10^{15}/3.08 \times 10^{19}) = 0.269$ eV. The intrinsic Fermi level in Fig. 7.3 is represented by E_i.

Initially, the interfaces and the oxide are considered to be ideal and to be free of charge. The energy-band diagram at initial contact before equilibrium can be established is shown in Fig. 7.4. It is clear that electrons in the metal have a higher Fermi level than for the semiconductor ($E_{f_{Al}} > E_{f_{Si}}$), and electrons will instantaneously transfer from the metal to the semiconductor to establish thermal equilibrium. In the MOS system, electrons cannot pass freely in either direction, and charge must redistribute by unintentional leakage paths or when connected into a circuit.

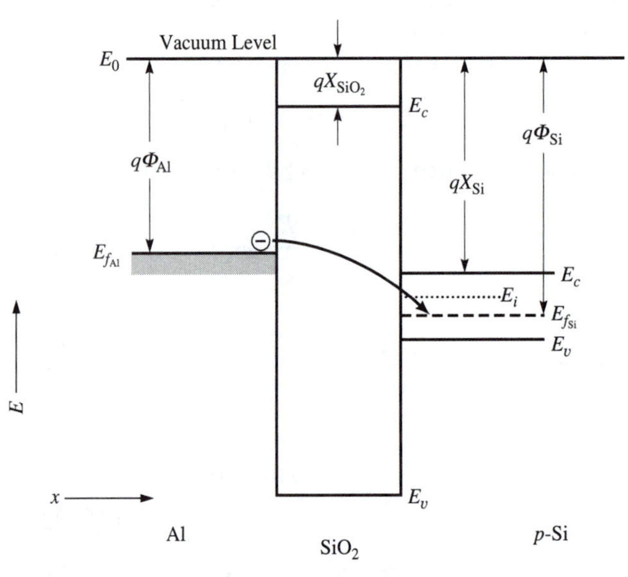

FIGURE 7.4 Energy-band diagram at initial contact for Al, SiO_2, and *p*-type Si before equilibrium. The representation of electrons in the metal at the SiO_2 interface flowing from the higher Fermi level to the lower Fermi level in the *p*-Si serves to indicate that the metal surface at the SiO_2 interface will lose negative charge. This loss of electrons results in positive charge on the metal at the SiO_2 interface, while the *p*-Si will gain excess negative charge at the SiO_2/*p*-Si interface at thermal equilibrium. Unless the SiO_2 is very thin, electrons cannot flow (tunnel) through the SiO_2, but redistribute by unintentional leakage paths or any electrical connection.

At thermal equilibrium, the Fermi levels in the Al and semiconductor must be equal and the vacuum level must be continuous. These two requirements determine a unique energy-band diagram for the ideal MOS capacitor as shown in Fig. 7.5 (*a*). With the loss of electrons, as illustrated in Fig. 7.4, the Al will have a thin layer of positive charge at the metal-oxide interface, while the *p*-type Si will have a negative space charge due to ionized acceptors N_a^- which are not neutralized by free holes. To accommodate the work function differences, the bands bend down. Note that this energy-band

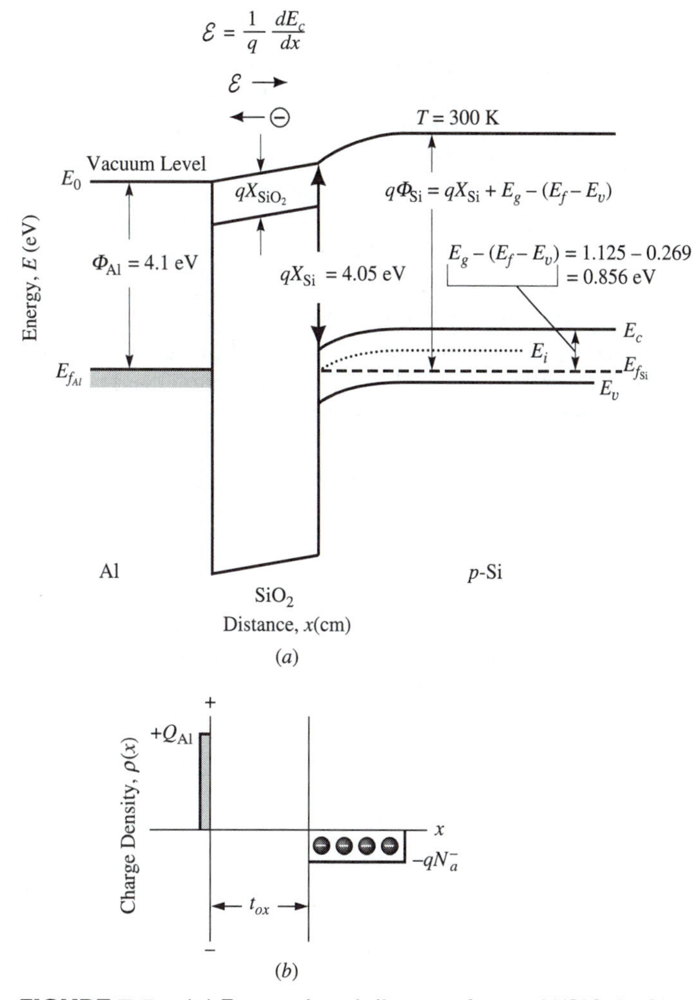

FIGURE 7.5 (*a*) Energy-band diagram for an Al/SiO_2/*p*-Si MOS capacitor at thermal equilibrium. The hole concentration in the *p*- type Si is 1×10^{15} cm^{-3}. (*b*) Charge distribution $\rho(x)$ vs. distance *x*.

diagram is similar to Fig. 6.3 for the Schottky barrier. The work function difference with the Si work function given by Eq. (7.2) is

$$\boxed{q\Phi_{\text{Al}} - q\Phi_{\text{Si}} = q\Phi_{\text{Al}} - [qX_{\text{Si}} + E_g - (E_f - E_v)]}. \quad (7.3)$$

For $p = 1 \times 10^{15}\ \text{cm}^{-3}$,

work function difference example

$$q\Phi_{\text{Al}} - q\Phi_{\text{Si}} = 4.1 - (4.05 + 1.125 - 0.269) = -0.806\ \text{eV}. \quad (7.4)$$

The difference in work functions, -0.806 eV, represents the amount of band bending at thermal equilibrium. The positive charge density due to the loss of electrons at the Al/SiO_2 interface and the negative charge due to depleted acceptors, N_a^-, is represented in Fig. 7.5 (*b*). As shown in Chapter 3 by Eq. (3.15), the electric field $\mathscr{E}$ in Fig. 7.5 (*a*) is given by $\mathscr{E} = (1/q)(dE_c/dx)$, which gives a positive electric field, and the force on an electron will be in the $-x$ direction as shown.

For the charge distribution shown in Fig. 7.5 (*b*), there is a depletion region in the Si at the SiO_2/Si interface, and this charge condition is called *depletion*. Depletion occurs in this case at thermal equilibrium. This charge condition, illustrated in Fig. 7.5 (*b*), is called *depletion* whether or not a bias voltage is required.

depletion condition

It is generally more convenient to use a modified energy-band diagram where the energies are measured to the top of the SiO_2 conduction band rather than to the vacuum level. Then the metal work function is replaced by a reduced work function $q\Phi'_M$ as

$$q\Phi'_M = q\Phi_M - qX_{\text{SiO}_2}, \quad (7.5)$$

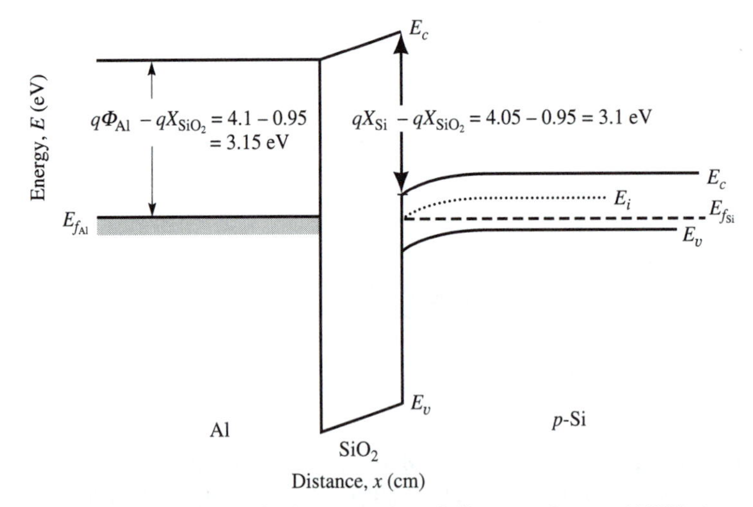

FIGURE 7.6 Simplified energy-band diagram for an Al/SiO_2/p-Si MOS capacitor at thermal equilibrium.

and the semiconductor electron affinity is replaced by a reduced electron affinity as

$$qX'_{\text{Si}} = qX_{\text{Si}} - qX_{\text{SiO}_2}. \tag{7.6}$$

The simplified energy-band diagram for Fig. 7.5 (*a*) is shown in Fig. 7.6.

7.2.2 Ideal Flat-band Voltage

flat-band condition

A quantity of frequent reference for MOS capacitors is the applied voltage which exactly compensates the difference in the work functions of the metal and the semiconductor. For this bias condition, the energy bands in the energy-band diagram are flat; i.e., do not vary with distance. This applied voltage to achieve flat bands is called the *flat-band voltage,* and is represented by V_{FB}° for the ideal MOS capacitor without oxide or interface charge. The polarity is selected to reduce the charge and electric fields to zero. As was done for the *p-n* junction and Schottky-barrier diode, it is helpful to use the *schematic space-charge model* to determine the polarity of the bias connected to the metal to achieve flat-band conditions. For the case shown in Fig. 7.5 (*b*) at thermal equilibrium, the charge on the metal plate is positive, and therefore a negative potential should be connected to the metal to eliminate the positive charge as illustrated in Fig. 7.7 (*a*). The energy-band diagram for the flat-band condition is shown in Fig. 7.7 (*b*). An important rule is illustrated by Eq. (7.4). *The sign of the difference of the metal and semiconductor work functions gives the polarity connected to the metal to obtain the flat-band condition.*

The flat-band voltage V_{FB}° is the voltage applied to achieve flat bands. The value of V_{FB}° varies with the free carrier concentration of the semiconductor and the gate material because they determine the semiconductor work function and the gate work function. The *ideal flat-band voltage* V_{FB}° is the work function difference given by Eq. (7.3):

ideal flat-band voltage

$$\boxed{V_{FB}^{\circ} = \Phi_M - \Phi_S = \{\Phi_M - [qX_S + E_g - (E_f - E_v)]/q\}}, \tag{7.7}$$

where $q\Phi_M$ is the metal gate work function, $q\Phi_S$ is the semiconductor work function, qX_S is the semiconductor electron affinity, E_g is the semiconductor energy gap, and $(E_f - E_v)$ is the position of the semiconductor Fermi level above the valence band in the neutral semiconductor bulk.

polycrystalline Si gate

As given in Eq. (7.7), the flat-band voltage depends on the gate work function. Often heavily doped polycrystalline Si is used for the gate electrode. It is convenient and accurate to assume that the Fermi level is at the conduction-band edge for heavily doped *n*-type polycrystalline Si and at the valence-band edge for heavily doped *p*-type polycrystalline Si.

7.2.3 Gate Bias Voltages for $|V_G| \neq V_{FB}^{\circ}$

Several useful definitions can be illustrated by considering the energy-band diagram and space-charge when $|V_G| \neq V_{FB}^{\circ}$. First, consider the case when the negative V_G

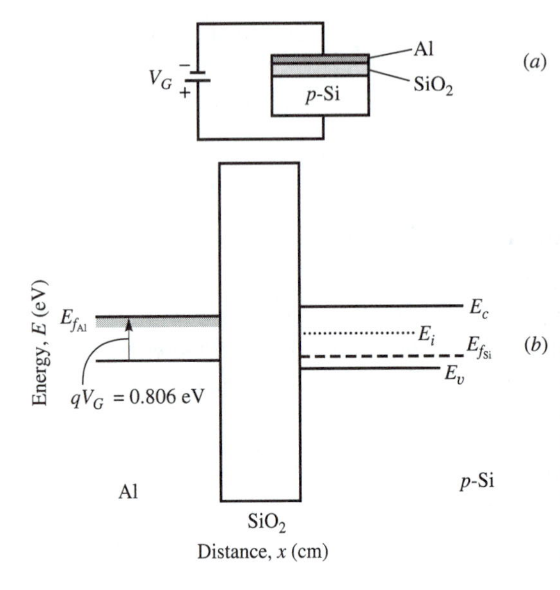

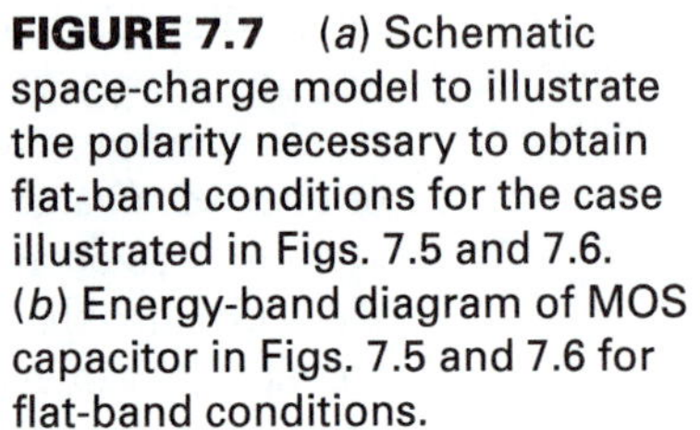

FIGURE 7.7 (*a*) Schematic space-charge model to illustrate the polarity necessary to obtain flat-band conditions for the case illustrated in Figs. 7.5 and 7.6. (*b*) Energy-band diagram of MOS capacitor in Figs. 7.5 and 7.6 for flat-band conditions.

shown in Fig. 7.7 (*a*) is increased so that $|V_G| > |V^\circ_{FB}|$. The schematic space-charge model is shown in Fig. 7.8 (*a*) and the energy-band diagram is shown in Fig. 7.8 (*b*). Since no current flows in the ideal MOS capacitor, the Fermi level in the semiconductor is constant with distance ($dE_f/dx = 0$). The hole concentration was written in Eq. (2.110) as

$$p = n_i \exp[(E_i - E_f)/kT]. \tag{7.8}$$

If the difference of E_i and E_f at the Si surface is labeled as $(E_i - E_f)_s$ and the difference of E_i and E_f in the neutral Si bulk is labeled as $(E_i - E_f)_b$, then Fig. 7.8 (*b*) shows that for the negative gate voltage $|V_G| > |V^\circ_{FB}|$,

$$(E_i - E_f)_s > (E_i - E_f)_b. \tag{7.9}$$

With the hole concentration at the Si surface represented by p_s [Fig. 7.8 (*c*)] and the hole concentration in the neutral Si bulk represented by p_b, then with p represented by Eq. (7.8) and $(E_i - E_f)_s > (E_i - E_f)_b$ as given in Eq. (7.9),

accumulation or enhancement $p_s > p_b$

$$p_s > p_b. \tag{7.10}$$

This case for a larger hole concentration at the surface than in the p-type bulk is called *accumulation* or *enhancement.* Similarly, for an n-type semiconductor, an electron concentration at the surface n_s greater than the electron concentration in the semiconductor

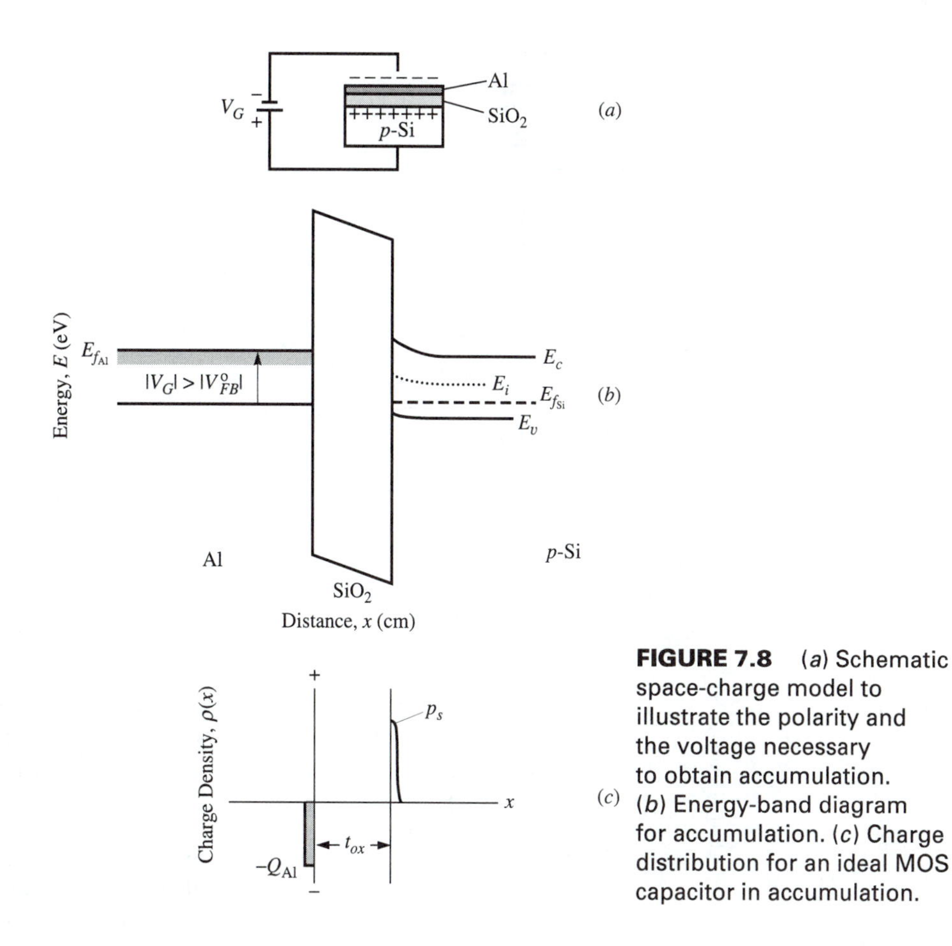

FIGURE 7.8 (*a*) Schematic space-charge model to illustrate the polarity and the voltage necessary to obtain accumulation. (*b*) Energy-band diagram for accumulation. (*c*) Charge distribution for an ideal MOS capacitor in accumulation.

bulk, $n_s > n_b$, is also *accumulation* or *enhancement.* For the case illustrated here, it was necessary to apply bias to achieve accumulation. It should be noted that accumulation refers to the charge condition and not the bias voltage. For a *p*-type semiconductor with $q\Phi_M > q\Phi_S$, accumulation occurs at thermal equilibrium.

For the next case, the polarity of the bias voltage is reversed so that the positive bias terminal is connected to the metal gate as illustrated in Fig. 7.9 (*a*). The energy-band diagram in Fig. 7.9 (*b*) shows that at the surface $(E_i - E_f)_s$ is negative, and by Eq. (7.8), $p_s < n_i$. The electron concentration was given in Eq. (2.109) as

$$n = n_i \exp[(E_f - E_i)/kT]. \tag{7.11}$$

In Fig. 7.9 (*b*), $(E_f - E_i)_s$ is positive and $n_s > n_i$. When $(E_f - E_i)_s$ is large enough to make the minority carrier concentration at the surface larger than the majority carrier concentration in the bulk, then $n_s > p_b$ and the surface is *inverted.* The inversion charge condition is illustrated in Fig. 7.9 (*c*).

inversion for $n_s > p_b$

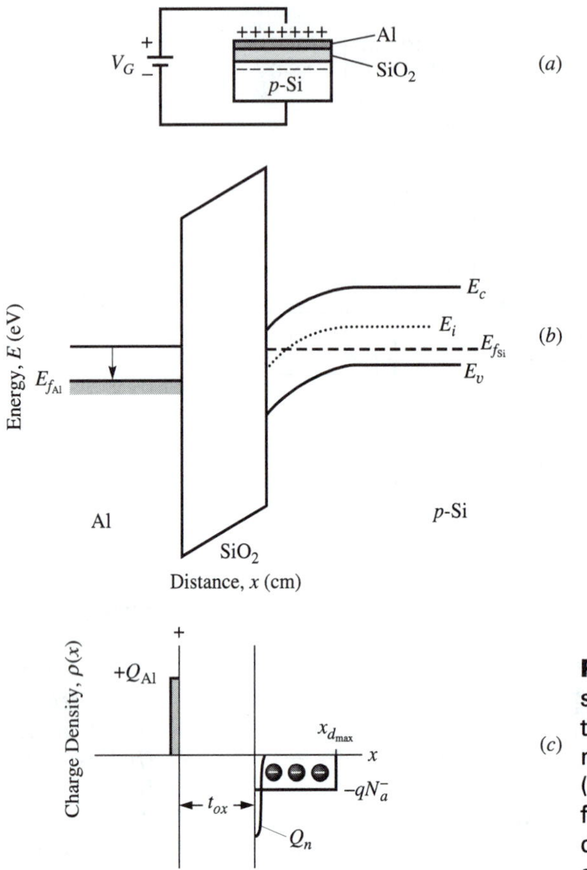

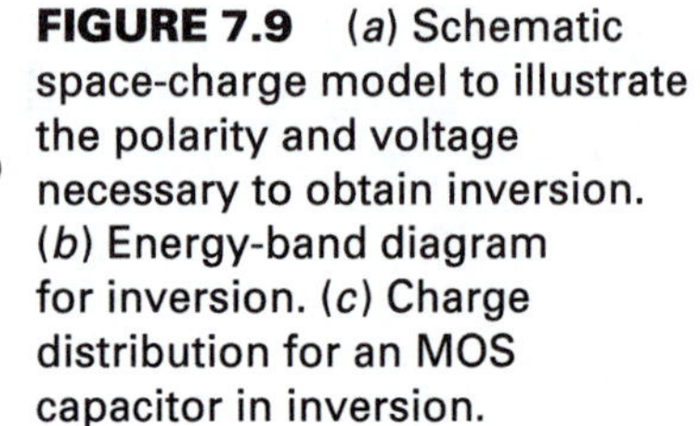

FIGURE 7.9 (*a*) Schematic space-charge model to illustrate the polarity and voltage necessary to obtain inversion. (*b*) Energy-band diagram for inversion. (*c*) Charge distribution for an MOS capacitor in inversion.

The charge components for inversion are labeled in Fig. 7.9 (*c*). The thin layer of positive charge on the surface of the Al at the Al/SiO_2 interface is Q_{Al}. With inversion, the negative mobile charge due to n_s is Q_n, which has a layer thickness typically between 10 and 100 Å, which is much less than the surface depletion layer (see Appendix C.1). As V_G increases to continue bending the bands from depletion to inversion, n_s exceeds n_i, and the surface depletion layer width reaches a maximum width, which is designated as $x_{d_{max}}$. The negative space charge in the semiconductor Q_{sc} is given at *maximum depletion* by

space-charge at maximum depletion

$$Q_{sc} = -qN_a^- x_{d_{max}} A, \tag{7.12}$$

where A is the cross-sectional area. Once the inversion layer is formed, Q_n grows exponentially with the band bending and x_d almost stops increasing as the gate voltage is increased.

7.2.4 Maximum Depletion Width $x_{d_{max}}$

The maximum depletion width $x_{d_{max}}$ is a frequently encountered quantity in the analysis of MOS devices. As the bands are pulled down far enough for inversion to occur, even a very small increase in band bending will result in a very large increase in charge contained within the inversion layer and a very small increase in depletion layer width. The change from rapid to only logarithmic increase in x_d with V_G in the p-type Si will occur when $n_s = p_b$ and may be obtained from Gauss's law which relates the electric field $\mathscr{E}$ to the charge density.

The potential $\psi(x)$ was initially defined in Chapter 3 by Eq. (3.17) as

$$\psi(x) = -E_i(x)/q, \tag{7.13}$$

where $E_i(x)$ is the intrinsic Fermi level, which may vary with position. In Chapter 4, the reference potential was selected as the Fermi level E_f to give Eq. (4.50) as

$$\psi(x) = (1/q)[E_f - E_i(x)]. \tag{7.14}$$

With Eq. (4.50) or (7.14) for potential, the hole concentration may be written with Eq. (2.110) as

$$p(x) = n_i \exp[(E_i(x) - E_f)/kT], \tag{7.15}$$

and the electron concentration may be written with Eq. (2.109) as

$$n(x) = n_i \exp\{[E_f - E_i(x)]/kT\}. \tag{7.16}$$

It has become common practice for analysis of MOS structures to label the potential given in Eq. (7.14) in the neutral semiconductor bulk as ψ_b. For the p-type semiconductor with $p = N_a^-$ and E_{i_b} as the intrinsic Fermi level in the neutral semiconductor bulk, Eqs. (7.14) and (7.15) give the hole concentration as

$$p = N_a^- = n_i \exp[(E_{i_b} - E_f)/kT] = n_i \exp(-q\psi_b/kT), \tag{7.17}$$

where

$$\boxed{\psi_b = (1/q)(E_f - E_{i_b}) = -(kT/q)\ln(N_a^-/n_i)}. \tag{7.18}$$

The potential ψ_b is illustrated in Fig. 7.10. It has also become common practice in MOS structures to redefine the potential in Eqs. (3.17) and (7.13) by adding the constant potential E_{i_b} as the reference potential for $E_i(x)$ to give

potential $\psi(x)$ with E_{i_b} as the reference potential

$$\boxed{\psi(x) = (1/q)[E_{i_b} - E_i(x)]}. \tag{7.19}$$

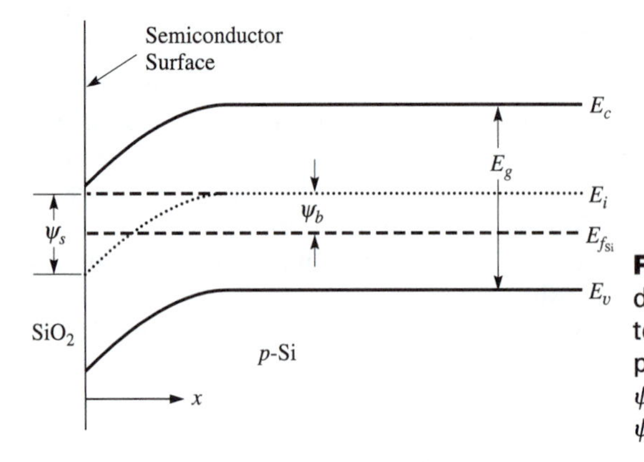

FIGURE 7.10 Energy-band diagram at the SiO_2/p-Si interface to illustrate the potential $\psi(x)$, the potential in the neutral p-type Si ψ_b, and the potential at the surface ψ_s.[13]

With this definition of potential, $\psi(x)$ goes to zero as $E_i(x)$ goes to E_{i_b} in the neutral semiconductor bulk. At the SiO_2/Si interface at $x = 0$, the potential is designated as

surface potential, ψ_s

$$\boxed{\psi_s = \psi(0) = (1/q)[E_{i_b} - E_i(0)]} \tag{7.20}$$

and is illustrated in Fig. 7.10. *This potential is called the* surface potential.

The hole concentration given in Eq. (7.15) may be written in a more convenient form for MOS structures by adding and subtracting E_{i_b} from $[E_i(x) - E_f]$ in Eq. (7.15) to give

$$\begin{aligned} p(x) &= n_i \exp\{[E_i(x) - E_f + E_{i_b} - E_{i_b}]/kT\} \\ &= n_i \exp[\underbrace{(E_{i_b} - E_f)}_{-q\psi_b}/kT] \exp\{-\underbrace{[E_{i_b} - E_i(x)]}_{q\psi(x)}/kT\}, \end{aligned}$$

which permits $p(x)$ to be written as

$$\boxed{p(x) = n_i \exp\{-q[\psi_b + \psi(x)]/kT\}}. \tag{7.21}$$

In a similar manner, the electron concentration given in Eq. (7.16) may be rewritten by adding and subtracting E_{i_b} from $[E_f - E_i(x)]$ to give

$$\boxed{n(x) = n_i \exp\{q[\psi_b + \psi(x)]/kT\}}. \tag{7.22}$$

For a p-type semiconductor, Eq. (7.18) gives a negative ψ_b, while Eq. (7.20) gives a negative ψ_s for accumulation and a positive ψ_s for depletion. Because ψ_s is the difference between the intrinsic Fermi level in the neutral bulk and the intrinsic Fermi level at the SiO_2/Si interface, *ψ_s is the voltage drop in the semiconductor, which is also* *the semiconductor band bending*. The variation of ψ_s for the designated conditions in a p-type semiconductor are:

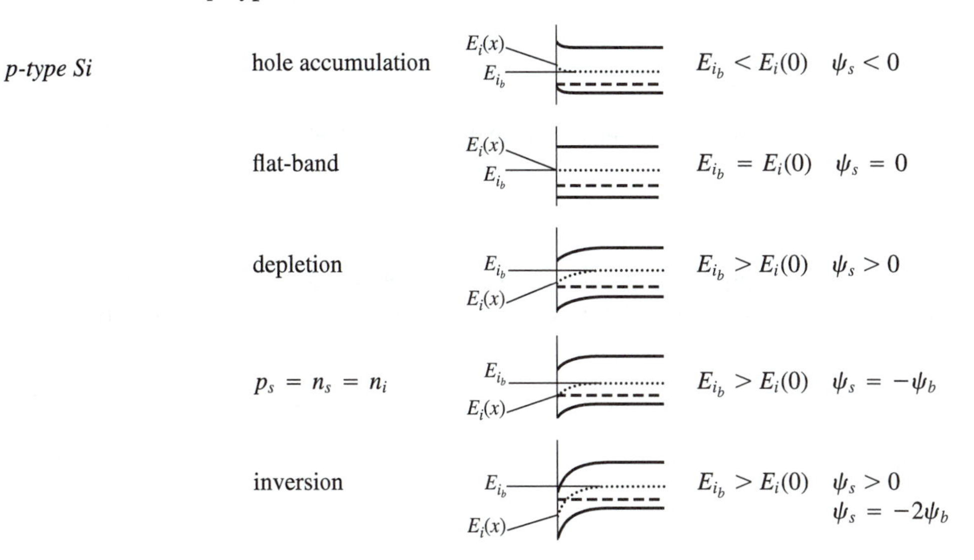

For an n-type semiconductor, $E_f > E_{ib}$ in the neutral semiconductor; therefore, ψ_b given by Eq. (7.14) will be positive. For accumulation of electrons, the intrinsic Fermi level in the bulk is greater than at the surface, $E_{i_b} > E_i(x)$, and Eq. (7.20) gives ψ_s as positive. The bands bend upward after the flat-band condition and depletion occurs. As the bands bend upward, $E_{i_b} < E_i(0)$ and ψ_s becomes negative. The variation in ψ_s for the designated conditions are:

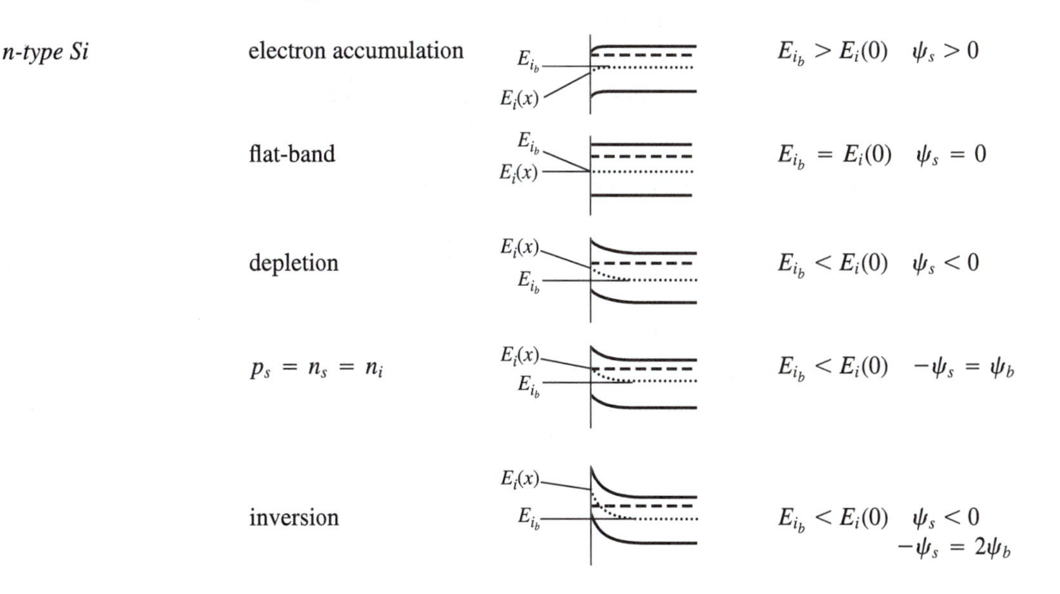

The potential as a function of distance $\psi(x)$ for depletion may be obtained from Gauss's law (Table 3.2):

$$\frac{d\mathscr{E}(x)}{dx} = \frac{\rho(x)}{\epsilon_{\text{Si}}} = -\frac{qN_a^-}{\epsilon_{\text{Si}}}, \tag{7.23}$$

where $\rho(x) = -qN_a^-$ is the space-charge density. Integration of Eq. (7.23) gives

$$\mathscr{E}(x) = -\frac{qN_a^-}{\epsilon_{\text{Si}}}x + C_1, \tag{7.24}$$

where C_1 is an arbitrary constant to be determined by the boundary condition. At the edge of the space-charge region at $x = x_d$, the field must be zero. This condition gives

$$C_1 = \frac{qN_a^-}{\epsilon_{\text{Si}}}x_d, \tag{7.25}$$

and

$$\mathscr{E}(x) = \frac{qN_a^-}{\epsilon_{\text{Si}}}(x_d - x). \tag{7.26}$$

For a static electric field, the potential $\psi(x)$ is related to the field by $\mathscr{E} = -d\psi/dx$ which gives

$$\frac{d\psi(x)}{dx} = -\frac{qN_a^-}{\epsilon_{\text{Si}}}(x_d - x). \tag{7.27}$$

Integration of Eq. (7.27) gives

$$\psi(x) = -\frac{qN_a^-}{\epsilon_{\text{Si}}}\left(x_d x - \frac{x^2}{2}\right) + C_2, \tag{7.28}$$

where the boundary condition of $\psi(x_d) = 0$ gives C_2 as

$$C_2 = \frac{qN_a^-}{2\epsilon_{\text{Si}}}x_d^2. \tag{7.29}$$

Equation (7.28) now becomes

$$\psi(x) = -\frac{qN_a^-}{2\epsilon_{\text{Si}}}x_d^2\left(\frac{2x}{x_d} - \frac{x^2}{x_d^2}\right) + \frac{qN_a^-}{2\epsilon_{\text{Si}}}x_d^2, \tag{7.30}$$

or,

$$\psi(x) = \frac{qN_a^-}{2\epsilon_{\text{Si}}}x_d^2\left(1 - \frac{x}{x_d}\right)^2. \tag{7.31}$$

The total potential at the semiconductor-oxide interface is given by Eq. (7.31) and represents the total band bending due to an applied voltage:

$$\psi(0) = \psi_s = \frac{qN_a^-}{2\epsilon_{\text{Si}}} x_d^2 \qquad \text{or} \qquad x_d = \sqrt{\frac{2\epsilon_{\text{Si}}\psi_s}{qN_a^-}}. \tag{7.32}$$

The surface is inverted whenever $n_s > p_b$, which requires $|\psi_s| > |\psi_b|$. *It has become common usage to assign the onset of strong inversion as when the electron concentration at the surface is equal to the hole concentration in the neutral bulk* p_b:

onset of strong inversion

$$n_s = p_b. \tag{7.33}$$

At the surface, $\psi(x)$ becomes ψ_s and Eq. (7.33) may be written with Eqs. (7.21) and (7.22) to give

$$\overbrace{n_i \exp[q(\psi_b + \psi_s)/kT]}^{n_s} = \overbrace{n_i \exp(-q\psi_b/kT)}^{p_b}, \tag{7.34}$$

which gives

$$\exp[q(2\psi_b + \psi_s)/kT] = 1. \tag{7.35}$$

Solving for ψ_s gives

ψ_s for inversion

$$\boxed{\psi_s = -2\psi_b}. \tag{7.36}$$

When $n_s = p_b$, then $\psi_s = -2\psi_b$, and the total band bending $\psi(0)$, as seen in Fig. 7.10, is $2|\psi_b|$. With the maximum depletion width $x_{d_{max}}$ taken as the onset of strong inversion, Eq. (7.32) for the total band bending may be written as

$$\psi_s = 2|\psi_b| = \frac{qN_a^-}{2\epsilon_{\text{Si}}} x_{d_{max}}^2, \tag{7.37}$$

or

$x_{d_{max}}$

$$\boxed{x_{d_{max}} = \sqrt{4\epsilon_{\text{Si}}|\psi_b|/qN_a^-}}. \tag{7.38}$$

The expression for $x_{d_{max}}$ is the same as the depletion width for the Schottky barrier given in Eq. (6.18), with $2|\psi_b|$ replacing the built-in potential V_{bi}. Variation of $x_{d_{max}}$ with the acceptor concentration N_a^- or donor concentration N_d^+ represented by N_B is shown in Fig. 7.11 for Si and GaAs. Note that $x_{d_{max}} \simeq 1\ \mu\text{m}$ for $N_B = 1 \times 10^{15}\ \text{cm}^{-3}$.

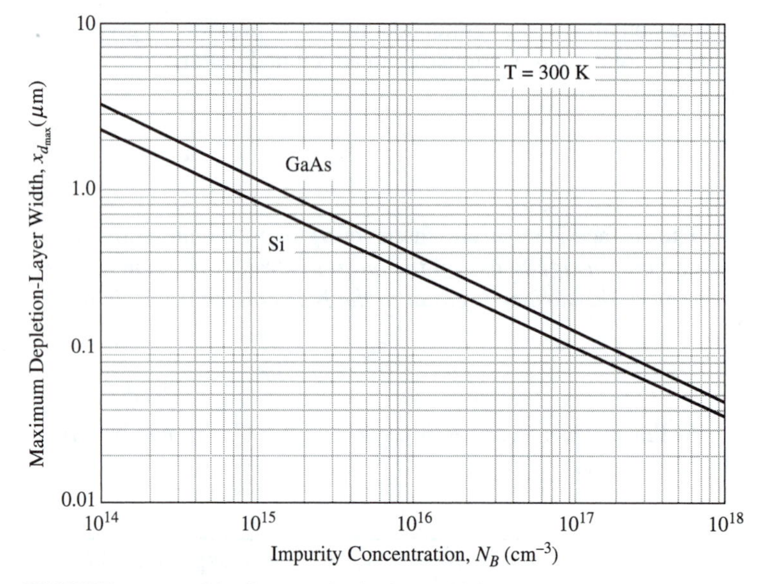

FIGURE 7.11 Maximum depletion width vs. impurity concentration for Si and GaAs for the condition of strong inversion.[14]

7.3 THRESHOLD VOLTAGE V_T

7.3.1 Threshold Condition

In MOS structures, as the gate bias is changed by an incremental voltage for the depletion condition, each charge added to the gate electrode causes an increase in the depletion-region width to balance the equal but opposite gate charge. For this condition, there are few mobile carriers contributing to the semiconductor charge which is made up of immobile ionized dopant atoms. Expansion of the depletion-region width continues to balance the addition of gate charge until inversion is reached with mobile minority carrier charge at the SiO_2/Si interface (inversion layer) and any additional gate charge is balanced by an increase in mobile charge carriers.

Brews[15] expresses this gate voltage when depletion ceases and inversion dominates as occurring when

threshold condition

$$\frac{dqN'_{inv}}{d\psi_s} = \frac{dQ'_{sc}}{d\psi_s}, \tag{7.39}$$

where ψ_s is the surface potential, qN'_{inv} is the inversion layer charge/unit area, and Q'_{sc} is the depletion-region charge/unit area. The inversion layer charge/unit area is given by

$$qN'_{inv} = qn_s x_c, \tag{7.40}$$

where x_c is an appropriate average thickness of the inversion layer (centroid of the inversion layer) given in Appendix C.1, and n_s is the electron concentration at the surface $x = 0$. The electron concentration at the surface is given by Eq. (7.22) as

$$\underbrace{n_s(0)}_{n_s} = n_i \exp\left\{ \frac{q}{kT} \underbrace{\left[-\frac{kT}{q} \ln \frac{N_a^-}{n_i} \right]}_{\psi_b} \right\} \exp[q \underbrace{\psi(0)}_{\psi_s} /kT], \tag{7.41}$$

where ψ_b was given in Eq. (7.18). Equation (7.41) may be written as

$$\boxed{n_s = \frac{n_i^2}{N_a^-} \exp(q\psi_s/kT)} \tag{7.42}$$

and demonstrates that $qN'_{inv} = qn_s x_c$ varies exponentially with ψ_s.

The depletion-region charge/unit area is given by

$$Q'_{sc} = -qN_a^- x_d, \tag{7.43}$$

with the depletion-region width given by Eq. (7.32) as

$$x_d = \sqrt{2\epsilon_{\text{Si}}\psi_s/qN_a^-}. \tag{7.44}$$

Therefore, with Eq. (7.44) for x_d in Eq. (7.43),

Q'_{sc}

$$\boxed{Q'_{sc} = -\sqrt{2q\epsilon_{\text{Si}}N_a^- \psi_s}} \tag{7.45}$$

and shows that Q'_{sc} varies as the square-root of ψ_s. According to Eq. (7.39), the *threshold* is taken as when the left and right sides of Eq. (7.39) are equal. As ψ_s is increased, both sides of Eq. (7.39) increase. However, the exponential increase in qN'_{inv} with ψ_s greatly exceeds the square-root dependence of Q'_{sc} with ψ_s. Therefore, as ψ_s increases, qN'_{inv} overtakes Q'_{sc}, and becomes the dominant response above threshold. The abruptness in this change of the dominant charge response is the reason this condition is termed the threshold.[15]

7.3.2 Threshold Surface Potential

The surface potential at threshold may be obtained by solving Eq. (7.39). This technique applies to both the uniformly doped case and for the nonuniformly doped case, such as for an ion-implanted layer. Only the uniformly doped case will be considered here.

From Eq. (C.1) in Appendix C at the end of this chapter for x_c and with Eq. (7.42) for n_s gives

$$qN'_{inv} = qn_sx_c = q\underbrace{\frac{n_i^2}{N_a^-}\exp(q\psi_s/kT)}_{n_s}\underbrace{\frac{\epsilon_{\text{Si}}kT}{qQ'_{sc}}}_{x_c}. \tag{7.46}$$

In Eq. (7.46), Q'_{sc} varies as $\sqrt{\psi_s}$ and the analysis can be simplified by neglecting the ψ_s dependence of Q'_{sc} as compared to the $\exp(q\psi_s/kT)$ term. Then

$$\frac{dqN'_{inv}}{d\psi_s} = q\frac{n_i^2}{N_a^-}\exp(q\psi_s/kT)\frac{q}{kT}\frac{\epsilon_{\text{Si}}kT}{qQ'_{sc}}. \tag{7.47}$$

The $(q/kT)(\epsilon_{\text{Si}}kT/qQ'_{sc})$ product in Eq. (7.47) may be rewritten as

$$\frac{q}{kT}\frac{\epsilon_{\text{Si}}kT}{qQ'_{sc}} = \frac{\epsilon_{\text{Si}}}{Q'_{sc}} = \frac{\epsilon_{\text{Si}}}{\sqrt{2q\epsilon_{\text{Si}}N_a^-\psi_s(qN_a^-/qN_a^-)}} = \frac{\epsilon_{\text{Si}}}{qN_a^-\underbrace{\sqrt{2\epsilon_{\text{Si}}\psi_s/qN_a^-}}_{x_d}}. \tag{7.48}$$

Then, the left side of Eq. (7.39) becomes

$$\frac{dqN'_{inv}}{d\psi_s} = \left(\frac{n_i}{N_a^-}\right)^2\frac{\epsilon_{\text{Si}}}{x_d}\exp(q\psi_s/kT). \tag{7.49}$$

The right side of Eq. (7.39) becomes

$$\begin{aligned}\frac{dQ'_{sc}}{d\psi_s} &= \frac{d}{d\psi_s}\sqrt{2q\epsilon_{\text{Si}}N_a^-\psi_s} \\ &= \sqrt{2q\epsilon_{\text{Si}}N_a^-}(\psi_s^{-1/2}/2) = \frac{\epsilon_{\text{Si}}}{\underbrace{\sqrt{2\epsilon_{\text{Si}}\psi_s/qN_a^-}}_{x_d}}.\end{aligned} \tag{7.50}$$

With Eqs. (7.49) and (7.50) in Eq. (7.39), the threshold condition becomes

$$\left(\frac{n_i}{N_a^-}\right)^2\frac{\epsilon_{\text{Si}}}{x_d}\exp(q\psi_s/kT) = \frac{\epsilon_{\text{Si}}}{x_d}. \tag{7.51}$$

Solving for ψ_s in Eq. (7.51) gives

ψ_s at threshold

$$\boxed{\psi_s = 2(kT/q)\ln(N_a^-/n_i)}. \tag{7.52}$$

With Eq. (7.18) for ψ_b, Eq. (7.57) gives

$$\psi_s = -2\psi_b, \tag{7.53}$$

which was given in Eq. (7.36). Equation (7.52) is a fundamental derivation of the relationship that $n_s = p_b$ at threshold. If the semiconductor had been an ion-implanted layer instead of uniformly doped, an additive term would result in Eq. (7.52) for ψ_s.[15]

7.3.3 Inversion by Carrier Generation

When the MOS capacitor is biased into inversion, a source for minority carriers is required. In the case of the capacitor, either generation of carriers by light or sufficient time must be allowed to permit thermal generation by the generation–recombination centers within the energy gap. The net nonradiative recombination rate U was given in Eq. (3.62) as

$$U = \frac{np - n_i^2}{\tau_{nr}(n + n_1) + \tau_{nr}(p + p_1)}. \tag{7.54}$$

Under conditions of depletion in the semiconductor, both n and p are less than the intrinsic concentration n_i. Also, the electron concentration n_1 or hole concentration p_1 (n and p when $E_f = E_i = E_t$) is n_i. Then, Eq. (7.54) reduces to

$$U = \frac{-n_i^2}{\tau_{nr}n_i + \tau_{nr}n_i} = -\frac{n_i}{2\tau_{nr}} = G, \tag{7.55}$$

where the minus sign signifies net generation G.

EXAMPLE 7.1 For an MOS capacitor on p-type Si with $p = N_a^- = 5\times10^{15}\ \mathrm{cm}^{-3}$ and a nonradiative lifetime $\tau_{nr} = 1\times10^{-8}$ s, how much time is required for generation of the electron concentration at the SiO_2/Si interface n_s to reach the hole concentration in the neutral bulk, $n_s = p_b$?

Solution The generation rate given by Eq. (7.55) when multiplied by the time t may be used to determine how long it takes for n_s to equal p_b:

$$Gt = n_s = p_b = n_i t/2\tau_{nr},$$

or

minority carrier generation time

$$t = 2\tau_{nr}p_b/n_i = 2\times1\times10^{-8}\times5\times10^{15}/1.0\times10^{10} \simeq 1\times10^{-2}\ \mathrm{s}.$$

This example demonstrates that times of the order of 10^{-2} s are required for generation of the minority carriers after the gate voltage V_G has been applied to create inversion. ∎

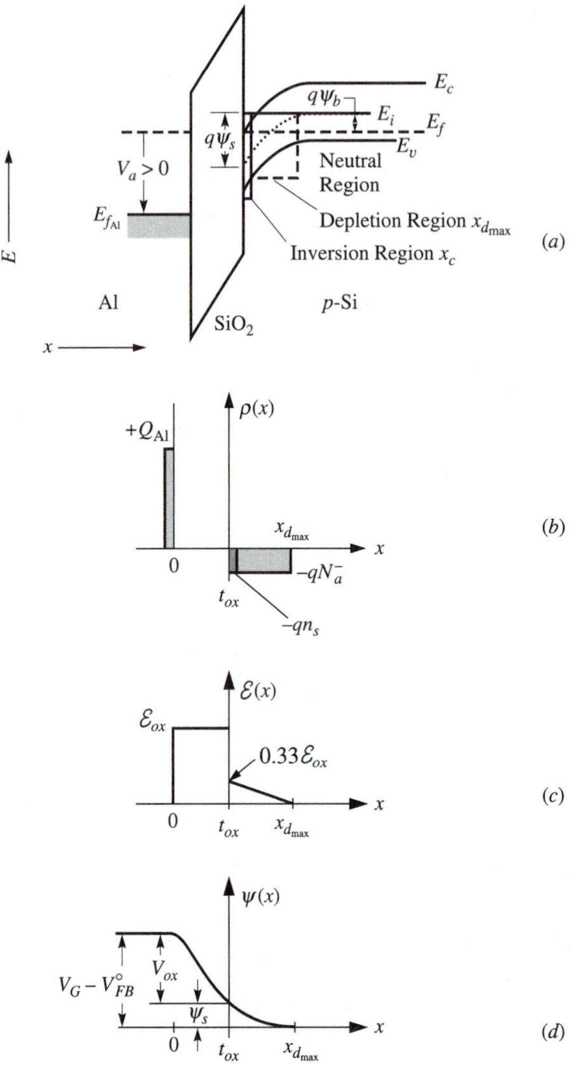

FIGURE 7.12 (*a*) Energy-band diagram for an ideal Al/SiO_2/*p*-Si MOS capacitor for the threshold bias condition. (*b*) Charge distribution for the MOS capacitor. (*c*) Electric-field distribution. (*d*) Potential distribution.

7.3.4 The Electric Field and Potential

The energy-band diagram for the Al/SiO_2/*p*-Si MOS capacitor biased for the condition of inversion is shown in Fig. 7.12 (*a*) and has the positive bias polarity connected to the Al gate, as was shown in Fig. 7.9 (*a*). The conditions have been chosen to give $t > 2\tau_{nr} p_b / n_i$ so that the inversion charge has reached the steady-state value represented by $-qN_{inv}$. The charge distribution is shown in Fig. 7.12 (*b*). As shown by Eq. (7.42), the free-electron concentration at the surface depends exponentially on the potential at the surface ψ_s, and $-qN_{inv}$ increases rapidly with V_G and the depletion width remains constant at $x_{d_{max}}$.

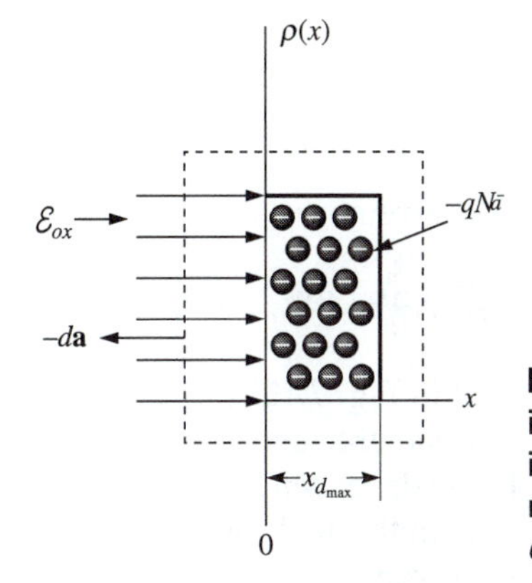

FIGURE 7.13 Determination of the electric field in the oxide due to the charge $Q'_{sc} = -qN_a^- x_{d_{max}}$ in the semiconductor. The dashed rectangular box represents the Gaussian surface enclosing Q'_{sc}, while $d\mathbf{a}$ is the incremental surface area.

The portion of the gate voltage across the Si at threshold is given by the potential $\psi_s = -2\psi_b$. The portion of the gate voltage across the oxide at threshold (V_{ox}) may be found from Gauss's law. Gauss's law in integral form states that the surface integral of the electric flux density, $\mathcal{D} = \epsilon\mathcal{E}$, is equal to the charge enclosed:

$$Q = \epsilon \oint \mathcal{E} \cdot d\mathbf{a}, \tag{7.56}$$

where $\oint$ is the surface integral. The negative charge in the p-type semiconductor at threshold is shown in Fig. 7.13. Only the charge due to the fixed ionized acceptors is considered because, as shown in Appendix C.1, the inversion layer charge/unit area qN'_{inv} is much less than the depletion-region charge/unit area $Q'_{sc} = -qN_a^- x_{d_{max}}$, and therefore, qN'_{inv} may be neglected in Eq. (7.56).

The negative charge per unit area Q'_{sc} is enclosed by the dashed surface illustrated in Fig. 7.13. The integral in Eq. (7.56) gives

$$Q_{sc} = -\epsilon_{ox}\mathcal{E}_{ox}A, \tag{7.57}$$

where A is the cross-sectional area of the MOS capacitor. The electric field in the oxide is

$$\mathcal{E}_{ox} = -Q_{sc}/\epsilon_{ox}A, \tag{7.58}$$

where $Q_{sc} = -qAN_a^- x_{d_{max}} = -A\sqrt{4q\epsilon_{\text{Si}}N_a^-|\psi_b|}$ at threshold with $x_{d_{max}}$ from Eq. (7.38). The dielectric constant for SiO_2 is given by $\epsilon_{ox} = 3.9\epsilon_0$.

At the SiO_2/Si interface, boundary conditions require that the electric flux density $\mathcal{D}$ must be continuous:

$$\mathcal{D}_{ox} = \mathcal{D}_{\text{Si}}, \tag{7.59}$$

or

$$\epsilon_{ox}\mathscr{E}_{ox} = \epsilon_{Si}\mathscr{E}_{Si}. \tag{7.60}$$

Then, the electric field at the Si surface is given by

Si surface electric field $\mathscr{E}_{Si}$

$$\mathscr{E}_{Si} = (\epsilon_{ox}/\epsilon_{Si})\mathscr{E}_{ox} = (3.9/11.7)\mathscr{E}_{ox} = 0.33\mathscr{E}_{ox}. \tag{7.61}$$

This electric field in the Si is represented in Fig. 7.12 (*c*). For larger ψ_s and $n_s > p_b$, the electric field in the Si is influenced by both the inversion layer charge qN_{inv} and the space charge Q_{sc}, and is not a simple linear dependence as shown in Fig. 7.12 (*c*). The detailed variation of $\mathscr{E}_{Si}(x)$ is not required in the analysis given in this chapter.

The potential $\psi_{ox}(x)$ in the SiO_2 is given by

$$\psi_{ox}(x) = -\int \mathscr{E}_{ox}dx = -\mathscr{E}_{ox}x + A_1, \tag{7.62}$$

where A_1 is an arbitrary integration constant determined by the boundary condition $\psi(t_{ox}) = \psi_s$. This boundary condition gives

$$\psi_{ox}(x) = \psi_s + \mathscr{E}_{ox}t_{ox} - \mathscr{E}_{ox}x \qquad \text{for } 0 < x < t_{ox}. \tag{7.63}$$

By Eq. (7.63), the potential across the oxide is the difference in $\psi(x)$ at $x = 0$ and $x = t_{ox}$,

$$V_{ox} = \psi_{ox}(0) - \psi_{ox}(t_{ox}) = \mathscr{E}_{ox}t_{ox}, \tag{7.64}$$

and is illustrated in Fig. 7.12 (*d*). By Eq. (7.58) for $\mathscr{E}_{ox}$ and $Q_{sc} = -qAN_a^- x_{d_{max}}$, V_{ox} becomes

V_{ox} at threshold

$$\boxed{V_{ox} = \frac{-Q_{sc}t_{ox}}{\epsilon_{ox}A} = \frac{\sqrt{4q\epsilon_{Si}N_a^-|\psi_b|}\; t_{ox}}{\epsilon_{ox}}}, \tag{7.65}$$

with $\psi_s = -2|\psi_b|$ at threshold. This expression for V_{ox} in Eq. (7.65) will be used in the next part of this section for the expression for threshold voltage.

7.3.5 Ideal Oxide Threshold Voltage V_T

The gate potential $V_G = V_T$ at threshold can be found by adding the potential drop across the semiconductor to the potential drop across the oxide and the work function difference for the ideal oxide:

$$V_T = V_G = \overbrace{\psi_T}^{\psi_s} + V_{ox} + V_{FB}^{\circ}. \tag{7.66}$$

The potential drop across the semiconductor at threshold, $\psi_T = \psi_s$, is given by Eq. (7.53) as $2|\psi_b|$ with ψ_b given by Eq. (7.18), and V_{ox} is given by Eq. (7.65). The threshold voltage for an n-channel MOS device on a p-type substrate is designated as $V_G = V_{T_n}$ and is given by

ideal oxide threshold voltage for n-channel on p-type Si

$$\boxed{V_{T_n} = V^{\circ}_{FB} + 2|\psi_b| + \sqrt{4q\epsilon_{\text{Si}}N_a^-|\psi_b|}/(\epsilon_{ox}/t_{ox})}. \tag{7.67}$$

Note that $\epsilon_{ox}/t_{ox} = C_{ox}/A = C'_{ox}$. The threshold voltage for the ideal SiO_2 given in Eq. (7.67) is one of the principal expressions used to describe MOSFETs.

When the work function of the p-Si is greater than the work function of the gate, the MOS capacitor is in depletion at thermal equilibrium, as shown in Fig. 7.5, and the flat-band voltage is negative. Therefore, the value of the gate voltage to go from flat-band conditions to inversion, which is the threshold voltage, is small because the MOS capacitor is already almost to inversion at thermal equilibrium. Equation (7.67) for the threshold voltage contains the *flat-band voltage* which is *negative,* plus the potential across the p-Si, $\psi_s = 2|\psi_b|$, and the potential across the oxide, $V_{ox} = Q'_{sc}/C'_{ox}$.

It is also important to note that the electron concentration at the SiO_2/Si interface n_s is given by Eq. (7.42) as $n_s = (n_i^2/N_a^-)\exp(q\psi_s/kT)$ while the minority carrier concentration at the edge of the depletion region n_p in a p-n junction is given by Eq. (4.85) as $n_p = (n_i^2/N_a^-)\exp(qV_a/kT)$. *In both cases, the minority carrier concentration varies exponentially with potential.* For the MOS case, ψ_s is the difference in the applied gate voltage V_G and the potential across the oxide V_{ox}, while for the p-n junction, V_a is the applied voltage.

For MOS devices prepared on n-type Si, the carriers in the surface inversion layer will be holes. This structure is designated a p-channel device. The threshold voltage may be derived in a manner analogous to the derivation given for the n-channel device in Eq. (7.67) and V_{T_p} for a p-channel device is given by

ideal oxide threshold voltage for p-channel on n-type Si

$$\boxed{V_{T_p} = V^{\circ}_{FB} - 2|\psi_b| - \sqrt{4q\epsilon_{\text{Si}}N_d^+|\psi_b|}/(\epsilon_{ox}/t_{ox})}. \tag{7.68}$$

EXAMPLE 7.2 An ideal MOS capacitor at 300 K has an Al gate with an oxide thickness of 500 Å (50 nm) on p-type Si with $p = N_a^- = 4\times 10^{15}$ cm^{-3}. Calculate the threshold voltage V_{T_n}.

Solution First, find V°_{FB}, ψ_b, and C'_{ox}.

$$V^{\circ}_{FB} = \Phi_{\text{Al}} - [X_{\text{Si}} + E_g - (E_f - E_v)]$$

$$V^{\circ}_{FB} = 4.1 - \{4.05 + 1.125 - [-0.026\ln(4\times 10^{15}/3.08\times 10^{19})]\} = \underline{-0.842\text{ V}}$$

$$\psi_b = -0.026\ln(N_a^-/n_i) = -0.026\ln(4\times 10^{15}/1.0\times 10^{10}) = \underline{-0.335\text{ V}}$$

$$C'_{ox} = \frac{\epsilon_{ox}}{t_{ox}} = \frac{3.9\times 8.85\times 10^{-14}}{500\times 10^{-8}} = \underline{6.903\times 10^{-8}\text{ F/cm}^2}$$

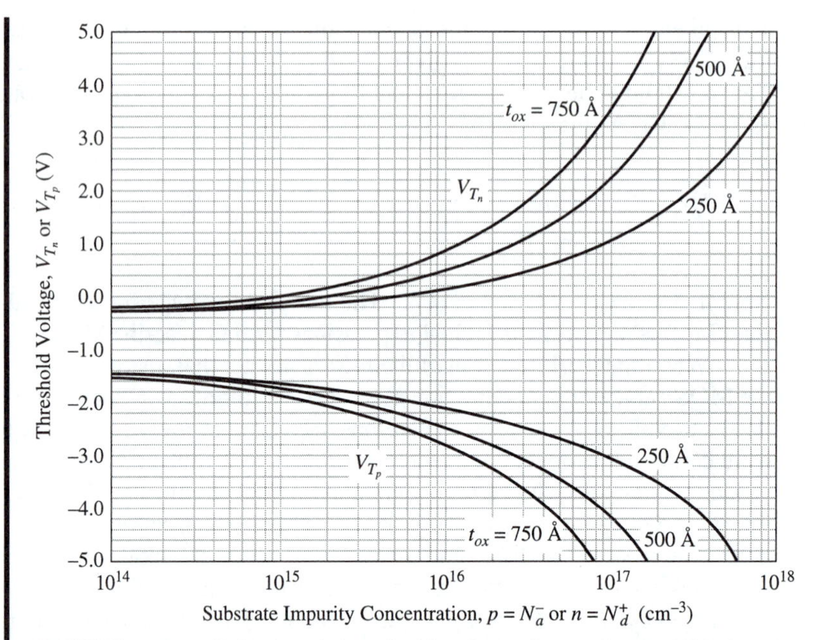

FIGURE 7.14 Calculated threshold voltage for *n*-channel (V_{T_n}) and *p*-channel (V_{T_p}) MOS capacitors as a function of substrate impurity concentration with an Al gate and an ideal oxide.

Note that typical values for C'_{ox} are $\sim 10^{-8}$ F/cm^2.

$$V_{T_n} = V^{\circ}_{FB} + 2|\psi_b| + \sqrt{4q\epsilon_{\mathrm{Si}}N_a^-|\psi_b|}/(\epsilon_{ox}/t_{ox})$$

$$V_{T_n} = -0.842 + 0.670 + \frac{\sqrt{4 \times 1.6 \times 10^{-19} \times 11.7 \times 8.85 \times 10^{-14} \times 4 \times 10^{15} \times 0.335}}{6.903 \times 10^{-8}}$$

$$V_{T_n} = -0.842 + 0.670 + 2.980 \times 10^{-8}/6.903 \times 10^{-8}$$

$$V_{T_n} = -0.842 + 0.670 + 0.432 = \underline{0.260 \text{ V}}$$

Note that $\sqrt{4q\epsilon_{\mathrm{Si}}N_a^-|\psi_b|} \simeq 10^{-8}$ and when divided by C'_{ox} the quotient is the order of 1 V.

The variation of the threshold voltage for both *n*-channel (V_{T_n}) and *p*-channel (V_{T_p}) MOS capacitors for three different oxide thicknesses as a function of the substrate doping is shown in Fig. 7.14. ■

7.3.6 Nonideal Oxide Threshold Voltage V_T

In an actual MOS capacitor or MOSFET, there may be charge within the oxide or at the amorphous SiO_2 and Si interface. At threshold, the concentration of electrons in *p*-type

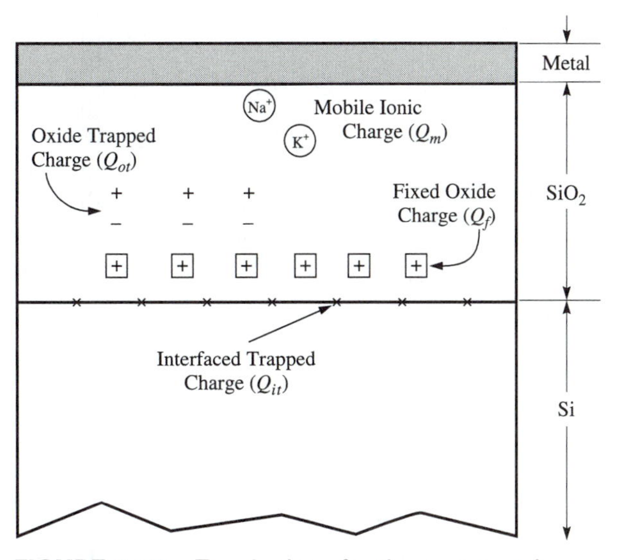

FIGURE 7.15 Terminology for the names and location of charges in thermally oxidized Si.[16]

Si will equal the ionized acceptor concentration. For the case of $p = N_a^- = 1 \times 10^{15}$ cm^{-3}, the area electron density per cm^2 will be $(1 \times 10^{15})^{2/3} = 1 \times 10^{10}$ cm^{-2}. The area density of Si atoms will be $(5 \times 10^{22})^{2/3} \simeq 10^{15}$ cm^{-2}. Therefore, an interface-charge density of 10^{10} cm^{-2}, which is 10^{-5} of the atom density, would be expected to affect a MOS device. One of the principal reasons Si became the preferred semiconductor for integrated circuits was that the SiO_2/Si interface can be prepared with charge densities of the order of 10^{10} cm^{-2} or less.

It has been established that four general types of charges are associated with the SiO_2/Si system.[16] The common terminology for oxide charges is summarized in Fig. 7.15. The charge is given by Q' coulombs/cm^2 and $|Q'/q| = N$, where N is the number/cm^2. A subscript is used to designate the type of oxide charge. The *fixed oxide charge*, Q'_f and N'_f, is a positive charge and is due primarily to structural defects (ionized Si) in the oxide layer less than 25 Å from the SiO_2/Si interface. The density of this charge is related to the oxidation process. The *interface trapped charge*, Q'_{it} and N_{it}, may be positive or negative charges and is due to structural, oxidation-induced defects, metal impurities, or defects caused by radiation or similar bond-breaking processes. This charge, unlike fixed charge or trapped charge, is in electrical communication with the underlying Si and can be charged or discharged. Most of the interface charge can be neutralized by annealing at 450°C in hydrogen. *This interface trapped charge has also been called surface states, fast states, or interface states.* The *oxide trapped charge*, Q'_{ot} and N_{ot}, may be positive or negative due to holes or electrons trapped in the bulk of the SiO_2. This trapped charge may result from ionizing radiation or avalanche injection. The *mobile ionic charge*, Q'_m and N_m, is due primarily to ionic impurities such as Li^+, Na^+, and K^+.

fixed oxide charge Q'_f

interface trapped charge or fast states Q'_{it}

oxide trapped charge Q'_{ot}

mobile ionic charge Q'_m

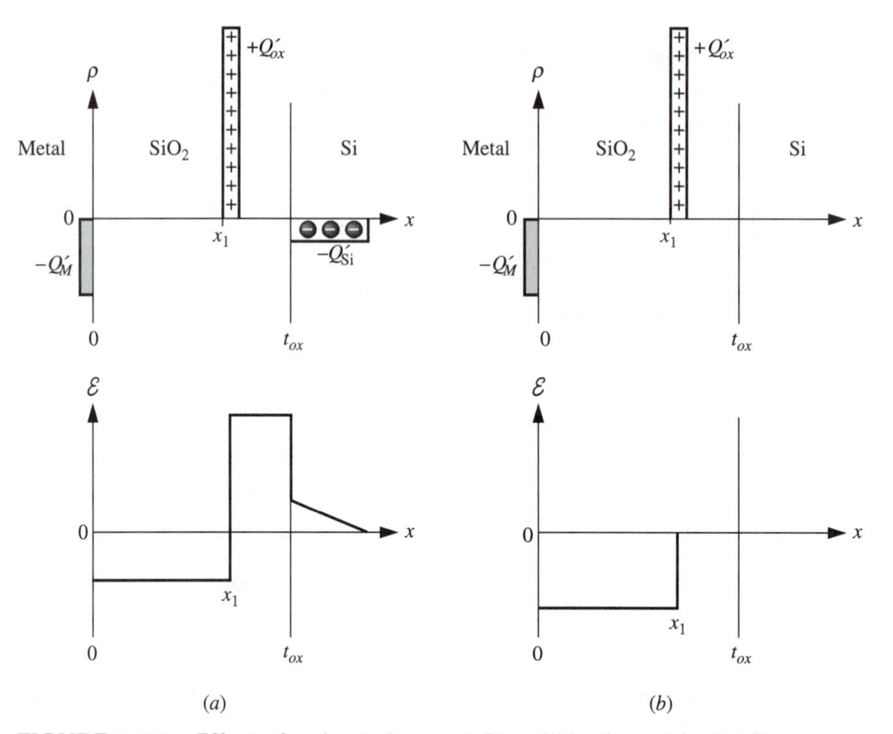

FIGURE 7.16 Effect of a sheet charge $+Q'_{ox}$ within the oxide. (*a*) Charge and electric field for $V_G = 0$. (*b*) Charge and electric field for flat-band conditions with $V_G \neq 0$.

The effect of a sheet charge $+Q'_{ox}$ on the MOS structure is illustrated in Fig. 7.16. For $V_G = 0$, the positive sheet charge will induce negative charges which are partly in the metal and partly in the semiconductor as illustrated in Fig. 7.16 (*a*). It will be shown that the closer the position x_1 of $+Q'_{ox}$ to the semiconductor, the greater the fraction of the negative charge induced in the semiconductor. In order to obtain the flat-band condition represented in Fig. 7.16 (*b*), *it is necessary to apply a negative voltage to the gate metal, which increases $-Q'_M$ and eliminates $-Q'_{\text{Si}}$*. For the flat-band condition, the field in the oxide is increased as shown in the lower part of Fig. 7.16 (*b*). Gauss's law has been applied to the charge on the metal gate as shown in Fig. 7.17, and gives

$$-Q'_M = Q'_{ox} = \frac{\epsilon_{ox}}{A} \oint \mathscr{E}_{ox} \cdot d\mathbf{a}, \tag{7.69}$$

which gives

$$-Q'_M = Q'_{ox} = -\epsilon_{ox}\mathscr{E}_{ox}, \tag{7.70}$$

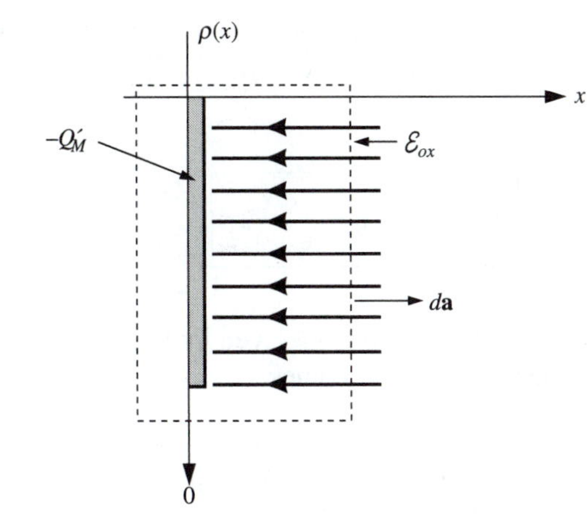

FIGURE 7.17 Determination of the electric field in the oxide due to the charge $-Q'_M$ on the metal electrode for $V_G \neq 0$ and $Q'_{Si} = 0$. The dashed rectangular box represents the Gaussian surface enclosing Q'_M, while $d\mathbf{a}$ is the incremental area.

or

$$\mathscr{E}_{ox} = -\frac{Q'_{ox}}{\epsilon_{ox}} \qquad \text{for } 0 < x < x_1. \tag{7.71}$$

The flat-band voltage shift ΔV_{FB} is given by the area of the $\mathscr{E}$ versus distance x plot:

$$\Delta V_{FB} = \psi(x) = -\int_0^{x_1} \mathscr{E}_{ox}\, dx = -(-\mathscr{E}_{ox})x_1 = \mathscr{E}_{ox}x_1. \tag{7.72}$$

With Eq. (7.71) for $\mathscr{E}_{ox}$ in Eq. (7.72),

$$\Delta V_{FB} = -\frac{Q'_{ox}x_1}{\epsilon_{ox}}. \tag{7.73}$$

With $C'_{ox} = \epsilon_{ox}/t_{ox}$ or $\epsilon_{ox} = C'_{ox}t_{ox}$, Eq. (7.73) becomes

$$\Delta V_{FB} = -\frac{Q'_{ox}x_1}{C'_{ox}t_{ox}}. \tag{7.74}$$

Equation (7.74) includes the amount of oxide charge and the effect of its location. The change in the flat-band voltage ΔV_{FB} is the difference between the flat-band voltage of the nonideal oxide and the flat-band voltage of the ideal oxide:

$$V_{FB} - V^{\circ}_{FB} = \Delta V_{FB}, \tag{7.75}$$

or

$$V_{FB} = V^{\circ}_{FB} + \Delta V_{FB} = V^{\circ}_{FB} - \frac{Q'_{ox}x_1}{C'_{ox}t_{ox}}. \tag{7.76}$$

By Eq. (7.76), Q'_{ox} will have little effect on V_{FB} when x_1 is very near the metal gate. The oxide charge Q'_{ox} will have maximum influence when Q'_{ox} is located at t_{ox}, and the flat-band voltage becomes

nonideal oxide flat-band voltage

$$\boxed{V_{FB} = V^{\circ}_{FB} - \frac{qN_f}{\epsilon_{ox}/t_{ox}}}, \tag{7.77}$$

when Q'_{ox} is due to fixed oxide charge. With V_{FB} given by Eq. (7.76) or Eq. (7.77), the threshold voltage for an n-channel device becomes

nonideal oxide threshold voltage for n-channel on p-Si

$$\boxed{V_{T_n} = V_{FB} + 2|\psi_b| + \sqrt{4q\epsilon_{\mathrm{Si}}N_a^-|\psi_b|/(\epsilon_{ox}/t_{ox})}}, \tag{7.78}$$

and for a p-channel device becomes

nonideal oxide threshold voltage for p-channel on n-Si

$$\boxed{V_{T_p} = V_{FB} - 2|\psi_b| - \sqrt{4q\epsilon_{\mathrm{Si}}N_d^+|\psi_b|/(\epsilon_{ox}/t_{ox})}}. \tag{7.79}$$

The oxide charges not only affect the threshold voltage, but also influence the capacitance-voltage behavior. Further discussion of the properties of the oxide charges are included in the next section.

7.4 DIFFERENTIAL MOS CAPACITANCE

7.4.1 Ideal Oxide MOS Capacitance-Voltage Behavior

Analysis of the behavior of the small-signal capacitance variation with bias voltage of an MOS capacitor provides further understanding of the electrical behavior of the MOS system. The static capacitance is defined as $C_{static} \equiv Q/V$, where Q is the total charge on the capacitor and V is the applied voltage. The differential capacitance is:

differential capacitance

$$C_{dif} \equiv \left|\frac{dQ_{\mathrm{Si}}}{dV_G}\right|. \tag{7.80}$$

The static and differential capacitances differ for the MOS capacitor because the charge on the MOS capacitor varies nonlinearly with the gate voltage. For the common parallel-plate capacitor there is a linear relationship between the charge and the voltage, and the static and differential capacitance will be the same. In this section, expressions for the differential capacitance will be derived for the conditions of accumulation, flat band, and depletion.

As illustrated in Fig. 7.12 (d) and Eq. (7.66), the gate voltage minus the flat-band voltage equals the voltage across the oxide and the Si, or

$$V_G = V_{FB}^{\circ} + V_{ox} + \psi_s. \tag{7.81}$$

The voltage across the oxide may be written as was given in Eq. (7.65) as

$$V_{ox} = -\frac{t_{ox} Q'_{\mathrm{Si}}}{\epsilon_{ox}}. \tag{7.82}$$

With Eq. (7.82) for V_{ox}, the gate voltage in Eq. (7.81) may be written as

$$V_G = V_{FB}^{\circ} - t_{ox} Q'_{\mathrm{Si}}/\epsilon_{ox} + \psi_s. \tag{7.83}$$

The differential capacitance per unit area from Eq. (7.80) with Eq. (7.83) for V_G now becomes

$$C'_{dif} = \left|\frac{dQ'_{\mathrm{Si}}}{dV_G}\right| = \left|\frac{dQ'_{\mathrm{Si}}}{-(t_{ox}/\epsilon_{ox})dQ'_{\mathrm{Si}} + d\psi_s}\right| = \frac{1}{1/C'_{ox} + 1/\left|\frac{dQ'_{\mathrm{Si}}}{d\psi_s}\right|}. \tag{7.84}$$

Equation (7.84) may be written as

$$\boxed{\frac{1}{C'_{dif}} = \frac{1}{C'_{ox}} + \frac{1}{C'_{\mathrm{Si}}}}, \tag{7.85}$$

with

Si differential capacitance C'_{Si}

$$\boxed{C'_{\mathrm{Si}} \equiv \left|\frac{dQ'_{\mathrm{Si}}}{d\psi_s}\right|}. \tag{7.86}$$

In the MOS literature, the oxide capacitance per unit area is written as C_{ox}. However, to be consistent with capacitances for the p-n junction and bipolar transistor where the capacitances C_j and C_s are not per unit area, C' will be used here to denote capacitances per unit area and capacitance $C = C'A$, where A is the area in cm^2.

It should be noted that the current charging the series capacitors C_{ox} and C_{Si} is the same and their charge therefore will be the same ($Q = I/t$). Then, $V_G = Q/C = Q/C_{ox} + Q/C_{\mathrm{Si}}$ or $1/C = 1/C_{ox} + 1/C_{\mathrm{Si}}$, and Eq. (7.85) represents this series combination. In Fig. 7.18, this series combination is represented as the fixed capacitance of the oxide and the variable capacitance of the Si which depends on the applied gate voltage through ψ_s.

total differential capacitance c'_{dif}

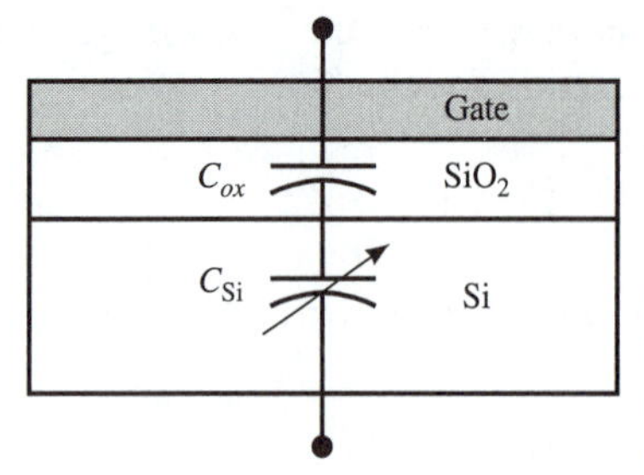

FIGURE 7.18 Schematic representation of the MOS capacitor as a series combination of the fixed oxide capacitor C_{ox} and the variable Si capacitor C_{Si}.

The evaluation of C'_{Si}, as given by Eq. (7.86), requires representing Q'_{Si} as a function of ψ_s. The potential ψ_s may be obtained from Poisson's equation, which includes the contribution of electrons and holes at the SiO_2/Si interface. As shown in Appendix C.2 at the end of this chapter, an integrating factor is used to make Poisson's equation an exact differential which permits integration. Use of boundary conditions at the SiO_2/Si interface and at the edge of the depletion region permit obtaining the electric field $\mathscr{E}_s$ at the SiO_2/Si interface. The charge in the Si per unit area is obtained from Gauss's law as $|Q'_{Si}| = \epsilon_{Si}\mathscr{E}_s$ [see Eq. (C.23)]. The capacitance per unit area for the Si in *accumulation* or *depletion* is given in Eq. (C.48) as

$$C'_{accum,dep} = \frac{\epsilon_{Si}}{\sqrt{2}L_D}\left|\frac{[1-\exp(-q\psi_s/kT)]}{\left[\frac{q\psi_s}{kT}+\exp(-q\psi_s/kT)-1\right]^{1/2}}\right| \text{ F/cm}^2, \quad (7.87)$$

where L_D is the Debye length for p Si with $p = N_a^-$ and is given by

Debye length L_D

$$\boxed{L_D = \sqrt{\frac{\epsilon_{Si}kT}{q^2N_a^-}}}. \quad (7.88)$$

The *Debye length* represents a distance over which the free carriers reduce (*screen*) the potential from the fixed impurity ions, and the reduced potential would decay as $\exp(-x/L_D)$. Evaluation of the numerical quantities in the prefactor $\epsilon_{Si}/\sqrt{2}L_D$ gives C'_{Si} as

semiconductor capacitance for accumulation or depletion $C'_{accum,dep}$

$$C'_{accum,dep} = 1.785 \times 10^{-15}\sqrt{N_a^-}\left|\frac{[1-\exp(-q\psi_s/kT)]}{\left[\frac{q\psi_s}{kT}+\exp(-q\psi_s/kT)-1\right]^{1/2}}\right| \text{ F/cm}^2. \quad (7.89)$$

inversion

For inversion, the complete expression given by Eq. (C.45) must be used for C'_{Si}.

For the flat-band condition, Q'_{Si} is zero, but as ψ_s changes from $\psi_s = 0$ at flat band, there will be a change in charge and hence a value for C'_{Si}. To evaluate Eq. (7.87) for small ψ_s, the exponential terms in the denominator requires a three-term series expansion to prevent the denominator from going to zero if only two terms are used. Therefore, $\exp(-q\psi_s/kT)$ becomes $[1-(q\psi_s/kT)+(1/2)(q\psi_s/kT)^2]$. The exponential in the numerator may be represented by the first two terms of the series expansion. The differential capacitance per unit area for $\psi_s \simeq 0$ is the flat-band capacitance C'_{FB} and is given by Eq. (7.87) as

semiconductor flat-band capacitance C'_{FB}

$$C'_{FB} = \frac{\epsilon_{\text{Si}}}{\sqrt{2}L_D} \frac{\dfrac{q\psi_s}{kT}}{\left[\dfrac{1}{2}\left(\dfrac{q\psi_s}{kT}\right)^2\right]^{1/2}} = \frac{\epsilon_{\text{Si}}\sqrt{2}}{\sqrt{2}L_D} = \frac{\epsilon_{\text{Si}}}{L_D} = 2.524 \times 10^{-15}\sqrt{N_a^-}\ \text{F/cm}^2. \tag{7.90}$$

As the applied voltage is increased for depletion and then inversion, the depletion width increases until strong inversion occurs for $\psi_s = 2|\psi_b|$. The maximum depletion width $x_{d_{max}}$ at the onset of strong inversion was given by Eq. (7.38) with $\psi_s = 2|\psi_b|$ as

$$x_{d_{max}} = \sqrt{\frac{2\epsilon_{\text{Si}}\psi_s}{qN_a^-}}. \tag{7.91}$$

The space charge per unit area Q'_{sc} due to the depletion region when $x = x_{d_{max}}$ is

$$Q'_{sc} = -qN_a^- x_{d_{max}} = -\sqrt{2q\epsilon_{\text{Si}}N_a^-\psi_s}. \tag{7.92}$$

The differential capacitance per unit area C'_{sc} at maximum depletion width may be obtained with Eq. (7.86) as

$$C'_{sc} = \left|\frac{dQ'_{sc}}{d\psi_s}\right| = \frac{1}{2}\sqrt{2q\epsilon_{\text{Si}}N_a^-}\,\psi_s^{-1/2}, \tag{7.93}$$

or

semiconductor capacitance at maximum depletion C'_{sc}

$$C'_{sc} = \epsilon_{\text{Si}}\underbrace{\sqrt{\frac{qN_a^-}{2\epsilon_{\text{Si}}\psi_s}}}_{1/x_{d_{max}}} = \frac{\epsilon_{\text{Si}}}{x_{d_{max}}}\ \text{F/cm}^2. \tag{7.94}$$

When the semiconductor is in accumulation, as illustrated in Fig. 7.8, E_i at $x = 0$ is larger than E_{i_b}, and ψ_s is negative. For $|\psi_s| > 3kT/q$, Eq. (7.87) becomes

semiconductor capacitance in accumulation C'_{accum}

$$C'_{accum} \simeq \frac{\epsilon_{\text{Si}}}{L_D}\left|\frac{-\exp[-q(-\psi_s)/kT]}{\sqrt{2}\exp[-q(-\psi_s)/2kT]}\right| \simeq C'_{FB}\frac{\exp[-q(-\psi_s)/2kT]}{\sqrt{2}}\ \text{F/cm}^2. \tag{7.95}$$

Because C'_{accum} increases as $\exp(q|\psi_s|/2kT)$, this capacitance increases very rapidly with $-\psi_s$.

The differential capacitance for the ideal MOS capacitor may conveniently be written from Eq. (7.85) as

$$\boxed{C'_{dif} = \frac{1}{\dfrac{1}{C'_{ox}} + \dfrac{1}{C'_{\text{Si}}}}}. \tag{7.96}$$

At $V_G = V_{FB}$, the differential MOS capacitance given by Eq. (7.96) with Eq. (7.90) for the *semiconductor flat-band capacitance* C'_{FB} and with $C'_{ox} = \epsilon_{ox}/t_{ox}$ becomes

MOS capacitance at flat-band C'_{dif}(FB)

$$\boxed{C'_{dif}(\text{FB}) = \frac{1}{\dfrac{t_{ox}}{\epsilon_{ox}} + \dfrac{L_D}{\epsilon_{\text{Si}}}} \text{ F/cm}^2}. \tag{7.97}$$

The semiconductor will be in *accumulation* for gate voltages more negative than the flat-band voltages for *p*-type Si. In accumulation, the differential capacitance given by Eq. (7.96) with Eq. (7.95) for the semiconductor accumulation capacitance C'_{accum} and with $C'_{ox} = \epsilon_{ox}/t_{ox}$ becomes

MOS capacitance in accumulation C'_{dif} (accum)

$$\boxed{C'_{dif}(\text{accum}) = \frac{1}{\dfrac{t_{ox}}{\epsilon_{ox}} + \dfrac{L_D}{\epsilon_{\text{Si}}} \dfrac{\sqrt{2}}{\exp[q|\psi_s|/2kT]}} \text{ F/cm}^2}. \tag{7.98}$$

For strong accumulation when $|\psi_s| > 6kT/q$, the second term in the denominator rapidly becomes smaller than t_{ox}/ϵ_{ox}, and

limiting MOS capacitance in accumulation C'_{dif}(accum)

$$\boxed{C'_{dif}(\text{accum}) = \frac{\epsilon_{ox}}{t_{ox}} = C'_{ox} \text{ F/cm}^2}. \tag{7.99}$$

When $V_G = V_T$, $\psi_s = 2|\psi_b|$ and the condition of *maximum depletion* gives the *space-charge capacitance,* and the differential MOS capacitance with Eq. (7.94) for C'_{sc} becomes

MOS capacitance at maximum depletion $C'_{dif}(x_{d_{max}})$

$$\boxed{C'_{dif}(x_{d_{max}}) = \frac{1}{\dfrac{t_{ox}}{\epsilon_{ox}} + \dfrac{x_{d_{max}}}{\epsilon_{\text{Si}}}} \text{ F/cm}^2}. \tag{7.100}$$

Equations (7.97), (7.99), and (7.100) may be used to sketch the C–V behavior of the ideal MOS capacitor from the three capacitances in limiting accumulation, flat-band, and maximum depletion. The ideal MOS capacitor *C–V* curve may readily be obtained by Eq. (7.96) for the differential MOS capacitance with Eq. (7.89) for C'_{Si} between the limiting accumulation capacitance and maximum depletion. The gate voltage V_G may be obtained from ψ_s and Q'_{Si}. As shown in Fig. 7.12 (*d*) for depletion and inversion, the difference $V_G - V^\circ_{FB}$ is the sum of the voltage across the oxide and the semiconductor, $V_{ox} + \psi_s$:

$$V_G - V^\circ_{FB} = V_{ox} + \psi_s. \tag{7.101}$$

The voltage across the oxide was given by Eq. (7.64) as

$$V_{ox} = \mathscr{E}_{ox} t_{ox}, \tag{7.102}$$

or

$$\mathscr{E}_{ox} = \frac{V_{ox}}{t_{ox}}. \tag{7.103}$$

In Eq. (7.57), the charge in the semiconductor was related to the electric field in the oxide as

$$Q_{\text{Si}} = -\epsilon_{ox} \mathscr{E}_{ox} A. \tag{7.104}$$

With Eq. (7.103) for $\mathscr{E}_{ox}$,

$$Q'_{\text{Si}} = \frac{Q_{\text{Si}}}{A} = -\epsilon_{ox} \mathscr{E}_{ox} = -\frac{\epsilon_{ox} V_{ox}}{t_{ox}}, \tag{7.105}$$

which gives

$$V_{ox} = -\frac{Q'_{\text{Si}}}{\epsilon_{ox}/t_{ox}} = -\frac{Q'_{\text{Si}}}{C'_{ox}}. \tag{7.106}$$

Then, Eq. (7.101) becomes

$$V_G = V^\circ_{FB} - \frac{Q'_{\text{Si}}}{C'_{ox}} + \psi_s, \tag{7.107}$$

with Q'_{Si} given by Eq. (C.47) for accumulation or depletion as

$$Q'_{\text{Si}} = \mp 9.282 \times 10^{-17} \sqrt{N_a^-} \left[\frac{q\psi_s}{kT} + \exp(-q\psi_s/kT) - 1 \right]^{1/2}. \tag{7.108}$$

For inversion, Q'_{Si} is given by Eq. (C.43), which includes the minority carriers at the SiO_2/Si interface and Q'_{Si} is negative. *Note that in Eq. (7.108) for depletion, the charge Q'_{Si} for a p-type semiconductor is due to negatively charged ionized acceptors N_a^- so that Q'_{Si}/C'_{ox} is negative.* The negative sign (−) in Eq. (7.108) applies for depletion. *For accumulation, the charge in the semiconductor is positive and the electric field in the oxide $\mathcal{E}_{ox}$ changes sign, and the positive sign (+) in Eq. (7.108) applies.* For depletion and inversion with *p*-type substrates,

$$V_G(\text{depletion and inversion}) = V_{FB}^{\circ} + \frac{|Q'_{\text{Si}}|}{C'_{ox}} + \psi_s, \tag{7.109}$$

and for accumulation with *p*-type substrates,

$$V_G(\text{accumulation}) = V_{FB}^{\circ} - \frac{|Q'_{\text{Si}}|}{C'_{ox}} + \psi_s, \tag{7.110}$$

7.4.2 Ideal Oxide MOS Capacitance-Voltage Behavior Examples

Measurements of the *C–V* behavior of MOS capacitors usually require comparison with the calculated *C–V* curve for the ideal oxide with a given gate metal, oxide thickness, and substrate carrier concentration. Extensive ideal oxide MOS *C–V* curves have been published by Goetzberger[17] and are a very complete compilation. An example calculation of the ideal MOS capacitance-voltage behavior illustrates evaluation of the equations given in the preceding part of this section.

EXAMPLE 7.3 An ideal MOS capacitor at 300 K has an Al gate with an oxide thickness of 500 Å (50 nm) on *p*-type Si with $p = N_a^- = 4 \times 10^{15}\ \text{cm}^{-3}$. Calculate the *C–V* variation for $-5\ \text{V} < V_G < +5\ \text{V}$. Normalize the capacitance by dividing the differential capacitance C'_{dif} by C'_{ox} and plot C'_{dif}/C'_{ox} vs. V_G.

Solution In Example 7.2 in Sec. 7.3.5, the flat-band voltage was given as $V_{FB}^{\circ} = -0.842$ V and the threshold voltage was given as $V_{T_n} = 0.260$ V. A sketch of the ideal MOS capacitor *C–V* behavior may be made from the differential MOS capacitance in the limiting accumulation [see Eq. (7.99)] of $C'_{dif}(\text{accum}) = C'_{ox}$ at $V_G = -5.0$ V, $C'_{dif}(\text{FB})$ at $V_G = V_{FB}^{\circ} = -0.842$ V, and $C'_{dif}(x_{d_{max}})$ at $V_G = V_{T_n} = 0.260$ V. For flat band,

$$L_D = \sqrt{\frac{\epsilon_{\text{Si}} kT}{q^2 N_a^-}} = \sqrt{\frac{6.47 \times 10^6 \times 0.026}{4 \times 10^{15}}} = 6.485 \times 10^{-6}\ \text{cm}$$

$$C'_{dif}(\text{FB}) = \frac{1}{\dfrac{t_{ox}}{\epsilon_{ox}} + \dfrac{L_D}{\epsilon_{\text{Si}}}} = \frac{1}{\dfrac{1}{6.903 \times 10^{-8}} + \dfrac{6.485 \times 10^{-6}}{11.7 \times 8.85 \times 10^{-14}}}$$

$$= 4.819 \times 10^{-8}\ \text{F/cm}^2$$

At the onset of strong inversion, $V_G = V_{T_n}$ and $x = x_{d_{max}}$:

$$x_{d_{max}} = 2L_D\sqrt{\ln(N_a^-/n_i)} = 2 \times 6.485 \times 10^{-6} \times \sqrt{\ln(4 \times 10^{15}/1.0 \times 10^{10})}$$

$$x_{d_{max}} = 4.658 \times 10^{-5} \text{ cm}$$

$$C'_{dif}(x = x_{d_{max}}) = \frac{1}{\dfrac{t_{ox}}{\epsilon_{ox}} + \dfrac{x_{d_{max}}}{\epsilon_{\text{Si}}}} = \frac{1}{1.449 \times 10^7 + \dfrac{4.658 \times 10^{-5}}{11.7 \times 8.85 \times 10^{-14}}}$$

$$C'_{dif}(x = x_{d_{max}}) = \underline{1.681 \times 10^{-8} \text{ F/cm}^2}$$

for accumulation

Next, the capacitances are normalized by dividing C'_{dif} by the oxide capacitance C'_{ox}:

$$\frac{C'_{dif}(V_G = -5 \text{ V})}{C'_{ox}} = \underline{1.0}$$

at flat band

$$\frac{C'_{dif}(V_G = -0.842 \text{ V})}{C'_{ox}} = \frac{4.819 \times 10^{-8}}{6.903 \times 10^{-8}} = \underline{0.698}$$

at $x_{d_{max}}$

$$\frac{C'_{dif}(V_G = 0.260 \text{ V})}{C'_{ox}} = \frac{1.681 \times 10^{-8}}{6.903 \times 10^{-8}} = \underline{0.244}$$

The resulting C–V sketch is given in Fig. 7.19 with the capacitance for accumulation designated by **A**, the capacitance at flat band designated by **B**, and the capacitance at the onset of strong inversion designated by **C**. If the AC measuring voltage has a frequency with a period much less than the time necessary for thermal generation of carriers in the space-charge region (see Example 7.1, $t = 2\tau_{nr}p_b/n_i$) while the DC gate voltage is slowly varying (such as a triangular ramp voltage), then the inversion layer

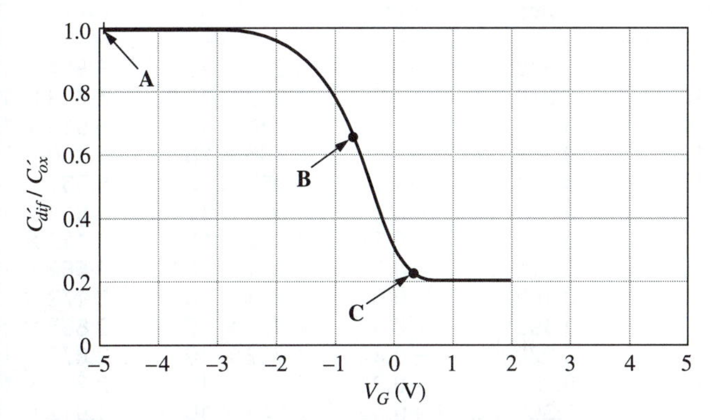

FIGURE 7.19 Sketch of the normalized capacitance C'_{dif}/C'_{ox} for the ideal oxide MOS capacitor on *p*-type Si in Example 7.3. The high-frequency measuring voltage is generally 1 MHz.

cannot respond to the measuring signal and the capacitance will remain constant for $V_G > V_T$. The 1-MHz measurement frequency is higher than the generation rate in the semiconductor space-charge layer and inversion will not be observed. For a very low measuring frequency or excitation by light, inversion can be observed for $V_G > V_T$ (see Sec. 7.4.3). ■

EXAMPLE 7.4 Extend the previous example to obtain the complete ideal oxide capacitor *C–V* behavior and plot the normalized capacitance C'_{dif}/C'_{ox} for V_G between -5.0 V and $+5.0$ V.

Solution First, Q'_{Si} is calculated from Eq. (7.108) for accumulation and depletion,

$$Q'_{\text{Si}} = \mp \underbrace{9.282 \times 10^{-17} \sqrt{N_a^-}}_{5.870\times10^{-9} \text{ for } N_a^- = 4\times10^{15}} \left[\frac{q\psi_s}{kT} + \exp(-q\psi_s/kT) - 1\right]^{1/2},$$

where the positive (+) sign applies for accumulation and the negative sign (−) applies for depletion. For $\psi_s = -0.200$ V,

$$Q'_{\text{Si}} = 5.870 \times 10^{-9}\left[\frac{-0.20}{0.026} + \exp(0.20/0.026) - 1\right]^{1/2},$$

or

$$5.870 \times 10^{-9}[-7.692 + 2.191 \times 10^3 - 1]^{1/2} = \underline{2.742 \times 10^{-7}} \text{ coulombs/cm}^2.$$

TABLE 7.1 Ideal Oxide MOS Capacitor *C–V* Behavior for Example 7.4.

ψ_s volts	Q'_{Si} coulombs/cm²	$\lvert Q'_{\text{Si}}\rvert/C'_{ox}$ volts	V_G volts	C'_{Si} F/cm²	C'_{dif} F/cm²	C'_{dif}/C'_{ox}
−0.20	2.742×10^{-7}	3.972	−5.01	5.292×10^{-6}	6.814×10^{-8}	0.987
−0.16	1.263×10^{-7}	1.830	−2.83	2.462×10^{-6}	6.713×10^{-8}	0.973
−0.12	5.733×10^{-8}	0.831	−1.79	1.156×10^{-6}	6.513×10^{-8}	0.943
−0.08	2.463×10^{-8}	0.357	−1.28	5.566×10^{-7}	6.140×10^{-8}	0.889
−0.04	8.544×10^{-9}	0.124	−1.01	2.837×10^{-7}	5.551×10^{-8}	0.804
0.0	0	0	−0.84	1.597×10^{-7}	4.819×10^{-8}	0.698
+0.04	5.094×10^{-9}	0.074	−0.73	1.022×10^{-7}	4.119×10^{-8}	0.597
+0.08	8.553×10^{-9}	0.124	−0.64	7.391×10^{-8}	3.569×10^{-8}	0.517
+0.12	1.118×10^{-8}	0.162	−0.56	5.871×10^{-8}	3.172×10^{-8}	0.460
+0.16	1.333×10^{-8}	0.193	−0.49	4.962×10^{-8}	2.886×10^{-8}	0.418
+0.20	1.519×10^{-8}	0.220	−0.42	4.362×10^{-8}	2.673×10^{-8}	0.387
+0.335	2.024×10^{-8}	0.293	−0.21	3.275×10^{-8}	2.221×10^{-8}	0.322
+0.50	2.506×10^{-8}	0.363	+0.02	2.644×10^{-8}	1.912×10^{-8}	0.277
+0.670	2.921×10^{-8}	0.423	+0.25	2.269×10^{-8}	1.707×10^{-8}	0.247

*$p = N_a^- = 4 \times 10^{15}$ cm^{-3} and $t_{ox} = 500$ Å (50 nm).

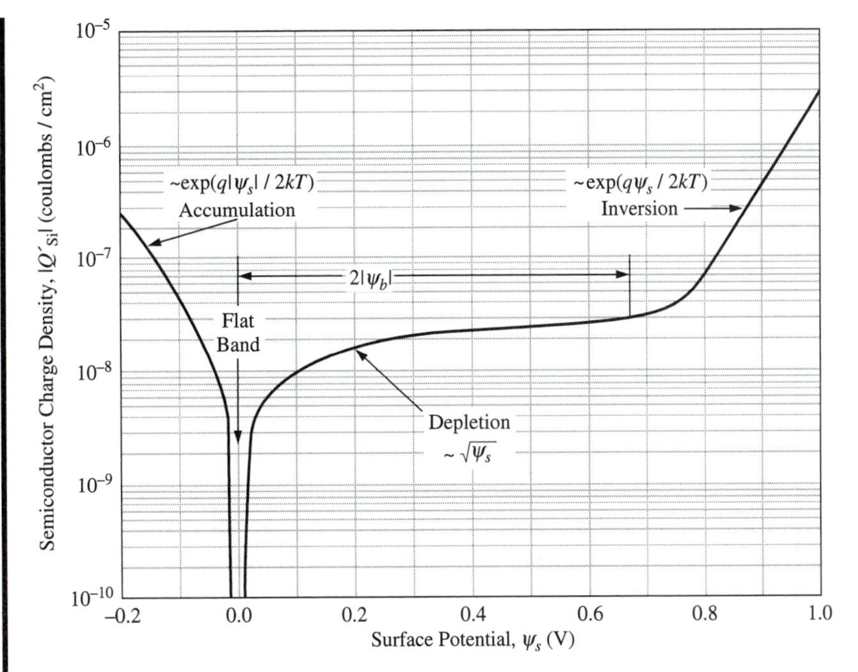

FIGURE 7.20 Variation of the magnitude of the charge density Q'_{Si} in the semiconductor as a function of the surface potential ψ_s for p-type Si with $p = N_a^- = 4 \times 10^{15}\ \text{cm}^{-3}$ at room temperature.

Values of Q'_{Si} as ψ_s is varied from accumulation $(-\psi_s)$ to flat band $(\psi_s = 0)$ to depletion $(+\psi_s)$ are given in Table 7.1. For inversion, $\psi_s > 2|\psi_b|$ $(\psi_b = -0.335\ \text{V})$, and Eq. (C.43) must be used for Q'_{Si}. A plot of $|Q'_{\text{Si}}|$ vs. ψ_s is shown in Fig. 7.20. Note the exponential variation of $|Q'_{\text{Si}}|$ for accumulation and for inversion when $\psi_s > 2|\psi_b|$.

The next quantity to be calculated is the gate voltage. In accumulation, V_G is given by Eq. (7.110) and for $\psi_s = -0.200$ V becomes

$$V_G = V_{FB}^{\circ} - \frac{|Q'_{sc}|}{C'_{ox}} + \psi_s = -0.842 - \frac{2.742 \times 10^{-7}}{6.903 \times 10^{-8}} - 0.200 = \underline{-5.01}\ \text{V}.$$

For depletion (and inversion), V_G is given by Eq. (7.109).

The capacitance per unit area for the Si in accumulation or depletion is given in Eq. (7.89) as

$$C'_{\text{Si}} = \underbrace{1.785 \times 10^{-15} \sqrt{N_a^-}}_{1.129\times10^{-7}} \left| \frac{[1 - \exp(-q\psi_s/kT)]}{\left[\dfrac{q\psi_s}{kT} + \exp(-q\psi_s/kT) - 1\right]^{1/2}} \right|,$$

and for $\psi_s = -0.200$ V,

$$C'_{\text{Si}} = 1.129 \times 10^{-7} \times \frac{2.190 \times 10^3}{4.672 \times 10^1} = \underline{5.292 \times 10^{-6}}\ \text{F/cm}^2.$$

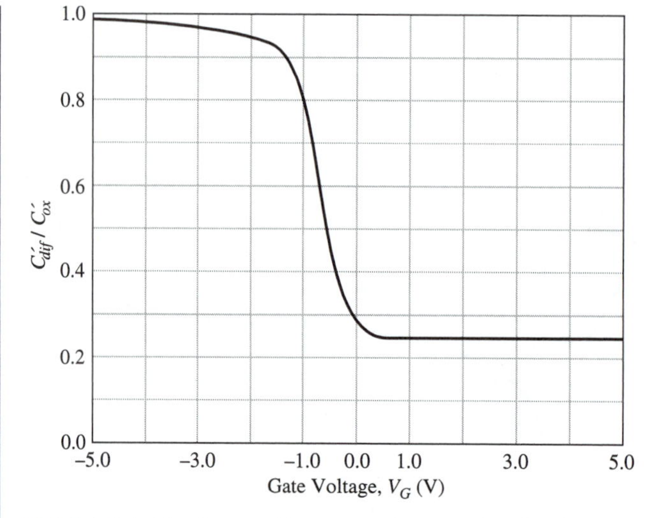

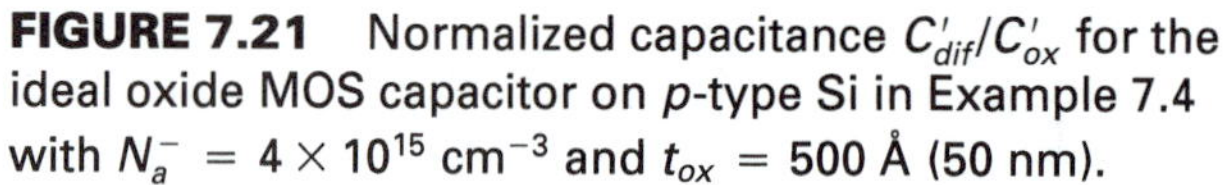
FIGURE 7.21 Normalized capacitance C'_{dif}/C'_{ox} for the ideal oxide MOS capacitor on *p*-type Si in Example 7.4 with $N_a^- = 4 \times 10^{15}$ cm^{-3} and $t_{ox} = 500$ Å (50 nm).

The MOS differential capacitance is given by Eq. (7.96) as

$$C'_{dif} = \frac{1}{1/C'_{ox} + 1/C'_{\text{Si}}},$$

and for $\psi_s = -0.200$ V and $t_{ox} = 500$ Å,

$$C'_{dif} = \frac{1}{1/6.903 \times 10^{-8} + 1/5.292 \times 10^{-6}} = \underline{6.814 \times 10^{-8}} \text{ F/cm}^2.$$

The normalized capacitance for $\psi_s = -0.200$ V is

$$C'_{dif}/C'_{ox} = 6.814 \times 10^{-8}/6.903 \times 10^{-8} = \underline{0.987}.$$

Capacitance values as a function of ψ_s are given in Table 7.1 and C'_{dif}/C'_{ox} is plotted in Fig. 7.21 as a function of V_G and is taken as constant for $\psi_s > 2|\psi_b|$.

To provide insight as to how the *C–V* curves vary with substrate carrier concentration and oxide thickness, the calculated *C–V* behavior is illustrated in Figs. 7.22 and 7.23. In Fig. 7.22, the *C–V* behavior is shown for a MOS capacitor with an Al gate, a 500-Å-thick SiO_2 layer, and substrate hole concentrations of 1×10^{17} cm^{-3}, 1×10^{16} cm^{-3}, and 1×10^{15} cm^{-3}. In Fig. 7.23, the oxide thickness is varied for a substrate hole concentration of 1×10^{16} cm^{-3}.

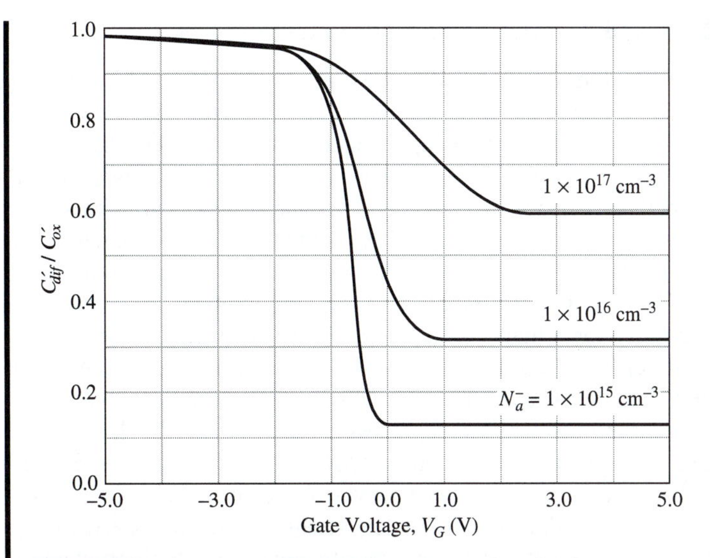

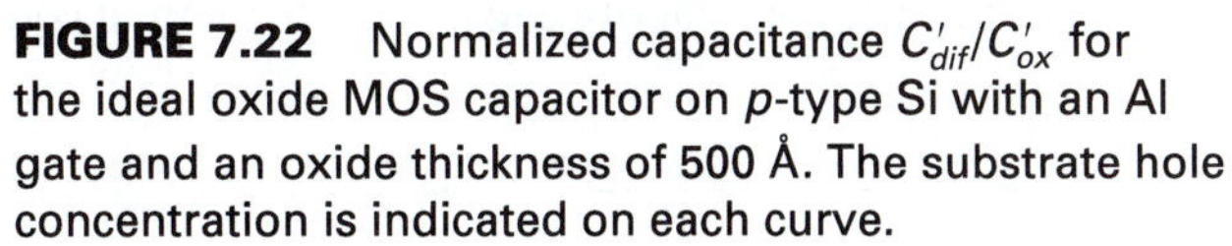

FIGURE 7.22 Normalized capacitance C'_{dif}/C'_{ox} for the ideal oxide MOS capacitor on *p*-type Si with an Al gate and an oxide thickness of 500 Å. The substrate hole concentration is indicated on each curve.

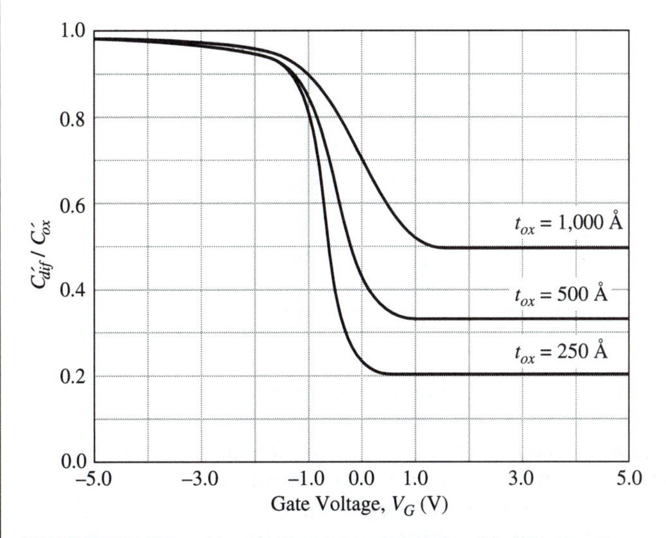

FIGURE 7.23 Normalized capacitance C'_{dif}/C'_{ox} for the ideal oxide MOS capacitor on *p*-type Si with an Al gate and a hole concentration of 1×10^{16} cm^{-3}. The oxide thickness is indicated on each curve.

■

7.4.3 MOS Capacitance Measurements

A variety of techniques are used to measure the properties of MOS capacitors.[18] These techniques have evolved from the use of L–C meters, and then lock-in amplifiers to instruments intended for MOS capacitance measurements such as the Hewlett-Packard 4284A Precision (LRC) Meter, which may be controlled by a personal computer to measure both the capacitance and conductance. These instruments use a linear voltage ramp to sweep the voltage over the bias range from accumulation to inversion (or from inversion to accumulation) at a specified bias ramp rate given in mV/s. A small sinusoidal AC voltage is superimposed on the ramp voltage to permit the measurement of the differential capacitance as $C_{dif} = dQ/dV$ or conductance as $dG = dI/dV$. This AC signal generally has a peak-to-peak voltage of less than 50 mV. The measurement frequency seldom exceeds the most common measurement frequency of 1 MHz, but measurement frequencies can be decreased to as low as 20 Hz.

The three principal C–V characteristics which are measured are illustrated by the curves shown in Fig. 7.24. These curves all occur in inversion for $V_G > V_T$ and are due to minority carrier effects. At low frequency for curve (a) in Fig. 7.24, the frequency is sufficiently low that the minority carriers are in thermal equilibrium with the

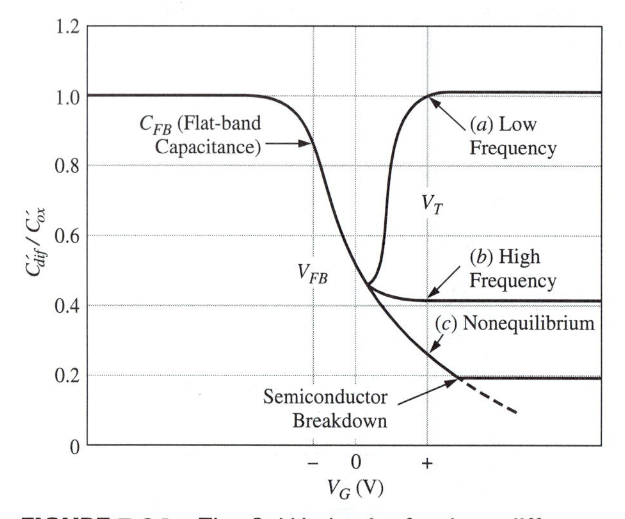

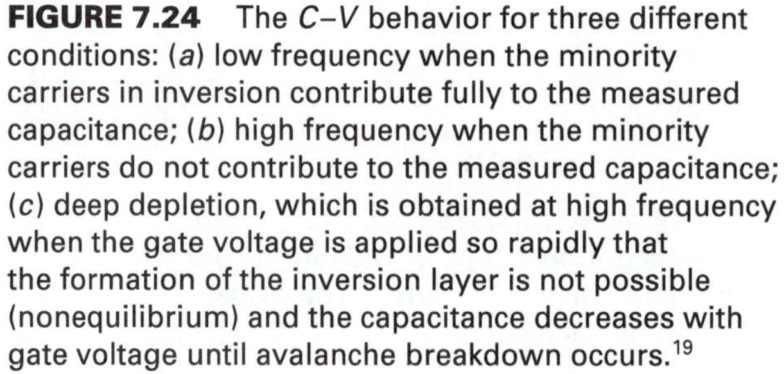

FIGURE 7.24 The C–V behavior for three different conditions: (a) low frequency when the minority carriers in inversion contribute fully to the measured capacitance; (b) high frequency when the minority carriers do not contribute to the measured capacitance; (c) deep depletion, which is obtained at high frequency when the gate voltage is applied so rapidly that the formation of the inversion layer is not possible (nonequilibrium) and the capacitance decreases with gate voltage until avalanche breakdown occurs.[19]

small-signal AC voltage. For the minority carriers to be in equilibrium with the small-signal AC voltage and to contribute to the capacitance, the measurement frequency period $1/f$ must be less than the minority carrier generation time as given by Eq. (7.55) and illustrated in Example 7.1. The capacitance for inversion is given by Eq. (C.45) in the Appendix at the end of this chapter. Curve (*b*) in Fig. 7.24 represents the high-frequency behavior where the minority carriers are in equilibrium with the slowly varying ramp voltage, but these minority carriers cannot vary with the more rapidly varying small-signal AC voltage. The minority carriers do pin the band bending so that x_d becomes fixed at $x_{d_{max}}$ [Eq. (7.38)] and independent of gate bias, and therefore, C_{dif} remains constant as V_G goes to larger values beyond V_T. Curve (*c*) in Fig. 7.24 is termed *deep depletion,* which requires further explanation.

deep depletion

In deep depletion, the gate bias ramp rate is too rapid to allow the inversion layer to form by thermal generation. Therefore, as V_G exceeds V_T, the absence of the inversion charge requires that the space charge in the semiconductor ($Q'_{sc} = -qN_a^- x_d$) must increase with $x_d > x_{d_{max}}$. This increase in x_d provides more negative charge due to the ionized acceptors in order to balance the larger positive gate charge. Further increases in gate voltage can result in avalanche breakdown as in a *p-n* junction.

Experimental MOS capacitor *C–V* measurements over a range of low frequencies and for 1 MHz are shown in Fig. 7.25. As described above, the variation in the measured *C–V* curves at different measurement frequencies occurs for gate voltages in the inversion region. At high frequencies (such as 1 MHz), the minority-carrier electrons at the SiO_2/Si interface are in equilibrium with the slowly varying voltage ramp, but the recombination-generation rate as given by Eq. (7.55) and evaluated as $\sim 1 \times 10^{-2}$ s in Example 7.1 is too slow to follow the AC measuring voltage. The minority-carrier electrons due to the slowly varying ramp voltage supply negative charge at the SiO_2/Si

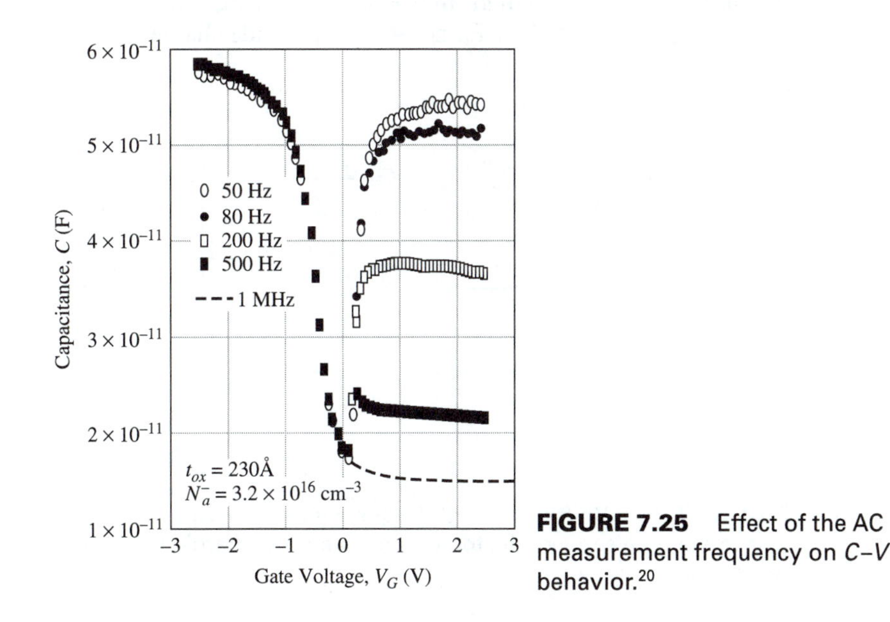

FIGURE 7.25 Effect of the AC measurement frequency on *C–V* behavior.[20]

interface so that the depletion layer width becomes fixed at $x_{d_{max}}$. For this reason, the high frequency capacitance becomes constant for $V_G > V_T$, as was represented by curve (*b*) in Fig. 7.24. As the frequency is decreased from 500 Hz to 50 Hz, the minority-carrier electrons approach equilibrium with both the slowly varying voltage ramp and the small-signal AC measuring voltage. In this case, the measured capacitance approaches the oxide layer capacitance C'_{ox} for $V_G > V_T$, as was represented by curve (*a*) in Fig. 7.24.

7.4.4 Nonideal Oxide Capacitance-Voltage Behavior

The nonideal oxide with charge within the oxide or at the SiO_2/Si interface was introduced in Sec. 7.3.6. The fixed oxide charge, $Q_f = qN_f$, is very near the SiO_2/Si interface and was shown in Eq. (7.77) to shift the flat-band voltage as $V_{FB} = V^\circ_{FB} - qN_f/(\epsilon_{ox}/t_{ox})$, where N_f was taken as a positive charge. Then the gate voltage given by Eq. (7.109) with Eq. (7.77) for V_{FB} becomes

$$V_G(\text{depletion and inversion}) = V^\circ_{FB} - \frac{qN_f}{\epsilon_{ox}/t_{ox}} + \frac{|Q'_{\text{Si}}|}{C'_{ox}} + \psi_s, \qquad (7.111)$$

where V_G is shifted from the ideal oxide case by $-qN_f/(\epsilon_{ox}/t_{ox})$. A similar equation for accumulation may be obtained with Eq. (7.77) in Eq. (7.110). The calculated *C–V* curve is shifted by the change in flat-band voltage due to the fixed oxide charge. Figure 7.26 illustrates for a *p*-type semiconductor the shift of the ideal *C–V* high-frequency curve along V_G. Part (*a*) is for positive fixed oxide charge $+Q_f$, and part (*b*) is for negative fixed oxide charge $-Q_f$. Figure 7.27 illustrates for an *n*-type semiconductor the shift of the ideal *C–V* high-frequency curve along V_G. Part (*a*) is for positive oxide charge $+Q_f$, and part (*b*) is for negative fixed oxide charge $-Q_f$.

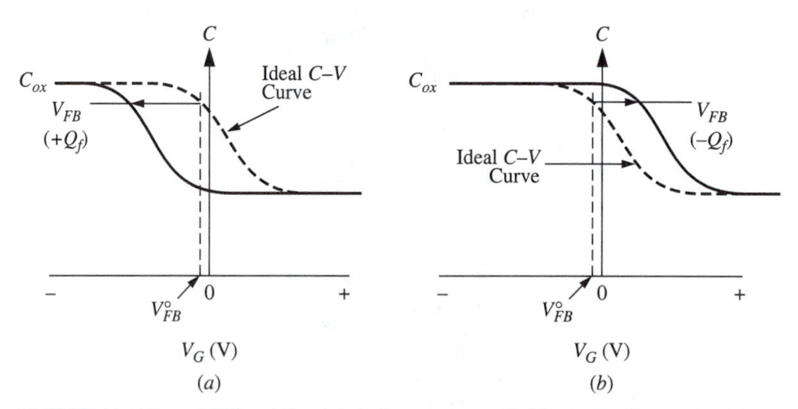

FIGURE 7.26 Shift of the high-frequency *C–V* curve along the V_G axis in a *p*-type semiconductor. (*a*) Positive fixed charge. (*b*) Negative fixed charge.

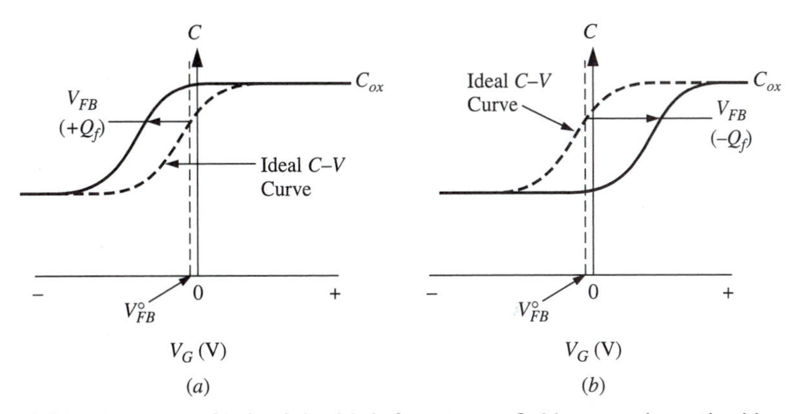

FIGURE 7.27 Shift of the high-frequency C–V curve along the V_G axis in an n-type semiconductor. (*a*) Positive fixed charge. (*b*) Negative fixed charge.

The interface trapped charge, $Q_{it} = qN_{it}$, is in electrical communication with the underlying semiconductor and can be charged or discharged. This interface trapped charge is often called *fast states* or *interface states,* and may be filled or empty depending whether these interface states are donor- or acceptor-type states and the position of the Fermi level at the semiconductor surface. As the gate voltage is varied, the position of the Fermi level at the surface will change. A donorlike state is illustrated in Fig. 7.28 for a p-type semiconductor. For accumulation shown in part (a), the Fermi level at the surface is below the interface donor level and the probability of being occupied by an electron, as given by Eq. (A.1) in Chapter 2, will be small. This case for an ionized donorlike surface state is represented by N_s^+. When the gate voltage bends the bands, as shown in part (b) for depletion and inversion, the Fermi level at the surface

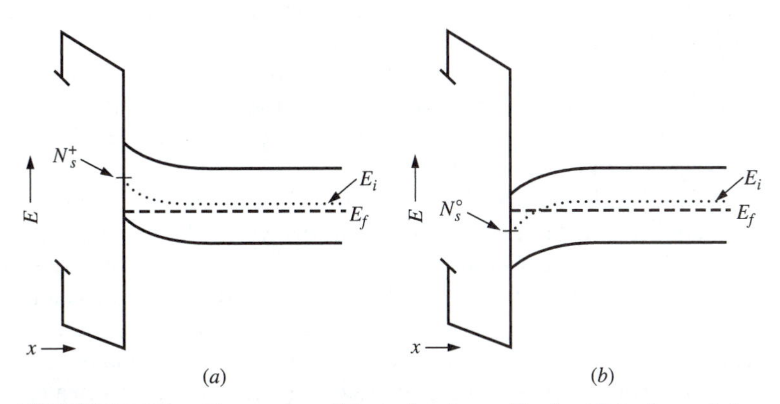

FIGURE 7.28 Charge condition of a donorlike fast interface state. (*a*) Positive-charge condition for accumulation. (*b*) Neutral-charge condition for depletion or inversion.

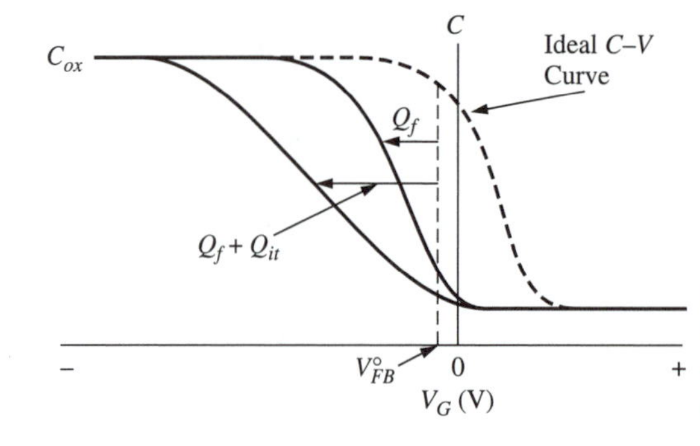

FIGURE 7.29 Distortion of the ideal oxide C–V curve due to fixed oxide charge Q_f and interface trapped charge Q_{it}.

is above the interface donor level, which will be occupied by an electron as a neutral donorlike surface state N_s°. Therefore, the charge of the fast states will vary with the band bending, which is represented by ψ_s. This charge will change the C–V behavior from the ideal oxide case by an amount which varies with ψ_s. For the more usual case of a continuum of interface states, the C–V curve tends to flatten out as illustrated in Fig. 7.29. Thus, a shift in position of the ideal oxide C– V curve is related to the fixed oxide charge Q_f, while departure from the shape of the ideal oxide C–V curve may be attributed to the fast interface states Q_{it}. The speed of response depends of the position of the trap with respect to the band edges; a detailed analysis may be found in Nicollian and Brews.[9]

7.5 CHARGE-COUPLED DEVICES

7.5.1 Basic Structure

There are numerous applications of MOS capacitors. These applications include but are not limited to charge storage as the memory element in the hundreds of millions of dynamic random access memory (DRAM) chips used in all types of computers, precision capacitors in switched-capacitor filters, analog-to-digital (A/D) converters, as well as the imager for high resolution and sensitivity astronomical observation and in rugged consumer camcorders. Only charge coupled devices (CCDs), which are used as imagers, will be considered here.

In 1970, Boyle and Smith[21] used MOS capacitors to form the CCDs which are now produced at the rate of 10 million per year. The essential idea for CCDs is to place MOS capacitors sufficiently close so that channel charge (electrons for p-Si) at the SiO_2/Si interface can be transferred to adjacent MOS capacitors by applying and removing the capacitor gate voltages. Close spacing means that the depletion regions of adjacent MOS capacitors overlap at the SiO_2/Si interface and hence the name, charge-coupled devices. Many adjacent capacitors (perhaps 1024) form a row while a comparable number of rows form a two-dimensional array. The focused optical image on the CCD generates the electrons which become an electronic representation of the optical image.

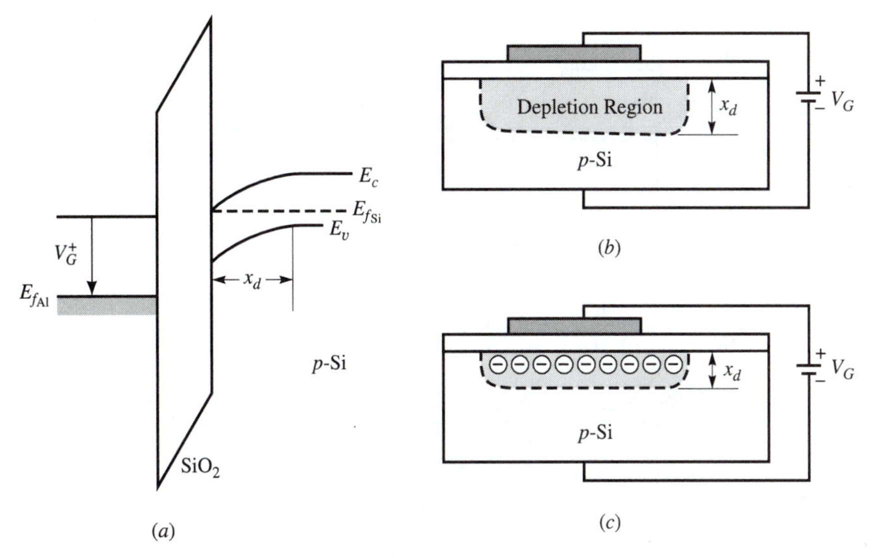

FIGURE 7.30 Representation of an MOS capacitor used in a CCD. (*a*) Energy-band diagram for an MOS capacitor in deep depletion. (*b*) Cross-sectional representation for part (*a*). (*c*) Cross-sectional representation of the MOS capacitor with electrons (channel charge) at the SiO_2/Si interface.

The basic operation of an individual capacitor in a CCD is illustrated in Fig. 7.30. When a positive gate voltage is applied to the gate of an MOS capacitor on *p*-Si, the time required for thermal generation to produce minority carriers, which was given in Example 7.1 with Eq. (7.55), can approach seconds for long-lifetime Si. With $V_G > V_T$, the MOS capacitor is initially driven into deep depletion where the depletion depth x_d is greater than the maximum depletion depth $x_{d_{max}}$, $x_d > x_{d_{max}}$, as described in Sec. 7.4.3 and illustrated in Fig. 7.30 (*a*). The MOS capacitor in deep depletion is illustrated by the cross-sectional representation in Fig. 7.30 (*b*). Channel charge Q_n due to electrons at the SiO_2/Si interface can be supplied by the absorption of incident light and is illustrated in Fig. 7.30 (*c*). The amount of charge is proportional to the light intensity. With channel charge Q_n, the depletion depth decreases with the total negative charge in the *p*-Si now composed of Q_n and qN_a^-.

The single MOS capacitor is combined with two closely spaced adjacent MOS capacitors as illustrated in the schematic representation of the CCD in Fig. 7.31. In Fig. 7.31 (*a*), the gate voltage V_2 is a sufficiently large positive voltage so that greater depletion occurs than for the adjacent capacitors. If minority carrier electrons are produced at the center capacitor, such as by the absorption of light, these electrons (channel charge) will reside in the potential well at the SiO_2/Si interface as was illustrated in Fig. 7.30 (*a*). The potentials V_1 on the left capacitor and V_3 on the right capacitor are insufficient (less than V_T) to permit minority-carrier electrons at the SiO_2/Si interface, although these capacitors would generally be in depletion.

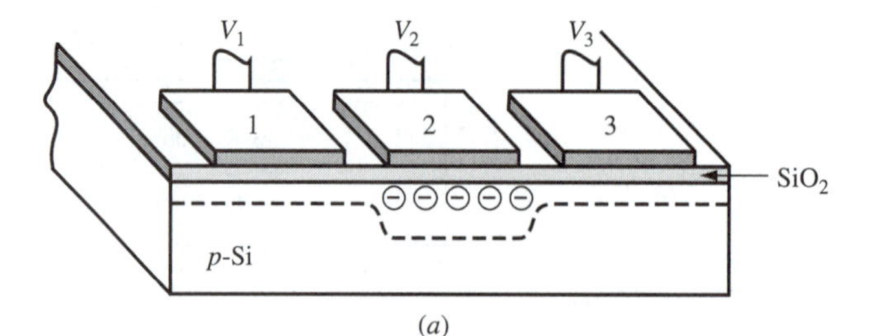

(a)

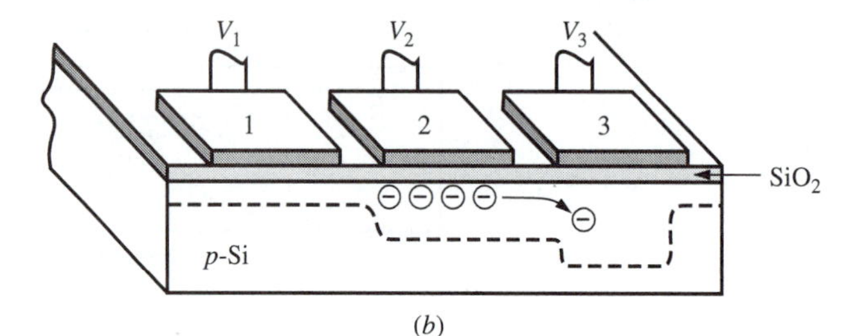

(b)

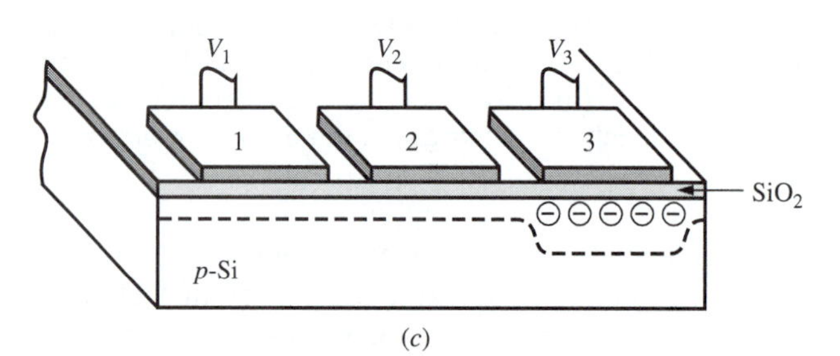

(c)

FIGURE 7.31 Cross-sectional representation of MOS capacitors illustrating channel charge transfer in CCDs. (a) $V_2 > V_T > V_1$ and V_3 to maintain channel charge under MOS capacitor 2. (b) $V_2 = V_3 > V_T > V_1$ to initiate channel charge transfer from MOS capacitor 2 to 3. (c) $V_3 > V_T > V_1$ and V_2 to maintain the channel charge under MOS capacitor 3.

To transfer the channel charge to the right, a larger positive potential V_3 would be applied to the right capacitor as shown in Fig. 7.31 (b), and then V_2 is reduced below V_T as shown in Fig. 7.31 (c). This two-stage charge-transfer process begins with the creation of the potential well (deep depletion) in capacitor 3, and is then followed by reduction of V_2 to empty the channel charge from capacitor 2 to capacitor 3. The charge is held and transferred at the MOS capacitors for times which are short compared to the thermal generation time.

For many applications, the three-phase operation, which has three MOS capacitors per stage, is typical. Every third gate electrode is connected to the same clock

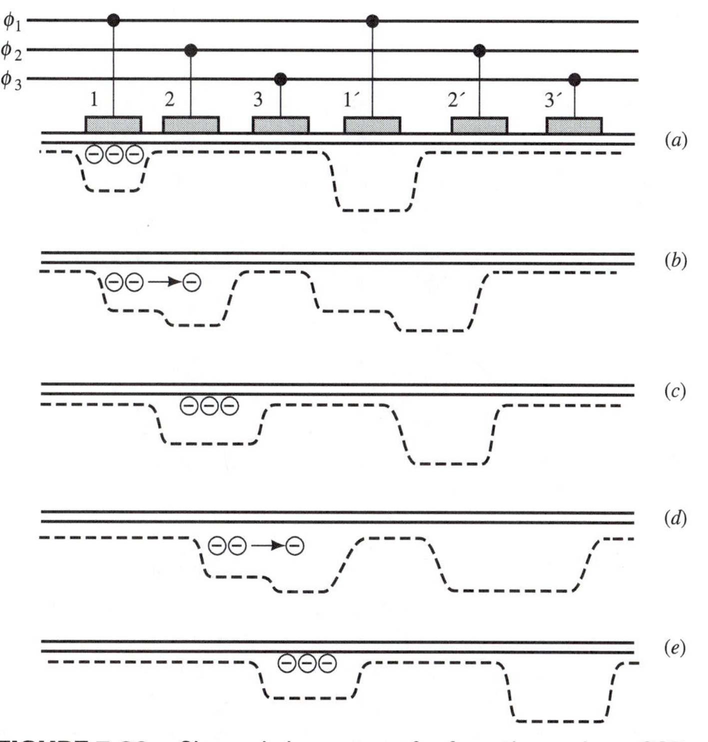

FIGURE 7.32 Channel charge transfer for a three-phase CCD with channel charge in capacitor 1 only. (*a*) $V_{G_1} = V_{G_{1'}} > V_T$. (*b*) $V_{G_1} = V_{G_{1'}} = V_{G_2} = V_{G_{2'}} > V_T$. (*c*) $V_{G_2} = V_{G_{2'}} > V_T$. (*d*) $V_{G_2} = V_{G_{2'}} = V_{G_3} = V_{G_{3'}} > V_T$. (*e*) $V_{G_3} = V_{G_{3'}} > V_T$.

voltage, as shown in Fig. 7.32. In Fig. 7.32 (*a*), capacitor 1 has channel charge, while capacitor 1′ has no channel charge, but capacitor 1 and 1′ have the same gate voltage $V_{G_1} = V_{G_{1'}} > V_T$. Initially, phase 1 ($\phi_1$) has $V_{G_1} > V_T$ and $V_{G_{1'}} > V_T$, while phase 2 (ϕ_2) and phase 3 (ϕ_3) are $V_{G_2} = V_{G_{2'}} = V_{G_3} = V_{G_{3'}} < V_T$. In Fig. 7.32 (*b*) both $V_{G_1} = V_{G_{1'}} > V_T$ and $V_{G_2} = V_{G_{2'}} > V_T$, while $V_{G_3} = V_{G_{3'}} < V_T$ and the charge in capacitor 1 transfers to capacitor 2 as $V_{G_1} = V_{G_{1'}} = V_{G_2} = V_{G_2'} > V_T$ and $V_{G_3} = V_{G_{3'}} < V_T$, as shown in Fig. 7.32 (*c*). This charge transfer was previously illustrated in Fig. 7.31. The channel charge in capacitor 2 is shifted to capacitor 3 in the same manner with $V_{G_3} = V_{G_{3'}} > V_T$, as shown in Fig. 7.32 (*d*) and (*e*).

For a complete clock cycle, each phase sequentially goes to $V_G > V_T$, and when V_1 again exceeds V_T, the charge originally in capacitor 1 has been transferred through capacitors 2 and 3 to capacitor 1′. In one complete clock cycle, the channel charge shifts to the same capacitor in the next stage, where a stage consists of three capacitors. More detailed consideration of CCDs may be found in the book by Séquin and Tompsett.[22]

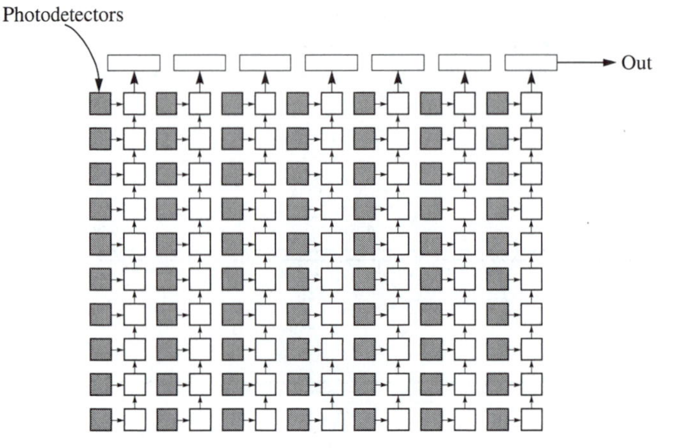

FIGURE 7.33 CCD imaging array with *p-n* junction photodetectors.[23]

7.5.2 Charge-Coupled Device Imaging

The CCD can serve as a photodetector, but other photodetectors such as *p-n* junctions may be used. A two-dimensional CCD imager is illustrated in Fig. 7.33 and uses *p-n* junctions as the photodetector. The whole CCD array is biased into deep depletion and a focused image is incident for a suitable time interval. The number of carriers generated at each pixel (picture element) depends on the intensity of the image at that location. The CCD transfers the charge line by line from the photodetectors to sense amplifiers shown at the top of the CCD. This analog signal is a serial format as a single output signal. The position of the pixel on the CCD occurs at a known time in the serial output signal which is synchronous with a master clock so that the image may be reconstructed at the original intensity and position of the image on the CCD.

The CCDs used in camcorders are 0.25 in. (3.6×2.7 mm^2) and have demanding performance although they must remain inexpensive. The image wavelength sensitivity is between 400 and 700 nm. For color cameras, a mosaic color filter pattern for red, green, and blue is used. Further description of CCD imagers may be found in the technical literature such as the special issue on solid-state image sensors.[24]

7.6 SUMMARY AND USEFUL EXPRESSIONS

This chapter introduced the metal-oxide-semiconductor capacitor.

- The applied voltage to achieve flat bands is given by the difference between the metal and semiconductor work functions.
- Conditions which give a larger majority carrier concentration at the semiconductor surface than in the neutral semiconductor bulk are called *accumulation* or *enhancement.*

- When conditions give a minority carrier concentration at the semiconductor surface larger than the majority carrier concentration in the semiconductor bulk, this condition is called *inversion.*
- Another frequently encountered quantity is the *maximum depletion width* $x_{d_{max}}$ which is the condition when further increases in gate voltage result in rapid increase in the inversion layer charge.
- The bias voltage necessary to give a minority carrier concentration at the Si surface equal to the majority carrier concentration in the neutral semiconductor bulk is termed the *threshold voltage.*
- The fixed-oxide charge, Q_f and N_f which are related as $|Q_f/q| = N_f$, is due primarily to structural defects (ionized Si) in the oxide layer less than 25 Å from the SiO_2/Si interface and is the most commonly encountered oxide charge.
- Oxide charge affects the flat-band voltage, and therefore the threshold voltage.
- The MOS capacitance is the series combination of the oxide and semiconductor capacitance.
- For the condition of maximum depletion, the semiconductor differential capacitance is $\epsilon_{Si}/x_{d_{max}}$, where $x_{d_{max}}$ is the maximum depletion width.
- At the flat-band voltage, the semiconductor differential capacitance per unit area is ϵ_{Si}/L_D, where L_D is the Debye length.
- In strong *accumulation*, the limiting MOS capacitance is the oxide capacitance, C'_{ox}(capacitance/unit area)$= \epsilon_{ox}/t_{ox}$.
- For nonideal oxides, the fixed oxide charge Q_f will shift the voltage at which the flat-band voltage occurs.
- The fast interface states will result in a less steep capacitance-voltage behavior as compared to the ideal oxide case.

The following expressions for MOS capacitors were obtained in this chapter.

Physical Parameters

$q\Phi_{Al} = 4.1$ eV $\qquad \epsilon_{ox} = 3.9\epsilon_0$

$qX_{ox} = 0.95$ eV $\qquad \epsilon_{Si} = 11.7\epsilon_0$

$qX_{Si} = 4.05$ eV $\qquad \epsilon_{Si}/q^2 = 6.47 \times 10^6$/eV-cm

Flat-Band Voltage

$qV^{\circ}_{FB} = q\Phi_{Al} - q\Phi_{Si} \qquad V_{FB} = V^{\circ}_{FB} - qN_f/(\epsilon_{ox}/t_{ox})$

Potential

$\psi(x) = (1/q)[E_{i_b} - E_i(x)] \qquad \psi_s = (1/q)[E_{i_b} - E_i(0)]$

p-type

$\psi_s < 0$ hole accumulation

$\psi_s > 0$ depletion and inversion

$\psi_s = 2|\psi_b|$ at threshold

$\psi_b = -(kT/q)\ln(N_a^-/n_i)$

n-type

$\psi_s > 0$ hole accumulation

$\psi_s < 0$ depletion and inversion

$\psi_s = 2|\psi_b|$ at threshold

$\psi_b = (kT/q)\ln(N_d^+/n_i)$

Carrier Concentration in *p*-Type Semiconductor

$$n(x) = (n_i^2/N_a^-)\exp[q\psi(x)/kT] = n_{po}\exp[q\psi(x)/kT]$$
$$p(x) = N_a^-\exp[-q\psi(x)/kT] = p_{po}\exp[-q\psi(x)/kT]$$

Carrier Concentration in *n*-Type Semiconductor

$$n(x) = N_d^+\exp[q\psi(x)/kT] = n_{no}\exp[q\psi(x)/kT]$$
$$p(x) = (n_i^2/N_d^+)\exp[-q\psi(x)/kT] = p_{no}\exp[-q\psi(x)/kT]$$

***n*-channel**

$$q\Phi_{\text{Si}} = qX_{\text{Si}} + E_g - (E_f - E_v) \qquad \psi_b = -(kT/q)\ln(N_a^-/n_i)$$
$$x_{d_{max}} = \sqrt{2\epsilon_{\text{Si}}(2|\psi_b|)/qN_a^-} \qquad Gt = N_a^- = (n_i/2\tau_o)t$$
$$\epsilon_{ox}\mathscr{E}_{ox} = \epsilon_{\text{Si}}\mathscr{E}_{\text{Si}} \qquad \mathscr{E}_{ox} = V_{ox}/t_{ox}$$
$$V_{ox} = \sqrt{4q\epsilon_{\text{Si}}N_a^-|\psi_b|}/(\epsilon_{ox}/t_{ox}) \text{ at threshold}$$
$$V_{T_n} = V_{FB} + 2|\psi_b| + \sqrt{4q\epsilon_{\text{Si}}N_a^-|\psi_b|}/(\epsilon_{ox}/t_{ox})$$
$$V_G(\text{depletion and inversion}) = V_{FB}^\circ + |Q'_{\text{Si}}|/C'_{ox} + \psi_s$$
$$V_G(\text{accumulation}) = V_{FB}^\circ - |Q'_{\text{Si}}|/C'_{ox} + \psi_s$$

***p*-channel**

$$q\Phi_{\text{Si}} = qX_{\text{Si}} + (E_c - E_f)$$
$$V_{T_p} = V_{FB} - 2|\psi_b| - \sqrt{4q\epsilon_{\text{Si}}N_d^+|\psi_b|}/(\epsilon_{ox}/t_{ox})$$

Capacitance

$$C'_{dif} \equiv \left|\frac{dQ'_{\text{Si}}}{dV_G}\right| = \frac{1}{1/C'_{ox} + 1/\left|\dfrac{dQ'_{\text{Si}}}{d\psi_s}\right|} = \frac{1}{[1/C'_{ox} + 1/C'_{\text{Si}}]}$$

$$C'_{\text{Si}} \equiv \left|\frac{dQ_{\text{Si}}}{d\psi_s}\right|$$

$$C'_{ox} = \epsilon_{ox}/t_{ox}$$
$$C'_{FB} = \epsilon_{\text{Si}}/L_D = 2.524 \times 10^{-15}\sqrt{N_a^- \text{ or } N_d^+}$$
$$L_D = \sqrt{(\epsilon_{\text{Si}}/q^2)(kT/(p \text{ or } n)} = \sqrt{6.47 \times 10^6 kT/(N_a^- \text{ or } N_d^+)}$$
$$C'_{accum,dep} = 1.785 \times 10^{-15}\sqrt{N_a^-}\left|\frac{[1 - \exp(-q\psi_s/kT)]}{\left[\dfrac{q\psi_s}{kT} + \exp(-q\psi_s/kT) - 1\right]^{1/2}}\right|$$
$$C'_{accum} = \epsilon_{ox}/t_{ox} = C'_{ox} \text{ (maximum value)}$$
$$C'_{sc} = \epsilon_{\text{Si}}/x_{d_{max}} \text{ (minimum value)}$$
$$C'_{dif}(\text{FB}) = 1/[(t_{ox}/\epsilon_{ox}) + (L_D/\epsilon_{\text{Si}})]$$
$$C'_{dif}(x_{d_{max}}) = 1/[(t_{ox}/\epsilon_{ox}) + (x_{d_{max}}/\epsilon_{Si})]$$

REFERENCES

1. C. J. Frosch and L. Derick, "Surface Protection and Selective Masking during Diffusion in Silicon," J. Electrochem. Soc. **104**, 547 (1957).

2. M. M. Atalla, E. Tannenbaum, and E. J. Scheibner, "Stabilization of Silicon Surfaces by Thermally Grown Oxides," Bell Syst. Tech. J. **38**, 749 (1959).

3. J. L. Moll, "Variable Capacitance with Large Capacity Change," Wescon Convention Record, Part 3, p. 32 (1959).

4. J. R. Ligenza and W. G. Spitzer, "The Mechanisms for Silicon Oxidation in Steam and Oxygen," Phys. Chem. Solids **14**, 131 (1960).

5. J. G. Ligenza, "Effect of Crystal Orientation on Oxidation Rates of Silicon in High Pressure Steam," J. Phys. Chem. **65,** 2011 (1961).

6. D. Kahng and M. M. Atalla, "Silicon-Silicon Dioxide Field Induced Surface Devices," *IRE-AIEE Solid-State Device Res. Conf.* (Carnegie Inst. of Technol., Pittsburgh, Pa. 1960).

7. L. M. Terman, "An Investigation of Surface States at a Silicon/Silicon Dioxide Interface Employing Metal-Oxide-Silicon Diodes," Solid-State Electron. **5**, 285 (1962).

8. A. S. Grove, B. E. Deal, E. H. Snow, and C. T. Sah, "Investigation of Thermally Oxidized Silicon Surfaces Using Metal-Oxide-Semiconductor Structures," Solid-State Electron. **8**, 145 (1965).

9. E. H. Nicollian and J. R. Brews, *MOS (Metal Oxide Semiconductor) Physics and Technology* (Wiley, New York, 1982).

10. S. M. Sze, *Semiconductor Devices: Physics and Technology* (Wiley, New York, 1985), p. 187.

11. R. S. Muller and T. I. Kamins, *Device Electronics for Integrated Circuits,* 2nd Ed. (Wiley, New York, 1977), p. 380.

12. H. Z. Massoud, "Charge-Transfer Dipole Moments at the Si-SiO_2 Interface," **63**, 2000 (1988).

13. Sze, *Semiconductor Devices,* p. 190.

14. Ibid., p. 192.

15. J. R. Brews, "The Metal-Oxide-Semiconductor Field Effect Transistor [MOSFET]," *The Electrical Engineering Handbook,* R. C. Dorf, Ed. (CRC Press, Boca Raton, Fla., 1993), p. 667.

16. B. E. Deal, "Standardized Terminology for Oxide Charges Associated with Thermally Oxidized Silicon," IEEE Trans. Electron Devices **ED-27**, 606 (1980).

17. A. Goetzberger, "Ideal MOS Curves for Silicon," Bell Syst. Tech. J. **45**, 1097 (1966).

18. Nicollian and Brews, *MOS,* Chap. 12.

19. A. Goetzberger and E. H. Nicollian, "MOS Avalanche and Tunneling Effects in Silicon Surfaces," J. Appl. Phys. **38,** 4582 (1967).

20. K. Henson, Private communication.

21. W. S. Boyle and G. E. Smith, "Charge Coupled Semiconductor Devices," Bell Syst. Tech. J. **49**, 587 (1970).

22. C. H. Séquin and F. T. Tompsett, *Charge Transfer Devices* (Academic Press, New York, 1975).

23. K. K. Ng, *Complete Guide to Semiconductor Devices* (McGraw-Hill, New York, 1995), pp. 132–142.

24. IEEE Trans. on Electron Devices **ED-32**, Aug. (1985).

PROBLEMS

All problems are for room temperature unless another temperature is specified.

7.1 A MOS capacitor at 300 K has an Al gate and a p-Si substrate with $p = N_a^- = 3 \times 10^{16}$ cm^{-3}. The oxide thickness is 550 Å (55 nm). There is no charge at the interface.

(a) What is the flat-band voltage V_{FB}°, and what polarity voltage must be connected to the Al gate to achieve flat-band conditions?

(b) Sketch the energy-band diagrams and charge distributions for the biasing conditions corresponding to thermal equilibrium, flat band, accumulation, depletion, and inversion. Specify the polarity of the voltage connected to the Al gate for each condition.

(c) Find the flat-band voltage when the Al gate is replaced by W, which has a work function of 4.5 eV.

(d) Find the flat-band voltage when the Al gate is replaced by n^+- Si, which has a work function of 3.83 eV.

7.2 For the MOS capacitor with an Al gate in Problem 7.1,

(a) Calculate the threshold voltage for $N_f = 0$.

(b) Calculate the threshold voltage for $N_f = 3 \times 10^{10}$ cm^{-2}.

(c) Describe how the threshold voltage would be expected to vary with p, N_f, and t_{ox}.

7.3 For the MOS capacitor with the Al gate in Problem 7.1, calculate C'_{ox}, C'_{dif}(FB), and $C'_{dif}(x_{d_{max}})$ with $N_f = 0$.

7.4 With the capacitance from Problem 7.3 normalized as C'_{dif}(FB)/C'_{ox} and $C'_{dif}(x_{d_{max}})/C'_{ox}$, sketch the high frequency C'_{dif}/C'_{ox} vs. V_G from -5 V to $+5$ V by plotting $C'_{ox}/C'_{ox} = 1.0$, C'_{dif}(FB)/C'_{ox}, and $C'_{dif}(x_{d_{max}})/C'_{ox}$ and then drawing a curve connecting these points.

(a) $N_f = 0$.

(b) $N_f = 5 \times 10^{10}$ cm^{-2}.

7.5 Repeat Problem 7.1, but replace the Al gate with a gold gate ($q\Phi_{\text{Au}} = 4.75$ eV) and let $p = N_a^- = 1 \times 10^{15}$ cm^{-3} for the p-Si.

7.6 For an ideal MOS capacitor with an Al gate and an n-Si substrate with $n = N_d^+ = 7 \times 10^{16}$ cm^{-3}:

(a) What is the flat-band voltage V_{FB}°, and polarity of the bias which must be connected to the Al gate to achieve flat-band conditions?

(b) Sketch the energy-band diagrams and charge distributions for biasing conditions corresponding to *accumulation, depletion,* and *strong inversion.* Specify the polarity of the bias connected to the Al gate.

7.7 For an ideal MOS capacitor with an Al gate and a p-Si substrate with $p = N_a^- = 2 \times 10^{16}$ cm^{-3}:

(a) What is the flat-band voltage V_{FB}° and polarity of the bias which must be connected to the Al gate to achieve flat-band conditions?

(b) Sketch the energy-band diagrams and charge distributions for biasing conditions corresponding to *accumulation, depletion,* and *strong inversion.* Specify the polarity of the bias connected to the Al gate.

7.8 An ideal Si MOS capacitor has a polycrystalline Si gate.

(a) With the Nilsson expression [Eq. (2.58)] for E_f (Sec. 2.6.2), what electron concentration gives $(E_c - E_f) = 0$ and what hole concentration gives $(E_f - E_v) = 0$?

(b) With this donor concentration for the gate "metal," sketch the thermal equilibrium energy-band diagram for a p-Si substrate with $p = N_a^- = 4 \times 10^{16}$ cm^{-3}.

(c) What is the flat-band voltage V_{FB}°, and what polarity bias must be connected to the gate for flat-band conditions?

7.9 Calculate the threshold voltage for an ideal MOS capacitor with an Al gate and p-Si substrate with $p = N_a^- = 3 \times 10^{16}$ cm^{-3}. The oxide thickness is 600 Å.

7.10 **(a)** Repeat Problem 7.9 for $p = N_a^- = 3 \times 10^{15}$ cm^{-3}.

(b) Repeat Problem 7.9 for $p = N_a^- = 3 \times 10^{17}$ cm^{-3}.

(c) Repeat Problem 7.9 for $t_{ox} = 1200$ Å.

7.11 Repeat Problem 7.9 for a fixed oxide charge $N_f = 1 \times 10^{11}$ cm^{-2}.

7.12 Repeat Problem 7.9 except for an n-Si substrate with $n = N_d^+ = 3 \times 10^{16}$ cm^{-3}.

7.13 Repeat Problem 7.9 for a Pt gate with a work function of 5.3 eV.

7.14 For the ideal MOS capacitor of Problem 7.9, sketch the C'_{dif}/C'_{ox} curve from -5 V to $+5$ V. To obtain this curve, first calculate C'_{ox}, C'_{dif}(FB), and $C'_{dif}(x_{d_{max}})$ and normalize to $C'_{ox} = 1.0$. Plot these points and sketch a curve connecting these points. Mark V_{FB}° and C'_{dif}(FB) on this curve.

7.15 **(a)** Find the numerical value for ϵ_{Si}/q^2 in units of eV^{-1}-cm^{-1}.

(b) Plot the Debye length on a log-log scale from a carrier concentration of 1×10^{15} cm^{-3} to 1×10^{19} cm^{-3} for 300 K.

7.16 The definition of the onset of strong inversion is that $n_s = N_a^-$ for p-Si. For this condition, show that $\psi_s = -\psi_p$ and that the total ψ is $2\psi_p$.

7.17 An ideal MOS capacitor has an Al gate, an oxide thickness of 800 Å, and the p-Si has $p = N_a^- = 4 \times 10^{15}$ cm^{-3}.

(a) Find the surface potential at the onset of strong inversion.

(b) What is the threshold voltage?

(c) What is the maximum depletion width for an ideal MOS capacitor?

(d) What is the threshold voltage for a fixed oxide charge $N_f = 9 \times 10^{10}$ cm^{-2}?

7.18 Consider the MOS capacitor in Problem 7.17 and calculate the major points on the C–V curve, such as:

(a) The maximum capacitance.

(b) The capacitance at flat band.

(c) The high-frequency capacitance at maximum depletion. Sketch the C'_{dif}/C'_{ox} curve from -5 V to $+5$ V. To obtain this curve, normalize to $C'_{ox} = 1$. Plot these points and sketch a curve connecting these points. Mark the flat-band voltage and C'_{dif}(FB) on this curve for both with and without the fixed oxide charge Q_f.

7.19 For the example problem summarized in Table 7.1, plot $|Q'_{Si}|$ on a log scale on the y-axis as a function of ψ_s with a linear scale on the x-axis. Indicate the accumulation, depletion, and inversion regions as well as flat band.

7.20 The variation of the electric field with distance at thermal equilibrium is shown below for an ideal MOS capacitor on a p-Si substrate.

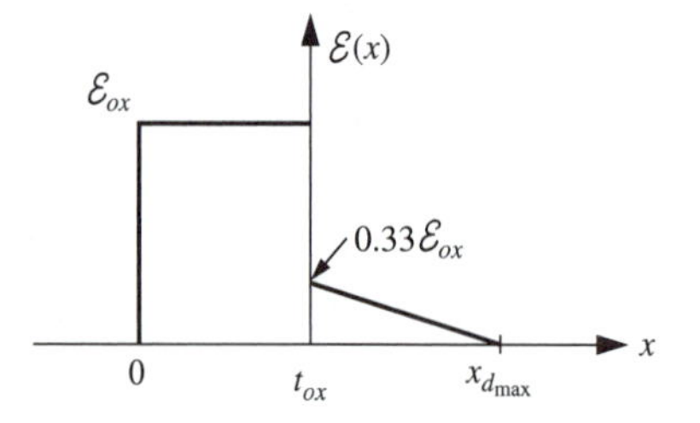

(a) Sketch the charge density $\rho(x)$ vs. distance x.

(b) Sketch the energy-band diagram corresponding to the electric field shown in the figure. Label the Fermi level and the intrinsic Fermi level.

(c) What polarity must be connected to the metal electrode to achieve the flat-band condition for the electric field (and your charge density) shown in the figure?

7.21 Find ψ_b for an n-type semiconductor.

7.22 Write a program to calculate and verify the quantities in Table 7.1 for $p = N_a^- = 4 \times 10^{15}$ cm^{-3} and $t_{ox} = 500$ Å at 300 K. Let $kT/q = 0.026$ V. Use $\psi = -0.20$ V to test your program and the additional value of $\psi = -0.30$ V to obtain the quantities in Table 7.1 to demonstrate you have a working program.

7.23 Write a program to calculate and then plot the normalized C–V behavior for an Al gate on an ideal SiO_2 layer on p-type InP with $p = N_a^- = 1 \times 10^{16}$ cm^{-3} and $t_{ox} = 400$ Å from V_G of -5.0 V to $+5.0$ V at room temperature. For InP, the electron affinity qX_{InP} is 4.4 eV, the energy gap E_g is 1.351 eV, the dielectric constant ϵ_{InP} is $12.4\epsilon_0$, the electron effective mass m_n^* is $0.077m_0$, and the effective hole mass is $m_p = 0.64m_0$. Also give your values for the effective density of states of electrons and holes and the intrinsic carrier concentration.

7.24 Although the dielectric breakdown of thermally grown SiO_2 often shows a wide variability and depends on the measurement technique, a maximum dielectric strength of defect-free SiO_2 would be $\sim 10 \times 10^6$ V/cm. If a maximum electric field is considered to be one-half this value, for a 5-V source, find:

(a) The minimum oxide thickness.

(b) The maximum capacitance for a 1-μm-diameter MOS capacitor.

7.25 For an oxide with a thickness of 300 Å (30 nm), plot ΔV_{FB} as the fixed oxide density is varied from 5×10^{10} cm^{-2} to 5×10^{12} cm^{-2}.

7.26 For an ideal MOS capacitor at 300 K with $p = N_a^- = 8 \times 10^{15}$ cm^{-3} and $t_{ox} = 350$ Å, find the voltage across the oxide and the electric field in the oxide at threshold.

7.27 Write equations for the gate voltage V_G for an n-type semiconductor in the form of Eqs. (7.109) and (7.110) which are for a p-type semiconductor.

7.28 Show that the maximum depletion width $x_{d_{max}}$ in a p-type semiconductor is given by $2L_D\sqrt{\ln(N_a^-/n_i)}$, where L_D is the Debye length.

7.29 Sketch the C–V behavior (similar to Fig. 7.19) for an ideal capacitor with an Al gate on n-type Si. Take the substrate doping as $n = N_d^+ = 4 \times 10^{15}$ cm^{-3} and the SiO_2 layer thickness $t_{ox} =$ 500 Å (50 nm).

7.30 Find the units for the prefactor expression $\sqrt{2\epsilon_{Si} kTN_a^-}$ for Q'_{Si} in Eq. (C.43) and rewritten in Eq. (C.46).

7.31 From Eq. (C.43), show that the minority carrier concentration in strong inversion for a p-type semiconductor varies as $\sim \exp(q\psi_s/2kT)$ where ψ_s is the surface potential.

7.32 Show that the majority carrier concentration in the accumulation layer for a p-type semiconductor varies as $\sim \exp(q|\psi_s|/2kT)$, where ψ_s is the surface potential.

7.33 What would be done to *prevent* inversion in Si under a metal interconnect over an oxide?

7.34 For the MOS capacitor in Problem 7.1, find V_G when

(a) $\psi_s = -0.1$ V.

(b) $\psi_s = 0$ V.

(c) $\psi_s = +0.1$ V.

7.35 The flat-band voltage V_{FB}° for a MOS capacitor may be obtained by fabricating and measuring a set of MOS capacitors with different SiO_2 layer thicknesses. With a known substrate doping concentration, the capacitance at flat band is given by Eq. (7.97). For a set of MOS capacitors, the voltage where the flat-band band capacitance occurs was found to be:

SiO_2 Layer Thickness, t_{ox} (Å)	400	600	800	1000
Flat-band Voltage, V_{FB}(V)	−2.15	−2.80	−3.44	−4.09

(a) Plot V_{FB} vs. t_{ox} and extrapolate to $t_{ox} = 0$ to obtain V_{FB}°.

(b) Find the fixed oxide charge density N_f.

7.36 For an ideal MOS capacitor with an Al gate, n-Si substrate with $n = N_d^+ = 7 \times 10^{15}$ cm^{-3}, and an oxide thickness of 500 Å, find:

(a) The oxide capacitance per unit area.

(b) The differential capacitance at flat band.

(c) The differential capacitance at threshold and plot as C'_{dif}/C'_{ox} from $V_G = -5$ V to $V_G = 5$ V.

7.37 For an ideal MOS capacitor with $p = N_a^- = 7 \times 10^{15}$ cm^{-3}, find the value of ψ_s when $n_s = 3p_b$.

7.38 In Fig. 7.22, C'_{dif}/C'_{ox} at $V_G = V_T$,where $x_d = x_{dmax}$ increases as N_a^- increases. Show that this behavior is expected from Eq. (7.100) for $C'_{dif}(x_{dmax})$ by showing that $C'_{dif}(x_{dmax})/C'_{ox}$ may be written as $1/[1 + C'_{ox}/(K\sqrt{N_a^-})]$.

7.39 Explain why the inversion charge Q_n can be neglected in the determination of the threshold voltage.

7.40 For an n-type semiconductor, find an expression for the gate voltage when biased for the condition of the onset of strong inversion, as was done for the p-type Si in Sec. 7.3.4 by finding:

(a) The electric field in the oxide $\mathscr{E}_{ox}$ in terms of the depletion space charge per unit area Q'_{sc} from Gauss's law.

(b) Find the potential $\psi_{ox}(x)$ in the oxide with the boundary condition $\psi_{ox}(t_{ox}) = -\psi_s$ at the SiO_2/Si interface.

(c) Find the potential difference across the oxide.

(d) Write an expression for the gate voltage V_G similar to Eq. (7.109).

APPENDIX C

C.1 INVERSION LAYER THICKNESS

As shown in Fig. 7.12 (*d*) for a *p*-type semiconductor, the potential $\psi(x)$ in the semiconductor is ψ_s at the SiO_2/Si interface and $\psi(x) = 0$ for $x > x_{d_{max}}$. It will be assumed here that the mobile charge in the inversion layer qN_{inv} at threshold is very small as compared to the immobile space charge Q_{sc}. Therefore, only Q_{sc} need be considered when determining the electric field $\mathcal{E}(x)$ and the potential $\psi(x)$ in the semiconductor from Gauss's law. The charge/unit area in the inversion layer can be represented as

$$qN'_{inv} = qn_s x_c, \tag{C.1}$$

where at threshold $n_s = p_b$, the hole concentration in the neutral semiconductor, and x_c is the centroid of the inversion layer charge. The depletion region space charge/unit area is given by Eq. (7.45) as

$$\begin{aligned} Q'_{sc} &= -qN_a^- x_{d_{max}} \\ &= -\sqrt{4q\epsilon_{ox}N_a^-|\psi_b|}, \end{aligned} \tag{C.2}$$

with ψ_b given by Eq. (7.18) and $\psi_s = 2|\psi_b|$ at threshold. With $n_s = p_b = N_a^-$ at threshold, the ratio of qN'_{inv} to Q'_{sc} is

$$\begin{aligned} \frac{qN'_{inv}}{Q'_{sc}} &= \frac{qn_s x_c}{qN_a^- x_{d_{max}}} \\ &= \frac{x_c}{x_{d_{max}}}. \end{aligned} \tag{C.3}$$

By determining x_c, it will be demonstrated that $x_c \ll x_{d_{max}}$, and therefore $qN'_{inv} \ll Q'_{sc}$.

*Analysis forAppendix C.1 was provided by J. R. Brews.

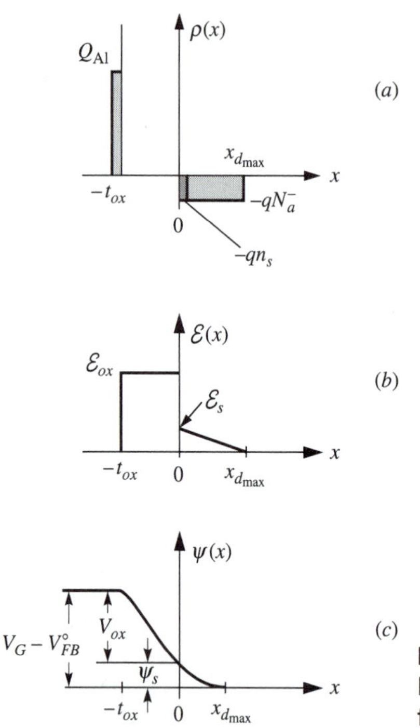

FIGURE C.1 *(a)* Charge distribution for the MOS capacitor with an ideal SiO_2. *(b)* Electric-field distribution. *(c)* Potential distribution.

The space-charge, electric field, and potential are replotted in Fig. C.1 where the x-axis has been relabeled with $x = 0$ at the SiO_2/Si interface to simplify the expressions used to calculate x_c. From Gauss's law, the electric field is given by

$$\frac{d\mathscr{E}(x)}{dx} = -\frac{qN_a^-}{\epsilon_{\text{Si}}}, \tag{C.4}$$

which gives

$$\mathscr{E}(x) = -\frac{qN_a^-}{\epsilon_{\text{Si}}}x + A_1. \tag{C.5}$$

With the boundary condition of $\mathscr{E}(x_{d_{max}}) = 0$, $A_1 = qN_a^- x_{d_{max}}/\epsilon_{\text{Si}}$, and

$$\mathscr{E}(x) = \frac{qN_a^-}{\epsilon_{\text{Si}}}(x_{d_{max}} - x). \tag{C.6}$$

The potential is given by

$$\psi(x) = -\int \mathscr{E}(x)\,dx = -\int \frac{qN_a^-}{\epsilon_{\text{Si}}}(x_{d_{max}} - x)\,dx, \tag{C.7}$$

which gives

$$\psi(x) = -\frac{qN_a^-}{\epsilon_{\text{Si}}}(x_{d_{max}}x - x^2/2) + B_2. \tag{C.8}$$

With the boundary condition of $\psi(x)$ at $x = 0$,

$$\psi(0) = \psi_s = B_2 \tag{C.9}$$

and

$$\psi(x) = \psi_s - \frac{qN_a^-}{\epsilon}(x_{d_{max}}x - x^2/2). \tag{C.10}$$

Because the inversion layer extends only a small distance from $x = 0$, the $x^2/2$ term will be neglected in Eq. (C.10) to give

$$\psi(x) = \psi_s - \frac{qN_a^- x_{d_{max}}}{\epsilon_{\text{Si}}}x \qquad \text{for } x \ll x_{d_{max}}. \tag{C.11}$$

From Eq. (C.6) for $\mathscr{E}(x)$ at the SiO_2/Si interface,

$$\mathscr{E}(0) = \mathscr{E}_s = \frac{qN_a^- x_{d_{max}}}{\epsilon_{\text{Si}}}. \tag{C.12}$$

By combining Eqs. (C.11) and (C.12), $\psi(x)$ becomes

$$\psi(x) = \psi_s - \mathscr{E}_s x \qquad \text{for } x \ll x_{d_{max}}. \tag{C.13}$$

From Eqs. (7.22) or (7.42), the electron concentration in the inversion layer of a p-type semiconductor varies as

$$n(x) \simeq \frac{n_i^2}{N_a^-}\exp[q\psi(x)/kT]. \tag{C.14}$$

With Eq. (C.13) for $\psi(x)$,

$$n(x) \simeq \underbrace{\frac{n_i^2}{N_a^-}\exp(q\psi_s/kT)}_{n_s}\exp(-q\mathscr{E}_s x/kT). \tag{C.15}$$

The centroid of the inversion layer x_c is obtained from

$$qN'_{inv}x_c = \int_0^\infty qxn(x)\,dx = qn_s\int_0^\infty x\exp(-q\mathscr{E}_s x/kT)\,dx, \tag{C.16}$$

which becomes

$$qn_s \int_0^\infty x \exp(-q\mathscr{E}_s x/kT)\, dx = qn_s \exp(-q\mathscr{E}_s x/kT) \left[\frac{x}{-q\mathscr{E}_s x/kT} - \frac{1}{(-q\mathscr{E}_s/kT)^2} \right] \Bigg|_0^\infty$$

$$= qn_s \left(\frac{kT}{q} \right)^2 \frac{1}{\mathscr{E}_s^2}. \tag{C.17}$$

Also, the inversion layer charge/unit area is given by

$$qN'_{inv} = \int_0^\infty qn(x)\, dx = qn_s \int_0^\infty \exp(-q\mathscr{E}_s x/kT)\, dx = qn_s \frac{kT}{q\mathscr{E}_s}. \tag{C.18}$$

With Eqs. (C.17) and (C.18), the centroid of the inversion layer is

$$x_c = \frac{N'_{inv} x_c}{N'_{inv}} = \frac{kT}{q\mathscr{E}_s}. \tag{C.19}$$

From Eq. (C.2) for Q'_{sc} and Eq. (C.12) for $\mathscr{E}_s$

$$Q'_{sc} = -qN_a^- x_{d_{max}} \frac{\epsilon_{\text{Si}}}{\epsilon_{\text{Si}}} = -\underbrace{\frac{qN_a^- x_{d_{max}}}{\epsilon_{\text{Si}}}}_{\mathscr{E}_s} \epsilon_{\text{Si}} = -\epsilon_{\text{Si}} \mathscr{E}_s, \tag{C.20}$$

which gives

$$\boxed{x_c = \frac{\epsilon_{\text{Si}} kT}{q|Q'_{sc}|}}. \tag{C.21}$$

A comparison of x_c and $x_{d_{max}}$ for a typical value of impurity concentration will demonstrate how much smaller $qN'_{inv}x_c$ is than Q'_{sc} at threshold. For example, with a substrate doping of $p_b = N_a^- = 1 \times 10^{16}\ \text{cm}^{-3}$,

$$\psi_b = -(kT/q)\ln(N_a^-/n_i) = -0.026 \ln(1 \times 10^{16}/1 \times 10^{10}) = -0.360\ \text{V},$$

$$Q'_{sc} = \sqrt{4q\epsilon_{\text{Si}} N_a^- |\psi_b|} = 4.879 \times 10^{-8}\ \text{coulombs/cm}^2,$$

numerical example for $x_c/x_{d_{max}}$

and

$$x_c = \frac{0.026 \times 11.7 \times 8.85 \times 10^{-14}}{4.879 \times 10^{-8}}$$

$$= 5.518 \times 10^{-7}\ \text{cm}.$$

With Eq. (7.38),

$$\begin{aligned} x_{d_{max}} &= \sqrt{4\epsilon_{\text{Si}}|\psi_b|/qN_a^-} \\ &= \sqrt{4 \times 11.7 \times 8.85 \times 10^{-14} \times 0.360/1.6 \times 10^{-19} \times 1 \times 10^{16}} \\ &= 3.053 \times 10^{-5} \text{ cm.} \end{aligned}$$

Then, with $n_s = p_b$, $qn_sx_c = 1.6 \times 10^{-19} \times 1 \times 10^{16} \times 5.518 \times 10^{-7} = 8.829 \times 10^{-10}$ coulombs/cm^2. Therefore, $qN_{inv} = 1.810 \times 10^{-2} Q'_{sc}$, which justifies ignoring the inversion layer charge at weak inversion or at threshold.

C.2 MOS CAPACITANCE

The most useful electrical measurement of a MOS structure is the small-signal capacitance. The differential capacitance of the semiconductor is given by Eq. (7.86) as

$$C'_{\text{Si}} = \left|\frac{dQ'_{\text{Si}}}{d\psi_s}\right|. \tag{C.22}$$

Therefore, Q'_{Si} in terms of the potential in the semiconductor must be known. From Eqs. (7.57) and (7.60) for Q'_{sc}:

$$Q'_{\text{Si}} = -\epsilon_{ox}\mathcal{E}_{ox} = -\epsilon_{\text{Si}}\mathcal{E}_{\text{Si}}, \tag{C.23}$$

which requires $\mathcal{E}_{\text{Si}}$ to obtain Q'_{Si}. The electric field $\mathcal{E}(x)$ and the potential $\psi(x)$ in the semiconductor as a function of distance may be obtained from the one-dimensional Poisson equation

$$\frac{d^2\psi(x)}{dx^2} = -\frac{\rho(x)}{\epsilon_{\text{Si}}}, \tag{C.24}$$

where $\rho(x)$ is the total space-charge density which is given by

$$\rho(x) = q[N_d^+ - N_a^- + p(x) - n(x)]. \tag{C.25}$$

The spatial variation of the hole concentration was given in Eq. (7.21) as

$$p(x) = n_i \exp\{-q[\psi_b + \psi(x)]/kT\} = \underbrace{n_i \exp(-q\psi_b/kT)}_{p_b = N_a^-} \exp[-q\psi(x)/kT], \tag{C.26}$$

or

$$p(x) = N_a^- \exp[-q\psi(x)/kT]. \tag{C.27}$$

The spatial variation of the electron concentration was given in Eq. (7.22) as

$$n(x) = n_i \exp\{q[\psi_b + \psi(x)]/kT\} = \underbrace{n_i \exp[q\psi_b/kT]}_{n_{po}=n_i^2/N_a^-} \exp[q\psi(x)/kT], \quad \text{(C.28)}$$

$$n(x) = (n_i^2/N_a^-)\exp[q\psi(x)/kT]. \quad \text{(C.29)}$$

For p-type Si, N_d^+ will be taken as zero.

Equation (C.24) with Eq. (C.25) becomes

$$\frac{d^2\psi(x)}{dx^2} = -\frac{q[-N_a^- + p(x) - n(x)]}{\epsilon_{\text{Si}}} = \frac{qN_a^-}{\epsilon_{\text{Si}}}\left[1 - \frac{p(x)}{N_a^-} + \frac{n(x)}{N_a^-}\right], \quad \text{(C.30)}$$

or with Eqs. (C.27) and (C.29),

$$\frac{d^2\psi(x)}{dx^2} = \frac{qN_a^-}{\epsilon_{\text{Si}}}\left[1 - \exp[-q\psi(x)/kT] + \left(\frac{n_i}{N_a^-}\right)^2 \exp[q\psi(x)/kT]\right]. \quad \text{(C.31)}$$

The coefficient on the right-hand side of Eq. (C.31) can be put in a more useful form by multiplying $(qN_a^-/\epsilon_{\text{Si}})$ by $(q/kT)^2(kT/q)^2$ to give

$$\frac{qN_a^-}{\epsilon_{\text{Si}}}\left(\frac{q}{kT}\right)^2\left(\frac{kT}{q}\right)^2 = \left(\frac{q^2N_a^-}{\epsilon_{\text{Si}}kT}\right)\frac{q}{kT}\left(\frac{kT}{q}\right)^2. \quad \text{(C.32)}$$

In this equation, the first quantity on the right is the reciprocal of the Debye length,

$$L_D = \sqrt{\frac{\epsilon_{\text{Si}}kT}{q^2N_a^-}}. \quad \text{(C.33)}$$

The Debye length L_D appears frequently in the following expressions and represents the screening of the potential of ionized acceptors (or donors) by the mobile free holes (or electrons). This variation is given as $\exp(-x/L_D)$. Note that the carrier concentration, as represented by $N_a^- = p$, appears in the denominator of L_D; therefore, for high carrier concentrations, the Debye length is small and the shielding is strong. Then, the expression on the right-hand side of Eq. (C.31) may be written as

$$\frac{qN_a^-}{\epsilon_{\text{Si}}} = \frac{1}{L_D^2}\left(\frac{kT}{q}\right)^2\frac{q}{kT} = \left(\frac{kT}{qL_D}\right)^2\frac{q}{kT}. \quad \text{(C.34)}$$

Poisson's equation may now be written as,

$$\frac{d^2\psi(x)}{dx^2} = \left(\frac{kT}{qL_D}\right)^2\frac{q}{kT}\left[1 - \exp[-q\psi(x)/kT] + \left(\frac{n_i}{N_a^-}\right)^2 \exp[q\psi(x)/kT]\right]. \quad \text{(C.35)}$$

Equation (C.35) can be made an exact differential by multiplying both sides by the integrating factor $d\psi(x)/dx$ to give

$$\frac{d\psi(x)}{dx}\left[\frac{d^2\psi(x)}{dx^2}\right] = \left(\frac{kT}{qL_D}\right)^2 \times\left[\frac{q}{kT} - \frac{q}{kT}\exp[-q\psi(x)/kT] + \frac{q}{kT}\left(\frac{n_i}{N_a^-}\right)^2\exp[q\psi(x)/kT]\right] \times\frac{d\psi(x)}{dx}. \tag{C.36}$$

With the identity,

$$\frac{1}{2}\frac{d}{dx}\left[\frac{d\psi(x)}{dx}\right]^2 = \frac{d\psi(x)}{dx}\left[\frac{d^2\psi(x)}{dx^2}\right],$$

Poisson's equation becomes,

$$\frac{1}{2}\frac{d}{dx}\left[\frac{d\psi(x)}{dx}\right]^2 = \left(\frac{kT}{qL_D}\right)^2 \times\frac{d}{dx}\left[\frac{q\psi(x)}{kT} + \exp[-q\psi(x)/kT] + \left(\frac{n_i}{N_a^-}\right)^2\exp[q\psi(x)/kT]\right]. \tag{C.37}$$

Integration of the exact differential in Eq. (C.37) from the SiO_2- semiconductor interface ($x = 0$) to the edge of the depletion region ($x = x_d$) gives

$$\left(\frac{d\psi(x)}{dx}\right)^2\bigg|_{x=x_d} - \left(\frac{d\psi(x)}{dx}\right)^2\bigg|_{x=0} = 2\left(\frac{kT}{qL_D}\right)^2 \times\left[\frac{q\psi(x)}{kT} + \exp[-q\psi(x)/kT] + \left(\frac{n_i}{N_a^-}\right)^2\exp[q\psi(x)/kT]\right]\Bigg|_{x=0}^{x=x_d}. \tag{C.38}$$

At the edge of the depletion region, both the potential and field are zero: at $x = x_d$, $\psi(x_d) = 0$ and $d\psi/dx = 0$. At the SiO_2/Si interface $x = 0$, $\psi(0) = \psi_s$. With these boundary conditions, Eq. (C.38) becomes

$$-\left(\frac{d\psi_s}{dx}\right)^2 = 2\left(\frac{kT}{qL_D}\right)^2\left\{\left[0 + 1 + \left(\frac{n_i}{N_a^-}\right)^2\right] - \left[\frac{q\psi_s}{kT} + \exp[-q\psi_s/kT] + \left(\frac{n_i}{N_a^-}\right)^2\exp[q\psi_s/kT]\right]\right\}, \tag{C.39}$$

or

$$-\left(\frac{d\psi_s}{dx}\right)^2 = 2\left(\frac{kT}{qL_D}\right)^2\left[1+\left(\frac{n_i}{N_a^-}\right)^2 - \frac{q\psi_s}{kT} - \exp[-q\psi_s/kT] - \left(\frac{n_i}{N_a^-}\right)^2 \exp[q\psi_s/kT]\right]. \tag{C.40}$$

Elimination of the minus sign on the left-hand side of Eq. (C.40) and collection of terms on the right-hand side gives

$$\left(\frac{d\psi_s}{dx}\right)^2 = 2\left(\frac{kT}{qL_D}\right)^2\left\{\left[\frac{q\psi_s}{kT} + \exp(-q\psi_s/kT) - 1\right] + \left(\frac{n_i}{N_a^-}\right)^2 \left[\exp(q\psi_s/kT) - 1\right]\right\}. \tag{C.41}$$

With

$$\mathscr{E}_s = -\frac{d\psi_s}{dx} \qquad \text{and} \qquad \mathscr{E}_s^2 = \left(\frac{d\psi_s}{dx}\right)^2,$$

Eq. (C.41) gives the electric field at the SiO_2/Si interface as

$$\mathscr{E}_s = \pm\sqrt{2}\left(\frac{kT}{qL_D}\right)\left\{\left[\frac{q\psi_s}{kT} + \exp(-q\psi_s/kT) - 1\right] + \left(\frac{n_i}{N_a^-}\right)^2 \left[\exp(q\psi_s/kT) - 1\right]\right\}^{1/2}. \tag{C.42}$$

With Eq. (C.23) for $\mathscr{E}_s$,

Q'_{Si} for p-type Si

$$Q'_{\mathrm{Si}} = -\epsilon_{\mathrm{Si}}\mathscr{E}_s$$

$$Q'_{\mathrm{Si}} = \mp\sqrt{2}\epsilon_{\mathrm{Si}}\left(\frac{kT}{qL_D}\right) \times \left\{\left[\frac{q\psi_s}{kT} + \exp(-q\psi_s/kT) - 1\right] + \left(\frac{n_i}{N_a^-}\right)^2 \left[\exp(q\psi_s/kT) - 1\right]\right\}^{1/2}. \tag{C.43}$$

For a p-type semiconductor, Q'_{Si} is positive for accumulation and negative for depletion and inversion. For an n-type semiconductor, Q'_{Si} is negative for accumulation and positive for depletion and inversion.

The differential capacitance per unit area of the semiconductor with Eq. (C.43) for Q'_{Si} is given by

$$C'_{\text{Si}} \equiv \left|\frac{dQ'_{\text{Si}}}{d\psi_s}\right| = \frac{\epsilon_{\text{Si}}}{\sqrt{2}}\left(\frac{kT}{qL_D}\right) \times \left|\frac{\left[\frac{q}{kT} - \frac{q}{kT}\exp(-q\psi_s/kT) + \frac{q}{kT}\left(\frac{n_i}{N_a^-}\right)^2 \exp(q\psi_s/kT)\right]}{\left\{\left[\frac{q\psi_s}{kT} + \exp(-q\psi_s/kT) - 1\right] + \left(\frac{n_i}{N_a^-}\right)^2 \left[\exp(q\psi_s/kT) - 1\right]\right\}^{1/2}}\right|. \tag{C.44}$$

First, the q/kT term is factored out of the numerator to give

semiconductor capacitance for p-type Si

$$C'_{\text{Si}} = \frac{\epsilon_{\text{Si}}}{\sqrt{2}L_D} \times \left|\frac{\left[1 - \exp(-q\psi_s/kT) + \left(\frac{n_i}{N_a^-}\right)^2 \exp(q\psi_s/kT)\right]}{\left\{\left[\frac{q\psi_s}{kT} + \exp(-q\psi_s/kT) - 1\right] + \left(\frac{n_i}{N_a^-}\right)^2 \left[\exp(q\psi_s/kT) - 1\right]\right\}^{1/2}}\right|. \tag{C.45}$$

Equation (C.45) gives the Si contribution to the capacitance as a function of ψ_s.

For numerical calculations, Eq. (C.43) for Q'_{Si} can be simplified for accumulation and depletion conditions. The ratio $(n_i/N_a^-)^2$ will be small so that the $(n_i/N_a^-)^2$ $[\exp(q\psi_s/kT) - 1]$ term for minority carriers may be neglected except during inversion. By replacing L_D in Eq. (C.43) with Eq. (C.33), the prefactor may be written as

$$\sqrt{2}\epsilon_{\text{Si}}\left(\frac{kT}{qL_D}\right) = \sqrt{2\epsilon_{\text{Si}}kTN_a^-} = 9.282 \times 10^{-17}\sqrt{N_a^-}. \tag{C.46}$$

Equation (C.43) now becomes

Q'_{Si} for p-type Si in accumulation or depletion

$$Q'_{\text{Si}} = \mp 9.282 \times 10^{-17}\sqrt{N_a^-}\left[\frac{q\psi_s}{kT} + \exp(-q\psi_s/kT) - 1\right]^{1/2}. \tag{C.47}$$

Equation (C.45) for C'_{Si} may also be simplified by neglecting the terms with $(n_i/N_a^-)^2$ for accumulation and depletion conditions. Evaluation of the numerical quantities in the prefactor $\epsilon_{\text{Si}}/\sqrt{2}L_D = 1.785 \times 10^{-15}\sqrt{N_a^-}$ gives

C'_{Si} for p-type Si in accumulation or depletion

$$C'_{\text{Si}} = 1.785 \times 10^{-15}\sqrt{N_a^-}\left|\frac{[1 - \exp(-q\psi_s/kT)]}{\left[\frac{q\psi_s}{kT} + \exp(-q\psi_s/kT) - 1\right]^{1/2}}\right|. \tag{C.48}$$

The complete expressions given by Eqs. (C.43) and (C.44) must be used for conditions resulting in inversion.

inversion

For n-type Si, the derivations for Q'_{Si} and C'_{Si} are the same as for p-type Si except that N_a^- in Eq. (C.25) is now taken as zero rather than N_d^+. The equation analogous to Eq. (C.47) for Q'_{Si} in n-type Si is

Q'_{Si} for n-type Si in accumulation or depletion

$$Q'_{\mathrm{Si}} = \mp 9.282 \times 10^{-17} \sqrt{N_d^+} \left[\frac{-q\psi_s}{kT} + \exp(q\psi_s/kT) - 1 \right]^{1/2}. \tag{C.49}$$

The equation analogous to Eq. (C.48) for C'_{Si} in n-type Si is

C'_{Si} for n-type Si in accumulation or depletion

$$C'_{\mathrm{Si}} = 1.785 \times 10^{-15} \sqrt{N_d^+} \left| \frac{[\exp(q\psi_s/kT) - 1]}{\left[\frac{-q\psi_s}{kT} + \exp(q\psi_s/kT) - 1 \right]^{1/2}} \right|. \tag{C.50}$$

For semiconductors other than Si, the numerical prefactors will have to be modified for the different dielectric constants of the semiconductors.

For the flat-band condition, Q'_{Si} is zero, but as ψ_s changes from $\psi_s = 0$ at flat band, there will be a change in charge, and hence a value for C_{Si}. To evaluate Eq. (C.45) for small ψ_s, the exponential terms in the denominator require a three-term series expansion to prevent the denominator from going to zero if only two terms are used. Therefore, $\exp(-\psi_s/kT)$ becomes $[1 - (q\psi_s/kT) + (1/2)(q\psi_s/kT)^2]$ and $\exp(\psi_s/kT)$ becomes $[1 + (q\psi_s/kT) + (1/2)(q\psi_s/kT)^2]$. The exponentials in the numerator may be represented by the first two terms of the series expansion. The differential capacitance per unit area for $\psi_s \simeq 0$ is the flat-band capacitance C'_{FB} and is given by Eq. (C.45) as

$$C'_{FB} = \frac{\epsilon_{\mathrm{Si}}}{\sqrt{2}L_D} \frac{\left\{ \frac{q\psi_s}{kT} \left[1 + \left(\frac{n_i}{N_a^-} \right)^2 \right] \right\}}{\left\{ \frac{1}{2} \left(\frac{q\psi_s}{kT} \right)^2 \left[1 + \left(\frac{n_i}{N_a^-} \right)^2 \right] \right\}^{1/2}}. \tag{C.51}$$

For p-type Si, $(n_i/N_a^-)^2 \ll 1$, and Eq. (C.51) becomes

C'_{Si} at flat band

$$C'_{FB} = \frac{\epsilon_{\mathrm{Si}}\sqrt{2}}{\sqrt{2}L_D} = \frac{\epsilon_{\mathrm{Si}}}{L_D}. \tag{C.52}$$

As the applied voltage is increased for depletion and then inversion, the depletion width increases until strong inversion occurs for $\psi_s = 2|\psi_b|$. The maximum depletion width $x_{d_{max}}$ at the onset of strong inversion was given by Eq. (7.38) as

$$x_{d_{max}} = \sqrt{\frac{2\epsilon_{\mathrm{Si}}\psi_s}{qN_a^-}}. \tag{C.53}$$

The semiconductor charge per unit area Q'_{Si} due to the depletion region at $x_{d_{max}}$ is

$$Q'_{\text{Si}} = Q'_{sc} = -qN_a^- x_{d_{max}} = -\sqrt{2q\epsilon_{\text{Si}} N_a^- \psi_s}. \tag{C.54}$$

The differential capacitance per unit area C'_{sc} at maximum depletion width may be obtained with Eq. (C.54) as

$$C'_{sc} = \left| \frac{dQ'_{\text{Si}}}{d\psi_s} \right| = \frac{1}{2}\sqrt{2q\epsilon_{\text{Si}} N_a^-}\,\psi_s^{-1/2}, \tag{C.55}$$

or

C'_{Si} in p-type Si at maximum depletion

$$C'_{sc} = \epsilon_{\text{Si}} \underbrace{\sqrt{\frac{qN_a^-}{2\epsilon_{\text{Si}}\psi_s}}}_{1/x_{d_{max}}} = \frac{\epsilon_{\text{Si}}}{x_{d_{max}}}. \tag{C.56}$$

In Eq. (C.51), ψ_s for the onset of strong inversion is $2|\psi_b|$, which was given in Eq. (7.18) as

$$\psi_b = -(kT/q)\ln(N_a^-/n_i), \tag{C.57}$$

or

$$\psi_s = 2|\psi_b| = 2(kT/q)\ln(N_a^-/n_i). \tag{C.58}$$

Replacement of ψ_s in Eq. (C.53) for $x_{d_{max}}$ with Eq. (C.58) gives

$$x_{d_{max}} = \sqrt{\left(\frac{\epsilon_{\text{Si}} kT}{q^2 N_a^-}\right) 4\ln(N_a^-/n_i)}, \tag{C.59}$$

where $\sqrt{\epsilon_{\text{Si}} kT/qN_a^-}$ was given in Eq. (C.33) as the Debye length. Therefore, $x_{d_{max}}$ may be written as

$$x_{d_{max}} = 2L_D\sqrt{\ln(N_a^-/n_i)}, \tag{C.60}$$

and

semiconductor capacitance at maximum depletion

$$C'_{sc} = \frac{\epsilon_{\text{Si}}}{2L_D\sqrt{\ln(N_a^-/n_i)}}. \tag{C.61}$$

CHAPTER 8

MOS FIELD-EFFECT TRANSISTORS

8.1 INTRODUCTION

The basic structure of a metal-oxide semiconductor field-effect transistor (MOSFET) is formed by adding two heavily doped n^+ regions to the MOS capacitor on p-type Si as shown in Fig. 8.1 (*a*). The MOSFET is also called the insulated gate FET (IGFET). A commonly used circuit symbol is shown in part (*b*). One n^+ region is designated the source, while the n^+ region where the carriers flow out of the MOSFET is the drain. Application of a positive gate bias produces the conductive path (by electrons) between the source and drain. The gate length is represented by L, while the gate width is represented by W. The bulk semiconductor is called the substrate or MOSFET body. The potential applied to the gate controls the flow of carriers from the source to the drain. Although MOSFET-based integrated circuits are now the dominant technology, numerous inventions and technology breakthroughs were required to overtake bipolar-junction transistor (BJT) integrated circuits.

initial MOSFET development

After the initial exclusive use of BJT integrated circuits in the early 1960s due to instabilities in the gate oxides, the MOSFET in the 1970s began to dominate integrated circuits in both unit volume and sales dollars.* Kahng and Atalla demonstrated the first Si MOSFET in 1960.[1] Two commercial MOSFETs were announced in late 1964, one by Fairchild and the other by RCA. In 1964, Snow et al.[2] showed that sodium ion drift in thermally grown SiO_2 was the principal cause of the threshold voltage instability, and Kerr et al.[3] found that phosphorus silicate glass getters sodium in SiO_2. Balk[4] suggested in 1965 that hydrogen could anneal out the interface traps by tying up the

*This evolution has been carefully documented by C. T. Sah in "Evolution of the MOS Transistor—From Conception to VLSI," Proc. IEEE **76**, 1280 (1988).

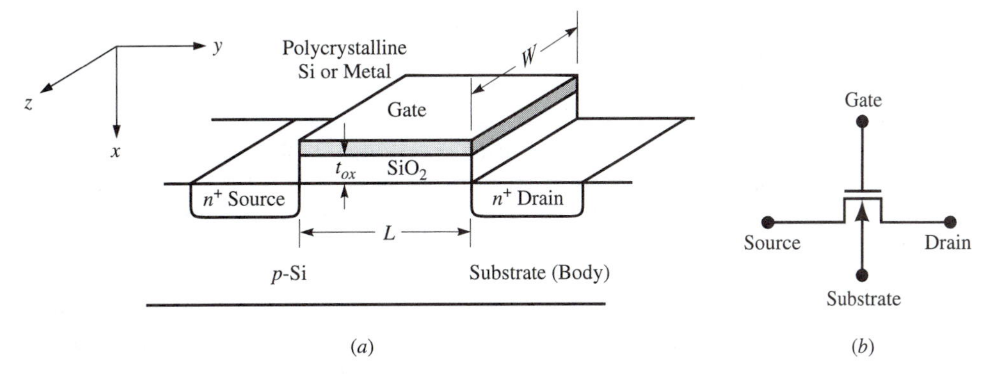

FIGURE 8.1 (*a*) Basic MOSFET structure. (*b*) MOSFET circuit symbol.

dangling Si and SiO_2 bonds. Later it was found that HCl added to oxygen during oxidation incorporates chlorine and apparently immobilizes sodium in SiO_2. In 1966, Sarace et al.[5] developed the polycrystalline Si gate which served as a self-aligning diffusion mask for the source and drain. This structure gave a self-aligned gate electrode and greatly simplified manufacturing processes and increased yield. With these technological advances, Intel developed the first large-volume MOS integrated circuit in 1970. This was a three-transistor, Si-gate, 1-kbit dynamic random-access memory (DRAM). Today DRAM chips have hundreds of millions of MOSFETs.

drain current behavior

The analysis of the MOSFET requires the derivation of the source-to-drain current (I_D) as a function of drain-to-source voltage (V_{DS}) for a given gate-to-source voltage (V_{GS}) when V_{GS} exceeds the threshold voltage (V_T). In Sec. 8.2, I_D is obtained for small V_{DS} to illustrate the analysis procedure. Next, the constraint of small V_{DS} is removed and I_D is obtained for the Shichman-Hodges model. A more complex representation is then given for I_D by the so-called variable-depletion charge model which takes into account the effect of the reverse-biased *p-n* junction space charge at the drain. These I_D vs. V_{DS} characteristics are derived for *p*-type substrates where the minority carriers for inversion are electrons. A qualitative description of drain current flow in the drain current saturation region is also given. Section 8.2 concludes with the use of substrate bias to adjust the threshold voltage.

n-channel
p-channel
enhancement mode
depletion mode

When electrons are the mobile charge carriers for inversion, the MOSFET is termed an *n-channel* MOSFET. Similarly for an *n*-type substrate with holes as the mobile carriers for inversion, the MOSFET is termed a *p-channel* MOSFET. When the MOSFET is off for $V_{GS} = 0$, it is termed an *enhancement-mode* MOSFET. When the MOSFET is on for $V_{GS} = 0$, it is termed a *depletion-mode* MOSFET. The I_D-V_{DS} characteristics for *n*-channel and *p*-channel enhancement- and depletion-mode MOSFETs are summarized in Sec. 8.3.

SPICE

The SPICE MOSFET DC model parameters are introduced and related to the device physics in Sec. 8.4. The element and model lines are first given and then the parameters used to represent the MOSFET are described. Because the description of the parasitic parameters is rather tedious, the SPICE resistance and charge-storage parameters, as well as the additional PSpice parameters, are given in Appendix D at the end of this chapter rather than in the text with the DC parameters. PSpice also includes the Berkeley short-channel IGFET model (BSIM). Parameters for the BSIM model are automatically extracted from the fabrication process.

small-signal AC model

The small-signal AC model is introduced in Sec. 8.5 to illustrate the parameters which influence high-frequency performance. The high-frequency figure of merit represented by the cutoff frequency f_T is also derived in this section.

subthreshold current

When the gate voltage is below the threshold voltage, but weakly inverted, the resulting drain current is called the subthreshold current. The conduction mechanism for the subthreshold current is given in Sec. 8.6 and is different from the conduction mechanism given in Sec. 8.2 for $V_{GS} > V_T$. Also, the theory for the I_D-V_{DS} characteristics presented in Sec. 8.2 does not account for two-dimensional effects when the gate length becomes comparable with the thickness of the source and drain depletion regions. These short-channel effects are briefly discussed in Sec. 8.7.

digital logic applications

CMOS

BiCMOS

Applications of MOSFETs are illustrated by consideration of digital logic circuits in Sec. 8.8. The inverter is first illustrated with an enhancement-mode *n*-channel MOSFET which is turned on when the input voltage exceeds the threshold voltage and is off when the input voltage is less than the threshold voltage. The load element is an *n*-channel depletion-mode MOSFET. Circuits with *n*-channel MOSFETs are called NMOS circuits. The combination of *n*-channel and *p*-channel enhancement-mode MOSFETs to form inverter circuits has become one of the most common circuit technologies due to the low power dissipation. This combination is called complementary MOS or CMOS, and the CMOS inverter is also presented. The combination of bipolar junction transistors (BJTs) with CMOS circuits is called BiCMOS and is also introduced in Sec. 8.8. The BJTs are used to provide a larger output current to drive capacitive loads than can be obtained with CMOS circuits. The concepts introduced in this chapter are summarized in Sec. 8.9 and the expressions used to represent the behavior of MOSFETs are listed. A more complete description of the MOSFET than is possible in this chapter may be found in the text by Tsividis.[6]

8.2 I_D-V_{DS} DC CHARACTERISTICS

8.2.1 Basic MOSFET Model

The basic MOSFET structure was shown in Fig. 8.1. Figure 8.2 illustrates the addition of the source and drain to a MOS capacitor to form the MOSFET. In part (*a*), a heavily doped n^+ region is shown adjacent to the MOS capacitor gate. When the gate voltage V_{GS} exceeds the threshold voltage V_T, then an inversion layer is formed at the SiO_2/Si interface. The n^+-source region can supply electrons to the inversion region without depending on the thermal generation rate as required for the MOS capacitor. In part (*b*), an n^+-drain region has been added so that electrons can flow from the source to the drain through the inversion layer when a positive drain voltage V_{DS} is applied to the drain. This electron flow constitutes the drain current I_D. Current into the drain is taken as a positive current. The positive gate voltage controls the number of electrons in the inversion layer and hence controls the drain current.

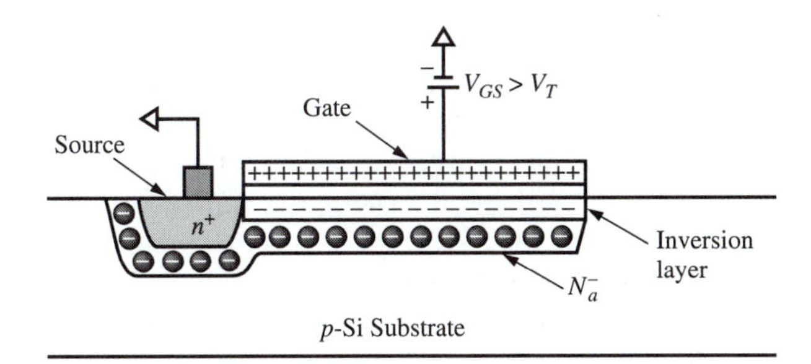

(a)

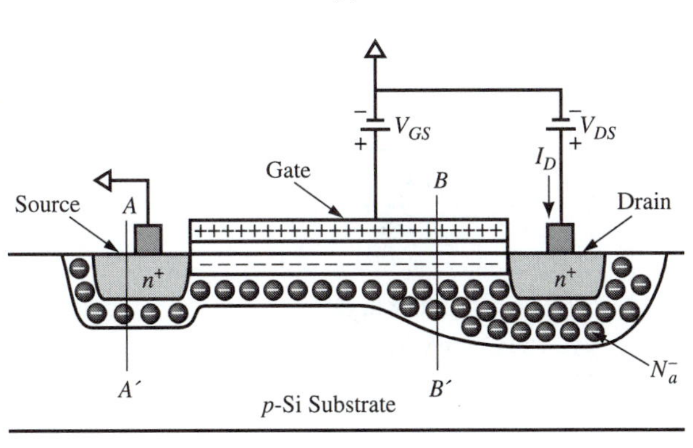

(b)

FIGURE 8.2 Formation of the MOSFET from a MOS capacitor. (*a*) Addition of an n^+-source region for the MOS capacitor biased to give an inversion layer. (*b*) Addition of an n^+-drain region which is biased for electron flow from source to drain. Cross sections in the source are represented by *A*-*A*′ and in the gate region by *B*-*B*′. These cross sections are shown in Figs. 8.3 and 8.4.

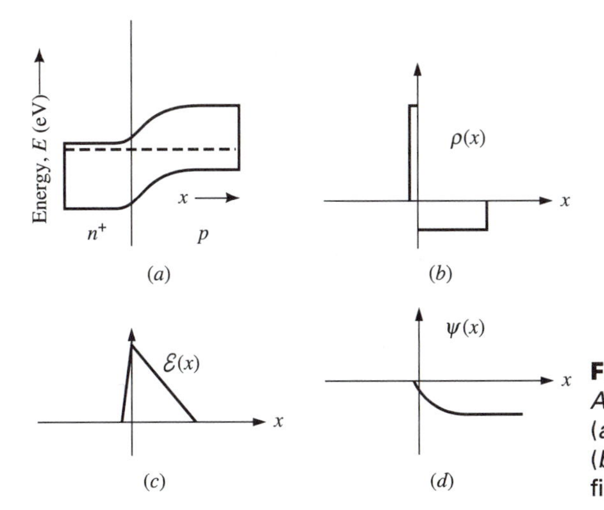

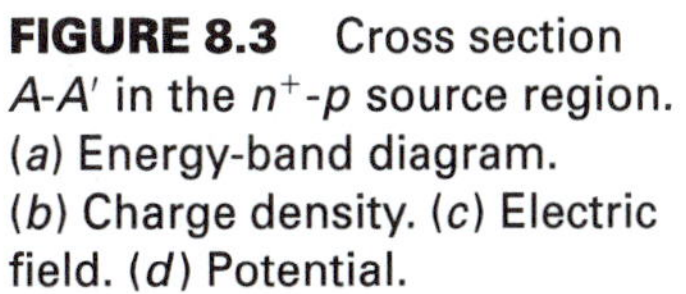

FIGURE 8.3 Cross section *A*-*A*′ in the n^+-p source region. (*a*) Energy-band diagram. (*b*) Charge density. (*c*) Electric field. (*d*) Potential.

Comparison of the energy-band diagram, charge density $\rho(x)$, electric field $\mathscr{E}(x)$ due to gate bias (perpendicular to the electron flow), and potential $\psi(x)$ in the source region marked by A-A' and the gate region marked by B-B' in Fig. 8.2 (b) is useful in describing MOSFET operation. The region A-A' represents an n^+-p junction at zero bias when the source is grounded and the energy-band diagram, $\rho(x)$, $\mathscr{E}(x)$, and $\psi(x)$ are shown in Fig. 8.3. Because the source is more heavily doped than the p-type substrate, most of the depletion region is on the substrate side (Sec. 4.4). The region B-B' is for a MOS capacitor in inversion, and the energy-band diagram, $\rho(x)$, $\mathscr{E}(x)$, and $\psi(x)$ are shown in Fig. 8.4; they were previously given in Fig. 7.12. The x-axis has been relabeled with $x = 0$ at the SiO_2/Si interface. Note that the direction of the perpendicular

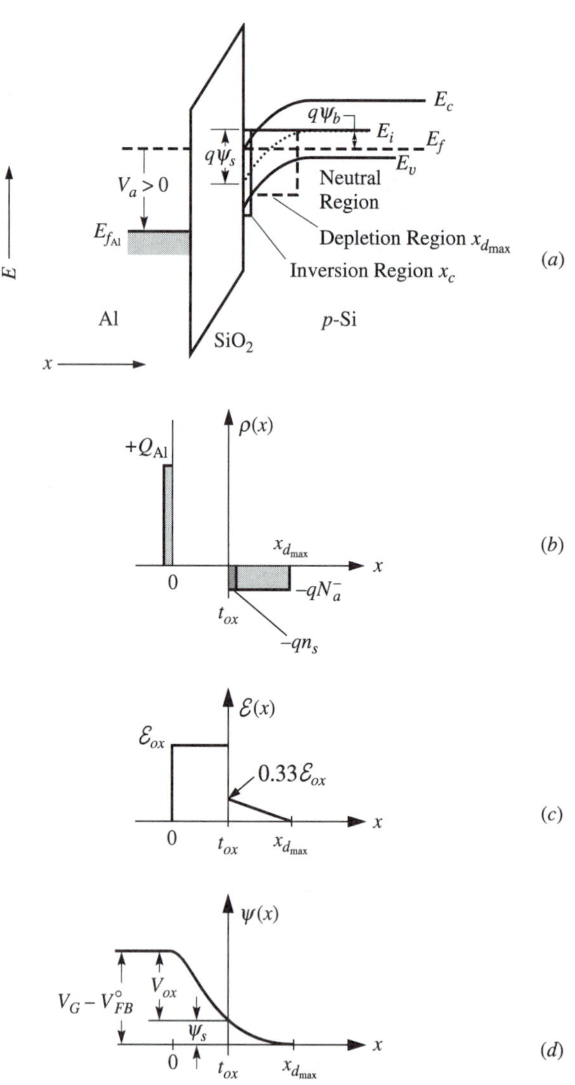

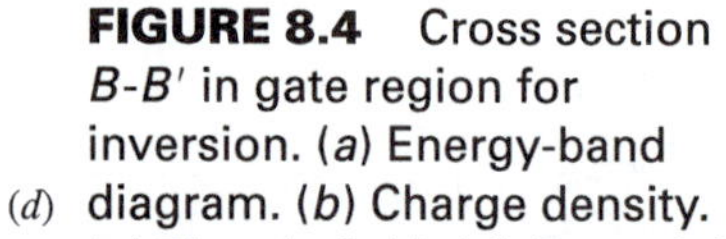
FIGURE 8.4 Cross section B-B' in gate region for inversion. (a) Energy-band diagram. (b) Charge density. (c) Electric field. (d) Potential.

electric field in the semiconductor is from the oxide toward the Si and gives a force on electrons ($F = -q\mathscr{E}$) toward the oxide which confines the electrons to the inversion layer. Aluminum has been used for the gate material for most of the MOSFETs considered in this chapter although the usual gate material is heavily doped polycrystalline Si because it can withstand the subsequent high temperature processing steps. With the p-type Si substrate, the gate material would be n^+-polysilicon. Also, to reduce the sheet resistance of the polysilicon, a refractory metal silicide ($TaSi_2$, $TaSi_2$, $MoSi_2$, or WSi_2) may be used on top of the polysilicon gate. Because the Fermi level, and therefore the work function of the gate, will depend on the impurity concentration of the polysilicon, it is not convenient to use a variable gate work function in a textbook, but V_{FB} can readily be assigned proper values when polysilicon gates are used.

The first derivation of drain current I_D as a function of drain voltage V_{DS} for a given gate voltage V_{GS} considers a *drain voltage that is small enough so that it does not affect the charge Q_n in the inversion layer.* The geometry for this derivation is illustrated in Fig. 8.5 with the y direction along the channel and the x direction perpendicular to the inversion layer as shown. The origin in the y direction for $y = 0$ is at the source, while the gate length is the distance from the source to the drain and is $y = L$. The origin in the x direction for $x = 0$ is at the SiO_2/Si interface, while the interface between the inversion layer and the space-charge region in the substrate is $x = x_i$. The incremental voltage $dV(y)$ across the incremental channel length dy at y, as illustrated in Fig. 8.5, becomes

$$dV(y) = I_D \, dR(y), \tag{8.1}$$

and the incremental resistance $dR(y)$ is

$$dR(y) = \frac{\rho}{W dx} dy, \tag{8.2}$$

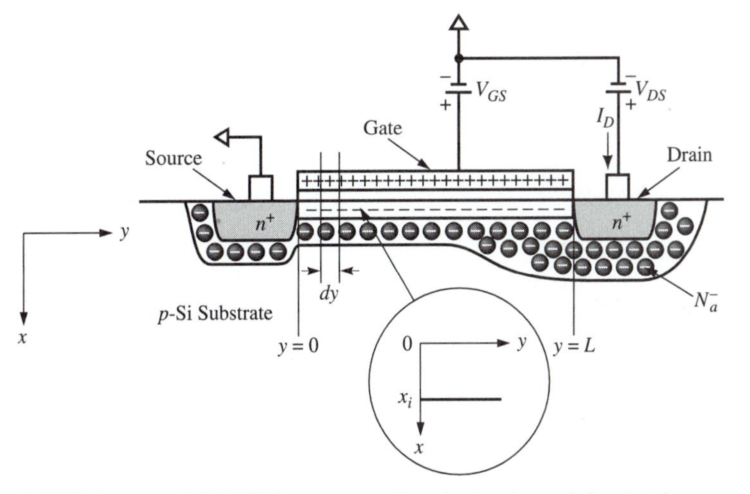

FIGURE 8.5 MOSFET geometry for derivation of the I_D-V_{DS} behavior.

where ρ is the resistivity, dy represents the incremental length, and the product of W and dx represents the cross-sectional area of the incremental volume. It is easier to manipulate the reciprocal of the incremental resistance, which becomes

$$\frac{1}{dR(y)} = \frac{1}{\frac{\rho\, dy}{W\, dx}} = \frac{q\mu_n n W\, dx}{dy}, \tag{8.3}$$

where the resistivity was given in Eq. (3.23). With the assumption of constant mobility in the inversion channel, Eq. (8.3) may be written as

$$\frac{1}{dR(y)} = \frac{W}{dy}\mu_n \int_0^{x_i} qn(x)\, dx, \tag{8.4}$$

$qN'_{inv} \equiv Q'_n$

where the integral represents the charge in the conducting inversion layer per unit area qN'_{inv}, which is generally written as Q'_n for MOSFET analysis. To avoid the complications of electron flow, which is opposite to conventional current flow, the sign of Q'_n will be neglected and Eq. (8.4) becomes

$$\frac{1}{dR(y)} = \frac{W}{dy}\mu_n |Q'_n|, \tag{8.5}$$

or

$$dR(y) = \frac{dy}{W\mu_n |Q'_n|}. \tag{8.6}$$

From Eq. (8.1),

$$dV(y) = I_D\, dR(y) = I_D \frac{dy}{W\mu_n |Q'_n|}, \tag{8.7}$$

or

$$I_D \int_0^L dy = W\mu_n \int_0^{V_{DS}} |Q'_n|\, dV(y). \tag{8.8}$$

Integration of the left-hand side gives the drain current as

basic equation for I_D-V_{DS} behavior

$$\boxed{I_D = \frac{W}{L}\mu_n \int_0^{V_{DS}} |Q'_n|\, dV(y)}. \tag{8.9}$$

Equation (8.9) is the basic equation for the I_D-V_{DS} behavior for various MOSFET models.

The difference between the expressions obtained for the I_D-V_{DS} behavior for (a) the constraint of small V_{DS}, (b) the Shichman-Hodges model, and (c) the variable depletion model is the expression used to represent $|Q'_n|$.

The electron mobility appears in Eq. (8.9), and this mobility for the carriers in the conducting channel between the source and drain is less than the bulk mobility given in Fig. 3.3. From both experimental and theoretical studies,[7] it has been found that the mobility for MOSFETs scales with an average or effective field perpendicular to Si/SiO_2 interface. This perpendicular field ($\mathscr{E}_\perp$) is related to the inversion charge (field at the Si/SiO_2 interface) and the field at the boundary of the inversion layer and the depletion layer. As the field $\mathscr{E}_\perp$ increases due to increases in the channel or depletion charge, the channel mobility decreases due to greater scattering at the Si/SiO_2 interface. The smaller mobility results in a smaller I_D. The mobility may be represented as

$$\mu = \frac{\mu_o}{1 + \Theta(V_{GS} - V_T)}, \tag{8.10}$$

where μ_o is the mobility at small $(V_{GS} - V_T)$ and Θ is an empirical parameter. This expression in Eq. (8.10) is used in SPICE (Sec. 8.4). For evaluation of the numerical expressions for I_D given in this section, the mobility will be taken as approximately μ(bulk)/2.

8.2.2 Small $V_{DS} \leq 0.1V$

An expression for $|Q'_n|$ may be obtained from the gate voltage V_{GS} for inversion, which was given in Eq. (7.109), and may be written as

$$V_{GS} = V_{FB} + \frac{|Q'_{Si}|}{C'_{ox}} + \psi_s = V_{FB} + \frac{|Q'_{sc} + Q'_n|}{C'_{ox}} + \psi_s, \tag{8.11}$$

where the total charge per unit area in the Si is the sum of the depletion-region space charge Q'_{sc} and the charge in the inversion layer Q'_n. The inversion layer charge may be written from Eq. (8.11) as

$$|Q'_n| = C'_{ox}\left\{V_{GS} - \left[V_{FB} + \frac{|Q'_{sc}|}{C'_{ox}} + \psi_s\right]\right\}. \tag{8.12}$$

To use this expression for Q'_n, it is necessary to assign a relationship between V_{GS} and ψ_s. With the plot of $|Q'_{Si}|$ from Fig. 7.20, V_{GS} for a given ψ_s may be obtained from Eq. (8.11). Table 7.1 gives ψ_s vs. V_{GS} for accumulation and depletion. For inversion, Q'_{Si} may be obtained from Eq. (C.43) and was plotted in Fig. 7.20. The resulting plot of ψ_s vs. V_{GS} is shown in Fig. 8.6.

It may be seen in Fig. 8.6 that ψ_s remains relatively constant with V_{GS} once inversion is reached. Therefore, in Eq. (8.12) it is convenient to let $\psi_s = 2|\psi_b|$, and $|Q'_n|$ may be written as

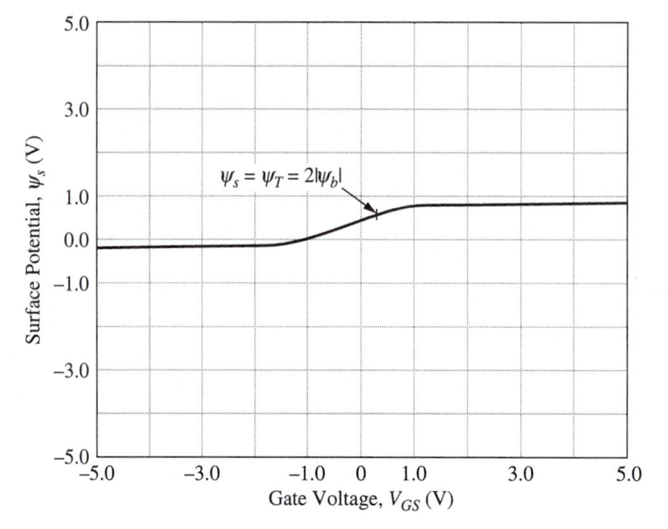

FIGURE 8.6 Variation of the surface potential with gate voltage for a substrate with $p = N_a^- = 4 \times 10^{15}$ cm^{-3} and SiO_2 thickness $t_{ox} = 500$ Å. The onset of strong inversion is indicated by $\psi_s = \psi_T = 2|\psi_b|$.

$$|Q_n'| = C_{ox}'\{V_{GS} - \underbrace{[V_{FB} + 2|\psi_b| + \sqrt{4q\epsilon_{\text{Si}}N_a^-|\psi_b|}/(\epsilon_{ox}/t_{ox})]}_{V_T}\}, \tag{8.13}$$

where Q_{sc}' was given in Eq. (7.45) and V_T is the threshold voltage given in Eq. (7.78). Equation (8.13) becomes

$$|Q_n'| = C_{ox}'(V_{GS} - V_T), \tag{8.14}$$

and Eq. (8.9) gives the drain current as

$$I_D = \frac{W}{L}\mu_n C_{ox}' \int_0^{V_{DS}} (V_{GS} - V_T)\, dV(y). \tag{8.15}$$

Integration of Eq. (8.15) gives

I_D-V_{DS} behavior for small V_{DS}

$$\boxed{I_D = (W/L)\mu_n C_{ox}'(V_{GS} - V_T)V_{DS}} \tag{8.16}$$

for the constraint that V_{DS} is sufficiently small that it does not influence $|Q_n'|$. For this case, the drain current varies linearly with the drain voltage, and the more the gate voltage exceeds the threshold voltage, the larger the drain current. This I_D-V_{DS} behavior is sketched in Fig. 8.7.

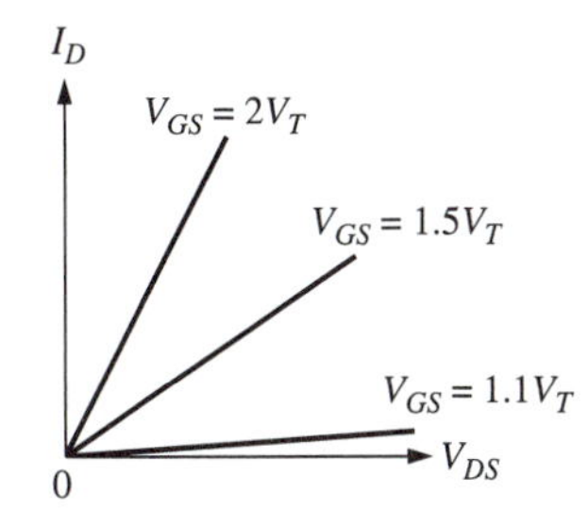

FIGURE 8.7 The I_D-V_{DS} behavior for the condition of small V_{DS}.

8.2.3 SPICE Level 1 MOSFET Model

In the previous part of this section, the I_D-V_{DS} characteristics were derived for small V_{DS}. "Small" V_{DS} means that V_{DS} is large enough for drain current flow, but small enough compared to V_{GS} so that V_{DS} does not affect the charge Q'_n in the inversion layer. The potential distribution along the channel $V(y)$ is linear and the drain current I_D varies linearly with V_{DS} as given by Eq. (8.16). As V_{DS} is increased, it begins to affect the channel charge Q'_n. The *Level 1 model*[8] used in SPICE considers the effect of V_{DS} on Q'_n and is frequently referred to as the Shichman and Hodges model.[9]

In Fig. 8.8 (*a*), an n^+-p junction such as the MOSFET drain is shown in reverse bias to emphasize that as V_{DS} is increased, the depletion-layer thickness will increase. This depletion tends to reduce the inversion due to V_{GS} so that Q'_n is no longer constant and decreases from the source to the drain. Part (*b*) of Fig. 8.8 illustrates the reduction of Q'_n between the source and drain due to the potential along the channel. The potential variation along the channel $V(y)$ is shown in Fig. 8.8 (*c*) and is defined as the difference between the potential at the semiconductor surface at the oxide and the potential V_{BS} at the substrate contact. At the source, $V(0) = V_S = 0$, while at the drain $V(L) = V_{DS} \pm V_{BS}$, which in this case is for $V_{BS} = 0$. This change in channel charge is represented by $C'_{ox}V(y)$, and the charge in the conducting channel previously given in Eq. (8.14) now becomes

substrate bias V_{BS}

$$|Q'_n(y)| = C'_{ox}[(V_{GS} - V_T) - V(y)]. \tag{8.17}$$

Note that current continuity requires that I_D is constant in the channel between the source and drain. The current in the conducting channel remains constant because as $|Q'_n(y)|$ decreases from the source to the drain, the longitudinal field $\mathscr{E}(y)$ due to V_{DS} increases as $|Q'_n(y)|$ decreases. The field $\mathscr{E}(y)$ is given by $-dV(y)/dy$ with $V(y)$ illustrated in Fig. 8.8 (*c*). In this manner, the drift current $[\propto |Q'_n(y)|\mathscr{E}(y)]$ remains constant between the source and drain.

The I_D-V_{DS} behavior may be found for this model with Eq. (8.17) in Eq. (8.9):

$$I_D = \frac{W}{L}\mu_n C'_{ox}\int_0^{V_{DS}} [(V_{GS} - V_T) - V(y)]\, dV(y), \tag{8.18}$$

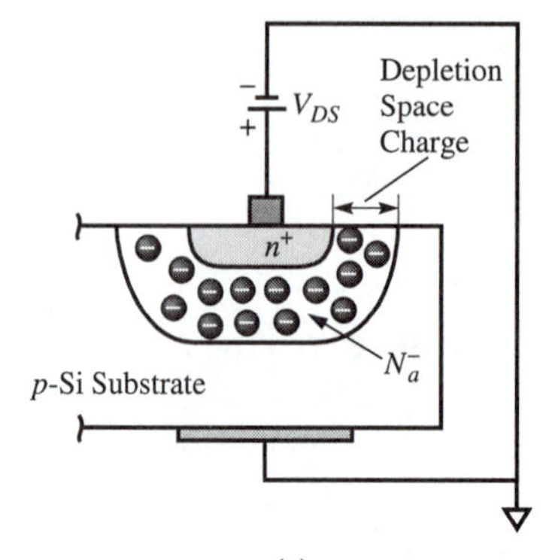

(a)

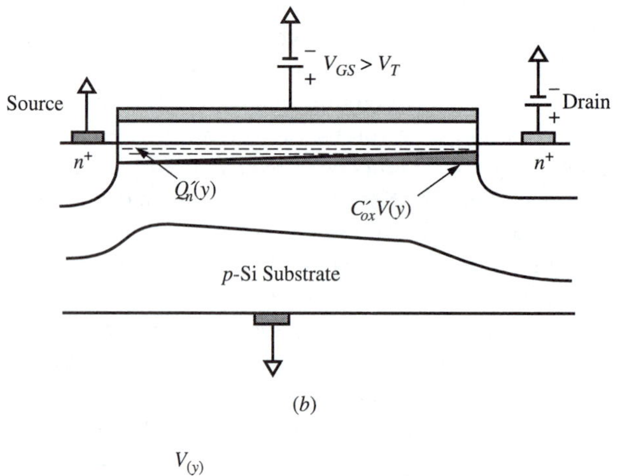

(b)

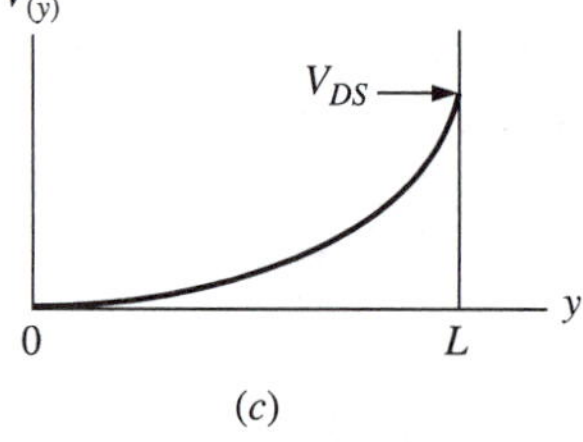

(c)

FIGURE 8.8 Effect of the drain voltage V_{DS} on the inversion channel charge Q_n'. (a) Depletion region for a reverse-biased n^+-p drain region. (b) Reduction in Q_n' due to the potential along the channel. (c) Variation of the drain voltage $V(y)$ along the inversion channel.

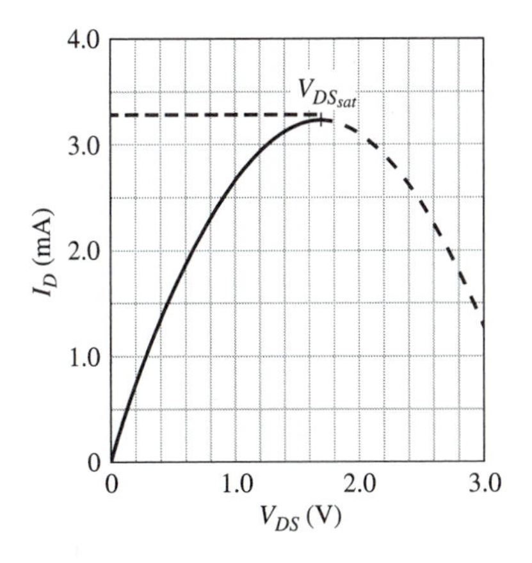

FIGURE 8.9 Drain current as a function of drain voltage given by Eq. (8.20), which is valid only up to the maximum indicated at $V_{DS} = V_{DS_{sat}}$.

which gives

$$I_D = \frac{W}{L}\mu_n C'_{ox}[(V_{GS} - V_T)V(y) - V(y)^2/2]\Big|_0^{V_{DS}}, \tag{8.19}$$

or

SPICE level 1 I_D-V_{DS} behavior

$$\boxed{I_D = \frac{W}{L}\mu_n C'_{ox}[(V_{GS} - V_T) - V_{DS}/2]V_{DS}}. \tag{8.20}$$

Equation (8.20) is only valid up to values of V_{DS} as represented by the solid line in Fig. 8.9 for I_D at $V_{DS} = V_{DS_{sat}}$, which occurs at the maximum of I_D where $dI_D/dV_{DS} = 0$. From Eq. (8.20),

$$\frac{dI_D}{dV_{DS}} = 0 = (W/L)\mu_n C'_{ox}\underbrace{[(V_{GS} - V_T) - V_{DS}]}_{0}, \tag{8.21}$$

which gives

SPICE level 1 drain saturation voltage $V_{DS_{sat}}$

$$\boxed{V_{DS_{sat}} = (V_{GS} - V_T)}, \tag{8.22}$$

and is known as the *drain saturation voltage* $V_{DS_{sat}}$. The current $I_{D_{sat}}$ at $V_{DS_{sat}}$ is given with Eq. (8.22) in Eq. (8.20) as

$$I_{D_{sat}} = (W/L)\mu_n C'_{ox}[(V_{GS} - V_T)(V_{GS} - V_T) - (V_{GS} - V_T)^2/2], \tag{8.23}$$

or

SPICE level 1 drain saturation current $I_{D_{sat}}$

$$\boxed{I_{D_{sat}} = \frac{W}{2L}\mu_n C'_{ox}(V_{GS} - V_T)^2}. \tag{8.24}$$

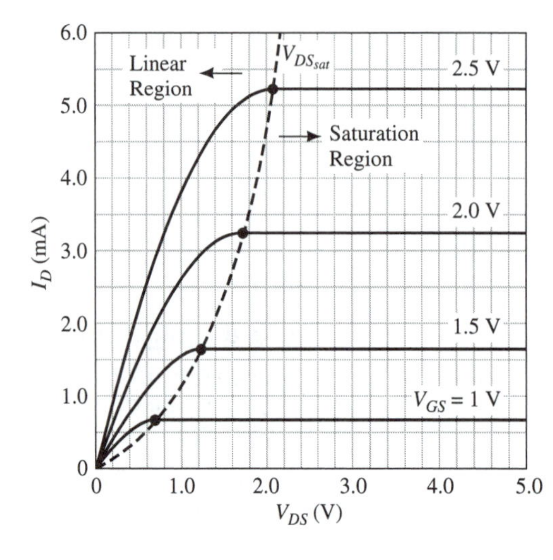

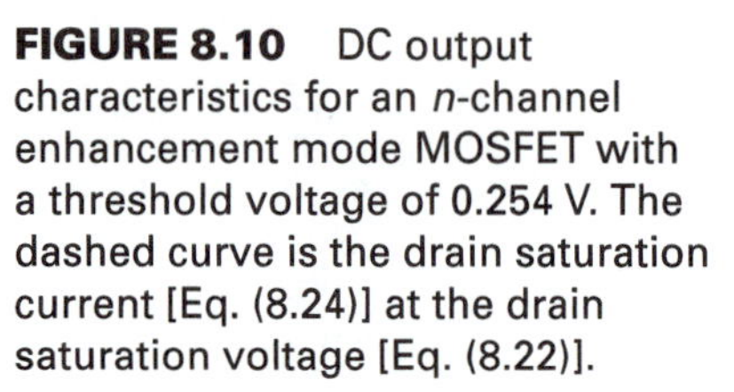
FIGURE 8.10 DC output characteristics for an n-channel enhancement mode MOSFET with a threshold voltage of 0.254 V. The dashed curve is the drain saturation current [Eq. (8.24)] at the drain saturation voltage [Eq. (8.22)].

linear region
saturation region
transfer characteristic

The complete I_D-V_{DS} DC output characteristic for the n-channel enhancement-mode MOSFET is illustrated in Fig. 8.10 with I_D calculated by Eq. (8.20) for $V_{DS} \leq V_{DS_{sat}}$ with $V_{DS_{sat}}$ given by Eq. (8.22). For $V_{DS} > V_{DS_{sat}}$, I_D has been taken constant at the value of $I_{D_{sat}}$. The dashed curve is $I_{D_{sat}}$ vs. $V_{DS_{sat}}$ as given by Eq. (8.24). The region for I_D with $V_{DS} < V_{DS_{sat}}$ is known as the *linear* region as indicated in Fig. 8.10, while the region for I_D with $V_{DS} > V_{DS_{sat}}$ is known as the *saturation* region where I_D becomes independent of drain bias. The relationship between the output drain current I_D and the input gate voltage V_{GS} has been called the *transfer characteristic* and is best represented by $\sqrt{I_{D_{sat}}}$ vs. V_{GS} because with $I_{D_{sat}}$ given by Eq. (8.24), $\sqrt{I_{D_{sat}}}$ goes to zero at $V_{GS} = V_T$. The plot of $\sqrt{I_{D_{sat}}}$ vs. V_{GS} is shown in Fig. 8.11 for the output characteristic given in Fig. 8.10.

The saturation condition of $V_{DS_{sat}} = (V_{GS} - V_T)$ represents the condition of the drain voltage becoming large enough to prevent inversion at the SiO_2/Si interface at

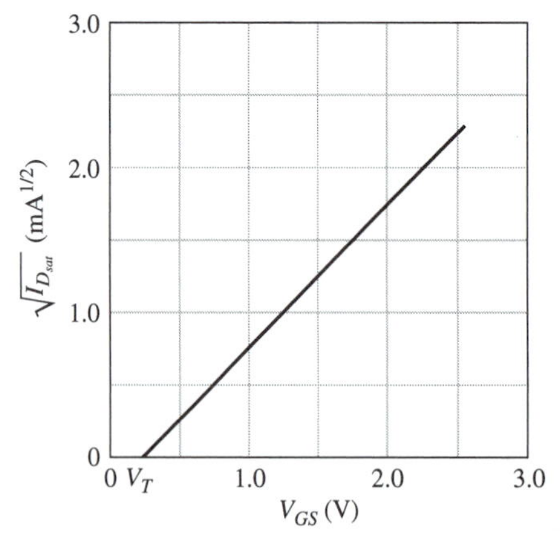

FIGURE 8.11 Transfer characteristic represented by $\sqrt{I_{D_{sat}}}$ vs. V_{GS} for the n-channel enhancement-mode MOSFET in Fig. 8.10 obtained from Eq. (8.24) for $V_T = 0.254$ V.

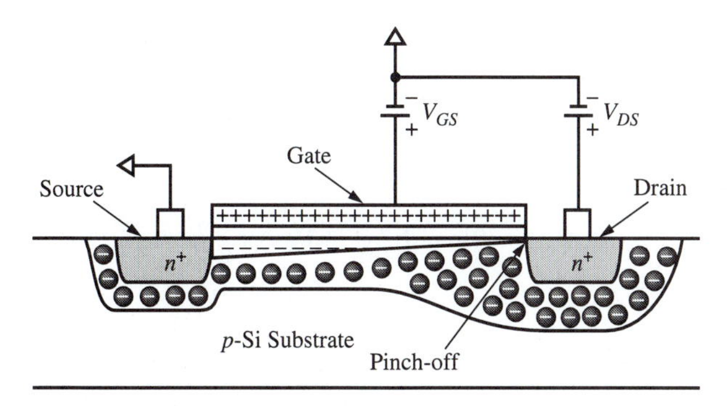

FIGURE 8.12 Illustration of pinch-off at the drain when $V_{DS} > V_{DS_{sat}} = (V_{GS} - V_T)$.

the drain. *Note that* $V_{DS_{sat}} = (V_{GS} - V_T)$ *is the value of* $V(y)$ *in Eq. (8.17) at which* $|Q'_n| = 0$. This effect is illustrated in Fig. 8.12 where the inversion layer is shown to diminish to zero at the drain; this condition is called *pinch-off*. The n^+-p junction at the drain is reverse biased and its depletion region opposes the inversion due to the gate voltage when $V_{DS} > (V_{GS} - V_T)$. An obvious question is, why does the drain current not go to zero for V_{DS} larger than pinch-off for $V_{DS} > V_{DS_{sat}}$? For a MOSFET operating in the saturation regime, the model used to obtain Eq. (8.20) becomes less valid near the drain where the potential distribution in the space-charge region becomes two-dimensional as a result of the interaction of the gate and drain electrodes. This effect is discussed further in Sec. 8.2.5. Not only is this a two-dimensional problem, but both carrier drift and diffusion must be considered.

pinch-off

8.2.4 SPICE Level 2 MOSFET Model

The SPICE level 1 MOSFET model given in Sec. 8.2.3 introduced the effect of V_{DS} on the mobile inversion layer charge Q'_n, but ignored the effect of V_{DS} on the space-charge Q'_{sc}, composed of ionized acceptors N_a^- for p-type Si. The SPICE level 2 MOSFET model includes the effect of V_{DS} on Q'_{sc} and is called the variable-depletion charge analysis. The resulting I_D-V_{DS} behavior was determined by Ihantola and Moll[10] and Meyer.[11] This model, which includes the effect of the variable space charge on I_D, could be referred to as the Ihantola and Moll model. Several assumptions are part of this model: (1) the effect of fixed oxide charge is contained in V_{FB}; (2) only drift current is considered; (3) the use of Eq. (8.9) assumes a constant mobility; (4) the impurity concentration is uniform in the channel region; (5) reverse n^+-p junction leakage current is negligibly small; and (6) the transverse field $\mathscr{E}_x$ (perpendicular to the current flow) in the channel is much larger than $\mathscr{E}_y$ (along the direction of current flow) in the channel. Note that $0 < V_{GS} < V_{DS}$, but x_i, the conducting channel thickness, is very small compared to the channel length L. Then, $\mathscr{E}_x \approx V_{GS}/x_i > V_{DS}/L \approx \mathscr{E}_y$. This condition on $\mathscr{E}_y$ applies to long-channel MOSFETs and is called the *gradual-channel approximation* and applies to both this model and the Shickman-Hodges model.

variable-depletion analysis

gradual-channel approximation

The potential as a function of distance $\psi(x)$ was obtained from Gauss's law for the MOS capacitor in Sec. 7.2.4. In Eq. (7.32), $\psi(0)$ at the SiO_2/Si interface for the MOSFET now includes the effect of V_{DS} and becomes $\psi_s + V(y) = 2|\psi_b| + V(y)$ to give

$$\psi(x=0) = 2|\psi_b| + V(y) = \frac{qN_a^-}{2\epsilon_{Si}} x_d^2, \tag{8.25}$$

and

$$x_d = \sqrt{2\epsilon_{Si}[2|\psi_b| + V(y)]/qN_a^-}. \tag{8.26}$$

The space charge, given for the MOS capacitor in Eqs. (7.43) and (7.45), for the MOSFET may now be written as

$$Q'_{sc} = -qN_a^- x_d = -\sqrt{2q\epsilon_{Si}N_a^-[2|\psi_b| + V(y)]}. \tag{8.27}$$

Equation (7.78) for the threshold voltage becomes

$$V_T = V_{FB} + 2|\psi_b| + \sqrt{2q\epsilon_{Si}N_a^-[2|\psi_b| + V(y)]}/(\epsilon_{ox}/t_{ox}), \tag{8.28}$$

and the channel charge given in Eq. (8.17) with V_T replaced by Eq. (8.28) becomes

$$Q'_n(y) = -\frac{\epsilon_{ox}}{t_{ox}}[(V_{GS} - V_{FB} - 2|\psi_b|) - V(y) - \sqrt{2q\epsilon_{Si}N_a^-[2|\psi_b| + V(y)]}/(\epsilon_{ox}/t_{ox})]. \tag{8.29}$$

It is more convenient to integrate $|Q'_n|$ when rewritten as

$$Q'_n(y) = -C'_{ox}[(V_{GS} - V_{FB} - 2|\psi_b|) - V(y) - \sqrt{2q\epsilon_{Si}N_a^-}\sqrt{2|\psi_b| + V(y)}/C'_{ox}]. \tag{8.30}$$

The I_D-V_{DS} behavior is given by Eq. (8.9) with Eq. (8.30) for $|Q'_n|$. Note that $\int x^{1/2}dx = (2/3)x^{3/2}$.

$$I_D = \frac{W}{L}\mu_n C'_{ox}\left\{(V_{GS} - V_{FB} - 2|\psi_b|)V(y) - \frac{V(y)^2}{2} - \frac{2}{3}\frac{\sqrt{2q\epsilon_{Si}N_a^-}}{C'_{ox}}\underbrace{[2|\psi_b| + V(y)]^{3/2}}_{(V_{DS}+2|\psi_b|)^{3/2}-(2|\psi_b|)^{3/2}}\Bigg|_0^{V_{DS}}\right\}. \tag{8.31}$$

The I_D-V_{DS} behavior for the SPICE level 2 model may be written as

SPICE level 2 I_D-V_{DS} behavior

$$I_D = \frac{W}{L}\mu_n C'_{ox}\left\{\left(V_{GS} - V_{FB} - 2|\psi_b| - \frac{V_{DS}}{2}\right)V_{DS} - \frac{2}{3}\frac{\sqrt{2q\epsilon_{Si}N_a^-}}{C'_{ox}}\left[(V_{DS} + 2|\psi_b|)^{3/2} - (2|\psi_b|)^{3/2}\right]\right\}. \tag{8.32}$$

Equation (8.32) for the SPICE level 2 model, which includes the effect of V_{DS} on both the mobile-channel charge and the space charge, has become significantly more complex than the SPICE level 1 model, which only considers the effect of V_{DS} on the mobile-channel charge.

Saturation for I_D is reached when the pinch-off condition occurs as illustrated in Fig. 8.12 and Q'_n at $y = L$ as given in Eq. (8.30) goes to zero for $V(y) = V(L) = V_{DS_{sat}}$:

$$Q'_n(L) = 0 = -C'_{ox}[(V_{GS} - V_{FB} - 2|\psi_b|) - V_{DS_{sat}}] + \sqrt{2q\epsilon_{Si}N_a^-(2|\psi_b| + V_{DS_{sat}})}. \tag{8.33}$$

This equation may be solved for $V_{DS_{sat}}$ by dividing through by C'_{ox} and squaring:

$$[\underbrace{(V_{GS} - V_{FB} - 2|\psi_b|)}_{A} - V_{DS_{sat}}]^2 = \underbrace{\frac{1}{C'^2_{ox}}[q\epsilon_{\mathrm{Si}}N_a^-}_{K^2}(4|\psi_b| + 2V_{DS_{sat}})], \tag{8.34}$$

which may be written in terms of A and K^2 as

$$(A - V_{DS_{sat}})^2 - K^2(4|\psi_b| + 2V_{DS_{sat}}) = 0. \tag{8.35}$$

Equation (8.35) may be arranged in the form of a quadratic equation:

$$V^2_{DS_{sat}} - 2(A + K^2)V_{DS_{sat}} + (A^2 - 4K^2|\psi_b|) = 0, \tag{8.36}$$

which gives,

$$V_{DS_{sat}} = (A + K^2) - \sqrt{(A + K^2)^2 - (A^2 - 4K^2|\psi_b|)}, \tag{8.37}$$

or

$$V_{DS_{sat}} = A + K^2 - K^2\sqrt{1 + \frac{2}{K^2}\underbrace{(A + 2|\psi_b|)}_{V_{GS}-V_{FB}-2|\psi_b|+2|\psi_b|}}. \tag{8.38}$$

Replacing A and K^2 gives

SPICE level 2 drain saturation voltage $V_{DS_{sat}}$

$$V_{DS_{sat}} = (V_{GS} - V_{FB} - 2|\psi_b|) + \frac{q\epsilon_{\text{Si}}N_a^-}{C_{ox}'^2}\left[1 - \sqrt{1 + \frac{2(V_{GS} - V_{FB})}{q\epsilon_{\text{Si}}N_a^-/C_{ox}'^2}}\right]. \quad (8.39)$$

A qualitative discussion of the drain current behavior in the saturation region is given in the next part of this section. In the saturation region, the two-dimensional solution of both Poisson's equation and the continuity equation for minority carriers with both drift and diffusion must be considered.

8.2.5 Drain Current in the Saturation Region

For a MOSFET operating in the saturation region, the gradual-channel approximation, which assumes $\mathscr{E}_y << \mathscr{E}_x$, becomes less valid as the drain region is approached in the conducting channel, and the potential distribution in the space-charge region becomes two dimensional as a result of the interaction of the gate and drain electrodes. The saturation voltage was given for the SPICE level 1 model in Eq. (8.22) and for the SPICE level 2 model in Eq. (8.39), and saturation at pinch-off occurs for $V_{DS} > V_{GS}$. For this condition, the *sign* of the oxide electric field normal to the SiO_2/Si interface will change at the pinch-off point. Also, for $V_{DS} > V_{DS_{sat}}$, the position of the pinch-off point will move toward the source as V_{DS} is increased.

Armstrong et al.[12] obtained the conducting channel potentials from an iterative solution of the current-continuity equation. Their results for the variation of the electric field $\mathscr{E}_{ox}(x)$ in the oxide and the potential along the conducting channel $V(y)$ are shown in Fig. 8.13 to illustrate this reversal in the direction of $\mathscr{E}_{ox}(x)$. Armstrong and

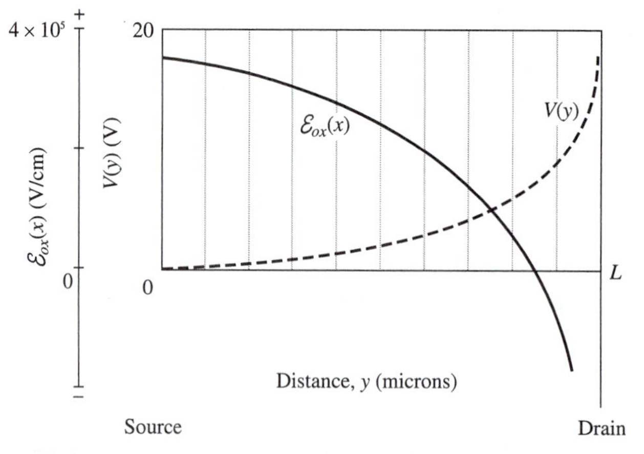

FIGURE 8.13 Variation of the channel voltage $V(y)$ and the normal component of the electric field $\mathscr{E}_{ox}(x)$ in the oxide between the source and drain.[12]

Magowan[13] showed that at the onset of pinch-off, the *transverse* drift and diffusion currents are approximately two orders of magnitude greater than the longitudinal drift current. Transverse current flow only becomes significant where the normal component of the substrate field inverts in sign. They obtained two-dimensional plots of the minority carrier concentrations at cross sections from the source to the drain and illustrated the absence of minority carriers at the SiO_2/Si interface, but significant minority carrier concentrations in the depletion regions between pinch-off and the drain.[14] Therefore, the two-dimensional continuity equation, which includes both drift and diffusion, is necessary in this region between the pinch-off point and the drain region.

A plot of current flow and the electric fields for conditions of saturation was obtained by Taylor[15] and is shown in Fig. 8.14. The distance y' is the distance from the source to the pinch-off point, and the distance the pinch-off point has moved toward the source is represented by ΔL. As the drain voltage increases, the position of the pinch-off point will move toward the source. Note in Fig. 8.14 that at pinch-off, the field in the oxide reverses and the current flows (by diffusion and drift) in the depletion region between pinch-off and the drain. This flow is not at the SiO_2/Si interface. El-Mansy and Boothroyd[16] found it convenient to analyze the MOSFET in saturation by considering the source section by the gradual-channel approximation and with a variable channel length obtained from a solution in the drain section. This model permitted determination of the I_D-V_{DS} behavior in the saturation region, which agreed with the experimental measurements. To analyze current flow in the pinch-off region, it is convenient to use

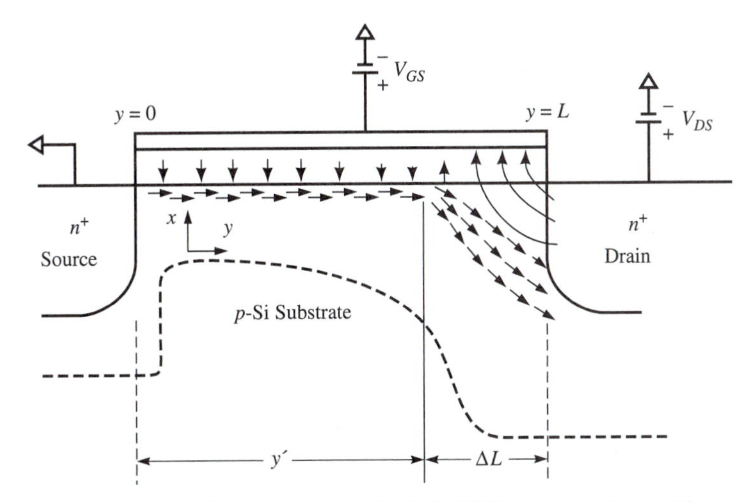

FIGURE 8.14 Cross section of a MOSFET in saturation with horizontal arrows representing the current flow and the vertical arrows representing the direction of the electric field. The dashed line represents the edge of the depletion region. The distance y' is the pinch-off point measured from the source and ΔL is the distance from the drain to the pinch-off point.[15]

two-dimensional device simulators, such as PISCES developed by Pinto et al.[17], which are now available.

8.2.6 Substrate Bias

In deriving the I_D-V_{DS} characteristics, the substrate bias V_{BS} has been taken as zero; however, V_{BS} may be included in the expressions for I_D.[18] Note that the substrate bias illustrated in Fig. 8.15 with the negative bias connected to the p-type substrate provides a reverse bias on the source n^+-p junction and an increased reverse bias on the drain n^+-p junction. A simplified representation of the effect of V_{BS} on the threshold voltage may be obtained by replacing $V(y)$ by V_{BS} in Eqs. (8.25)–(8.28) to give

$$V_T = V_{FB} + 2|\psi_b| + \sqrt{2q\epsilon_{\mathrm{Si}}N_a^-(2|\psi_b| + V_{BS})}/C'_{ox}. \tag{8.40}$$

Then, the change in threshold voltage due to the substrate bias is

$$\Delta V_T = \frac{\sqrt{2q\epsilon_{\mathrm{Si}}N_a^-}}{C'_{ox}}\left(\sqrt{2|\psi_b| + V_{BS}} - \sqrt{2|\psi_b|}\right). \tag{8.41}$$

body effect

The influence of substrate bias is like a second gate and is called the *body effect*.

With the level 2 SPICE model, the I_D-V_{DS} behavior is more conveniently obtained with SPICE than by numerical evaluation of Eq. (8.32), as was illustrated in Fig. 8.10 for the level 1 SPICE model with Eq. (8.20). The effect of substrate bias can also more conveniently be determined by SPICE than by adding more terms to already lengthy expressions. The expression of ΔV_T in Eq. (8.41) does illustrate that the threshold voltage can be adjusted by V_{BS}, and V_T varies approximately as $\sqrt{V_{BS}}$, when V_{BS} becomes larger than $2|\psi_b|$.

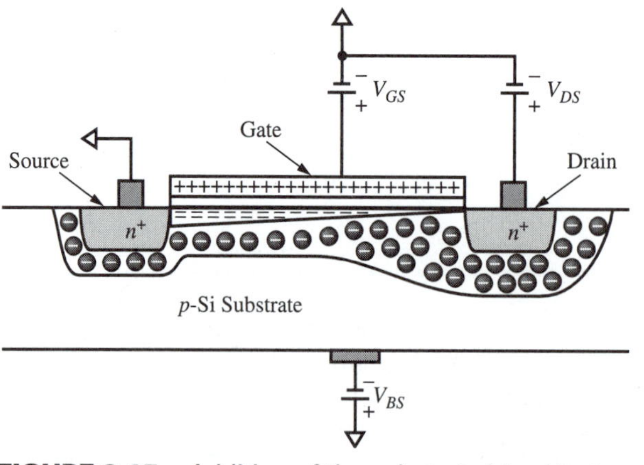

FIGURE 8.15 Addition of the substrate bias V_{BS} to the MOSFET.

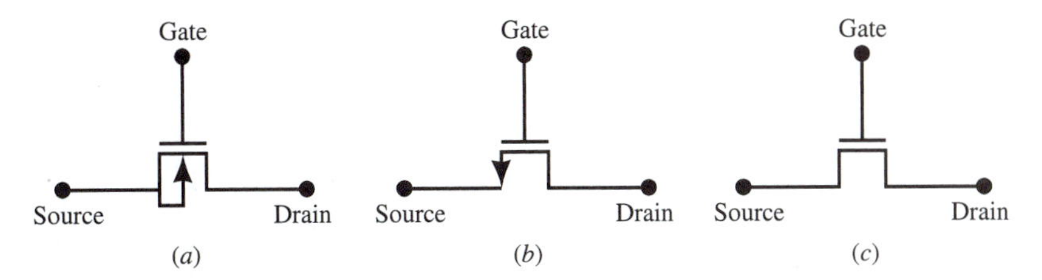

FIGURE 8.16 Circuit symbols for the *n*-channel enhancement-mode MOSFET, which differ from the circuit symbol given in Fig. 8.1. (*a*) The arrow indicates the substrate is *p*-type. (*b*) Arrow indicates direction of current flow. (*c*) Simplified circuit symbol.

8.3 TYPES OF MOSFETS

As presented in Sec. 8.2 when the mobile charge carriers were electrons, the transistor is termed an *n-channel* MOSFET. When the MOSFET is off for $V_{GS} = 0$, it is termed an *enhancement-mode* (normally off) MOSFET. When the MOSFET is on for $V_{GS} = 0$, it is termed a *depletion-mode* (normally on) MOSFET. For an *n*-type substrate with holes as the mobile carriers, the MOSFET is termed a *p-channel* MOSFET. In this section, the schematic symbol, the output characteristics, and the transfer characteristics are summarized.

For the *n*-channel enhancement-mode MOSFET, the circuit symbol was given in Fig. 8.1, the DC output characteristics were given in Fig. 8.10, and the transfer characteristic with $\sqrt{I_D}$ vs. V_{GS} was given in Fig. 8.11. Several other circuit symbols are used to represent this device and are shown in Fig. 8.16.

The *n*-channel depletion-mode MOSFET is normally on for $V_{GS} = 0$ and is generally obtained by implanting *n*-type impurities into the *p*-type substrate to give a threshold voltage in the range from -1 V to -4 V. Ion implantation[19] provides controllable and reproducible impurity distributions in thin layers. For ion implantation, acceptor elements such as B or donor elements such as As are vaporized and ionized, accelerated, and directed at the Si substrate. The energetic impurity ions enter the Si lattice, collide with Si atoms, gradually lose energy, and come to rest within the lattice. The average depth is determined by the acceleration voltage. The amount of impurities, called the *dose,* may be controlled by measuring the ion current during implantation. Subsequent heat treatment removes the lattice damage caused by the ion collisions. Doses are near 10^{12} ions/cm^2 for threshold adjustment. If Q_c' is the electrically active implanted charge per unit area in the channel region due to the implant, the threshold voltage is shifted by approximately $-Q_c'/C_{ox}'$ for ionized positive donors:

$$\Delta V_T \simeq -\frac{Q_c'}{C_{ox}'}. \tag{8.42}$$

Because of the complicated shape of the implant profile, the detailed analysis is very involved. Further discussion has been given by Tsividis.[20] The schematic symbol for depletion-mode devices is often represented by a heavier line, as shown in Fig. 8.17.

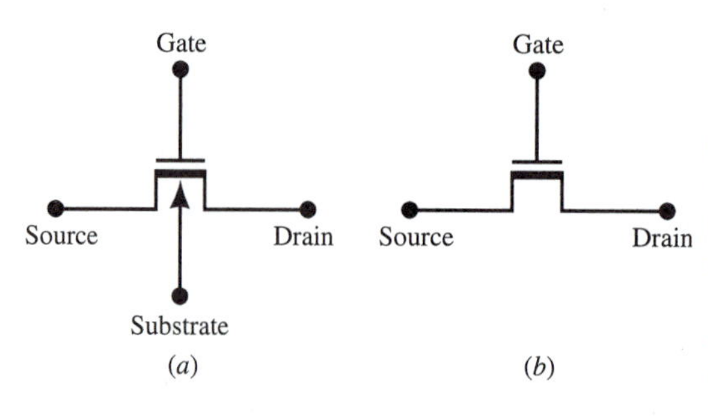

FIGURE 8.17 Circuit symbols for the *n*-channel depletion-mode MOSFET. The arrow indicates that the substrate is *p*-type. (*a*) Circuit symbol with a contact to the substrate. (*b*) Simplified symbol with wider line only.

MOSFETs prepared on *n*-type substrates have minority-carrier holes as the mobile charge carriers for inversion. The threshold voltage for the *p*-channel device was given in Eq. (7.79) and is negative for the normally off enhancement-mode MOSFET. The normally on depletion-mode *p*-channel MOSFET will have a thin implanted layer of acceptors and a threshold voltage between +1 and +3 V.

The MOSFET cross sections, circuit symbols, output characteristics (I_D vs. V_{DS}), and transfer characteristics (I_D vs. V_{GS}) are summarized in Fig. 8.18. The arrow for the

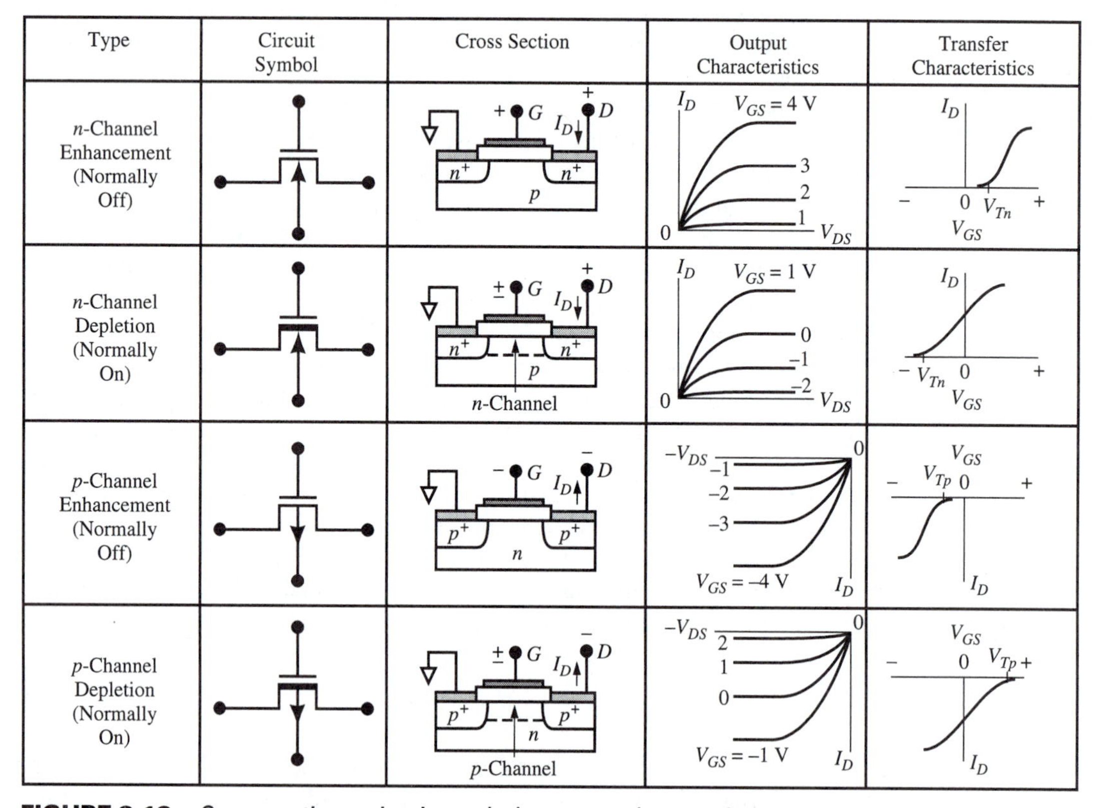

FIGURE 8.18 Cross sections, circuit symbols, output characteristics, and transfer characteristics for the four types of MOSFETs.[21]

substrate is directed to the gate for the n-channel MOSFET with a p-type substrate and away from the gate for the p-channel MOSFET with an n-type substrate. The drain current I_D is taken as positive when I_D flows into the drain. For the transfer characteristics shown in Fig. 8.18, the source-to-drain voltage V_{DS} has been selected so that for smaller gate-to-source voltages V_{GS}, the drain current I_D is in the linear region and in the saturation region at higher values of V_{GS}. For these conditions, the S-shaped transfer characteristic as shown for the n-channel enhancement-mode MOSFET is obtained. For a normally on depletion-mode n-channel MOSFET, current flows at $V_{GS} = 0$, and the current may be increased or decreased by varying the gate voltage.

8.4 SPICE MODEL

8.4.1 Element and Model Lines

A very complete text on MOSFET modeling with SPICE by Foty[22] provides an in-depth presentation and should be consulted for further description. That text includes HSPICE, which is a high-performance version run with UNIX. Also, the SPICE parameters are grouped as *physical* parameters and *electrical* parameters. The physical parameters are related to SPICE parameters such as the substrate doping concentration and the oxide thickness. The electrical parameters, such as the threshold voltage, are obtained from test-device measurements. Helpful PSpice descriptions have been given in texts such as those by Rashid[23] and by Banzhaf.[24] Here, the use of SPICE for the MOSFET will follow the description given by Banzhaf.[24] The general form of the *element line* for a MOSFET is

`MOSFET` *element line*

```
MXXXXXX ND NG NS NB MODNAME <L=VAL> <W=VAL> <AD=VAL>
+<AS=VAL> <PD=VAL> <PS=VAL> <NRD=VAL> <NRS=VAL> <OFF>
+<IC=VDS,VGS,VBS>
```

where `MXXXXXX` is the name of the MOSFET, `ND`, `NG`, `NS`, and `NB` are the drain, gate, source, and bulk (or substrate) nodes, respectively. `MODNAME` is the model name which is used in an associated `.MODEL` control line. These items are required in the MOSFET element line. The optional parameters are the quantities in $< \cdots >$, and an element line may be continued by entering a `+` sign at the start of the next line. The meanings of the optional parameters are

`L` length of the channel, in meters.
`W` width of the channel, in meters.
`AD` area of the drain diffusion, in square meters.
`AS` area of source diffusion, in square meters.

If any of these parameters are not specified, default values are used. The remaining optional parameters and their definitions are

PD	perimeter of drain junction, in meters.
PS	perimeter of source junction, in meters.
NRD	number of squares for the drain region.
NRS	number of squares for the source region.
OFF	indicates an initial condition for DC analysis.
IC=	sets the initial conditions for V_{GS}, V_{DS}, and V_{BS} for use with a transient analysis.

NRD and NRS multiply the sheet resistance RSH specified on the .MODEL line to accurately represent the source and drain resistance.

The model form or the *model lines* for the MOSFET are:

MOSFET .MODEL *line*

```
.MODEL MODNAME NMOS <(PAR1=PVAL1 PAR2=PVAL2 ...)>
.MODEL MODNAME PMOS <(PAR1=PVAL1 PAR2=PVAL2 ...)>
```

MODNAME is the model name given to a MOSFET in the element line, and NMOS or PMOS denotes whether the MOSFET is an *n*-channel or a *p*-channel device. PAR is the parameter name of one of the optional parameters listed in Table 8.1. PVAL is the value of the designated parameter. Care must be taken to assign the correct units which are also designated in these tables.

8.4.2 SPICE2 and PSpice DC Parameters

The model of Shichman and Hodges presented in Sec. 8.2.3 had simple enough expressions to permit determination of the I_D-V_{DS} characteristics by hand calculations and is the model designated the *SPICE level 1* model. The variable-depletion charge analysis for the I_D-V_{DS} characteristics given in Sec. 8.2.4 resulted in considerably more complex expressions. This model is the *SPICE level 2* model. The *SPICE level 3* model was developed to simulate short-channel MOSFETs.[25] The equations for the level 3 model are formulated in the same way as for the level 2 model; however, simplifications have been made to allow more manageable equations. Many of the equations used in this model are empirical. The books by Foty[22] and by Massobrio and Antognetti[26] should be consulted for further description of the MOSFET SPICE parameters.

Presentation of the MOSFET SPICE parameters has been broken into four separate parts due to many possible parameter choices. The first group (Table 8.1) begins with the designation of level 1, 2, or 3 for the MOSFET and includes those parameters necessary to obtain the DC characteristics. Although necessary for circuit simulation, the description of the parasitic parameters is very tedious, and therefore, these additional 25 parameters are listed and described in Appendix D at the end of this chapter. Table D.1 lists the parasitic parameters, Table D.2 gives the two noise parameters, and Table D.3 gives the additional parasitic parameters used in PSpice.

The parameters necessary for the DC characteristics are given in Table 8.1. Parameter no. 1 is the LEVEL which must be specified in the .MODEL line or the default level 1 model will be used. The DC characteristics are given by the device parameters VTO, GAMMA, PHI, KP, and LAMBDA. The second parameter VTO in Table 8.1

TABLE 8.1 SPICE2 and PSpice MOSFET DC Model Parameters.

No.	Text Symbol	SPICE Keyword	Level	Parameter Name	Default Value	Units
1	—	LEVEL	1–3	SPICE model 1, 2 or 3	1	—
2	V_T	VTO	1–3	Zero-bias threshold voltage	0.0	V
3	γ	GAMMA	1–3	Bulk space-charge parameter	0.0	$V^{0.5}$
4	ψ_s	PHI	1–3	Surface potential	0.6	V
5	KP	KP	1–3	Transconductance parameter	2.0E-5	A/V^2
6	λ	LAMBDA	1, 2	Channel-length modulation	0	V^{-1}
7	t_{ox}	TOX	1–3	Gate-oxide thickness	1.0E-7	meter
8	N_b	NSUB	1–3	Substrate doping	0.0	cm^{-3}
9	N_f	NSS	2, 3	Fixed oxide charge	0.0	cm^{-2}
10	N_{it}	NFS	2, 3	Interface-trapped charge	0.0	cm^{-2}
11	—	TPG	2, 3	Type of gate material +1 opp. to substrate −1 same as substrate 0 Al gate	1	—
12	μ	UO	1–3	Surface mobility	600	cm^2/Vs
13	U_c	UCRIT	2	Critical electric field for mobility	1E4	V/cm
14	U_e	UEXP	2	Exponential coefficient for mobility	0.0	—
15	U_t	UTRA	2	Transverse field coefficient	0.0	—
16	x_j	XJ	2, 3	Source or drain junction depth	0.0	meters
17	x_{jl}	LD	1–3	Lateral diffusion	0.0	meters
18	v_{max}	VMAX	2, 3	Maximum carrier drift velocity	0.0	meters/s
19	N_{eff}	NEFF	2	Total channel charge coefficient	1	—
20	δ	DELTA	2, 3	Width effect on threshold voltage	0.0	—
21	η	ETA	3	Static feedback on threshold voltage	0.0	—
22	V_{bi}	PB	1–3	Source and drain junction built-in potential	0.80	V
23	θ	THETA	3	Mobility modulation	0.0	—
24	κ	KAPPA	3	Saturation field factor	0.2	—

is the threshold voltage without substrate bias as presented in Sec. 7.3.6. The parameter VTO is positive (negative) for enhancement mode and negative (positive) for depletion mode n-channel (p-channel) devices. With substrate bias, the threshold voltage includes the expression from Eq. (8.41):

VTO

$$V_T = V_{TO} + \frac{\sqrt{2q\epsilon_{\mathrm{Si}}N_a^-}}{C'_{ox}}\left(\sqrt{2|\psi_b| + V_{BS}} - \sqrt{2|\psi_b|}\right). \tag{8.43}$$

The third parameter in the table is GAMMA and is the coefficient in Eq. (8.43):

GAMMA

$$\gamma = \frac{\sqrt{2q\epsilon_{\mathrm{Si}}N_a^-}}{C'_{ox}}. \tag{8.44}$$

Parameter no. 4 also appears in Eq. (8.43) as

PHI

$$\psi_s = 2|\psi_b|, \tag{8.45}$$

which was given in Sec. 7.2.4 as the surface potential for the onset of strong inversion. Parameters 5 and 6 can be illustrated by rewriting Eq. (8.24) as

$$I_{D_{sat}} = KP\frac{W}{2L}(V_{GS} - V_T)^2(1 + \lambda V_{DS}), \tag{8.46}$$

where

KP

$$KP = \mu C'_{ox}. \tag{8.47}$$

LAMBDA

The quantity $(1 + \lambda V_{DS}D)$ is an empirical correction which gives a finite slope (conductance) to the I_D-V_{DS} characteristic due to channel length modulation (see Figs. 8.26 and 8.29) in the saturation region and is represented by parameter no. 6, `LAMBDA`.

TOX

NSUB

NSS & NFS

TPG

UO

SPICE will compute the device parameters `VTO`, `GAMMA`, `PHI`, and `KP` if the process parameters nos. 7–12 are given. However, user-specified values for `VTO`, `GAMMA`, `PHI`, and `KP` always override the computed quantities. Note that for parameter no. 7 for the oxide thickness t_{ox} is specified in meters. Parameter no. 8 is the substrate doping N_a^- for *n*-channel devices and N_d^+ for *p*-channel devices. The fixed oxide charge (parameter no. 9) and the interface trapped charge (parameter no. 10) were presented in Sec. 7.3.6 and influence the threshold voltage. Parameter no. 11 permits three choices for specification of the gate material which determines the flat-band voltage and hence the threshold voltage. The transconductance parameter [`KP`$=\mu(\epsilon_{ox}/t_{ox})$] may be obtained by specification of channel (surface) mobility `UO`, which is parameter no. 12, and the oxide thickness (parameter no. 7).

The next three parameters (13–15) only apply to the level 2 model and represent the effect of the electric field perpendicular to the conducting channel on the saturation of the carrier mobility as described for Eq. (8.10). This effect is included in SPICE by modifying the `KP` parameter by the expression:

$$KP' = KP\left(\frac{\epsilon_{Si}}{\epsilon_{ox}}\frac{U_c t_{ox}}{V_{GS} - V_T - U_t V_{DS}}\right)^{U_e}. \tag{8.48}$$

UCRIT

UEXP

UTRA

In Eq. (8.48), U_c is the gate-to-channel critical field represented as `UCRIT` in parameter no. 13, and has a typical value of 1×10^4 V/cm. The term $(V_{GS} - V_T - U_t V_{DS})/t_{ox}$ represents the average field perpendicular to the channel. The exponent U_e, which has typical values of 0.1, is parameter no. 14, `UEXP`. The term U_t (parameter no. 15) represents the effect of the drain voltage on the perpendicular electric field and has values between 0 and 0.5.

XJ

LD

In processing integrated circuits, even with a self-aligned gate technology, the formation of the source and drain regions by diffusion or ion implantation can often result in the lateral extension of these regions under the gate electrode during processing at elevated temperatures, as illustrated in Fig. 8.19. The source and drain junction depth is x_j, and the lateral diffusion of the source or drain under the gate electrode is x_{jl}. Thus,

$$L_{eff} = L - 2x_{jl}. \tag{8.49}$$

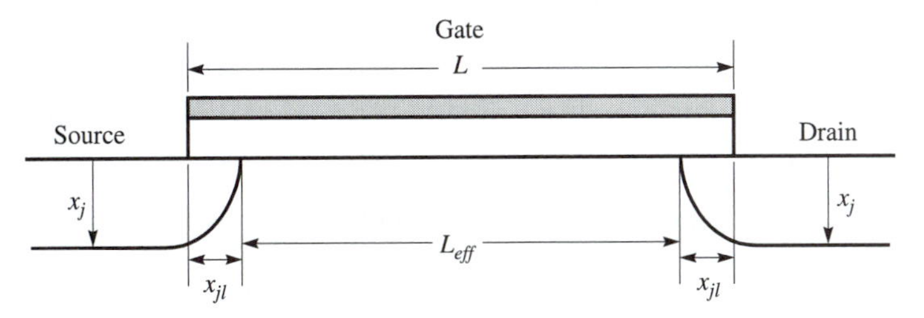

FIGURE 8.19 Representation of the overlap of the source and drain regions with the gate electrode.

In the previous expressions for I_D in Secs. 8.2.3 and 8.2.4, the quantity L for the gate length must be replaced by L_{eff}. These parameters, XJ and LD are parameters no. 16 and no. 17 in Table 8.1.

Several remaining parameters are SPICE modifications to the basic equations. The electric field in the depletion region at the drain can be sufficiently high for the carriers to reach their maximum velocity v_{max}, as presented in Sec. 3.8, and v_{max} is parameter 18, VMAX. This parameter affects the channel charge and the channel length modulation.[27] The next parameter, N_{eff}, is used as a fitting parameter to modify N_a^- by replacing N_a^- by $N_a^- N_{eff}$ in expressions related to depletion-space charge.[27]

VMAX

NEFF

When the gate width W is less than 4 or 5 μm, the threshold voltage is larger than the value given by the usual expression in Eq. (7.78) due to the two-dimensional distribution of the charge at the edges of the channel. An additive term which includes the parameter δ is added to the usual expression for threshold voltage,[28] and is parameter no. 20, DELTA. The effect of the channel *width* on the threshold voltage for the level 2 or 3 model is expressed as

DELTA

$$V_T = V_{FB} + 2|\psi_b| + \frac{\sqrt{2q\epsilon_{\text{Si}} N_a^-(2|\psi_b| + V_{BS})}}{C'_{ox}} + \frac{\epsilon_{\text{Si}}\delta\pi}{4C'_{ox}W}(2|\psi_b| + V_{BS}). \quad (8.50)$$

The effect of the channel *length* on the threshold voltage for the level 3 model is given by an equation proposed by Dang,[25,29]

$$V_T = V_{FB} + 2|\psi_b| - \sigma V_{DS} + F_s \frac{\sqrt{2q\epsilon_{\text{Si}} N_a^-(2|\psi_b| + V_{BS})}}{C'_{ox}} + \frac{\epsilon_{\text{Si}}\delta\pi}{2C'_{ox}W}(2|\psi_b| + V_{BS}), \quad (8.51)$$

where σ is given by

ETA

$$\sigma = \eta \frac{8.15 \times 10^{-22}}{C'_{ox} L^3_{eff}}. \quad (8.52)$$

Parameter no. 21 in Table 8.1 is η and would have a typical value of 1. The correction factor F_s represents the effect of the short channel and is given as

$$F_s = 1 - \frac{x_j}{L_{eff}}\left[\frac{x_{jl} + W_c}{x_j}\sqrt{1 - \left(\frac{W_p}{x_j + W_p}\right)^2} - \frac{x_{jl}}{x_j}\right], \tag{8.53}$$

where W_p is the depleted region on the bottom of the source junction:

$$W_p = \sqrt{\frac{2\epsilon_{Si}(V_{bi} + V_{BS})}{qN_a^-}}. \tag{8.54}$$

In Eq. (8.54), V_{bi} is the built-in potential of the source n^+-p junction, and is parameter no. 22. Equation (4.60) represented the built-in potential as

PB

$$V_{bi} = (kT/q)\ln(N_a^- N_d^+ / n_i^2). \tag{8.55}$$

The quantity W_c in Eq. (8.53) is given by the empirical expression[29]

$$\frac{W_c}{x_j} = 0.06313292 + 0.8013292\frac{W_p}{x_j} - 0.01110777\left(\frac{W_p}{x_j}\right)^2. \tag{8.56}$$

The quantities x_j and x_{jl} were illustrated in Fig. 8.19.

The next parameter in Table 8.1, no. 23, also applies to the level 3 model. The empirical equations used in the level 3 model are intended to improve the precision of the model and to shorten the computer computation time. The current in the linear region is given by[29]

$$I_D = \mu C'_{ox}\frac{W}{L_{eff}}\left(V_{GS} - V_T - \frac{1 + F_B}{2}V_{DS}\right)V_{DS}, \tag{8.57}$$

where

$$F_B = \frac{\sqrt{2\epsilon_{Si} q N_a^-}}{C'_{ox}} \times \frac{F_s}{4\sqrt{2|\psi_b| + V_{BS}}} + \frac{\epsilon_{Si}\delta\pi}{2C'_{ox}W}, \tag{8.58}$$

and the correction factor F_s was given in Eq. (8.53). The mobility dependence on the perpendicular electric field for the level 3 model was given in Eq. (8.10) as

THETA

$$\mu = \frac{\text{UO}}{1 + \theta(V_{GS} - V_T)}, \tag{8.59}$$

where θ is the mobility modulation parameter and is parameter no. 23 in Table 8.1.

One difference between the SPICE level 1 and level 2 models occurs when the gate length becomes less than 10 μm. For these shorter gate lengths, the pinch-off point at saturation moves toward the source for $V_{DS} > V_{DS_{sat}}$ and reduces the channel length

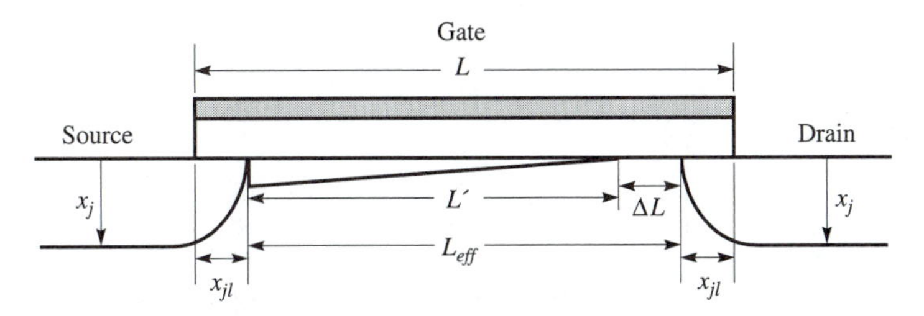

FIGURE 8.20 Reduction in channel length for a MOSFET when the drain voltage moves the pinch-off point toward the source by the amount ΔL. This effect can be expected for gate lengths less than 10 μm.

from L_{eff} by ΔL as illustrated in Fig. 8.20. Model parameter no. 6 in Table 8.1, LAMBDA, may be specified to give L' in the level 2 model as

$$L' = L_{eff} - \texttt{LAMBDA} \times V_{DS}. \tag{8.60}$$

The channel-length modulation parameter λ typically has values of 0.05 to 0.005 V^{-1} and represents the effect of V_{DS} on I_{DS} as $(1 + \lambda V_{DS})$ in Eq. (8.46).

Variation of the channel length in the saturation region for the level 3 model is simulated with the expression[30]

$$L_{eff} - L' = \sqrt{\left(\frac{\mathscr{E}_p \epsilon_{\text{Si}}}{qN_a^-}\right)^2 + \kappa \frac{2\epsilon_{\text{Si}}}{qN_a^-}(V_{DS} - V_{DS_{sat}})} - \frac{\mathscr{E}_p \epsilon_{\text{Si}}}{qN_a^-}. \tag{8.61}$$

In Eq. (8.61), L_{eff} was given in Eq. (8.49), L' was given in Eq. (8.60), and the lateral electric field at the pinch-off point is

$$\mathscr{E}_p = \kappa \frac{I_{D_{sat}}}{g_{D_{sat}} L_{eff}}, \tag{8.62}$$

KAPPA

where KAPPA is parameter no. 24, and $g_{D_{sat}}$ is the output conductance given by $g_{D_{sat}} \equiv \partial I_D / \partial V_{DS}$ at a fixed V_{GS}.

Examples are useful to illustrate the use of SPICE to obtain the MOSFET DC characteristics. Additional SPICE2 examples for MOSFETs have been given by Pulfrey and Tarr.[31]

EXAMPLE 8.1 An n-channel enhancement mode MOSFET has the following parameters:

substrate: $p = N_a^- = 4 \times 10^{15}$ cm^{-3}
oxide thickness: $t_{ox} = 50$ nm
channel mobility: $\mu_n = 600$ cm^2/Vs
Al gate

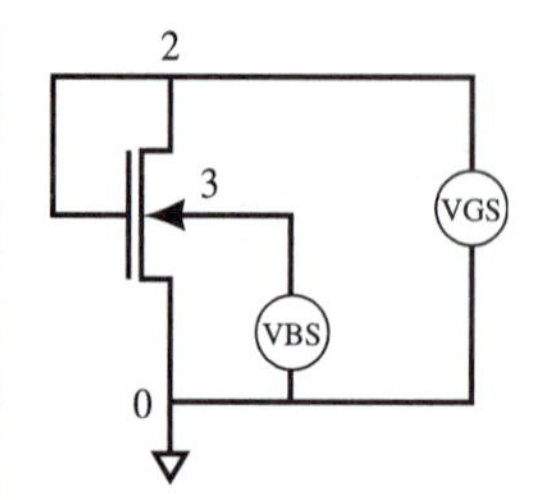

FIGURE 8.21 Circuit for Example 8.1.

gate length: $L = 3\ \mu\text{m}$

gate width: $W = 12\ \mu\text{m}$.

With the level 1 model, obtain V_T with PSpice in the saturation region by plotting $\sqrt{I_D}$ as V_{GS} is varied from 0 V to +2 V in 0.01 V-steps.

Solution To ensure saturation, which was given by Eq. (8.22) as $V_{DS_{sat}} = (V_{GS} - V_T)$, the gate is connected to the drain so that $V_{DS} > V_{DS_{sat}}$. Note that $\sqrt{I_{D_{sat}}}$ given in Eq. (8.24) goes to zero for $V_{GS} = V_T$. The circuit diagram with the node numbers is shown in Fig. 8.21. The PSpice input file is

```
*MOSFET THRESHOLD VOLTAGE
VGS 2 0
VBS 3 0 DC 0
MOS1 2 2 0 3 MOSVT L=3U W=12U
.MODEL MOSVT NMOS LEVEL=1 NSUB=4E15
+U0=600 TOX=50E-9 TPG=0
.DC VGS 0 2 0.01
.PROBE
.END
```

The plot of $\sqrt{I_D}$ as a function of V_{GS} is shown in Fig. 8.22. Inclusion of the substrate bias permits observing the effect of V_{BS} on the threshold voltage as given in Sec. 8.2.6.

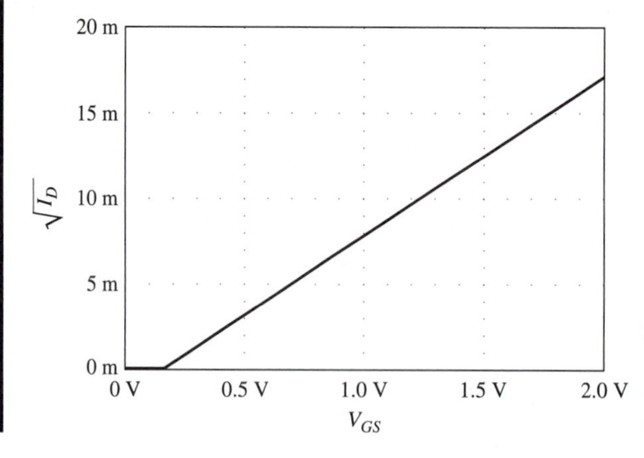

FIGURE 8.22 Variation of $\sqrt{I_D}$ with V_{GS} to illustrate that the $\sqrt{I_D} = 0$ intercept is V_T in Example 8.1.

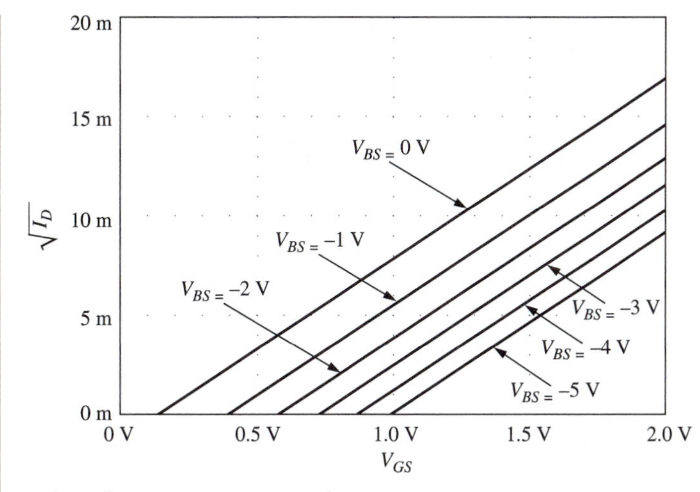

FIGURE 8.23 Variation of V_T with the substrate bias V_{BS}.

A family of curves with V_{BS} as the parameter may be obtained with PSpice by changing the `.DC` control line to `.DC VGS 0 2 0.01 VBS 0 -5 -1`. The plot of $\sqrt{I_D}$ for V_{BS} varied from 0 to -5 V in steps of 1 V is shown in Fig. 8.23 and illustrates the variation of V_T with V_{BS}. ■

Another example illustrates the differences in the DC I_D-V_{DS} characteristics obtained for the level 1 and level 2 SPICE models. For channel lengths near 10 μm and larger, the I_D-V_{DS} characteristics are represented by Eq. (8.20) for the level 1 model and Eq. (8.32) for the level 2 model. In this long-channel case, the drain current in the saturation region is less for the level 2 model than for the level 1 model. Examination of Eq. (8.29) for the channel charge Q_n' shows that for the same V_{GS}, Q_n' becomes less (and hence smaller I_D) because the space charge Q_{sc}' in Eq. (8.27) increases with $V(y)$. Therefore, more of the negative charge in the Si is now space charge rather than channel charge.

EXAMPLE 8.2 Compare the I_D-V_{DS} characteristics for the level 1 and level 2 SPICE models for the long-channel case. Use the device parameters from Example

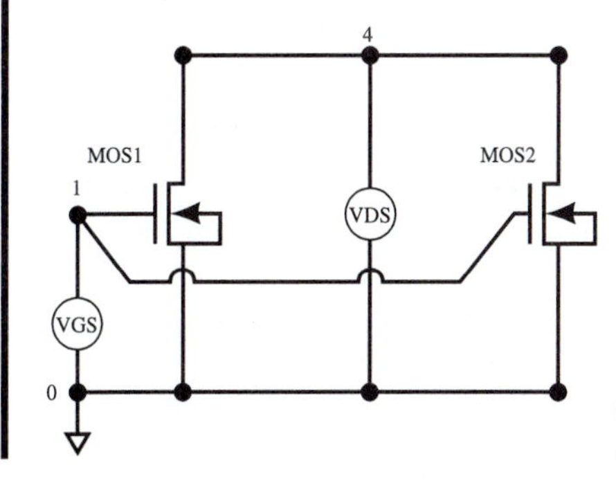

FIGURE 8.24 Circuit for the I_D-V_{DS} comparison for the SPICE level 1 and level 2 models.

8.1, except let the gate length $L = 10\ \mu$m and the gatewidth $W = 40\ \mu$m. Let V_{DS} vary from 0 to 5 V and V_{GS} vary from 0 to 5 V in 1-V steps.

Solution The circuit diagram with the node numbers is shown in Fig. 8.24. This SPICE input file will generate the desired I_D-V_{DS} characteristics.

```
*MOSFET I-V CHARACTERISTIC LEVEL 1 AND 2 MODEL
VDS 4 0
VGS 1 0
MOS1 4 1 0 0 NMOS1 L=10U W=40U
MOS2 4 1 0 0 NMOS2 L=10U W=40U
.MODEL NMOS1 NMOS LEVEL=1 NSUB=4E15
+UO=600 TOX=50E-9 TPG=0
.MODEL NMOS2 NMOS LEVEL=2 NSUB=4E15
+UO=600 TOX=50E-9 TPG=0
.DC VDS 0 5 0.1 VGS 0 5 1
.PROBE
.END
```

The resulting I_D-V_{DS} characteristics are shown in Fig. 8.25 and illustrate the effect of including the variable-space charge due to V_{DS} in the level 2 model. The drain current in the saturation region is greater for the level 1 model than for the level 2 model: $I_{D_{sat}}$(level 1)$>$ $I_{D_{sat}}$(level 2).

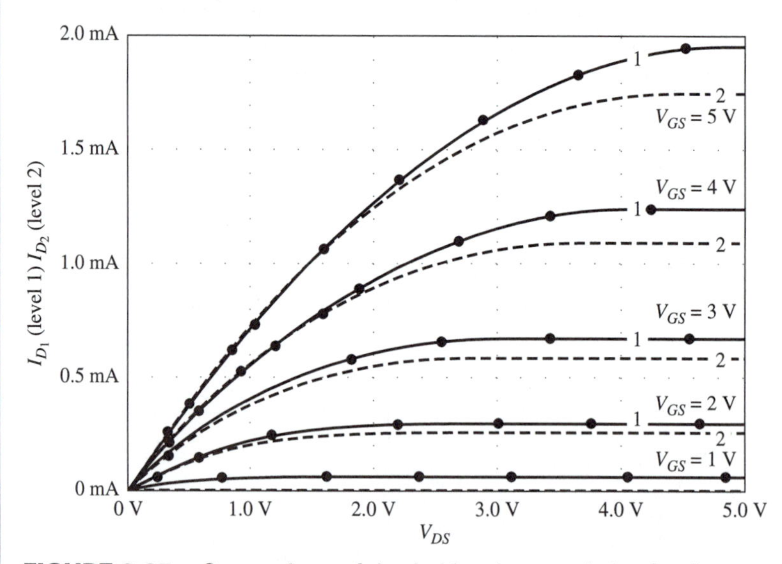

FIGURE 8.25 Comparison of the I_D-V_{DS} characteristics for the SPICE level 1 and level 2 models for long-channel lengths. The I_D-V_{DS} characteristics are labeled with 1 for the level 1 model and with 2 for the level 2 model.

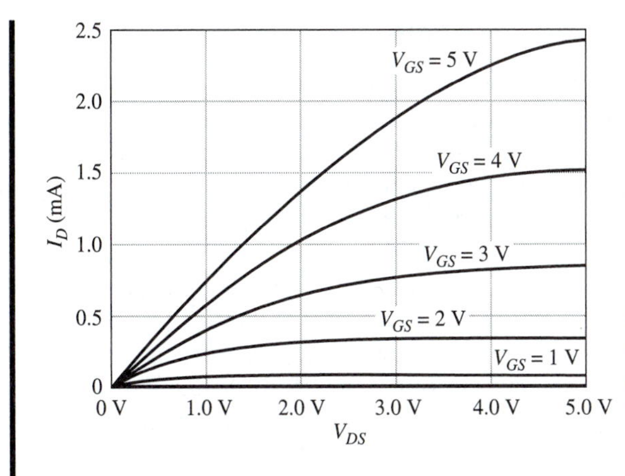

FIGURE 8.26 Effect of the channel-length modulation parameter λ (`LAMBDA`) on the I_D-V_{DS} characteristics of the level 1 example in Fig. 8.25.

The effect of assigning $\lambda = 0.05\ \text{V}^{-1}$ on the level 1 I_D-V_{DS} characteristics example in Fig. 8.25 is shown in Fig. 8.26. However, for the level 2 model, if `LAMBDA` is not included in the `.MODEL` line as a parameter, then L', which was given in Eq. (8.60), will be computed by the SPICE program and I_D does not saturate for $V_{DS} > V_{DS_{sat}}$. If the gate length in this example is reduced to $L = 1\ \mu$m and the W/L ratio is retained as four so that $W = 4\ \mu$m, the I_D-V_{DS} characteristics shown in Fig. 8.27 are obtained. The dashed line is for the level 1 model, and the I_D-V_{DS} characteristic remains the same as in Fig. 8.25. The solid line is for the level 2 model, and the smaller L' results in a much larger I_D than was obtained in Fig. 8.25, and I_D does not saturate.

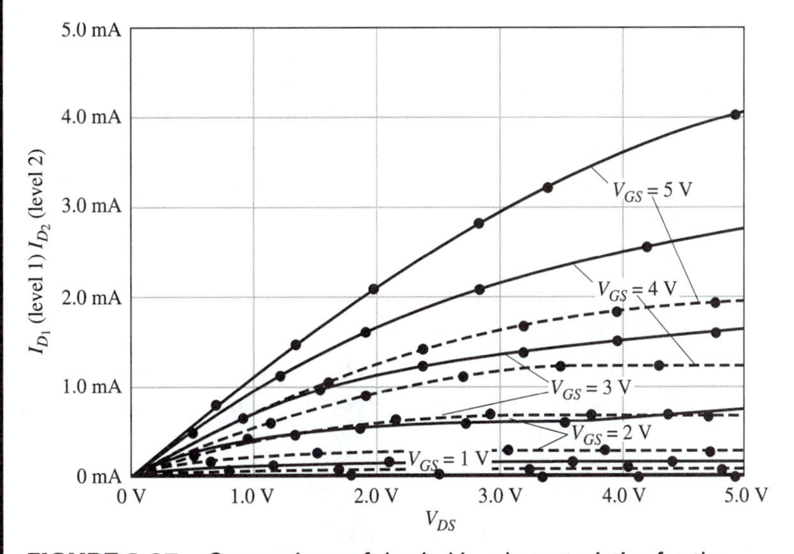

FIGURE 8.27 Comparison of the I_D-V_{DS} characteristics for the level 1 model (dashed line) and the level 2 model (solid line) with $L = 1\ \mu$m and $W = 4\ \mu$m. The level 2 model drain current has significantly increased because $W/L' > W/L$.

■

8.4.3 PSpice BSIM Models

As pointed out by Foty,[22] the level 1, level 2, and level 3 models were based on the simple physical MOSFET models. The next models are based on the Berkeley Short-channel IGFET Model (BSIM) model and introduce a semiempirical approach for the electrical parameters of small-geometry MOSFETs.[32–35] With these models, the emphasis shifted to the circuit design user. The parameters are extracted by measurement of MOSFETs representative of the process used in the integrated circuit fabrication. The measurements use a personal computer or workstation to analyze the data acquired with a semiconductor parameter analyzer such as the Hewlett-Packard 4145B, on a test MOSFET accessed with a probe station.[35] For circuit design, only the layout geometries of the MOSFETs and the parasitic elements are required for circuit simulation with BSIM. There are no default parameters and all parameters must be specified.

The BSIM1 model is the PSpice level 4 model and is suitable for representing MOSFETs with limiting parameters of $L \simeq 1$ μm and $t_{ox} \simeq 15$ nm. The parameter list for BSIM1 may be found in Ref. 32 and the parameters are described in Refs. 32 and 33.

A BSIM1-based submicron model, BSIM2, is the level 5 PSpice model.[33,36] It includes submicron behavior such as threshold voltage reduction, nonuniform doping for ion-implanted devices, mobility reduction due to the transverse field, depletion charge sharing by drain and source, carrier velocity saturation, drain-induced barrier lowering, channel-length modulation, hot-carrier-induced output resistance reduction, subthreshold conduction, source-drain parasitic resistance, drain voltage reduction in the light-doped drain (LDD) structure (see Sec. 8.7.4).[33,35]

The most recent models, such as BSIM3, were influenced by the need for suitable models for analog circuit design and the desire to reduce the number of model parameters used in BSIM1 and BSIM2. These models also returned to more physical meaning for the parameters. The BSIM3 model includes the major effects that are important in modeling deep-submicrometer MOSFETs. Although BSIM3 development is continuing at the University of California at Berkeley, the preliminary release of this model, Version 1.0, is the PSpice level 5 model. The current Version 2.0 adds several new parameters and corrects several errors and problems with Version 1.0. This Version 2.0 is the PSpice level 6 model.

8.4.4 Other SPICE Models

SPICE3 provides four MOSFET models.[37] In addition to the level 1, 2, and 3 models given in SPICE2, the level 4 model gives the BSIM1 model. Some versions of SPICE3 include the BSIM2 model.

HSPICE contains both public domain and proprietary models given in the `.MODEL` line.[38,39] The level 1, 2, and 3 models are the same as for SPICE2. The level 4 model is the same as for the level 2 model in SPICE2 except that there are no narrow channel effects ($\eta = 1$), no short-channel effects, and the gate type defaults to an Al gate. The level 5 model uses units in microns rather than meters, and includes both enhancement and depletion modes by specifying `ZENH=1` and `ZENH=0`, respectively. The depletion model is approximated by a step profile for the ion-implanted channel layer. The level 13 model is an adaptation of the BSIM model. The level 27 model is for the silicon on sapphire FET (SOSFET).

HSPICE runs faster for large circuits and has better convergence than many other versions of SPICE. To use HSPICE, a good understanding of SPICE is necessary because no texts similar to those for PSpice are currently available. Therefore, a good knowledge of SPICE2 or PSpice will greatly assist in the use of the more advanced simulators, such as HSPICE.

8.5 SMALL-SIGNAL AC MODEL

8.5.1 Equivalent Circuit

Many applications of MOSFETs are for high switching speed in digital logic circuits or for operation at high frequencies. Small signal means that the AC voltages and currents are small compared to the DC values and that the AC currents are proportional to the AC voltages. A small-signal equivalent circuit will be obtained in this part of Sec. 8.5. Then, to evaluate the ability of a device to operate at high frequencies, the common figure of merit, the cutoff frequency f_T, will be derived in Sec. 8.5.2. The cutoff frequency f_T is the frequency at which the current gain in a common-source configuration becomes unity for the output shorted. It should be noted that f_T will be derived for the bipolar-junction transistor (BJT) in Sec. 9.6.2.

An amplifying circuit is shown in Fig. 8.28 (*a*) for an *n*-channel enhancement-mode MOSFET in the common-source configuration with a load resistance R_L and a

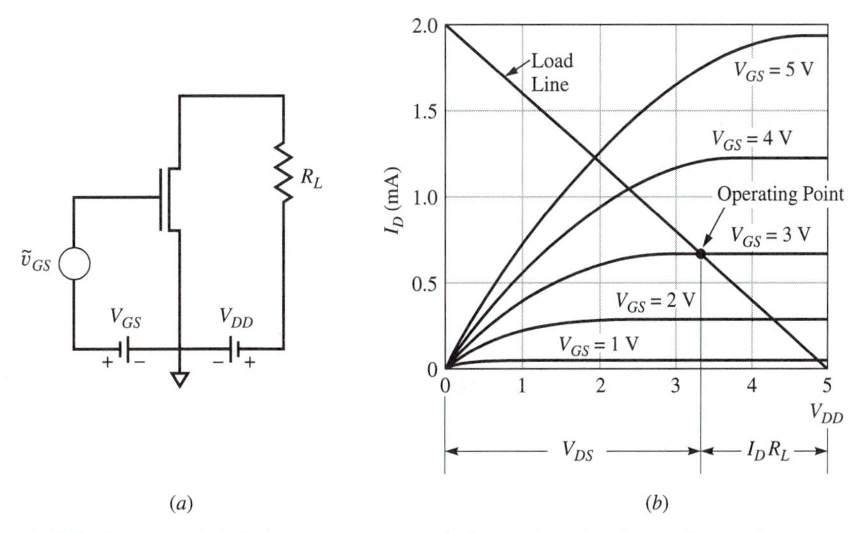

FIGURE 8.28 (*a*) Common-source configuration for the *n*-channel enhancement-mode MOSFET with the AC input voltage $\tilde{v}_{GS}$. (*b*) Output I_D-V_{DS} characteristics with the load line and operating point.

supply voltage V_{DD}. The small-signal operation of the MOSFET is illustrated by the load line drawn on the I_D-V_{DS} output characteristics shown in Fig. 8.28 (*b*). The DC voltage V_{DS} across the MOSFET and the voltage across R_L must equal V_{DD}:

$$V_{DD} = V_{DS} + I_D R_L. \tag{8.63}$$

The load line is for $V_{DS} = 0$ and $I_D = V_{DD}/R_L$. The operating point is determined by V_{GS}. For small variations around the operating point, the nonlinear I–V and C–V characteristics can be linearized and incremental currents become proportional to the incremental voltages when the voltage variations are smaller than kT/q. The *small-signal equivalent circuit* is used to design amplifier circuits.

The Shichman-Hodges model [Eq. (8.24)] is used to represent the current in the saturation region because it is a simple enough equation to permit obtaining useful expressions for the desired elements in the equivalent circuit.[40] The saturation current given in Eq. (8.24) is used with the empirical correction term $(1 + \lambda V_{DS})$, where λ is parameter no. 6 in Table 8.1 and represents the effect of channel-length modulation by V_{DS}:

$$I_{D_{sat}} = \frac{W}{2L}\mu_n C'_{ox}(V_{GS} - V_T)^2(1 + \lambda V_{DS}). \tag{8.64}$$

The transconductance (g_m) is the change in the output current (I_D) with respect to the change in the input voltage (V_{GS}) and is given by

$$g_m \equiv \left.\frac{\partial I_D}{\partial V_{GS}}\right|_{V_{DS}} = \frac{W}{L}\mu_n C'_{ox}(V_{GS} - V_T)(1 + \lambda V_{DS}). \tag{8.65}$$

With the approximation $\lambda V_{DS} < 1$, the transconductance in the saturation region becomes

transconductance

$$\boxed{g_m = \left.\frac{\partial I_D}{\partial V_{GS}}\right|_{V_{DS}} = \frac{W}{L}\mu_n C'_{ox}(V_{GS} - V_T) = \sqrt{2\frac{W}{L}\mu_n C'_{ox} I_{D_{sat}}}}. \tag{8.66}$$

The transconductance represents the effectiveness of the gate voltage in controlling the drain current. To compare transconductance for various devices, it has become standard practice to normalize g_m by dividing through by the gatewidth W in mm. Common values of g_m range from 10 to 100 millisiemens per millimeter (mS/mm).

The transconductance given in Eq. (8.66) ignores the velocity saturation effects on the mobility as introduced with Eq. (8.10). As L decreases, velocity saturation is achieved at smaller values of V_{DS}. The spacing of the I_D-V_{DS} curves is not according to the square law (see Sec. 8.7 and Fig. 8.38) given by Eq. (8.24) and the spacing depends almost *linearly*[41] on $(V_{GS} - V_T)$. Therefore, velocity saturation will reduce the g_m as represented in Eq. (8.66).

The output conductance is the slope of the I_D-V_{DS} output characteristic in the saturation region and is given by

$$g_d \equiv \left.\frac{\partial I_D}{\partial V_{DS}}\right|_{V_{GS}} = \frac{1}{r_o}, \tag{8.67}$$

where r_o is the output resistance. The variation of I_D with V_{DS} is related to the movement of the pinch-off point toward the source as illustrated in Fig. 8.14 with $y' = L_{eff}$. The effective channel length is taken as

$$L_{eff} = L - x_d, \tag{8.68}$$

where x_d is the drain depletion depth which varies with V_{DS} and is represented by ΔL in Fig. 8.14. With $I_{D_{sat}}$ taken as Eq. (8.24), the drain current is written as

$$I_{D_{sat}} = \frac{W}{2L_{eff}}\mu_n C'_{ox}(V_{GS} - V_T)^2, \tag{8.69}$$

where the effect of the $(1 + \lambda V_{DS})$ term in Eq. (8.64) is given by L_{eff}. Note that for $\Delta L/L << L$, $1/L_{eff} = 1/(L - \Delta L) \simeq (1/L)(1 + \Delta L/L)$, where $\Delta L/L$ becomes λV_{DS}. Then with Eq. (8.69) in Eq. (8.67),

$$g_d = \left.\frac{\partial I_{D_{sat}}}{\partial V_{DS}}\right|_{V_{GS}} = -\frac{W}{2L_{eff}^2}\mu_n C'_{ox}(V_{GS} - V_T)^2 \frac{dL_{eff}}{dV_{DS}}. \tag{8.70}$$

With L_{eff} from Eq. (8.68), $dL_{eff}/dV_{DS} = -dx_d/dV_{DS}$, and

$$g_d = \frac{\partial I_{D_{sat}}}{\partial V_{DS}} = \frac{I_{D_{sat}}}{L_{eff}}\frac{dx_d}{dV_{DS}}. \tag{8.71}$$

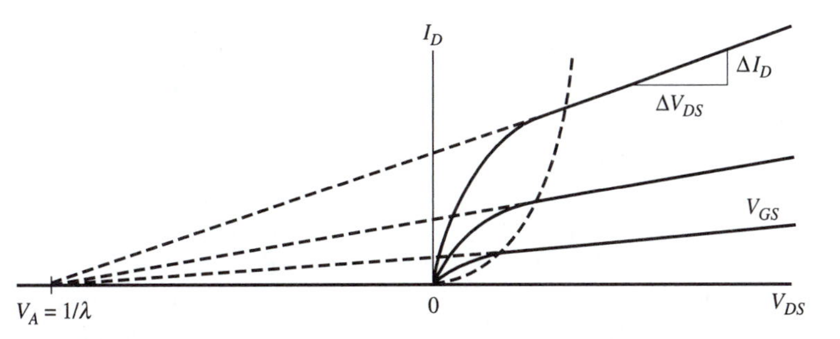

FIGURE 8.29 Output characteristic for an *n*-channel enhancement-mode MOSFET with drain current slope $\partial I_{D_{sat}}/\partial V_{DS}$ to illustrate the voltage V_A, which represents changes in channel length due to V_{DS}.

The output conductance can be obtained from the similar triangles shown in Fig. 8.29, which gives

$$\frac{\partial I_{D_{sat}}}{\partial V_{DS}} = \frac{I_{D_{sat}}}{|V_A|}. \tag{8.72}$$

Comparison of Eqs. (8.71) and (8.72) gives

$$|V_A| = L_{eff}\left(\frac{dx_d}{dV_{DS}}\right)^{-1}, \tag{8.73}$$

so that the output conductance in Eq. (8.71) may be written as

$$g_d = \frac{I_{D_{sat}}}{|V_A|}. \tag{8.74}$$

Comparison of Eq. (8.64) and (8.74) with $\partial I_{D_{sat}}/\partial V_{DS} = I_{D_{sat}}\lambda$ from Eq. (8.64) permits writing the output resistance as

output resistance

$$\boxed{r_o = g_d^{-1} = \frac{|V_A|}{I_{D_{sat}}} = \frac{1}{\lambda I_{D_{sat}}}}. \tag{8.75}$$

Also, the variation of the substrate bias V_{BS} can affect the drain current and this effect can be represented by a transconductance[40] g_{mb}, which is in parallel to g_m. This transconductance is given by

$$g_{mb} \equiv \left.\frac{\partial I_D}{\partial V_{BS}}\right|_{V_{DS}}, \tag{8.76}$$

where, as previously shown in Eq. (8.41), V_T has a dependence on V_{BS}. Considerations of variations in I_D due to variations in the source-to-substrate (or body) voltage will be neglected in the equivalent circuit considered here.

The intrinsic capacitances are illustrated in Fig. D.1. The capacitance C_{GB}, between the gate metal (or polysilicon) and the substrate outside the active device includes the capacitance due to the metal (or polysilicon) interconnects and the underlying substrate. This capacitance is taken as constant and will depend on the particular fabrication process. The capacitances C_{GS} and C_{GD} are related to the gate capacitance C_{ox}. In the saturation region, the channel is very thin near the drain and V_{DS} has little effect on the channel or gate charge and the intrinsic part of C_{GD} may be neglected. The parasitic capacitance represented in Fig. D.3 (*b*) by the overlap of the gate over the drain is taken as C_{GD}.

The value of C_{GS} in the saturation region is obtained from the total channel charge Q_T, which is obtained by integrating $|Q_n'(y)|$ given in Eq. (8.17):

$$Q_T = WC'_{ox}\int_0^L [(V_{GS} - V_T) - V(y)]\, dy. \tag{8.77}$$

The variable dy may be replaced by $dV(y)$ from the relationship given in Eq. (8.7) and $|Q'_n|$ from Eq. (8.17) to give

$$dy = \frac{W\mu_n|Q'_n|}{I_D}\, dV(y) = \frac{W\mu_n C'_{ox}[(V_{GS} - V_T) - V(y)]}{I_D}\, dV(y). \tag{8.78}$$

With dy from Eq. (8.78), the total channel charge may be written as

$$Q_T = \frac{W^2 C'^{\,2}_{ox}\mu_n}{I_D}\int_0^{V_{GS}-V_T} [(V_{GS} - V_T) - V(y)]^2\, dV(y). \tag{8.79}$$

Integration and replacement of I_D with the saturation current given in Eq. (8.24) gives

$$Q_T = -\frac{2}{3}WLC'_{ox}(V_{GS} - V_T) = -\frac{2}{3}C_{ox}(V_{GS} - V_T). \tag{8.80}$$

The gate-to-source capacitance is

$$C_{GS} = \left|\frac{\partial Q_T}{\partial V_{GS}}\right| = \frac{2}{3}C_{ox}. \tag{8.81}$$

This value for C_{GS} is given in Fig. D.2 (*c*) in Appendix D at the end of this chapter. The gate-to-source overlap capacitance C_{GSO} was given in Fig. D.3 (*a*) and in Eqs. (D.3) and (D.4). This overlap capacitance is added to (2/3)C_{ox} given in Eq. (8.81) to give the total gate-to-source capacitance. Expressions similar to Eqs. (D.3) and (D.4) give the drain-to-gate capacitance C_{GD}, which is the overlap capacitance with ΔL_S replaced by ΔL_D.

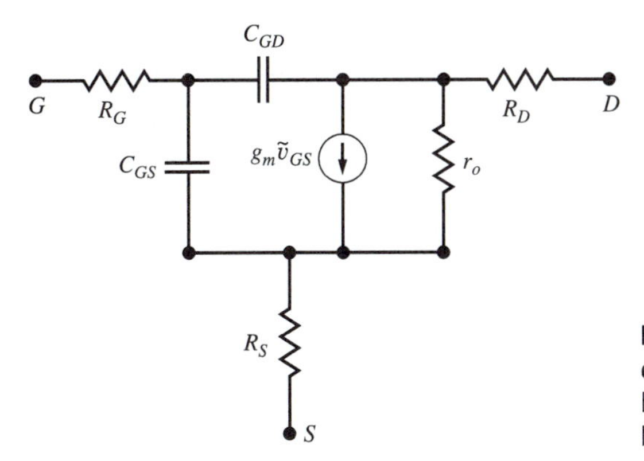

FIGURE 8.30 Small-signal equivalent circuit for an *n*-channel MOSFET. Both g_{mb} and C_{GB} have been omitted.

The small-signal equivalent circuit also has series resistances for each of the terminals: the source resistance R_S, the gate resistance R_G, and the drain resistance R_D. These parasitic resistances together with the transconductance and the source and drain capacitances give the equivalent circuit shown in Fig. 8.30. Parasitic capacitances between the substrate, drain, and source regions may also be included in the small-signal equivalent circuit,[40,42] but have not been included in Fig. 8.30. The capacitance of the drain region is given in Eq. (D.9). The gate-to-substrate capacitance varies with V_{GS} and is given in Fig. D.2 (b).

The effect of the parasitic resistances is to reduce the intrinsic transconductance. To illustrate the reduction of g_m, the equivalent circuit in Fig. 8.30 is considered at low frequency so that C_{GS} and C_{GD} may be neglected. Also, R_G, R_D, and r_o are neglected to give a simplified equivalent circuit which helps emphasize the effect of R_S. Only the source resistance is retained in Fig. 8.31. The AC voltage across the gate-to-source terminals is given by

$$\tilde{v}_{GS} = \tilde{v}'_{GS} + R_S\tilde{i}_D. \tag{8.82}$$

The drain current is given by

$$\tilde{i}_D = g_m\tilde{v}'_{GS}. \tag{8.83}$$

Replacement of $\tilde{v}'_{GS}$ in Eq. (8.82) with $\tilde{v}'_{GS}$ from Eq. (8.83) gives

$$\tilde{i}_D = \left[\frac{g_m}{1 + R_S g_m}\right]\tilde{v}_{GS}. \tag{8.84}$$

Thus, the transconductance in the presence of a source resistance becomes

$$g'_m = \left[\frac{g_m}{1 + R_S g_m}\right], \tag{8.85}$$

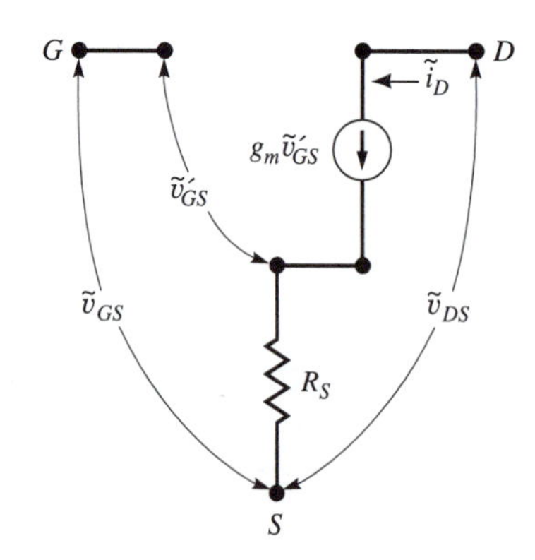

FIGURE 8.31 Low-frequency MOSFET equivalent circuit to illustrate the effect of R_S on the intrinsic tranconductance.

which shows that the intrinsic transconductance is degraded as $R_S g_m$ becomes of significant value as compared to 1.0.

8.5.2 Cutoff Frequency f_T

The cutoff frequency f_T is often used to specify the MOSFET high-frequency performance and is the frequency at which the current gain in the common-source configuration is unity for the output shorted for an AC signal. Measurements of f_T are used to compare the performance of MOSFETs for new Si fabrication technologies. The shorted output condition eliminates the output resistance r_o and connects C_{GD} in parallel with C_{GS}.

From Fig. 8.32, the input current is given by

$$\tilde{i}_{in} = j\omega(C_{GS} + C_{GD})\tilde{v}_{GS}. \tag{8.86}$$

The magnitude of the current gain is unity at f_T:

$$\left|\frac{\tilde{i}_{out}}{\tilde{i}_{in}}\right| = \frac{g_m}{\omega(C_{GS} + C_{GD})} = 1, \tag{8.87}$$

or

$$f_T = \frac{g_m}{2\pi(C_{GS} + C_{GD})}. \tag{8.88}$$

Note that the source resistance would reduce f_T through the reduced transconductance given in Eq. (8.85). For g_m represented by Eq. (8.66), and $C_{GS} + C_{GD}$ represented by $C'_{ox}WL$, the cutoff frequency is approximated by

cutoff frequency f_T

$$f_T \simeq \frac{(W/L)\mu_n C'_{ox}(V_{GS} - V_T)}{2\pi(2/3)C'_{ox}WL} = \frac{3\mu_n(V_{GS} - V_T)}{4\pi L^2}. \tag{8.89}$$

These equations show that for high-speed operation, a large g_m is desired together with a small capacitance. Although only an approximation, Eq. (8.89) helps to illustrate that

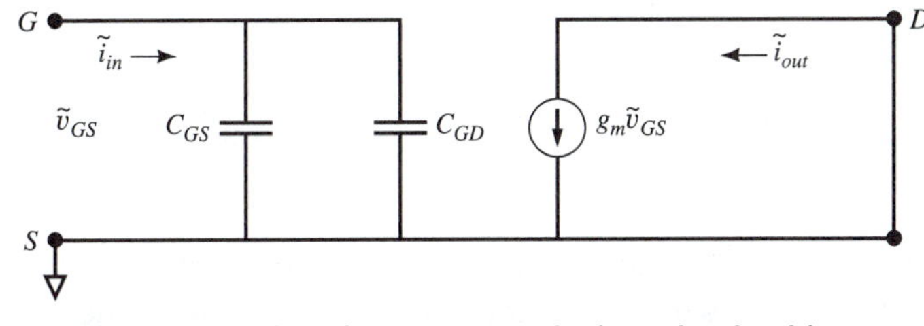

FIGURE 8.32 High-frequency equivalent circuit with the drain shorted to the source (output shorted) for the determination of f_T.

higher speed implies shorter gate length. The highest speed MOSFETs that have been fabricated and measured have gate lengths near 0.1 μm and f_T near 100 GHz.[43] The cutoff frequency for the BJT will be given in Eq. (9.86) for the *pnp* BJT as $f_T = D_p/\pi W_n^2$, where W_n is the base thickness. Therefore, comparison of Eqs. (9.86) and (8.89) shows that f_T for the MOSFET depends directly on the carrier mobility and inversely on the channel length squared, while f_T for the BJT varies directly with the carrier diffusivity and inversely as the base thickness squared.

8.6 SUBTHRESHOLD CURRENT BEHAVIOR

8.6.1 Subthreshold Drain Current

As shown in Fig. 8.33 (*a*), the measured drain current gradually goes to zero and the threshold voltage is taken as the linear extrapolation of $\sqrt{I_D}$ to zero as given by the Shichman-Hodges model (see Sec. 8.2.3). The drain current for $V_{GS} < V_T$ is known as the subthreshold current. The semilogarithmic plot of the measured subthreshold current shown in Fig. 8.33 (*b*) demonstrates an exponential variation with gate voltage V_{GS} when V_{GS} is less than V_T. The previously introduced Shichman-Hodges model and the variable depletion model for the MOSFET, which represent the level 1 and level 2 SPICE models, respectively, do not correctly represent the subthreshold region.

Subthreshold currents become an important consideration for low leakage applications such as wristwatches and dynamic memories. It also becomes a problem as tens of millions of MOSFETs are used in an IC and, as the gate length decreases into the submicron region, the turn-off further degrades.

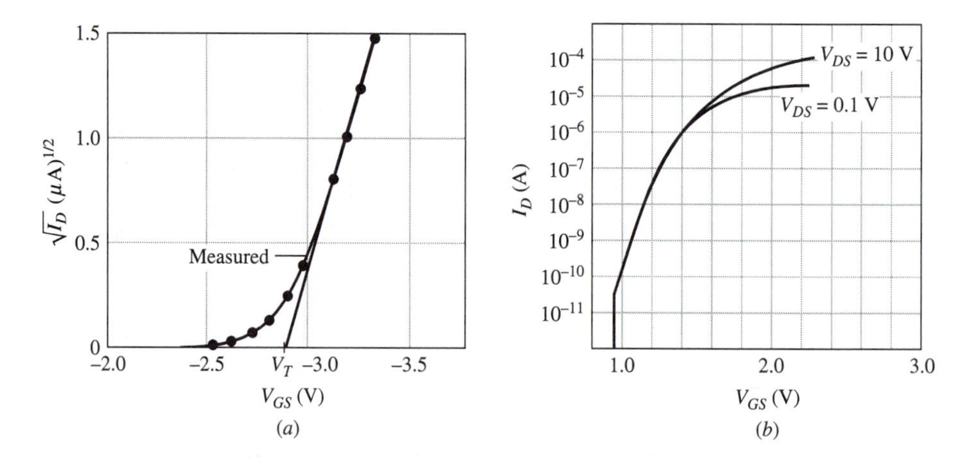

FIGURE 8.33 (*a*) The variation of $\sqrt{I_D}$ vs. V_{GS} to illustrate that the threshold voltage V_T is taken as the extrapolation of $\sqrt{I_D}$ to zero while the measured I_D gradually decreases as shown by the dashed line.[44] (*b*) Experimental I_D in the subthreshold region for a long channel device (L = 15.5 μm) and V_T = 1.3 V.[45]

Several analyses for the subthreshold region have been given.[44–53] The charge-sheet model,[51–53] which compresses the inversion layer into a conducting plane of zero thickness, will be presented here. The more general charge-sheet model applies from subthreshold to saturation conditions and to both uniformly and nonuniformly doped MOSFETs,[52] but requires a more complex analysis than for the models given in Sec. 8.2. The long-channel case is considered here and means that the source-to-drain distance is much greater than the source and drain depletion layer widths.

charge-sheet model

In the charge-sheet model, the inversion layer carrier density per unit area, N_I', is needed to describe the MOSFET. As shown in Appendix C.1 at the end of Chapter 7, the electrons (for p-Si) in the inversion layer are within 30–300 Å of the SiO_2/Si interface. This inversion layer is taken as a charge sheet of infinitesimal thickness.

The total charge per unit area, Q_{Si}', in the Si was derived in Appendix C.2 at the end of Chapter 7 and was given in Eq. (C.43). The charge per unit area in the depletion region, Q_{sc}', was given in Eq. (7.43) as $-qN_a^- x_d$, where x_d is the depletion width given by Eq. (7.32) as $\sqrt{2\epsilon_{\mathrm{Si}}\psi_s/qN_a^-}$. Then the charge in the inversion layer is the difference between the total charge and the charge in the depletion region:

$$-qN_I' = -Q_{\mathrm{Si}}' - (-qN_a^- x_d), \tag{8.90}$$

which is

$$\begin{aligned} qN_I' = {} & \sqrt{2}\epsilon_{\mathrm{Si}}\left(\frac{kT}{qL_D}\right) \\ & \times \left\{\left[\frac{q\psi_s}{kT} + \exp(-q\psi_s/kT) - 1\right] + \left(\frac{n_i}{N_a^-}\right)^2 \left[\exp(q\psi_s/kT) - 1\right]\right\}^{1/2} \\ & - qN_a^- \sqrt{2\epsilon_{\mathrm{Si}}\psi_s/qN_a^-}, \end{aligned} \tag{8.91}$$

where L_D is the Debye length given by $\sqrt{(\epsilon_{\mathrm{Si}}/q^2)(kT/N_a^-)}$. For weak inversion, $|\psi_b| < \psi_s < 2|\psi_b|$, where $\psi_b = -(kT/q)\ln(N_a^-/n_i)$, which permits neglecting the $[\exp(-q\psi_s/kT) - 1]$ term compared to $q\psi_s/kT$, and the -1 term may be neglected compared to $\exp(q\psi_s/kT)$. By multiplying both sides by $1/q$, Eq. (8.91) now may be written as

$$N_I' = \sqrt{2}\frac{\epsilon_{\mathrm{Si}}}{q}\left(\frac{kT}{qL_D}\right)\left[\frac{q\psi_s}{kT} + \left(\frac{n_i}{N_a^-}\right)^2 \exp(q\psi_s/kT)\right]^{1/2} - N_a^- \sqrt{2\epsilon_{\mathrm{Si}}\psi_s/qN_a^-}. \tag{8.92}$$

The coefficient $\sqrt{2}(\epsilon_{\mathrm{Si}}/q)(kT/qL_D)$ may be written in a more convenient form by multiplying by N_a^-/N_a^- to give

$$\sqrt{2}\frac{\epsilon_{\mathrm{Si}}}{q}\left(\frac{kT}{qL_D}\right)\frac{N_a^-}{N_a^-} = \sqrt{2}\sqrt{\frac{\epsilon_{\mathrm{Si}}}{q^2}\frac{kT}{N_a^-}}N_a^- = \sqrt{2}N_a^- L_D. \tag{8.93}$$

The last term in Eq. (8.92) may be written in a similar form by multiplying by $\sqrt{(q/kT)(kT/q)}$ to give

$$N_a^- \sqrt{\frac{2\epsilon_{\mathrm{Si}}\psi_s}{qN_a^-}} = N_a^- \sqrt{\frac{2\epsilon_{\mathrm{Si}}\psi_s}{qN_a^-}} \sqrt{\frac{q}{kT}\frac{kT}{q}} = \sqrt{2}N_a^- L_D \sqrt{\frac{q\psi_s}{kT}}. \tag{8.94}$$

With these two expressions, Eq. (8.92) may be written as

$$N_I' = \sqrt{2}N_a^- L_D \left\{ \left[\frac{q\psi_s}{kT} + \left(\frac{n_i}{N_a^-}\right)^2 \exp(q\psi_s/kT) \right]^{1/2} - \left[\frac{q\psi_s}{kT}\right]^{1/2} \right\}. \tag{8.95}$$

In Eq. (8.95), the $[(q\psi_s/kT) + (n_i/N_a^-)^2 \exp(q\psi_s/kT)]^{1/2}$ term can be expanded in the first two terms of a Taylor's series $[(1 + x)^{1/2} \simeq 1 + \frac{1}{2}x]$ to give

$$\sqrt{q\psi_s/kT} + \frac{1}{2}\frac{1}{\sqrt{q\psi_s/kT}}\left(\frac{n_i}{N_a^-}\right)^2 \exp(q\psi_s/kT), \tag{8.96}$$

so that Eq. (8.95) becomes

inversion layer charge density

$$\boxed{N_I' = N_a^- L_D (n_i/N_a^-)^2 \exp(q\psi_s/kT)(2q\psi_s/kT)^{-1/2}} \tag{8.97}$$

for weak inversion in the subthreshold region. Note that the inversion layer charge per unit area N_I' in Eq. (8.97) has dimensions of $N_a^- L_D$ which is cm^{-2}.

In the charge-sheet model, the current flow is partly due to drift by the electric fields and partly due to diffusion. The carrier densities adjust so that at each location in the channel the current continuity is maintained. This accommodation of carrier densities to the current flow is expressed by an adjustment of the quasi-Fermi level (see Sec. 4.6.1), which varies from point to point in the channel. The MOSFET I_D-V_{DS} output characteristics are obtained with Eq. (4.74), which gives J_n as $\mu_n n(dE_{f_n}/dy)$, where E_{f_n} is the electron quasi-Fermi level. The charge-sheet model is a more rigorous model than the Shichman-Hodges model or the Ihantola-Moll model and may be extended for analysis of the subthreshold current. However, the charge-sheet model does not permit description of the SPICE level 1 and level 2 models.

For n-channel devices on p-Si, the electron current density J_n was given by Eq. (4.69) as

$$J_n = q\mu_n n\mathscr{E} + qD_n\frac{dn}{dy}, \tag{8.98}$$

with $\mathscr{E} = -d\psi/dy$. As pointed out by Brews,[53] the small value of N_I' in weak inversion has two consequences. First, ψ_s is found to be the saturation value ψ_{sat} obtained for the pinch-off condition. Second, ψ_s is very nearly ψ_{sat} throughout the channel length and is given later in Eq. (8.111). Therefore, $-d\psi/dy = \mathscr{E} = 0$ and there is *no drift current in the subthreshold region.* The diffusion current is a result of the gradient in N_I'.

The drain current is obtained from Eqs. (8.97) and (8.98) with $\mathscr{E} = 0$ to give

$$I_D = qAD_n\frac{dn}{dy} = qWD_n\frac{dN_I'}{dy}, \tag{8.99}$$

where W is the gatewidth and the other dimension in the cross-sectional area A is the infinitesimal charge layer thickness contained in N_I'. Note, that in Eq. (8.97) for N_I' a thickness dimension is given by the Debye length L_D. Because the current must be the same at any point along the channel, dN_I'/dy must be a constant, which permits writing Eq. (8.99) as

$$I_D = qWD_n\frac{N_I'(\text{drain}) - N_I'(\text{source})}{y(\text{drain}) - y(\text{source})} = -qW\left(\mu_n\frac{kT}{q}\right)\frac{N_I'(\text{source}) - N_I'(\text{drain})}{L}, \tag{8.100}$$

where the Einstein relation $D = \mu kT/q$ has been used to replace the diffusivity D_n. In Sec. 8.2, drain current flowing into the drain (electron current flowing out of the drain) was taken as positive; therefore, the negative sign in Eq. (8.100) will be omitted.

Consideration of the terms containing ψ_s in N_I'(source) and N_I'(drain), which are used in Eq. (8.100) to evaluate I_D, is a laborious task. First, consider Poisson's equation given by Eqs. (C.30) with the variation of the electron concentration in the conducting channel represented by $n(x)$, which becomes $(n_i^2/N_a^-)^2 \exp[q\psi(x)/kT]$ in Eq. (C.31). This expression for $(n_i^2/N_a^-)^2 \exp[q\psi(0)/kT]$ is $(n_i^2/N_a^-)^2 \exp(q\psi_s/kT)$ at the source end ($y = 0$) because $\psi(0) = \psi_s$. At the drain, $\psi(0) = (\psi_s - V_{DS})$ at $y = L$, which gives $(n_i^2/N_a^-)^2 \exp[q(\psi_s - V_{DS})/kT]$. Also, the $(2q\psi_s/kT)^{-1/2}$ term entered Eq. (8.97) from the charge in the depletion region, as given in Eqs. (8.90) and (8.91), and will be the same for $y = 0$ and $y = L$. Therefore,

$$N_I'(\text{source}) = N_a^- L_D(n_i/N_a^-)^2 \exp(q\psi_s/kT)(2q\psi_s/kT)^{-1/2}, \tag{8.101}$$

and

$$N_I'(\text{drain}) = N_a^- L_D(n_i/N_a^-)^2 \exp[q(\psi_s - V_{DS})/kT](2q\psi_s/kT)^{-1/2}, \tag{8.102}$$

where V_{DS} is the drain voltage. This drain current in Eq. (8.100) with Eqs. (8.101) and (8.102) becomes

$$I_D = q(W/L)\mu_n(kT/q)N_a^- L_D(n_i/N_a^-)^2 \exp(q\psi_s/kT) \times [1 - \exp(-qV_{DS}/kT)](2q\psi_s/kT)^{-1/2}. \tag{8.103}$$

In Eq. (8.103) for $V_{DS} > 3kT/q$, the $[1 - \exp(-qV_{DS}/kT]$ term may be taken as 1.0. Equation (8.103) may now be written as

subthreshold current

$$\boxed{I_D = q\left(\frac{W}{L}\right)\mu_n\left(\frac{kT}{q}\right)N_a^- L_D\left(\frac{n_i}{N_a^-}\right)^2 \exp\left(\frac{q\psi_s}{kT}\right)\left(\frac{2q\psi_s}{kT}\right)^{-1/2}}. \tag{8.104}$$

To evaluate Eq. (8.104), the relationship between the applied gate voltage V_{GS} and the surface potential ψ_s must be obtained. Previously, this relationship was given by Eq. (7.109) for p-Si in depletion or inversion:

$$V_{GS} = V_{FB}^\circ + \frac{|Q'_{\mathrm{Si}}|}{C'_{ox}} + \psi_s, \tag{8.105}$$

or

$$C'_{ox}[(V_{GS} - V_{FB}^\circ) - \psi_s] = |Q'_{\mathrm{Si}}| = q(N'_I + qN_a^- x_d), \tag{8.106}$$

where the charge in the Si is represented as the sum of the inversion layer charge and the depletion-layer charge. The depletion-layer charge is $-qN_a^- x_d$ with x_d given in Eq. (7.32) as used previously for Eq. (8.90) to give

$$[(V_{GS} - V_{FB}^\circ) - \psi_s] = \frac{1}{C'_{ox}}\sqrt{2q\epsilon_{\mathrm{Si}}N_a^-\psi_s}, \tag{8.107}$$

where the inversion-layer charge has been neglected compared to the depletion-layer charge. By squaring Eq. (8.107)

$$(V_{GS} - V_{FB}^\circ)^2 - 2(V_{GS} - V_{FB}^\circ)\psi_s + \psi_s^2 = (1/C'_{ox})^2 2q\epsilon_{\mathrm{Si}}N_a^-\psi_s. \tag{8.108}$$

By multiplying the right-hand side of Eq. (8.108) by $(\epsilon_{\mathrm{Si}}/\epsilon_{\mathrm{Si}})(q/kT)(kT/q)$ gives

$$\begin{aligned}(1/C'_{ox})^2 2q\epsilon_{\mathrm{Si}}N_a^-\psi_s &= 2(1/C'_{ox})^2 \underbrace{\left(\frac{q^2}{\epsilon_{\mathrm{Si}}}\frac{N_a^-}{kT}\right)}_{1/L_D^2}\epsilon_{\mathrm{Si}}^2\frac{kT}{q}\psi_s \\ &= \underbrace{2\left(\frac{1}{C'_{ox}}\right)^2\frac{\epsilon_{\mathrm{Si}}^2}{L_D^2}}_{a^2}\frac{kT}{q}\psi_s.\end{aligned} \tag{8.109}$$

Then, Eq. (8.108) becomes

$$\psi_s^2 - [a^2(kT/q) + 2(V_{GS} - V_{FB}^\circ)]\psi_s + (V_{GS} - V_{FB}^\circ)^2 = 0. \tag{8.110}$$

Solving the quadratic equation for ψ_s gives

$$\boxed{\psi_s = (V_{GS} - V_{FB}^{\circ}) + (a^2/2)(kT/q) - a\sqrt{(V_{GS} - V_{FB}^{\circ})(kT/q) + (a^2/4)(kT/q)^2}}\,. \tag{8.111}$$

Brews[53] showed that ψ_s in Eq. (8.111) is the same expression for ψ_s at pinch-off for $V_{GS} > V_T$. Therefore, in the subthreshold region $\psi_s = \psi_{sat}$.

Unfortunately, ψ_s given in Eq. (8.111) is not a simple expression, and because ψ_s appears as $\exp(q\psi_s/kT)$, approximation of ψ_s leads to significant errors for I_D in Eq. (8.104). It should be noted that the subthreshold current usually is plotted as a function of V_{GS} or $(V_{GS} - V_T)$. To evaluate the subthreshold current, it is necessary to assign a value for V_{GS} and calculate ψ_s with Eq. (8.111). Then, with this ψ_s, the subthreshold drain current I_D is calculated with Eq. (8.104).

8.6.2 Gate Voltage Swing Parameter *S*

Subthreshold drain current behavior has become characterized by the gate voltage needed to reduce I_D by one decade. For example, in Fig. 8.34, I_D decreases from 1×10^{-4} A to 1×10^{-5} A as the gate voltage is decreased by 85 mV. This gate voltage swing is given by

$$S \equiv V_{GS}(I_D') - V_{GS}(0.1I_D') = \frac{V_{GS}(I_D') - V_{GS}(0.1I_D')}{\ln I_D' - \ln 0.1I_D'} \times \ln 10, \tag{8.112}$$

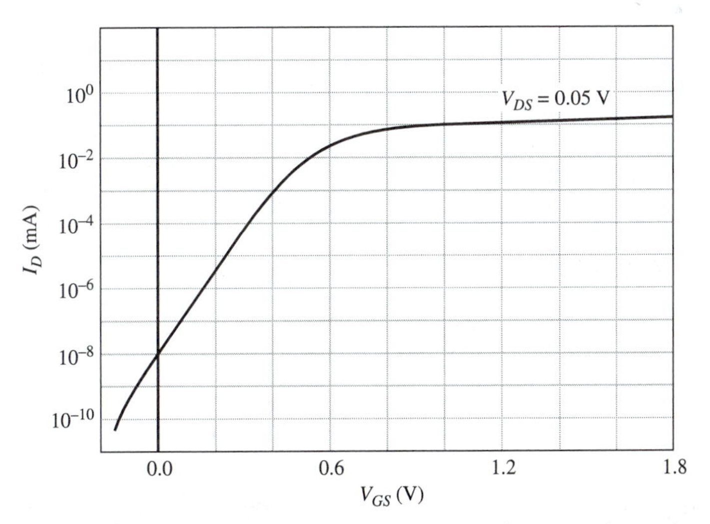

FIGURE 8.34 Subthreshold drain current for an *n*-channel MOSFET with a gate voltage swing parameter $S = 85$ mV/decade.[54]

where I'_D is a selected current in the exponential region with a gate voltage $V_{GS}(I'_D)$. The second form for S in Eq. (8.112) has been multiplied by $\ln 10/(\ln(I'_D - \ln 0.1I'_D)$ to give S in a form suitable for evaluation with I_D given by the charge-sheet model and is[52]

$$S \equiv \frac{\ln 10}{d \ln I_D / dV_{GS}}, \tag{8.113}$$

or

$$S \equiv \ln 10 \times \left\{\left[\frac{d \ln I_D}{d\psi_{sat}}\right]\left[\frac{d\psi_{sat}}{dV_{GS}}\right]\right\}^{-1} = \ln 10 \times \left(\frac{dV_{GS}/d\psi_{sat}}{d \ln I_D / d\psi_{sat}}\right). \tag{8.114}$$

From Eqs. (8.108) and (8.109) with $\psi_s = \psi_{sat}$,

$$[(V_{GS} - V^{\circ}_{FB}) - \psi_{sat}]^2 = a^2 (kT/q)\psi_{sat}. \tag{8.115}$$

Taking the square root of both sides of Eq. (8.115) and rearranging gives

$$V_{GS} = V^{\circ}_{FB} + \psi_{sat} + a\sqrt{(kT/q)\psi_{sat}}. \tag{8.116}$$

Then, at a constant temperature,

$$\frac{dV_{GS}}{d\psi_{sat}} = 1 + \frac{a}{2}\left(\frac{q}{kT}\psi_{sat}\right)^{-1/2}. \tag{8.117}$$

The other derivative in Eq. (8.114) is obtained by taking the ln of Eq. (8.104) for $V_{DS} > 3kT/q$:

$$\ln I_D = \ln\left[q\frac{W}{L}\mu_n\left(\frac{kT}{q}\right)L_D N_a^- \left(\frac{n_i}{N_a^-}\right)^2\right] + \frac{q\psi_{sat}}{kT} - \frac{1}{2}\ln\left(\frac{2q\psi_{sat}}{kT}\right). \tag{8.118}$$

Then, at constant temperature,

$$\begin{aligned}\frac{d \ln I_D}{d\psi_{sat}} &= \frac{q}{kT} - \frac{1}{2\psi_{sat}} \\ &= \frac{q}{kT}\left[1 - \frac{1}{2(q/kT)\psi_{sat}}\right].\end{aligned} \tag{8.119}$$

Equations (8.117) and (8.119) are most useful written in terms of the oxide capacitance per unit area $C'_{ox} = \epsilon_{ox}/t_{ox}$, the flat-band capacitance per unit area given by Eq. (C.52) as $C'_{FB} = \epsilon_{Si}/L_D$, and the depletion capacitance per unit area $C'_D(\psi_{sat})$ given by Eq. (C.45) when $(n_i/N_a^-)^2 \exp(q\psi_{sat}) << q\psi_{sat}/kT$ and $\exp(-q\psi_{sat}/kT) <$

1 as

$$C'_D(\psi_{sat}) = \frac{1}{\sqrt{2}} \underbrace{\frac{\epsilon_{\text{Si}}}{L_D}}_{C'_{FB}} \left(\frac{q\psi_{sat}}{kT}\right)^{-1/2}, \tag{8.120}$$

or

$$\frac{1}{\sqrt{2(q/kT)\psi_{sat}}} = \frac{C'_D(\psi_{sat})}{C'_{FB}}. \tag{8.121}$$

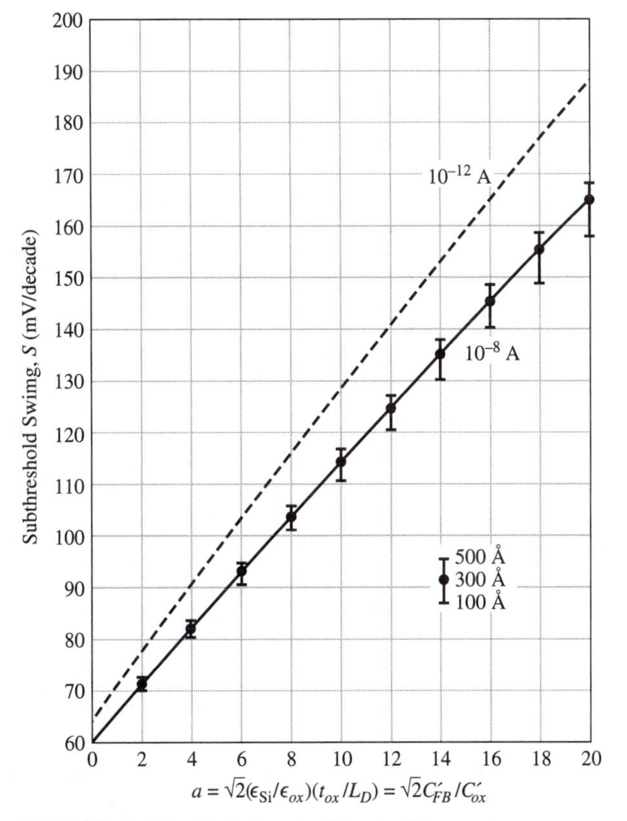

FIGURE 8.35 Subthreshold gate bias swing S required to reduce subthreshold current one decade versus the parameter a with $V_{DS} > 3kT/q$. The vertical bars indicate the variation of S at a constant value of a when the oxide thickness is varied from 500 to 100 Å. The curve was calculated for an oxide thickness of 300 Å, gate width to length ratio $W/L = 10$, and a drain current I_D of 10^{-8} A.[52]

Equation (8.117) with $a = \sqrt{2}(C'_{FB}/C'_{ox})$ and Eq. (8.121) becomes

$$\frac{dV_{GS}}{d\psi_{sat}} = 1 + \frac{C'_{FB}}{C'_{ox}} \frac{1}{\sqrt{2(q/kT)\psi_{sat}}}$$
$$= 1 + \frac{C'_{FB}}{C'_{ox}} \frac{C'_D(\psi_{sat})}{C'_{FB}} = 1 + \frac{C'_D(\psi_{sat})}{C'_{ox}}. \tag{8.122}$$

Equation (8.119) with $(2/a^2)(C'_{FB}/C'_{ox})^2 = 1$ and Eq. (8.121) now becomes

$$\frac{d \ln I_D}{d\psi_{sat}} = \frac{q}{kT}\left[1 - \left(\frac{C'_D(\psi_{sat})}{C'_{FB}}\right)^2 \frac{2}{a^2}\left(\frac{C'_{FB}}{C'_{ox}}\right)^2\right]$$
$$= \frac{q}{kT}\left[1 - \frac{2}{a^2}\left(\frac{C'_D(\psi_{sat})}{C'_{ox}}\right)^2\right]. \tag{8.123}$$

With Eqs. (8.122) and (8.123), Eq. (8.114) for S becomes

$$\boxed{S = (kT/q) \ln 10[1 + C'_D(\psi_{sat})/C'_{ox}]/\{1 - (2/a^2)[C'_D(\psi_{sat})/C'_{ox}]^2\}}. \tag{8.124}$$

As for other results for MOS devices such as threshold voltage, the parameter S given by Eq. (8.124) must be plotted to gain insight into the dependence on oxide thickness and doping level. To compare S values for different devices, ψ_{sat} should be chosen to correspond to the same current in both devices being compared. Therefore, the plot of S vs. a shown in Fig. 8.35 is given for two current levels.[52] Because a depends on t_{ox} and L_D, the subthreshold swing S will vary at a constant a as the impurity concentration changes in L_D when t_{ox} is varied. This variation is represented by the error bars in Fig. 8.35. A plot of constant values of $a = \sqrt{2}C'_{FB}/C'_{ox}$ as a function of oxide thickness t_{ox} and substrate doping N_a^- is shown in Fig. 8.36. The values of a and S are indicated on the curves. Values of a in Fig. 8.35 give corresponding values of S. Because $a = \sqrt{2}(\epsilon_{Si}/\epsilon_{ox})(t_{ox}/L_D)$, the parameter a depends on both t_{ox} and N_a^-. For $a = 3.336$, S from Fig. 8.35 is 77 mV/decade and $t_{ox} = 500$ Å and $N_a^- = 5 \times 10^{15}$ cm^{-3} from Fig. 8.36.

In some cases, the expression for S in Eq. (8.124) may be simplified. First, let $C'_D(\psi_{sat})$ be taken as $C'_{Si} = \epsilon_{Si}/x_{d_{max}}$. Also, the $\{1 - (2/a^2)[C'_D(\psi_{sat})/C'_{ox}]^2\}$ term may be taken as 1.0 with $a = \sqrt{2}C'_{FB}/C'_{ox} = \sqrt{2}\epsilon_{Si}t_{ox}/L_D\epsilon_{ox} > 1$, because $L_D \simeq t_{ox}$ and $\epsilon_{Si} = 3.0\epsilon_{ox}$ and $C'_D(\psi_{sat})/C'_{ox} < 1$. Therefore, Eq. (8.124) may be written as

$$\boxed{S \simeq (kT/q) \ln 10[(1 + C'_{Si}(x_{d_{max}})/C'_{ox})]}, \tag{8.125}$$

where $C'_{Si}(x_{d_{max}})/C'_{ox} = \epsilon_{Si}t_{ox}/\epsilon_{ox}x_{d_{max}}$. Equation (8.125) is a commonly used expression for S.

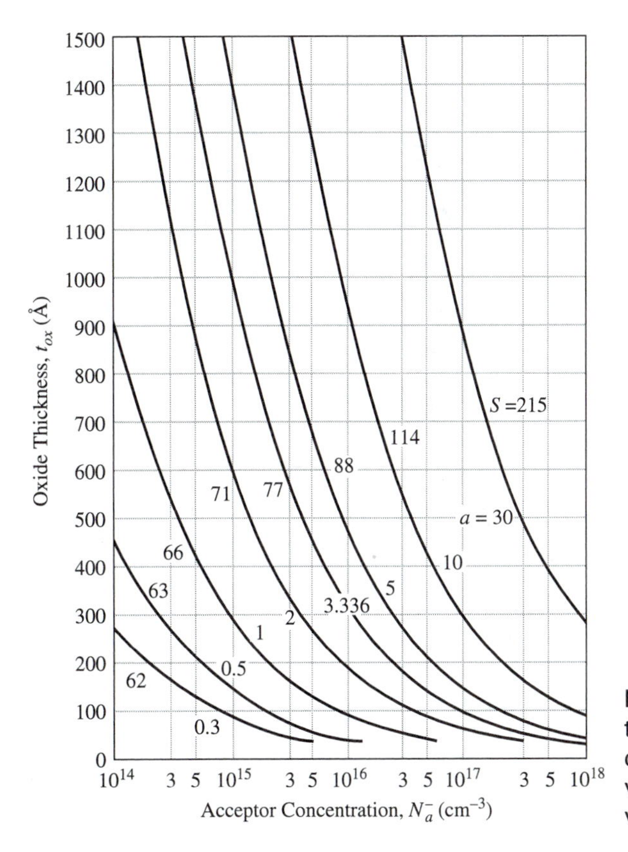

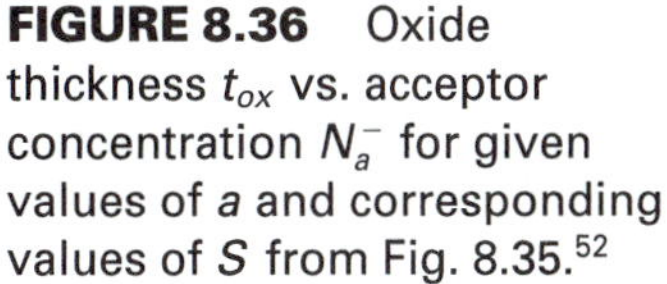

FIGURE 8.36 Oxide thickness t_{ox} vs. acceptor concentration N_a^- for given values of a and corresponding values of S from Fig. 8.35.[52]

8.7 MOSFET MINIATURIZATION

8.7.1 General Comments

As integrated circuit chips incorporate more circuits and are designed for higher speed operation, the MOSFET feature sizes such as gate length continue to shrink. The three basic variables of miniaturization are oxide thickness, source and drain junction depth, and source and drain depletion depth. As the feature sizes shrink, the one-dimensional analysis for the long-channel device is no longer adequate to represent the short-channel device. The boundary between the long-channel and the short-channel device is when the drain current no longer varies as $1/L$, the threshold voltage decreases, and the gate voltage necessary to reduce the subthreshold current becomes larger. As shown in Fig. 8.37 for a fixed oxide thickness, doping, source and drain junction depth, and V_{DS}, long-channel behavior is observed for $L = 5$ μm, but as L is reduced, the threshold voltage begins to shift to smaller values. As the gate length is further decreased, much larger changes in gate voltage are necessary to reduce the subthreshold current.

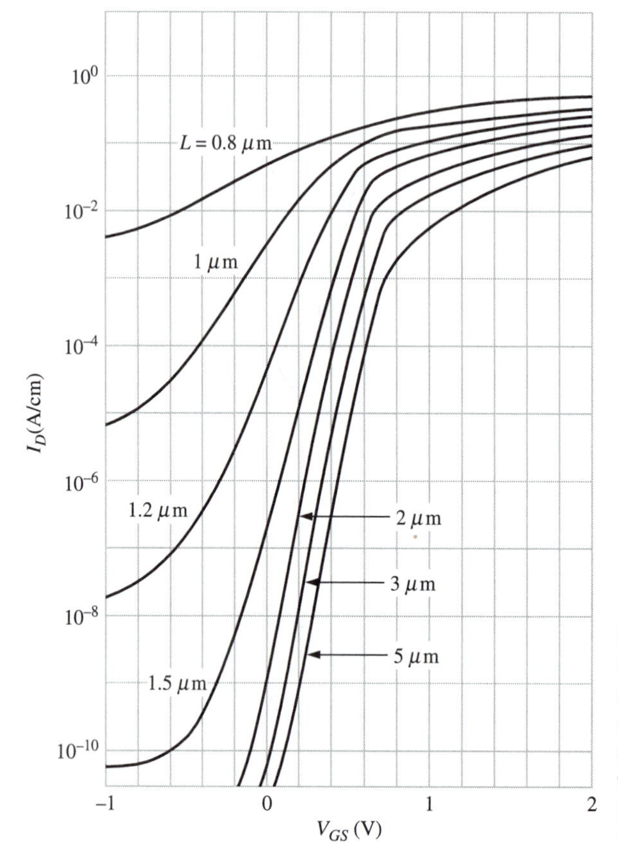

FIGURE 8.37 Subthreshold current behavior as a function of gate length. These curves are from two-dimensional computer calculations with $N_a^- = 1 \times 10^{15}$ cm^{-3}, $t_{ox} = 0.5$ μm, $r_j = 0.33$ μm, $V_{DS} = 2.0$ V, and $V_{BS} = 0$ V.[55]

As the gate length becomes smaller, the field increases and the carrier velocity can saturate (Sec. 3.8). The simulated drain current with a constant mobility is shown in Fig. 8.38 (*a*), and with a field-dependent mobility is shown in Fig. 8.38 (*b*).[56] In this figure, the drain saturation current in Fig. 8.38 (*b*) is greatly reduced and has an almost linear dependence on gate voltage (as shown by the spacing of the curves) rather than the nearly quadratic dependence shown in Fig. 8.38 (*a*). Further discussion of short-channel effects may be found in other texts.[57–59]

One approach to avoid short-channel effects has been to use a successful long-channel device from an existing fabrication generation and scale the device in dimension, doping, and voltage.[49] The linear dimensions such as t_{ox} and L are reduced by the unitless scaling factor κ, the drain voltage is also reduced by κ, and the substrate doping is increased by the same factor κ. Scaling will be presented in Sec. 8.7.2, but scaling has limitations when physical mechanisms do not scale with field, such as in the subthreshold region where current is due to diffusion and not drift. A simple empirical formula by Brews et al.,[60] for the minimum gate length to maintain long-channel behavior takes

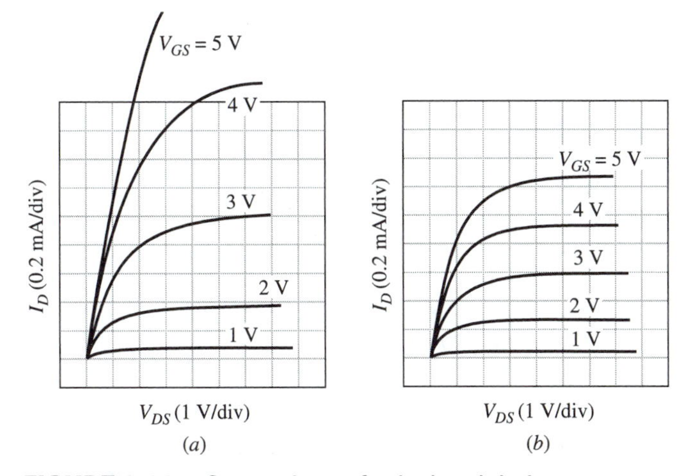

FIGURE 8.38 Comparison of calculated drain current for (*a*) constant mobility (no velocity saturation) and (*b*) velocity saturation.[56]

into account lower voltages, shallower source-drain junctions, thinner oxides, and heavier doping as the MOSFET dimensions are reduced. However, this approach does not require reduction of all dimensions by the same factor, and is summarized in Sec. 8.7.3. Other techniques, such as the lightly-doped drain (LDD)[61] to decrease the high electric field at pinch-off near the drain, are considered in Sec. 8.7.4.

8.7.2 Scaling

As fabrication technology improves, effort is made to further miniaturize the MOSFET to increase speed and circuit complexity. As the channel length is decreased, the source and drain depletion regions can meet, which results in punch-through. The depletion widths can be reduced by increasing the substrate doping concentration and decreasing the drain voltage. As shown in Fig. 7.14, the increased substrate doping increases the threshold voltage, but can be reduced by decreasing the oxide thickness. Denard et al.,[49] suggested a technique that decreases the MOSFET gate length by scaling down a successful long-channel device. This approach assumes that the two devices will behave the same and avoid the short-channel effects if the internal fields are the same, because effects such as carrier multiplication and oxide breakdown depend upon the field.

In this approach, the large device is scaled in gate length ($L^* = L/\kappa$), oxide thickness ($t_{ox}^* = t_{ox}/\kappa$), and junction depth ($x_d^* = x_d/\kappa$) by a unitless scaling factor $1/\kappa$ with $\kappa > 1$, where the $*$ parameter refers to the new scaled-down device. The depletion region of the drain p-n^+ junction was given in Eq. (4.79) with $N_d^+ >> N_a^-$ as

$$x_d = \sqrt{\frac{2\epsilon_{\text{Si}}(V_{bi} - V_{DS})}{qN_a^-}}. \tag{8.126}$$

If N_a^- is increased by κN_a^- and V_{DS} is decreased by V_{DS}/κ, x_d then decreases approximately as $\sqrt{1/\kappa^2}$ or $1/\kappa$. The scaled-gate capacitance is given as

$$C_{ox}^* = \frac{\epsilon_{ox}}{t_{ox}/\kappa} \frac{W}{\kappa} \frac{L}{\kappa}, \tag{8.127}$$

and therefore the gate capacitance scales as $1/\kappa$. With the Shichman-Hodges expression in Eq. (8.20) for the drain current, the scaled drain current is found to scale as I_D/κ:

$$I_d^* = \frac{W/\kappa}{L/\kappa} \mu_n \frac{\epsilon_{ox}}{t_{ox}/\kappa} \left[\frac{(V_{GS} - V_T) - V_{DS}}{\kappa} \right] \frac{V_{DS}}{\kappa} = I_d/\kappa, \tag{8.128}$$

where the decrease in mobility due to the higher impurity concentration is ignored.

With the drain current scaled by $1/\kappa$ and the gate capacitance also scaled by $1/\kappa$, the rate of change for charging the capacitance, $dV/dt = I_d/C_{ox}$, does not scale. However, the capacitance is now only charged to voltages scaled to $1/\kappa$, and therefore the charging time scales as $1/\kappa$ and the speed increases by κ. The power dissipation IV scales as $1/\kappa^2$ and the power-delay product scales as $1/\kappa^3$. Also, the device areas have been scaled by $1/\kappa^2$ so that power per unit chip area is not scaled. The subthreshold turn-off behavior of the MOSFET was shown in Eq. (8.125) to depend on the ratio of the depletion layer capacitance $[C'_{Si}(x_{d_{max}}) = \epsilon_{Si}/x_{d_{max}}, x^*_{d_{max}} = x_{d_{max}}/\kappa]$ to the oxide capacitance $(C'_{ox} = \epsilon_{ox}/t_{ox}, t^*_{ox} = t_{ox}/\kappa)$, which is unchanged by scaling, so the same voltage swing is needed to reduce I_D by one decade as for the larger device. Also, scaling may not be meaningful when the physical mechanism limiting device size does not scale.

8.7.3 Empirical Formula

It was observed by Brews et al.[60] that the minimum channel length $L \geq L_{min}$ to preserve long-channel behavior could be represented by

$$L_{min}(\mu\text{m}) = 0.41(\text{Å})^{-1/3}\{r_j(\mu\text{m})t_{ox}(\text{Å})[\text{W}_\text{S}\,(\mu\text{m}) + \text{W}_\text{D}\,(\mu\text{m})]^2\}^{1/3}, \tag{8.129}$$

where r_j is the source and drain junction depth, and W_S and W_D are the source and drain depletion widths, respectively. Equation (8.129) was established for 100 Å$\leq t_{ox} \leq$ 1000 Å, $10^{14} \leq N_a^- \leq 10^{17}$ cm^{-3}, $0.3 \leq r_j \leq 1.5$ μm by comparison to experimental measurements and computer simulations. A comparison of experimental and computed L_{min} with the predictions of Eq. (8.129) is shown in Fig. 8.39.

A more recent empirical formula for the deep-submicron region from 0.5 μm down to sub-0.1 μm was given by Hu et al.[62] as

$$L_{min}(\mu\text{m}) = 2.2\mu\text{m}^{-2}\left(\frac{\delta V_T}{\delta V_{DS}}\right)^{-0.37} \times (t_{ox} + 0.012\mu\text{m})(W_S + W_D + 0.15\mu)(x_j + 2.9\mu\text{m}), \tag{8.130}$$

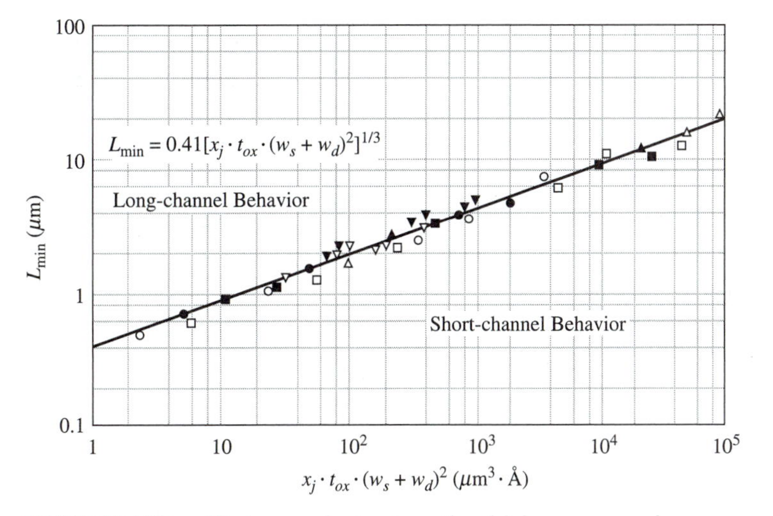

FIGURE 8.39 Minimum device length which preserves long-channel behavior.[60]

where $\delta V_T/\delta V_{DS}$ is obtained from the parallel shift of $\log I_D$ vs. V_{GS} at a given drain current level, corresponding approximately to $V_{GS} = V_T$ in the subthreshold region. *N*-Channel MOSFETs with channel lengths down to sub-0.1 μm were fabricated and evaluated. It was found that the trade-off between device performance (represented by g_m/WC'_{ox}) and the short-channel effect (represented by $\delta V_T/\delta V_{DS}$) is dominated by source-drain parameters x_j, total source and drain parasitic resistance R_{SD}, and junction abruptness, rather than channel parameters t_{ox}, V_T, and channel doping profile.

8.7.4 Other Miniaturization Techniques

A commonly used structure in submicron MOSFETs is the LDD.[61] In the LDD structure, self-aligned n^- regions are introduced between the channel and n^+-source-drain regions to spread the high field at the drain pinch-off region and therefore reduce the maximum electric field. This structure increases breakdown voltage and reduces impact ionization by spreading the high electric field at the drain pinch-off into the n^- region and allows a reduction in channel length or an increase in V_{DS}. The LDD structure is illustrated in Fig. 8.40 (*a*) and compared to the conventional structure in Fig. 8.40 (*b*). The reduction in electric field due to the LDD structure is illustrated by the two-dimensional simulation shown in Fig. 8.41.

The reduction in MOSFET gate length with year for laboratory demonstration structures is shown in Fig. 8.42.[63] As shown in this figure, the gate length was reduced to 1 μm in 1974 by Denard et al.[49] Reduction to 0.1 μm was reported by Sai-Halasz et al.[64] in 1987. In 1995, Ono et al.[63] reported an *n*-channel MOSFET with a 40-nm (0.04-μm) gate length and a maximum transconductance g_m of 428 mS/mm. Table 8.2

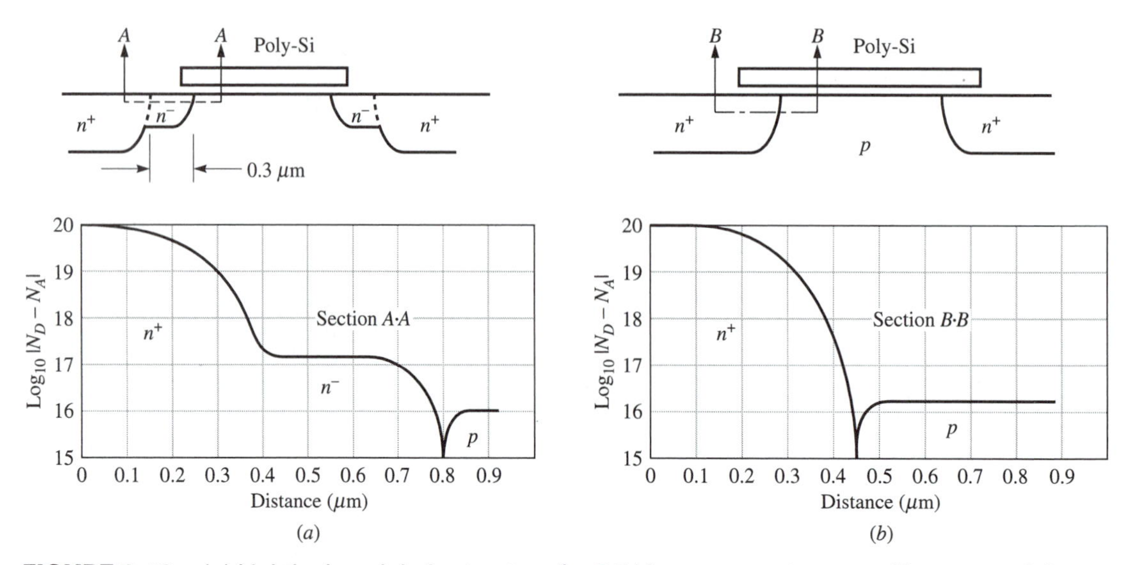

FIGURE 8.40 (*a*) Lightly doped drain structure for 8.5-V power supply, $t_{ox} = 45$ nm, $x_j = 0.3\ \mu$m (n^-), $L = 1.2\ \mu$m, substrate resistivity $\rho_S = 5$ ohm–cm. (*b*) Conventional structure for 5-V power supply with design parameters the same as in part (*a*) except $L = 1.5\ \mu$m and $\rho_s = 15$ ohm–cm.[61]

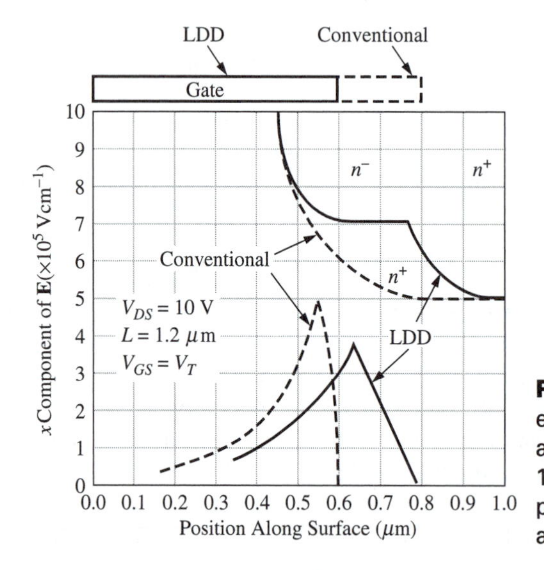

FIGURE 8.41 Magnitude of the electric field at the Si/SiO_2 interface as a function of distance with $L = 1.2\ \mu$m, $V_{DS} = 8.5$ V, $V_{GS} = V_T$. The physical geometries for both devices are shown above the plot.[61]

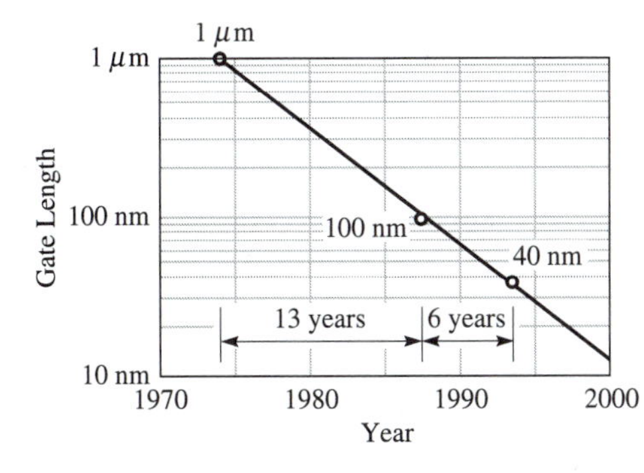

FIGURE 8.42 Trends in the fabrication of small geometry MOSFETs.[63]

TABLE 8.2 Parameter Scaling for 40-nm Gate-length MOSFET.

Parameter	Conventional 0.1-μm MOSFET	Scaling Factor	40-nm Gate-Length MOSFET
L	100 nm	1/2.5	40 nm
t_{ox}	3 nm	1/1	3 nm
N_a^-	1×10^{18} cm^{-3}	1/1	1×10^{18} cm^{-3}
x_j	40 nm	1/4	10 nm
V_{DS}	1.5 V	1/1	1.5 V

summarizes the parameters used in this device. Because the oxide thickness at 3–4 nm has reached the tunneling leakage current limit, the doping is 1×10^{18} cm^{-3}, and $V_{DS} =$ 1.5 V, these quantities were not changed from the $L = 0.1$-μm gate-length MOSFET. Special techniques were used to form the ultrashallow source and drain junctions of 10 nm. The ultrathin junctions were formed by solid-state diffusion from phosphorus phosphosilicate glass (PSG) which is illustrated in Fig. 8.43 (*a*). Arsenic was implanted outside of the gate with an energy of 30 keV to create the deeper source and drain regions. The conventional structure is shown in Fig. 8.43 (*b*).

8.7.5 Reliability

The push for further miniaturization of MOSFET feature sizes and larger numbers of devices on integrated-circuit chips makes reliability a significant part of MOSFET device design. Reliability refers to the time an integrated circuit continues to operate after it has passed initial testing. The common unit of failure rate is the failure unit (FIT), which is the number of units that fail in 10^9 device hours (1 failure in 10^5 in 10^3 h $=$ 10 FIT). A useful summary of reliability is chapter 5 in the book by Pimbley et al.[65]

As the channel length becomes submicron, the substrate doping is increased and the oxide thickness decreased, which can result in oxide reliability problems. When

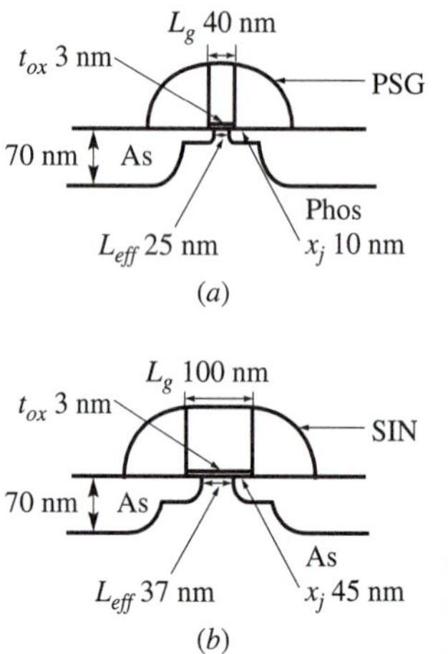

FIGURE 8.43 Schematic cross sections of *n*-channel MOSFETs. (*a*) $L = 40$ nm MOSFET. (*b*) Conventional $L = 0.1\ \mu$m MOSFET.[63]

the electrons in the conducting channel are accelerated to velocity saturation ($\sim 1 \times 10^7$ cm/s), the electron distribution will have more high-energy electrons than for a thermal distribution. The energetic electrons are referred to as "hot" electrons. These hot electrons can be injected over the $\sim$ 3.1-eV barrier at the Si/SiO_2 interface (see Fig. 7.5). Some of these injected electrons can be trapped in the oxide, which leads to threshold voltage drift and degradation of the transconductance. High-field effects at the drain were decreased by the use of the lightly doped drain introduced in the previous part of this section. Another reliability concern is oxide wear-out, where the oxide abruptly ruptures after many hours of operations at fields in the range of 2–6 MV/cm. This failure mode has been suggested to be due to subtle defects within the oxide.

Electromigration is a failure mechanism related to the small size of the metal interconnects such as Al and is the motion of ions of a conductor due to the passage of current. This migration leads to accumulation of vacancies within the metal which become a void and then an open circuit. Continued work on reliability will be required as feature sizes decrease and chip complexity continues to grow.

8.8 MOSFET INTEGRATED-CIRCUIT SPICE EXAMPLES

8.8.1 NMOS Inverter

The *n*-channel NMOS inverter shown in Fig. 8.44 (*a*) is the basic building block for digital integrated circuits. The CMOS inverter, which is presented in Sec. 8.8.2, has now become the preferred digital technology because of its low power requirements. However, an understanding of the NMOS inverter is a useful first step before considering the CMOS inverter. The output characteristics for the enhancement-mode MOSFET

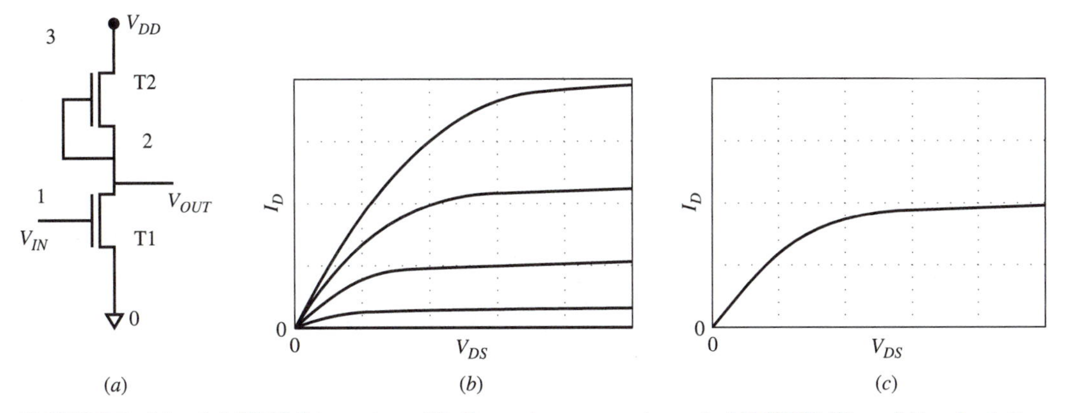

FIGURE 8.44 (*a*) NMOS inverter with the enhancement mode MOSFET T1 and the depletion mode MOSFET T2. (*b*) Output characteristics of the enhancement mode MOSFET T1. (*c*) Output characteristics of the depletion-mode MOSFET T2.

T1 are shown in Fig. 8.44 (*b*). The enhancement-mode MOSFET with a resistive load was presented in Sec. 8.5.1 and shown in Fig. 8.28, but the depletion-mode MOSFET T2 is used in the inverter to replace the load resistor because the MOSFET takes up less chip area and gives more favorable voltage transfer characteristics. The depletion-mode MOSFET with the gate connected to the source means that the MOSFET is always on with $V_{GS} = 0$ V, as shown by the output characteristics in Fig. 8.44 (*c*). The operation of the NMOS inverter will be illustrated by an example.

EXAMPLE 8.3 Find the voltage transfer characteristic for an NMOS inverter [Fig. 8.44 (*a*)] with the following properties:

Property	MOSFET T1	MOSFET T2
L	4.0 μm	4.0 μm
W	4.0 μm	4.0 μm
V_T	0.83 V	−3.0 V
t_{ox}	445 Å	445Å
μ_n	600 cm^2/Vs	600 cm^2/Vs
N_a^-	5.5×10^{15} cm^{-3}	5.5×10^{15} cm^{-3}

Solution The PSpice input file is

```
NMOS Inverter voltage transfer characteristic
vin 1 0
vdd 3 0 dc 5.0
mT1 2 1 0 0 mosT1 L=4U W=4U
```

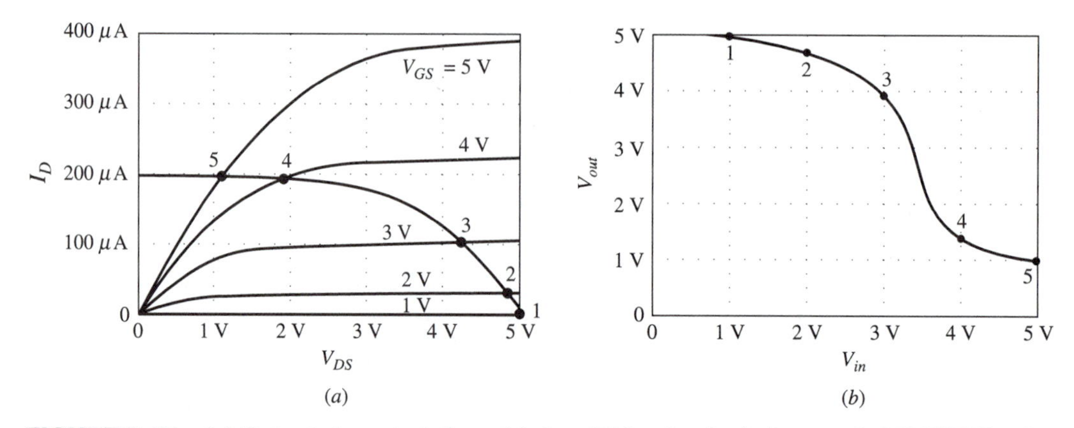

FIGURE 8.45 (*a*) Output characteristics with $L = W$ for the depletion-mode MOSFET load. The values of V_{GS} are given by the labels 1 to 5. (*b*) Voltage transfer characteristic.

```
mT2 3 2 2 0 mosT2 L=4U W=4U
.model mosT1 nmos level=2 vto=0.83 tox=445E-10
+nsub=5.5E15 uo=600
.model mosT2 nmos level=2 vto=-3.0 tox=445E-10
+nsub=5.5E15 uo=600
.dc vin 0 5 0.02
.probe
.end
```

The output characteristics with the depletion-mode MOSFET as the load are shown in Fig. 8.45 (*a*), and the voltage transfer characteristic is shown in Fig. 8.45 (*b*). The intersections of the depletion-mode load with the output characteristics of the enhancement-mode MOSFET are labeled 1–5, where the number corresponds to the value of V_{GS}. The corresponding values of $V_{in} = V_{GS}$ are labeled in Fig. 8.45 (*b*). ■

The operation of the inverter in Fig. 8.44 (*a*) can be illustrated with a capacitive load C_L connected across T1 at VOUT. When the input voltage VIN is ~ 0 V, the enhancement mode MOSFET T1 is off and the load capacitor charges to ~VDD through the depletion-mode MOSFET T2. When the input voltage VIN is ~VDD, the enhancement-mode MOSFET T1 is on, and the load capacitor discharges through both MOSFETs.

noise margin

The measure of the desirability of a particular voltage transfer characteristic is the noise margin which represents the uncertainty in whether an input voltage magnitude will turn the inverter from off to on or from on to off. The inverter voltage transfer characteristic in Fig. 8.45 (*b*) is repeated in Fig. 8.46, where the -1 slope points define the high (output near VDD, logic 1) and the low (output near 0, logic 0) logic levels which represent the two binary logic values. As illustrated in Fig. 8.46, if the output of the driving inverter is at logic 1 at V_{OH}, and V_{IH} is the smallest input which can be interpreted as a logic 1, then there is a margin of safety equal to

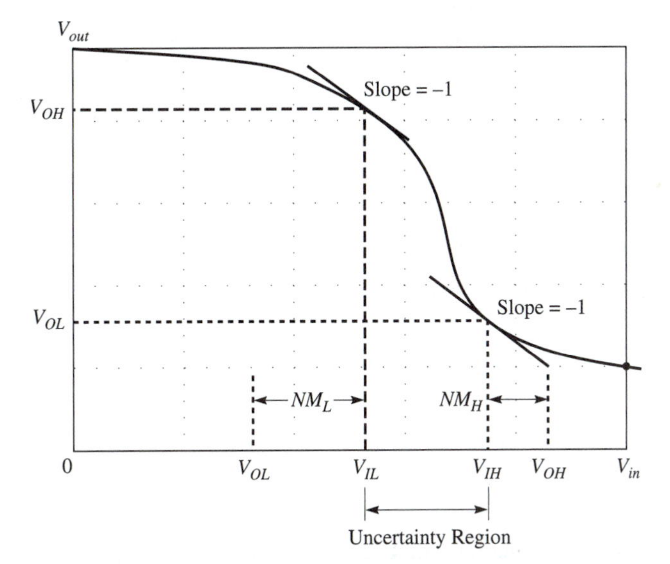

FIGURE 8.46 Voltage transfer characteristic from Fig. 8.45 (*b*) with the points at which the slope is −1 define the high and low logic levels.

$(V_{OH} - V_{IH}) = 4.3 - 3.8$. This difference is called the noise margin for the high input, NM_H, and is given as

$$NM_H = V_{OH} - V_{IH} \tag{8.131}$$

If the output of the driving inverter is at logic 0 at V_{OL}, and V_{IL} is the largest input which can be interpreted as a logic 0, then there is a margin of safety equal to $(V_{IL} - V_{OL}) = 2.6 - 1.6$. This difference is called the noise margin for the low input, NM_L, and is given as

$$NM_L = V_{IL} - V_{OL}. \tag{8.132}$$

The noise margins are a measure of the ability of the inverter to tolerate input voltage variations due to transistor characteristic variations and to be able to restore the signal to logic 0 and logic 1 while the input varies within the noise margins. Changes in the input voltages within the noise margins do not affect the inverter performance.

The desirable voltage transfer characteristic will have large noise margins and minimize the uncertainty region $V_{IH} - V_{IL}$. By increasing the gate length of the depletion-mode MOSFET in Example 8.3 from 4 μm to 20 μm, the output characteristic with the increased gate-length depletion-mode MOSFET is shown in Fig. 8.47 (*a*). The transfer voltage characteristic shown in Fig. 8.47 (*b*) is obtained and the

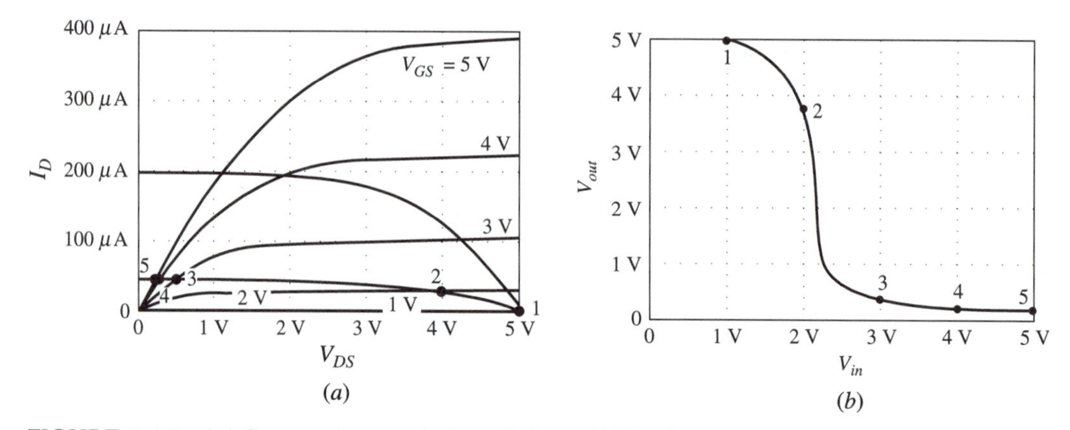

FIGURE 8.47 (*a*) Output characteristic with $L = 4W$ for the depletion-mode MOSFET load. (*b*) Voltage transfer characteristic.

region of uncertainty is decreased. The disadvantage of an increased L in the depletion mode MOSFET is the decrease in the capacitive load charging current when the input MOSFET is turned off. A more complete discussion of noise margin has been given by Hauser.[66]

8.8.2 CMOS Inverter

As feature sizes became micron and submicron, the complementary MOS CMOS technology with both n-channel enhancement-mode MOSFETs (NMOS) and p-channel enhancement-mode MOSFETs (PMOS) became the dominate technology. With NMOS inverters, power is dissipated when the input MOSFET is on, which limits the number of MOSFETs that can be integrated on a single chip due to excessive heating. With CMOS inverters, power is only dissipated during switching from on to off or from off to on, which permits millions of MOSFETs to be integrated on a single chip. A CMOS inverter is shown in Fig. 8.48 with the drain of the NMOS transistor connected to the

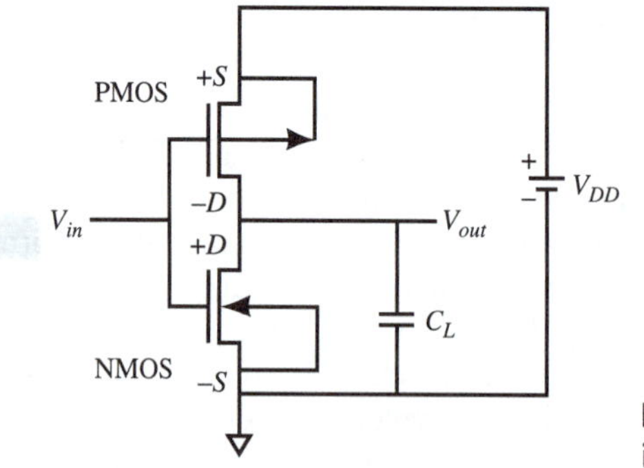

FIGURE 8.48 Circuit for the CMOS inverter with the capacitive load C_L.

drain of the PMOS transistor. Note that the terminal where the carriers flow out is designated the drain. The supply voltage V_{DD} is connected to the PMOS transistor source. The gates of both transistors are connected together so that the input V_{in} drives both MOSFETs. The output load is represented as the capacitance C_L, and the output voltage V_{out} is the voltage at the drains of both MOSFETs. The book by Chen presents both the devices and technology for CMOS[67], while the book by Kang and Leblebici[68] covers MOS integrated circuits.

When $V_{in} = 0$, the NMOS transistor is off, but for the PMOS transistor the gate-to-source voltage $V_{GS_p} = -V_{DD}$ and therefore is on. The PMOS transistor conducts current only while the capacitive load charges to V_{DD}. When $V_{in} = V_{DD}$, the gate-to-source voltage for the NMOS transistor is $V_{GS_n} = V_{DD}$ and therefore is on. The NMOS transistor only conducts current while the charged capacitive load discharges through the conductive path to ground provided by the NMOS transistor. With $V_{in} = V_{DD}$, the PMOS transistor is off with $V_{GS_p} = 0$. Also, the CMOS inverter has a very favorable noise margin because $V_{OH} = V_{DD}$ and $V_{OL} = 0$.

Load line plots for the CMOS inverter are not as straightforward as for the NMOS inverter shown in Figs. 8.45 (*a*) and 8.47 (*a*). The output characteristics for the NMOS and PMOS transistors were summarized in Fig. 8.18. Note that the drain current for the NMOS transistor I_{D_n} flows into the drain (electrons flow out of the drain), while the drain current for the PMOS transistor I_{D_p} flows out of the drain:

$$-I_{D_p} = I_{D_n}. \tag{8.133}$$

The gate-to-source voltage for the NMOS transistor, V_{GS_n}, is V_{in}:

$$V_{GS_n} = V_{in} \tag{8.134}$$

while the gate-to-source voltage for the PMOS transistor is V_{GS_p}:

$$V_{GS_p} = -(V_{DD} - V_{in}) = V_{in} - V_{DD}, \tag{8.135}$$

where V_{GS_p} is negative [see Fig. 8.49 (*a*)]. The drain-to-source voltage for the NMOS transistor V_{DS_n} is V_{out},

$$V_{DS_n} = V_{out}. \tag{8.136}$$

The drain-to-source voltage for the PMOS transistor V_{DS_p} is:

$$V_{DS_p} = -(V_{DD} - V_{out}) = V_{out} - V_{DD} \tag{8.137}$$

or

$$V_{out} = V_{DD} + V_{DS_p}, \tag{8.138}$$

where V_{DS_p} varies between 0 and $-V_{DD}$.

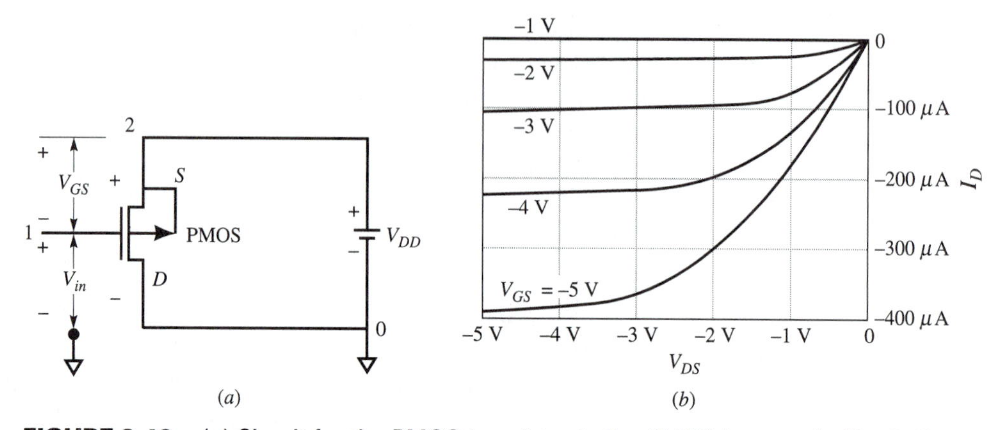

FIGURE 8.49 (*a*) Circuit for the PMOS transistor in the CMOS inverter in Fig. 8.48. (*b*) Output characteristics for the PMOS transistor.

EXAMPLE 8.4 Find the output characteristics for the PMOS transistor in the CMOS inverter configuration shown in Fig. 8.48 with the following properties similar to the NMOS transistor in Example 8.3: $L = 4\ \mu$m, $W = 8\ \mu$m, $V_T = -0.83$ V, $t_{ox} = 445$ Å, $\mu_p = 300$ cm^2/Vs, and $N_d^+ = 5 \times 10^{15}$ cm^{-3}. Note that with $\mu_p = 0.5\mu_n$, the gatewidth for the PMOS transistor is increased to 8 μm to give the same saturation current as for the NMOS transistor in Example 8.3. Let $V_{DD} = 5$ V, and V_{GS_p} vary from 0 to -5 V in -1 V steps.

Solution The circuit diagram is shown in Fig. 8.49 (*a*). The PSpice input file is

```
vds 2 0
vgs 1 2
mos1 0 1 2 2 pmos1 L=4E-6 W=8E-6
.model pmos1 pmos level=2 vto=-0.83 tox=445E-10
+nsub=5.5E15 uo=300
.dc vds 0 5 0.05 vgs 0 -5 -1
.probe
.end
```

The output characteristics for the PMOS transistor in the configuration represented in Fig. 8.49 (*a*) are given in Fig. 8.49 (*b*). ■

EXAMPLE 8.5 Convert the PMOS output characteristics to have the gate voltage in terms of V_{in} and the drain voltage in terms of V_{out}.

Solution The PMOS output characteristics are first transformed to I_{D_n} by Eq. (8.133), and then the gate voltage is written in terms of the input voltage, with Eq. (8.135) to give the output characteristics shown in Fig. 8.50 (*a*). Then the output characteristics are transformed with Eq. (8.138) in Fig. 8.50 (*b*) to permit plotting as the load characteristics for the NMOS transistor. ■

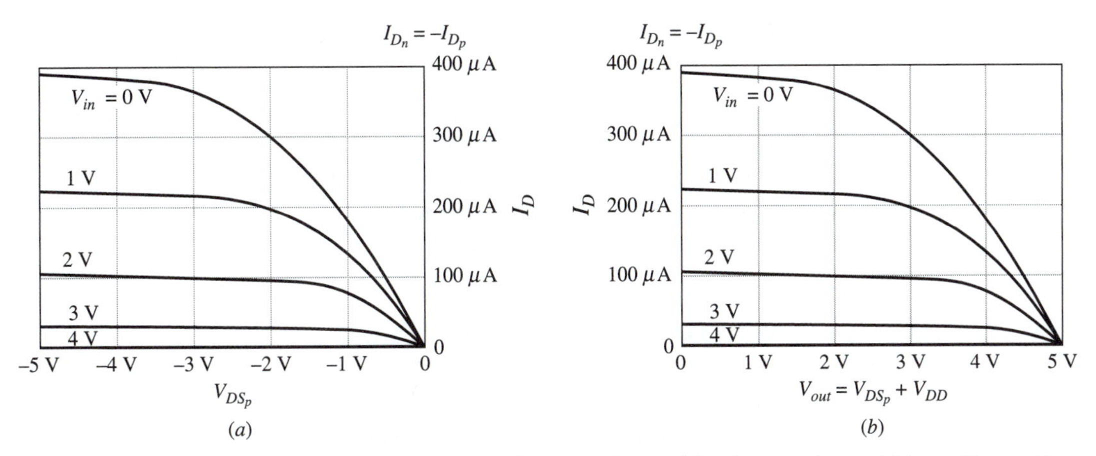

FIGURE 8.50 (*a*) PMOS output characteristics transformed by $I_{D_n} = -I_{D_p}$ and $V_{in} = V_{GS_p} + V_{DD}$. (*b*) PMOS output characteristics transformed from part (*a*) by $V_{out} = V_{DS_p} + V_{DD}$.

EXAMPLE 8.6 Find the load curves for the NMOS output characteristics given in Example 8.3 with the transformed PMOS characteristics given in Fig. 8.50 (*b*), and with this plot, find the voltage transfer characteristic (see Problem 8.28).

Solution The load curves for the NMOS and PMOS transistors are shown in Fig. 8.51 (*a*). The operating points occur where the current through both transistors is equal, which is the intersection at the same V_{in} for both transistors, as indicated in the figure. The resulting voltage transfer characteristic is shown in Fig. 8.51 (b). Note that all the operating points are all near $V_{out} = V_{DD}$ or $V_{out} = 0$. Therefore, the CMOS inverter has a very narrow transition region. ■

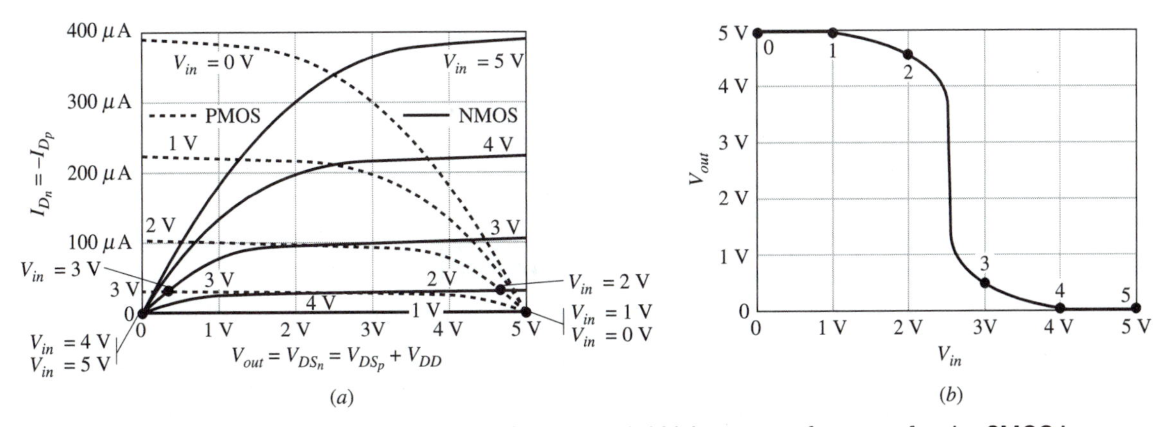

FIGURE 8.51 (*a*) Load curves for the CMOS inverter. (*b*) Voltage transfer curve for the CMOS inverter.

8.8.3 BiCMOS

When it is necessary to drive long interconnects or chip output pads, CMOS circuits are slower than bipolar circuits. Bipolar transistors can provide large driving currents, but have high power dissipation. The BiCMOS technology uses low-power CMOS digital circuits with high-speed bipolar transistors. Also, BiCMOS can use CMOS for digital circuits and bipolar for analog circuits on the same chip. An introduction to BiCMOS inverters and applications has been given by Chen.[67]

8.9 SUMMARY AND USEFUL EXPRESSIONS

This chapter extended the behavior of the two-terminal metal-oxide semiconductor capacitor to the three-terminal metal-oxide semiconductor field-effect transistor (MOSFET).

- The basic I_D-V_{DS} behavior of the MOSFET is represented by

$$I_D = \frac{W}{L}\mu_n \int_0^{V_{DS}} |Q_n'|\, dV(y),$$

  where $|Q_n'|$ represents the charge per unit area in the conducting inversion layer between the source and drain.
- The Shichman-Hodges model (SPICE level 1 model) accounts for the decrease in $|Q_n'|$ from source to drain due to V_{DS}.
- In the "linear" region, $V_{DS} < V_{DS_{sat}}$.
- In the "saturation" region, $V_{DS} > V_{DS_{sat}}$.
- The variable-depletion charge model (SPICE level 2 model) includes the effect of V_{DS} on the depletion-layer charge in the representation of $|Q_n'|$.
- The condition where the charge in the inversion layer diminishes to zero at the drain is called pinch-off, and the pinch-off point moves toward the source as V_{DS} is increased further.
- Current flows by transverse drift and diffusion current in the depletion region between the pinch-off point and the drain.
- There are four basic types of MOSFETs.

1. When the substrate is p-type and the mobile charge carriers are electrons, the transistor is termed an n-channel MOSFET.
2. When the substrate is n-type and the mobile charge carriers are holes, the MOSFET is termed a p-channel MOSFET.
3. When the transistor is off for $V_{GS} = 0$, it is termed an enhancement-mode (normally off) MOSFET.

4. When the MOSFET is on for $V_{GS} = 0$, it is termed a depletion-mode (normally on) MOSFET.

- The transconductance, $g_m = \partial I_D / \partial V_{GS}$, represents the effectiveness of the gate voltage in controlling the drain current.
- The most common figure of merit for operation at high frequencies is the frequency f_T at which the current gain in a common-source configuration becomes unity for the output shorted.
- Subthreshold currents become important, as micron and submicron gate lengths permit millions of MOSFETs on single chips.
- The subthreshold current is represented by the gate voltage swing parameter S, which is the gate voltage change needed to reduce I_D by one decade.
- MOSFET miniaturization can result in short-channel effects where the drain current no longer scales as $1/L$, the threshold voltage decreases, and the gate voltage necessary to reduce the subthreshold current becomes larger.
- A simple empirical formula for the minimum gate length to maintain long-channel behavior takes into account shallower source-drain junctions, thinner oxides, and heavier doping as the MOSFET dimensions are reduced.

The following expressions for the MOSFET were obtained in this chapter.

General Expressions

$$I_D = \frac{W}{L}\mu_n \int_0^{V_{DS}} |Q'_n| \, dV(y)$$

$$g_m \equiv \left. \frac{\partial I_D}{\partial V_{GS}} \right|_{V_{DS}} \qquad f_T = \frac{g_m}{2\pi(C_{GS} + C_{GD})}$$

SPICE Level 1 Model (Shichman-Hodges Model)

$$I_D = (W/L)\mu_n C'_{ox}[(V_{GS} - V_T) - (V_{DS}/2)]V_{DS}$$

$$V_{DS_{sat}} = (V_{GS} - V_T)$$

$$I_{D_{sat}} = (W/2L)\mu_n C'_{ox}(V_{GS} - V_T)^2$$

SPICE Level 2 Model (Variable-depletion Model)

$$I_D = (W/L)\mu_n C'_{ox}\{[V_{GS} - V_{FB} - 2|\psi_b| - V_{DS}/2]V_{DS} - (2/3)(\sqrt{2q\epsilon_{\mathrm{Si}}N_a^-}/C'_{ox})[(V_{DS} + 2|\psi_b|)^{3/2} - (2|\psi_b|)^{3/2}]\}$$

Element Line

```
MXXXXXX ND NG NS NB MODNAME <L=VAL> <W=VAL> <AD=VAL>
+<AS=VAL> <PS=VAL> <PD=VAL> <NRD=VAL> <NRS=VAL> <OFF>
+<IC=VDS,VGS,VBS>
```

Model Line

```
.MODEL MODNAME NMOS <(PAR1=PVAL1 PAR2=PVAL2 . . .)>
.MODEL MODNAME PMOS <(PAR1=PVAL1 PAR2=PVAL2 . . .)>
```

Subthreshold Gate Voltage Swing Parameter

$S \simeq (kT/q) \ln 10 \times [(1 + C'_{\text{Si}}(x_{d_{max}})/C'_{ox})]$

MOSFET Miniaturization

$L_{\text{min}}(\mu\text{m}) = 0.41(\text{Å})^{-1/3}\{r_j(\mu\text{m})t_{ox}(\text{Å})[W_S(\mu\text{m}) + W_D(\mu\text{m})]^2\}^{1/3}$

REFERENCES

1. D. Kahng and M. M. Atalla, "Silicon-Silicon Dioxide Field Induced Surface Devices," *IRE-AIEE Solid-State Device Res. Conf.* (Carnegie Inst. of Tech., Pittsburgh, Pa. 1960).
2. E. H. Snow, B. E. Deal, A. S. Grove, and C. T. Sah, "Ion Transport Phenomena in Insulating Films Using the MOS Structure," J. Appl. Phys. **36**, 1664 (1965).
3. D. R. Kerr, J. S. Logan, P. J. Burkhardt, and W. A. Pliskin, "Stabilization of SiO_2 Passivated Layers with P_2O_5," IBM J. Res. Develop. **8**, 376 (1964).
4. J. B. Balk, "Effects of Hydrogen Annealing on Silicon Surfaces," Electrochem. Soc. Meeting (San Francisco, Calif.), 1965.
5. J. C. Sarace, R. E. Kerwin, D. L. Klein, and R. Edwards, "Metal-Nitride-Silicon Field-Effect Transistors with Self-Aligned Gates," Solid-State Electron. **11**, 653 (1968).
6. Y. P. Tsividis, *Operation and Modeling of the MOS Transistor* (McGraw-Hill, New York, 1987).
7. D. K. Ferry, L. A. Akers, and E. W. Greeneich, *Ultra Large Scale Integrated Electronics* (Prentice Hall, Englewood Cliffs, N.J., 1988), p. 38.
8. G. Massobrio and P. Antignetti, *Semiconductor Device Modeling with SPICE,* 2nd Ed. (McGraw-Hill, New York, 1993), p. 174.
9. H. Shichman and D. A. Hodges, "Modeling and Simulation of Insulated-Gate Field-Effect Transistors Switching Circuits," IEEE J. Solid-State Circuits **SC-3**, 285 (1968).
10. H. K. J. Ihantola and J. L. Moll, "Design Theory of a Surface Field-Effect Transistor," Solid-State Electron. **7**, 423 (1964).
11. J. E. Meyer, "MOS Models and Circuit Simulation," RCA Rev. **32**, 42 (1971).
12. G. A. Armstrong, J. A. Magowan, and W. D. Ryan, "Two-Dimensional Solution of the D.C. Characteristics for the M.O.S.T." Electron. Lett. **5**, 406 (1969).
13. G. A. Armstrong and J. A. Magowan, "Pinch-Off in Insulated Gate Field-Effect Transistors," Solid-State Electron. **14**, 760 (1971).
14. G. A. Armstrong and J. A. Magowan, "The Distribution of Mobile Carriers in the Pinch-Off Region of an Insulated-Gate Field-Effect Transistor and Its Influence on Device Breakdowns," Solid-State Electron. **14**, 723 (1971).
15. G. W. Taylor, "Velocity-Saturated Characteristics of Short-Channel MOSFETs," AT&T Bell Lab. Tech. J. **63**, 1325 (1984).
16. Y. A. El-Mansy and A. R. Boothroyd, "A Simple Two-Dimensional Model for IGFET Operation in the Saturation Region," IEEE Trans. Electron Devices **ED-24**, 254 (1977).
17. M. R. Pinto, C. S. Rafferty, H. R. Yeager, and R. W. Dutton, "PISCES-IIB," Supplementary Report, Stanford Electronics Laboratory, Department of Electrical Engineering, Stanford University, Stanford, Calif. 94305 (1985).
18. R. S. Muller and T. I. Kamins, *Device Electronics for Integrated Circuits,* 2nd ed. (Wiley, New York), p. 436.

19. M. D. Giles, "Ion Implantation" *VLSI Technology,* 2nd ed., S. M. Sze, Ed.(McGraw-Hill, New York, 1983), Chap. 8.

20. Tsividis, *Operation and Modeling,* Chap. 6.

21. S. M. Sze, *Semiconductor Devices: Physics and Technology* (Wiley, New York, 1985), p. 209.

22. D. Foty, *MOSFET Modeling with SPICE* (Prentice Hall, Saddle River, N.J., 1997).

23. M. H. Rashid, *SPICE for Circuits and Electronics Using PSpice* (Prentice Hall, Saddle River, N.J., 1995).

24. W. Banzhaf, *Computer-Aided Circuit Analysis using PSpice,* 2nd ed. (Regents/Prentice Hall, Englewood Cliffs, N.J., 1992), p. 117.

25. L. M. Dang, "A Simple Current Model for Short-Channel IGFET and Its Application to Circuit Simulation," IEEE J. Solid-State Circuits **SC-14**, 358 (1979).

26. Massobrio and Antignetti, *Semiconductor Device Modeling,* p. 161.

27. Ibid., p. 192.

28. Ibid., p. 195.

29. Ibid., pp. 202–207.

30. Ibid., p. 209.

31. D. L. Pulfrey and N. G. Tarr, *Introduction to Microelectronic Devices* (Prentice Hall, Englewood Cliffs, N.J., 1989), Chap. 7.

32. Foty, *MOSFET Modeling,* pp. 213–276.

33. Massobrio and Antignetti, *Semiconductor Device Modeling,* pp. 216–240.

34. J. Y. Chen, *CMOS Devices and Technology for VLSI* (Prentice Hall, Englewood Cliffs, N.J., 1990) pp. 74–77.

35. B. J. Sheu, D. L. Scharfetter, P.-K. Ko, and M.-C. Jeng,"BSIM: Berkeley Short-Channel IGFET Model for MOS Transistors," IEEE J. Solid-State Circuits **SC-22**, 558 (1987).

36. Foty, *MOSFET Modeling,* pp. 318–378.

37. Massobrio and Antignetti, *Semiconductor Device Modeling,* p. 240.

38. Ibid., p. 242.

39. Foty, *MOSFET Modeling,* pp. 277–317.

40. P. R. Gray and R. G. Meyer, *Analysis and Design of Analog Integrated Circuits,* 3rd ed. (Wiley, New York, 1993), p. 67.

41. Tsividis, *Operation and Modeling,* pp. 175–181.

42. Massobrio and Antignetti, *Semiconductor Device Modeling,* p. 213.

43. R.-H. Yan, K. F. Lee, D. Y. Jeon, Y. O. Kim, B. G. Park, M. R. Pinto, C. S. Rafferty, D. M. Tennant, E. H. Westerwick, G. M. Chin, M. D. Morris, K. Early, P. Mulgrew, W. M. Mansfield, R. K. Watts, A. M. Voshchenkov, J. Bokor, R. G. Swartz, and O. Ourmazd, "89-GHz f_T Room-Temperature Silicon MOSFETs," IEEE Electron Devices Lett. **EDL-13**, 256 (1992).

44. M. B. Barron, "Low Level Currents in Insulated Gate Field Effect Transistors," Solid-State Electron. **15**, 293 (1972).

45. R. R. Troutman, "Subthreshold Design Considerations for Insulated Gate Field-Effect Transistors," IEEE J. Solid-State Circuits **SC-9**, 55 (1974).

46. R. M. Swanson and J. D. Meindl, "Ion-Implanted Complementary MOS Transistors in Low-Voltage Circuits," IEEE J. Solid-State Circuits **SC-7**, 146 (1992).

47. R. R. Troutman and S. N. Chakravarti, "Subthreshold Characteristics of Insulated-Gate Field-Effect Transistors," IEEE Trans. Circuit Theory **CT-20**, 659 (1973).

48. R. J. Van Overstraeten, G. J. DeClerck, and P. A. Muls, "Theory of the MOS Transistor in Weak Inversion-New Method to Determine the Number of Surface States," IEEE Trans. Electron Devices **ED-22**, 282 (1975).

49. R. H. Denard, F. H. Gaensslen, H. N. Yu, V. L. Rideout, E. Bassous, and A. R. LeBlanc, "Design of Ion-Implanted MOSFET's with Very Small Physical Dimensions," IEEE J. Solid-State Circuits **SC-9**, 256 (1974).

50. G. W. Taylor, "Subthreshold Conduction in MOSFET's," IEEE Trans. Electron Devices **ED-25**, 337 (1978).

51. J. R. Brews, "A Charge-Sheet Model of the MOSFET," Solid-State Electron. **21**, 345 (1978).

52. J. R. Brews, "Subthreshold Behavior of Uniformly and Nonuniformly Doped Long-Channel MOSFET," IEEE Trans. Electron Devices **ED-26**, 1282 (1979).

53. J. R. Brews, *Physics of the MOS Transistor,* D. Kahng, Ed. (Academic Press, New York, 1981) pp. 1–120.

54. J. E. Chung, M.-C. Jeng, J. E. Moon, P.-K. Ko, and C. Hu, "Performance and Reliability Design Issues for Deep-Submicrometer MOSFETs," IEEE Trans. Electron Devices **ED-38**, 545 (1991).

55. N. Kotani and S. Kawazu, "Computer Analysis of Punch-Through in MOSFETs," Solid-State Electron. **22**, 63 (1979).

56. K. Yamaguchi, "Field-Dependent Mobility Model for Two-Dimensional Numerical Analysis of MOSFETs," IEEE Trans. Electron Devices **ED-26**, 1068 (1973).

57. Brews, *Physics of the MOS Transistor,* pp. 60–108.

58. J. R. Brews, "The Submicron MOSFET," *High-Speed Semiconductor Devices,* S. M. Sze, Ed. (Wiley, New York, 1990), Chap. 3.

59. Tsividis, *Operation and Modeling,* pp. 181–202.

60. J. R. Brews, W. Fichtner, E. H. Nichollian, and S. M. Sze, "Generalized Guide for MOSFET Miniaturization," IEEE Electron Devices Lett. **EDL-1**, 2 (1980).

61. S. Ogura, P. J. Tsang, W. W. Walker, D. L. Critchlow, and J. F. Sheppard, "Design and Characterization of Lightly Doped Drain-Source (LDD) Insulated Gate Field-Effect Transistors," IEEE Trans. Electron Devices **ED-27**, 1359 (1980).

62. H. Hu, J. B. Jacobs, L. T. Su, and D. A. Antoniadis, "A Study of Deep-Submicron MOSFET Scaling Based on Experiment and Simulation," IEEE Trans. Electron Devices **ED-42**, 669 (1995).

63. M. Ono, M. Saito, T. Yoshitomi, C. Fiegna, T. Ohguro, and H. Iwari, "A 40 nm Gate Length *n*-MOSFET," IEEE Trans. Electron Devices **ED-42**, 1822 (1995).

64. G. Sai-Halasz, M. Wordeman, D. Kern, E. Ganin, S. Rishton, D. Zicherman, H. Schmid, M. Polcari, H. Ng, P. Restle, T. Chang, and R. Denard, "Design and Experimental Technology for 0.1 μm Gate-Length Low-Temperature Operation of FETs," IEEE Electron Devices Lett. **EDL-8**, 463 (1987).

65. J. M. Pimbley, M. Ghezzo, H. G. Parks, and D. M. Brown, *Advanced CMOS Process Technology,* Vol. 19, *Electronics Microstructure Science*, N. G. Einspruch, Ed. (Academic Press, San Diego, 1989), pp. 181–225.

66. J. R. Hauser, "Noise Margin Criteria for Digital Logic Circuits," IEEE Trans. Educ. **E-36**, 363 (1993).

67. Chen, *CMOS Devices,* p. 92.

68. S.-M. Kang and Y. Leblebici, *CMOS Digital Integrated Circuits, Analysis and Design,* (McGraw-Hill, New York, 1996) p. 175.

69. Ibid., p. 150.

70. Tsividis, *Operation and Modeling,* p. 310.

71. Massobrio and Antignetti, *Semiconductor Device Modeling,* pp. 195–202.

72. D. E. Ward and R. W. Dutton, "A Charge-Oriented Model for MOS Transistor Capacitances," IEEE J. Solid-State Circuits **SC-13**, 703 (1978).

73. S. Y. Oh, D. E. Ward, and R. W. Dutton, "Transient Analysis of MOS Transistors," IEEE Trans. Electron Devices **ED-27**, 1571 (1980).

74. D. A. Hodges and H. G. Jackson, *Analysis and Design of Digital Integrated Circuits,* 2nd. ed. (McGraw-Hill, New York, 1988) pp. 41 and 73.

75. Banzhaf, *Computer-Aided Circuit Analysis,* p. 120.

PROBLEMS

All problems are for room temperature unless another temperature is specified.

8.1 Consider an n-channel enhancement-mode MOSFET with the following parameters: $W = 4\ \mu$m, $L = 2\ \mu$m, $\mu_n = 675$ cm²/V-s, $t_{ox} = 445$ Å, and $V_T = 0.86$ V. With the Shichman-Hodges model,

(a) Calculate C_{ox} and C'_{ox}.

(b) For $V_{GS} = 1, 2, 3, 4$, and 5 V, calculate $V_{DS_{sat}}$ and $I_{D_{sat}}$.

(c) For each V_{GS}, plot $I_{D_{sat}}$ vs. $V_{DS_{sat}}$. Use a scale of 0 to 1×10^{-3} A for I_D and a scale of 0 to 5 V for V_{DS}.

(d) Sketch the I_D-V_{DS} characteristic for the conditions of part (c). No detailed calculations are necessary.

8.2 Consider an n-channel enhancement-mode MOSFET with the following parameters: $\mu_n = 550$ cm²/V-s, $W = 350\ \mu$m, $L = 5\ \mu$m, $t_{ox} = 800$ Å, and $V_T = 1.2$ V. With the Shichman-Hodges model,

(a) Calculate C_{ox} and C'_{ox}.

(b) Sketch I_D vs. V_{DS} for $V_{DS} = 0$ to 5 V with $V_{GS} = 2, 3, 4$, and 5 V. Give the value of $I_{D_{sat}}$ for each V_{GS}.

8.3 Consider an n-channel enhancement-mode MOSFET with the following parameters: substrate with $p = N_a^- = 5 \times 10^{16}$ cm^{-3}, channel mobility of $\mu_n = 550$ cm²/V-s, $t_{ox} = 40$ nm, Al gate, $L = 3\ \mu$m, $W = 25\ \mu$m, and no interface charge. With the PSpice level 2 model, plot $\sqrt{I_D}$ with $V_{DS} = 5$ V vs. V_{GS}.

(a) Find the threshold voltage V_T by plotting $\sqrt{I_D}$ as V_{GS} is varied from -1 V to $+2.0$ V in 0.1-V steps.

(b) Find V_T as in part (a), but with $N_f = 8 \times 10^{10}$ cm^{-2}.

(c) Find V_T as in part (a), but with $p = N_a^- = 5 \times 10^{15}$ cm^{-3}, $\mu_n = 700$ cm^2/V-s, and $N_f = 8 \times 10^{10}$ cm^2.

(d) Find V_T as in part (a), but with $t_{ox} = 60$ nm, and $N_f = 8 \times 10^{10}$ cm^2.

(e) Find V_T as in part (a), but use an n^+ Si gate.

Please note how you can vary the parameters to change V_T.

8.4 For the parameters of Problem 8.3, part (b), use PSpice to plot the I_D-V_{DS} characteristics for V_{DS} from 0 to +5 V in 0.2-V steps. Set the I_D scale from 0 to 3.0 mA. Let V_{GS} vary from 0 to 5 V in 1-V steps.

(a) Level-1 model.

(b) Level-2 model.

(c) What are the significant differences in the I_D-V_{DS} curves for the level 1 and level 2 models?

8.5 The threshold voltage for an n-channel MOSFET (MOSFET 1) with an Al gate is 0.782 V. The threshold voltage for an identical MOSFET (MOSFET 2) except with a different gate material is found to have a threshold voltage of 0.480 V. What is the work function for MOSFET 2 gate material?

8.6 Selected parameters for an n-channel enhancement-mode MOSFET with 2-μm feature size which were provided by the Si foundry program managed by MOSIS and are summarized below:

1. LEVEL=2	2. LD=0.25U	3. TOX=445E-10
4. NSUB=5.5E+15	5. VTO=0.86	6. KP=5.25E-05
7. GAMMA=0.548	8. PHI=0.6	9. UO=675
10. UEXP=0.230	11. UCRIT=80560	12. DELTA=1.0E-06
13. VMAX=87100	14. XJ=0.25U	15. LAMBDA=2.89E-02
16. RSH=19.6	17. CGDO=2.9E-10	18. CGSO=2.9E-10
19. CGBO=3.8E-10	20. CJ=9.46E-05	21. MJ=0.67
22. CJSW=4.5E-10	23. MJSW=0.10	24. PB=0.80

In words or by a sketch, describe what each parameter represents.

8.7 An n-channel enhancement mode Si MOSFET with an Al gate has the following properties at room temperature: gate length $= 2.5\mu$m, gatewidth $= 300\mu$m, channel mobility $= 575$ cm^2/V-s, oxide thickness $= 500$ Å, Debye length $= 5.80 \times 10^{-6}$ cm, fixed oxide charge density $= 4 \times 10^{10}$ cm^{-2}, maximum depletion width $= 4.19 \times 10^{-5}$ cm, depletion-layer charge at maximum depletion $= 3.352 \times 10^{-8}$ coulombs/cm^2, Fermi level from conduction band edge $= 0.898$ eV (neutral bulk), potential ψ_b in the neutral Si bulk $= 0.339$ V. *Find the threshold voltage.*

8.8 Design an Al gate n-channel MOSFET with an SiO_2 layer 1000 Å thick and a gate length of 5 μm.

(a) Designate the substrate impurity concentration to give a threshold voltage of +1.0 V.

(b) Designate the gatewidth to give a saturated source-to-drain current I_D of 1.0 mA for $V_{GS} = 3$ V and $V_{DS} = 5$ V.

Use the Shichman-Hodges MOSFET model and assume that the carrier mobility in the channel is one-half the bulk value.

8.9 The mobile charge in the channel of a MOSFET may be represented by

$$C'_{ox}(V_{GS} - V_T)[1 + V(y) - \delta V(y)^2],$$

where δ is an empirical constant. Find the expression for the drain current I_D as a function of the drain-to-source voltage V_{DS}.

8.10 A Si MOSFET ($L = 5$ μm and $W = 40$ μm) with an n^+ polycrystalline Si gate and a p-type substrate has an ideal flat-band voltage of -0.292 V and the following SPICE parameters:

```
LEVEL=2, VTO=0.86, GAMMA=0.550, PHI=0.687, KP=5.25E-05,
TOX=445E-10, NSUB=5.5E+15, NSS=1.0E+10, and TPG=1.0.
```

(a) Find C'_{ox}.

(b) Find the threshold voltage from the other parameters just specified and compare to `VTO`.

(c) Find the threshold voltage V_T by plotting $\sqrt{I_D}$ with $V_{DS} = 5$ V vs. V_{GS} with PSpice as V_{GS} is varied from -1.0 V to $+2.0$ V in $+0.05$ V steps.

8.11 When $|Q'_n| = C'_{ox}\{V_{GS} - [V_{FB} + 2|\psi_b| + \sqrt{4q\epsilon_{\mathrm{Si}}N_a^-|\psi_b|}/(\epsilon_{ox}/t_{ox})]\}$, what assumption has been made regarding the relationship between V_{GS} and ψ_s?

8.12 What conditions are set equal to zero in the SPICE level 1 and 2 models to obtain expressions for the saturation value of V_{DS}?

8.13 Why does the drain current continue to flow after pinch-off is reached?

8.14 For the MOSFET parameters given in Problem 8.6, with $L = 2$ μm, $W = 4$ μm, NFS = 5.55E + 12, NEFF = 1, NSS = 1.0E + 10, and TPG = 1.0:

(a) To observe the subthreshold current, plot I_D from 1×10^{-7} A to 1×10^{-4} A as V_{GS} is varied from 0 to 5 V in 0.1 V steps and with $V_{DS} = 0.1$ V.

(b) With $V_{DS} = 0.1$ V, find the threshold voltage V_T by plotting I_D vs. V_{GS} as V_{GS} is varied from -1.0 V to $+2.0$ V in 0.1-V steps.

(c) Repeat part (a) with $V_{DS} = 5.0$ V and comment on the differences with part (a).

(d) Plot the I_D-V_{DS} characteristics for V_{GS} from 0 to +5 V in 1-V steps and V_{DS} from 0 to 5 V in 0.2-V steps. Set the I_D scale to vary from 0 to 1.0 mA.

8.15 Read the paper by M. Ono, M. Saito, T. Yoshitomi, C. Fiegna, T. Ohguro, and H. Iwai, "A 40 nm Gate Length n-MOSFET," *IEEE Transactions on Electron Devices,* **ED-42,** pp. 1822–1830, Oct. 1995, and write a one-page summary of what you think are the most important results. Please use a word processing system such as LATEX.

8.16 For an n-channel MOSFET with $t_{ox} = 400$ Å and $N_a^- = 6.6 \times 10^{15}$ cm^{-3}:

(a) With $\psi_s = 2|\psi_b|$, compare qN'_I obtained with the complete expression to qN'_I obtained with the approximate expression.

(b) Compare qN'_I with Q'_{sc} for $\psi_s = 2|\psi_b|$.

(c) How much would the threshold voltage change if qN'_I was included with Q'_{sc} for Q'_{Si}?

8.17 For an n-channel MOSFET with $t_{ox} = 400$ Å and $N_a^- = 6.6 \times 10^{15}$ cm^{-3}, plot qN_I' and Q_{sc}' (on a semi-log scale) for ψ_s between $|\psi_b|$ and $2|\psi_b|$. Use the approximate expression for qN_I'.

8.18 For the subthreshold current given in Eq. (8.122), show that the expression reduces to units of Amperes.

8.19 This problem investigates the transconductance g_m and the cutoff frequency f_T based on the parameters from previous examples. From Example 7.2 for an Al gate on p Si, the hole concentration was given as $p = N_a^- = 4 \times 10^{15}$ cm^{-3}, and the oxide thickness was given as $t_{ox} =$ 500 Å (50 nm). The flat-band voltage V_{FB}° was found to be -0.842 V, the capacitance per unit area was $C_{ox}' = 6.903 \times 10^{-8}$ F/cm^2, and the threshold voltage was $V_T = 0.230$ V. In Example 7.3, the Debye length L_D was found to be 6.485×10^{-6} cm. The mobility μ_n was given as 600 cm^2/Vs in Example 8.1, and in Example 8.2 the gate dimensions were $L = 10$ μm and $W =$ 40 μm.

(a) As a first step, find $V_{DS_{sat}}$ and $I_{D_{sat}}$ with the Shichman-Hodges model for $V_{GS} = 1$, 2, 3, 4, and 5 V.

(b) For the small-signal model, find g_m for V_{GS} values in part (a) and normalize by dividing by the gatewidth W in units of mm to give g_m in mS/mm.

(c) For g_m at $V_{GS} = 4$ V, find the value of the source resistance R_s that reduces g_m to $0.85g_m$.

(d) As g_m gets larger, what influence does R_s have?

(e) Find an approximate value for the cutoff frequency f_T for the V_{GS} values in part (a).

(f) If the gate length is reduced from 10 μm to 2 μm, how much does f_T increase?

8.20 Consider an n-channel MOSFET with the following parameters: $V_{FB}^\circ = -0.842$ V, $V_{T_n} = 0.260$ V, $\mu_n = 600$ cm^2/Vs, $t_{ox} = 500$ Å, $W = 40$ μm, $L = 10$ μm, $p = N_a^- = 4 \times 10^{15}$ cm^{-3}, $C_{ox}' = 6.903 \times 10^{-8}$ F/cm^2, $L_D = 6.485 \times 10^{-6}$ cm, $C_{FB}' = 1.597 \times 10^{-7}$ F/cm^2, and $|\psi_b| =$ 0.335 V. Calculate the subthreshold drain current with the expressions in Sec. 8.6.1 for V_{GS} from -0.2 V to the threshold value of 0.260 V and plot I_D vs. V_{GS}. Let V_{DS} be greater than $3kT/q$.

8.21 This problem uses the level 1 SPICE model to determine the transconductance. The MOSFET on p-type Si has $p = N_a^- = 7 \times 10^{15}$ cm^{-3}, an oxide thickness of 35 nm, an Al gate with $L = 10$ μm and $W = 45$ μm, and the electron channel mobility of 600 cm^2/Vs.

(a) Use the SPICE level 1 model to obtain the I_D-V_{DS} characteristics. Let V_{GS} vary from 0 to 5 V in 1-V steps with V_{DS} between 0 and 5 V.

(b) From the results given in the *filename.out* file, obtain the threshold voltage.

(c) With the equations for the Shichman-Hodges model, calculate $I_{D_{sat}}$ and then g_m for each V_{GS} value.

(d) Use PSpice to obtain the variation of the drain current with the gate voltage for drain voltages of $V_{DS} = 1$, 2, 3, 4, and 5 V.

(e) From the results in part (d), find the transconductance g_m. In probe, after Trace has been selected and you have clicked on Add, the derivative is obtained by entering in the Trace Command: dialog box the expression `d(y)/d(x)`, where y is your symbol for the drain current [such as I(d)] and x is your symbol for the gate voltage.

(f) Normalize the transconductance in part (c) by dividing by the gatewidth in mm to give the transconductance in units of mS/mm.

8.22 For an n-channel MOSFET with $p = N_a^- = 5.5 \times 10^{15}$ cm^{-3} and an oxide thickness of 44.5 nm, find the voltage swing parameter S by the approximate expression.

8.23 **(a)** For an n-channel MOSFET with $p = N_a^- = 6.6 \times 10^{15}$ cm^{-3} and an oxide thickness of 40 nm, find the voltage swing parameter S by the approximate expression.

(b) Compare the result in part (a) with S obtained from Fig. 8.40.

8.24 For the MOSFET given in Problem 8.6, find the minimum channel length to obtain long-channel behavior at $V_{DS} = 5$ V.

8.25 Compare the oxide thickness given in Problem 8.6 with the oxide thickness obtained from the empirical formula in Sec. 8.7.3 for $L = 2\ \mu$m and $V_{DS} = 5$ V.

8.26 **(a)** For the NMOS inverter in Example 8.3, plot the voltage transfer characteristic for the depletion-mode device with $L = 4\ \mu$m, $L = 6\ \mu$m, and $L = 8\ \mu$m.

(b) Which gate length gives the more desirable transfer characteristic?

8.27 For the depletion-mode gate lengths given in Problem 8.26 for the NMOS inverter in Example 8.3, find the uncertainty region $V_{IH} - V_{IL}$ as shown in Fig. 8.51.

8.28 Write a PSpice file for the CMOS inverter circuit shown in Fig. 8.53 and obtain the voltage transfer characteristic for $V_{DD} = 5$ V. Use the NMOS parameters from Example 8.3 and the PMOS parameters from Example 8.4.

8.29 For the load curves and the voltage transfer curve in Fig. 8.56, determine whether each transistor at the input voltage values ($V_{in} = 0, 1, 2, 3, 4,$ and 5 V) is in the saturation, linear region or turned off.

8.30 A CMOS inverter with a second stage as the load is shown in the figure. Find the pulse response for the first stage of the CMOS inverter with $V_{DD} = 5$ V and the input pulse given by:

```
vin 1 0 pulse 0 5 0.5ns 0.1ns 0.1ns 3ns 10ns.
```

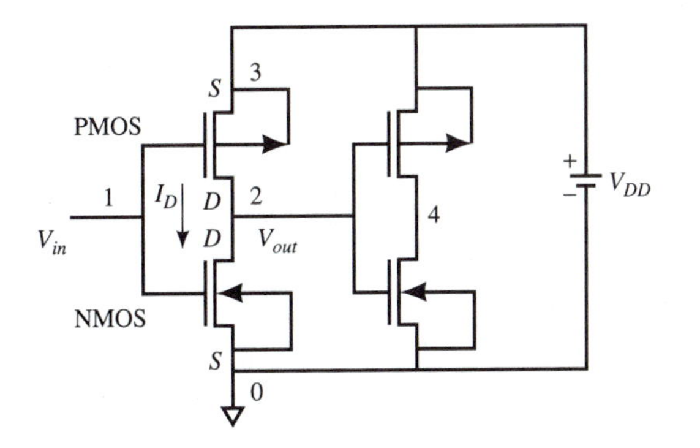

Plot both V_{in} and V_{out} on the same plot with a time scale from 0 to 10 ns.

The NMOS MOSFET has $L = 5\ \mu$m and $W = 5\ \mu$m, and the PMOS MOSFET has $L = 5\ \mu$m and $W = 15\ \mu$m. The parameters for the MOSFETs are for a 2-μm process from MOSIS and are:

```
nmos LEVEL=2 PHI=0.600000 TOX=4.1300E-08 XJ=0.200000U
+ TPG=1 VTO=0.7108 DELTA=4.8120E+00 LD=2.9230E-07
+ KP=4.7115E-05 UO=563.5 UEXP=1.5690E-01
+ UCRIT=1.0980E+05 RSH=2.6430E+01 GAMMA=0.5617
+ NSUB=6.6450E+15 NFS=2.060E+11 VMAX=6.4920E+04
+ LAMBDA=3.2380E-02 CGDO=3.6659E-10 CGSO=3.6659E-10
+ CGBO=3.7314E-10 CJ=1.0789E-04 MJ=0.6654 CJSW=4.5280E-10
+ MJSW=0.310750 PB=0.800000
pmos LEVEL=2 PHI=0.600000 TOX=4.1300E-08 XJ=0.200000U
+ TPG=-1 VTO=-0.7905 DELTA=2.7300E+00 LD=2.8650E-07
+ KP=2.1087E-05 UO=252.2 UEXP=2.6920E-01
+ UCRIT=4.6950E+04 RSH=7.3710E+01 GAMMA=0.6379
+ NSUB=8.5700E+15 NFS=2.770E+11 VMAX=9.9990E+05
+ LAMBDA=4.4130E-02 CGDO=3.5932E-10 CGSO=3.5932E-10
+ CGBO=4.3195E-10 CJ=2.5057E-04 MJ=0.5508 CJSW=2.8373E-10
+ MJSW=0.273554 PB=0.800000
```

APPENDIX **D**

D.1 RESISTANCE AND CHARGE-STORAGE PARAMETERS

RS & RD

RSH

The parasitic source and drain series resistance can be specified by two separate methods. In one, the source and drain resistances are specified in the .MODEL line as RS and RD by parameters 25 and 26 in Table D.1. The other method uses parameter 27 to give RSH, which is the source and drain sheet resistance [Eq. (3.33)] in ohms per square. The source and drain resistances are then obtained by SPICE from RSH and from the number of squares for the source and drain which are given in the MOSFET element line by NRS for the source and NRD for the drain.

TABLE D.1 SPICE MOSFET Parasitic Parameters.

No.	Text Symbol	SPICE Keyword	Level	Parameter Name	Default Value	Units
25	R_S	RS	1–3	Source ohmic resistance	0.0	Ω
26	R_D	RD	1–3	Drain ohmic resistance	0.0	Ω
27	R_{sh}	RSH	1–3	Source and drain sheet resistance	0.0	Ω/□
28	—	QXC	2,3	Coefficient of channel charge share	0.0	—
29	C_{GSO}	CGSO	1–3	Gate-source overlap capacitance per meter	0.0	F/m
30	C_{GDO}	CGDO	1–3	Gate-drain overlap capacitance per meter	0.0	F/m
31	C_{GBO}	CGBO	1–3	Gate-bulk overlap capacitance per meter	0.0	F/m
32	C_{j0}	CJ	1–3	Zero-bias bulk-bottom capacitance per m^2	0.0	F/m^2
33	M_j	MJ	1–3	Bulk-bottom capacitance grading coefficient	0.5	—
34	C_{jsw0}	CJSW	1–3	Zero-bias perimeter capacitance per m	0.0	F/m
35	M_{jsw}	MJSW	1–3	Perimeter capacitance grading coefficient	0.5	—
36	—	FC	1–3	Coefficient for forward-bias source and drain depletion capacitance formula	0.5	—
37	C_{bs}	CBS	1–3	Zero-bias bulk-source capacitance	0.0	F
38	C_{bd}	CBD	1–3	Zero-bias bulk-drain capacitance	0.0	F
39	J_s	JS	1–3	Source or drain *p-n* junction saturation current density per m^2	1×10^{-8}	A/m^2
40	I_s	IS	1–3	Source or drain *p-n* junction saturation current	1×10^{-14}	A

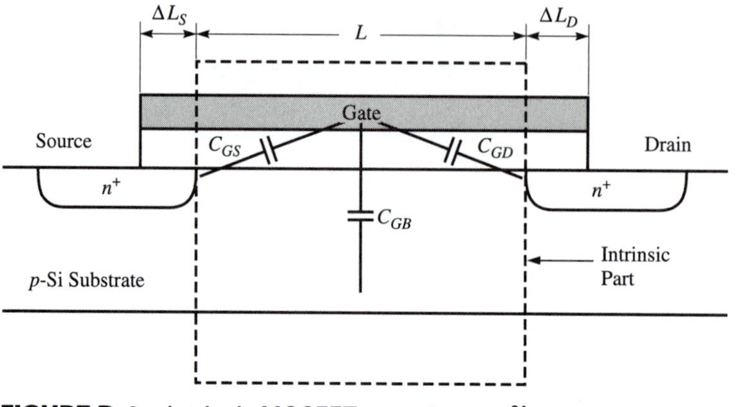

FIGURE D.1 Intrinsic MOSFET capacitances.[31]

The charge-storage parameters for large-signal transient analysis, such as for digital logic circuits, are much more complex than the small-signal capacitances which are introduced in Sec. 8.5, because the MOSFET is operated for input voltages which drive the device from cutoff conditions to saturation. For SPICE circuit applications, the charge-storage parameters are separated into the *intrinsic* capacitances for the gate region between the source and drain as indicated by the region inside the dashed line in Fig. D.1 and the *extrinsic* capacitances. The extrinsic capacitances are associated with the overlap of the gate over the source and drain as indicated by ΔL_S and ΔL_D in Fig. D.1 and with the junction capacitances between the n^+-source and -drain regions with the p-type substrate. Only the capacitances of the overlap of the gate over the source and drain can be taken as voltage independent due to the heavy doping of the source and drain regions which makes the n^+ regions "metallic like."

gate capacitance

Division of the gate capacitance C_G into the three voltage-dependent capacitances C_{GB}, C_{GS}, and C_{GD} associated with the intrinsic part of the MOSFET is rather complex and depends on whether the MOSFET is in cutoff, the linear, or saturation regions. The SPICE circuit analysis program calculates the appropriate capacitances from the oxide layer thickness, device dimensions, and the electrode voltages with a gate capacitance model similar to that of Meyer.[11] In this model, the gate voltage is considered to vary between $-V_{DS} < V_{GS} < +V_{DS}$ and the capacitances are represented by the change in gate charge Q_G due to changes in V_{BS}, V_{GS}, and V_{DS}. The three capacitances C_{GB}, C_{GS}, and C_{GD} are represented at a given V_{DS}, as V_{GS} is varied between $-V_{DS}$ and $+V_{DS}$. This variation with V_{DS} is represented in the I_D-V_{DS} characteristics of Fig. D.2 (*a*), where $V_{DS} = 2$ V and V_{GS} is varied from $-V_{DS}$ to V_T, with V_T represented by $V_1 = V_T$ in Fig. D.2 (*a*). As V_{DS} goes from $V_1 = V_T$ to V_2, I_D is in the saturation region, and as V_{GS} goes from V_2 to V_3, I_D is in the linear region. With $-V_{DS} < V_{GS} < V_T$, the MOSFET is biased so that the Si at the SiO_2/Si interface is in accumulation or depletion. The capacitance C_{GB}, defined as $C_{GB} = \partial Q_G / \partial V_{BS}$, is the capacitance of the MOS capacitor as given in Sec. 7.4. Just as for the MOS capacitor when the inversion layer is formed for $V_{GS} > V_T$, the semiconductor fixed space charge due to ionized impurities becomes constant and Q_G ceases to respond to changes in V_{BS} above

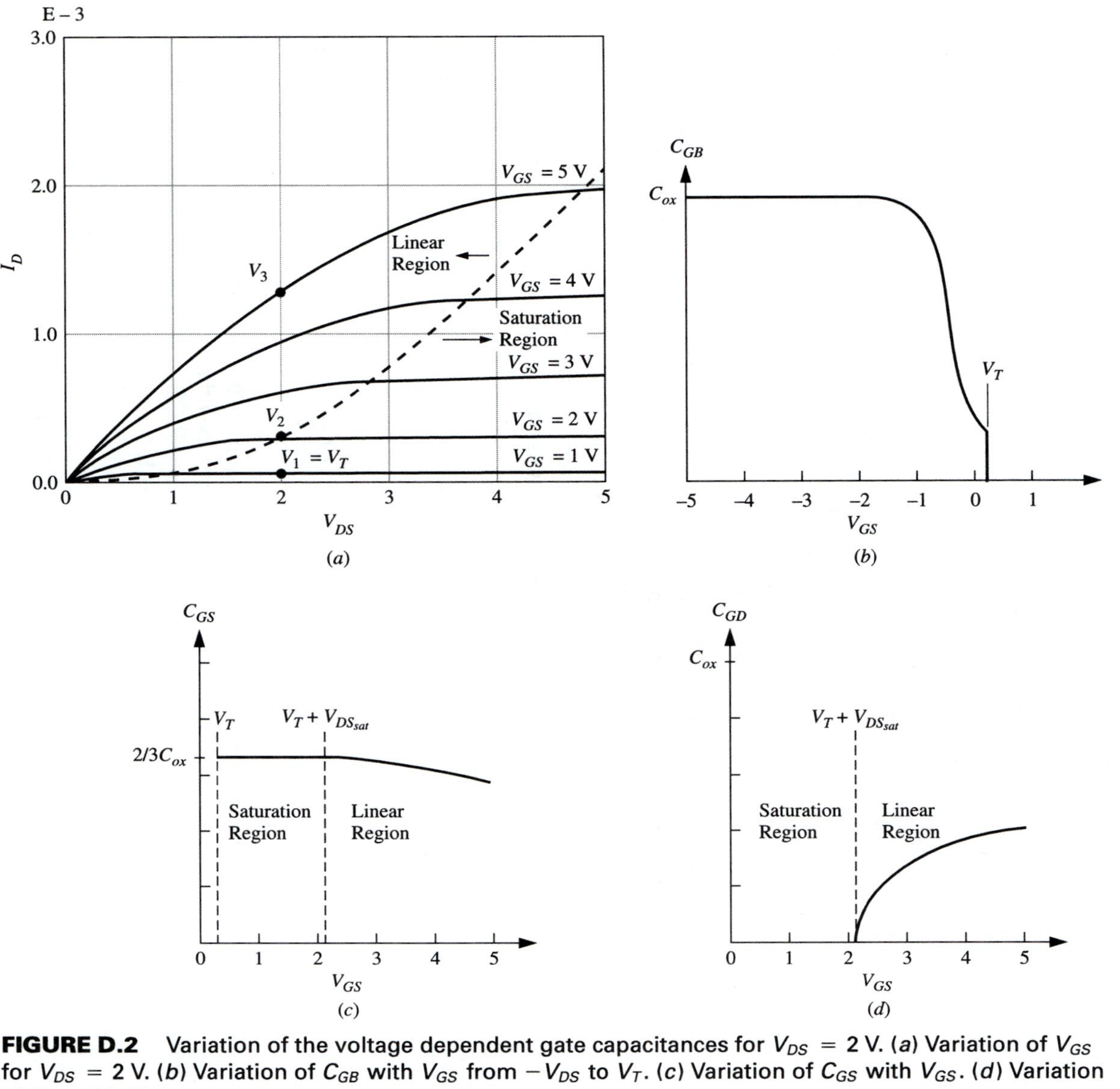

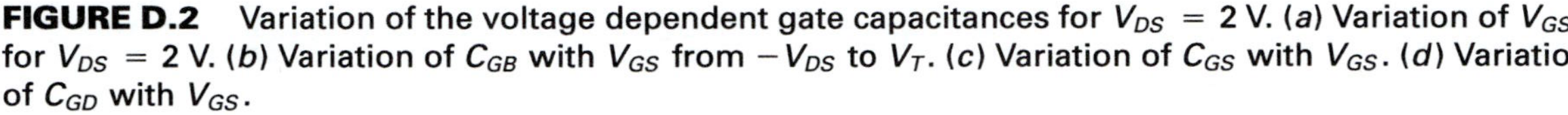

FIGURE D.2 Variation of the voltage dependent gate capacitances for $V_{DS} = 2$ V. (*a*) Variation of V_{GS} for $V_{DS} = 2$ V. (*b*) Variation of C_{GB} with V_{GS} from $-V_{DS}$ to V_T. (*c*) Variation of C_{GS} with V_{GS}. (*d*) Variation of C_{GD} with V_{GS}.

threshold and C_{GB} becomes zero when the MOSFET is on. The variation of C_{GB} with V_{GS} is illustrated in Fig. D.2 (*b*).

The models used for C_{GS} and C_{GD} are known as quasi-static models. In this case, the n^+-source region can supply the necessary carriers for the conducting channel in the MOSFET, and it is assumed that the terminal voltages change slowly enough so that the carrier distributions are always close to their steady-state distributions. Derivation of expressions for C_{GS} and C_{GD} have been given by Pulfrey and Tarr[31] and Tsividis.[70] Massobrio and Antognetti[71] describe the use of these capacitances in SPICE. Note that for the boundary between the saturation and linear regions, which is represented by the dashed line in Fig. D.2 (*a*), $V_{DS_{sat}} = (V_{GS} - V_T)$ by Eq. (8.22) or $V_{GS} = (V_{DS_{sat}} + V_T)$. Therefore, in the saturation region for V_{GS} between V_T and $(V_{DS_{sat}} + V_T)$, C_{GS} is taken as $(2/3)C_{ox}$ (see Sec. 8.5) and C_{GD} is taken as zero. In the linear region, Meyer's model[11] is used for C_{GS} and C_{GD}. These values for C_{GS} and C_{GD} are shown in Fig. D.2 (*c*) and (*d*).

While the Meyer model works well in many cases, it fails for circuits where charge storage is important. The charge-control model of Ward and Dutton[72] and Oh et al.[73] has been introduced into SPICE to avoid this problem. In this model, the transient currents are found directly from the charge distributions in the MOSFET rather than from capacitances and can be used with the level 2 and level 3 SPICE models. When `XQC`, parameter 28, is assigned a value between 0.0 and 0.5, the Ward and Dutton model is used in the SPICE circuit simulation. This parameter determines the fraction of the total mobile electron channel charge Q_n that is partitioned between the source charge Q_{nS} and the drain charge Q_{nD} as

`XQC`

$$Q_{nD} = \texttt{XQC} \times Q_n, \tag{D.1}$$

and

$$Q_{nS} = (1 - \texttt{XQC}) \times Q_n. \tag{D.2}$$

A commonly used value for `XQC` is 0.5.

The gate-source and gate-drain overlap capacitances due to ΔL_S and ΔL_D were illustrated in Fig. D.1. This overlap of the gate over the source and drain is represented by the extrinsic capacitances specified in the SPICE models by the parameters `CGSO` and `CGDO`, parameters 29 and 30 in Table D.1, and are normally identical. The overlap capacitance C_{GSO} is represented in Fig. D.3 (*a*) and is obtained from the parameter

`CGSO`

$$\texttt{CGSO} = \frac{\epsilon_{ox}}{t_{ox}} \Delta L_S, \tag{D.3}$$

where ϵ_{ox} is in F/m, t_{ox} is in m, and ΔL_S, which is x_{jl} in Fig. 8.19, is in m to give `CGSO` in F/m. The overlap capacitance is given by

$$C_{GSO} = \texttt{CGSO} \times W, \tag{D.4}$$

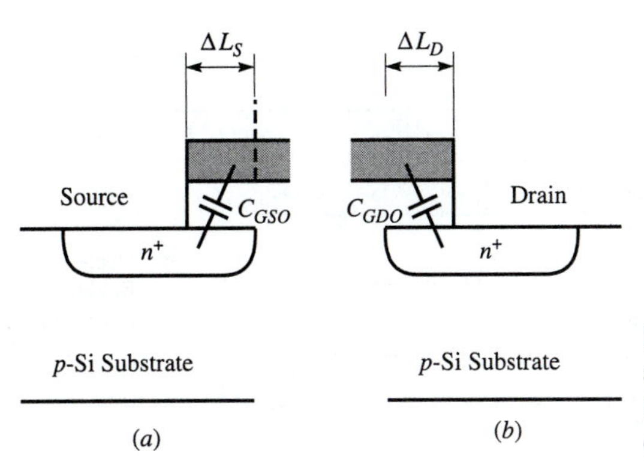

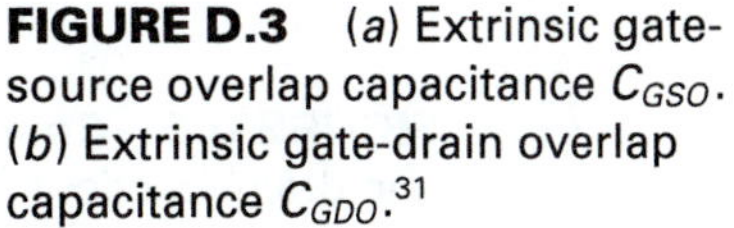
FIGURE D.3 (*a*) Extrinsic gate-source overlap capacitance C_{GSO}. (*b*) Extrinsic gate-drain overlap capacitance C_{GDO}.[31]

where W is the gatewidth in m and is given in the element line. Similar expressions apply to C_{GDO}, which is illustrated in Fig. D.3 (*b*).

field oxide

The SiO_2 in the regions between transistors where the polysilicon or metal is patterned to provide the interconnections between transistors is much thicker than the gate oxide and is called the *field oxide*. As illustrated in Sec. 7.3, the threshold voltage increases with oxide thickness. Therefore, the field oxide has sufficient thickness to raise the threshold voltage for the interconnections to prevent inversion between transistors. The overlap of the polysilicon or metal gate over the field oxide gives the extrinsic capacitance C_{GBO} shown in Fig. D.4. This gate-substrate overlap capacitance per unit gate length is the SPICE parameter `CGBO`, parameter 31 in Table D.1, and is calculated from

`CGBO`

$$\texttt{CGBO} = \frac{\epsilon_{ox}}{t_{ox_{field}}} \Delta W, \tag{D.5}$$

where $t_{ox_{field}}$ is the field oxide thickness in meters and ΔW is the overlap of the gate over the field oxide in meters as shown in Fig. D.4. The gate-substrate capacitance

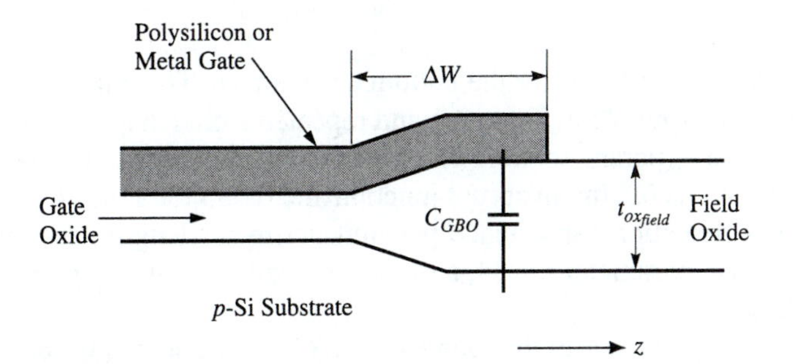

FIGURE D.4 The extrinsic capacitance C_{GBO} due to the polysilicon or metal gate overlap of the field oxide.[31]

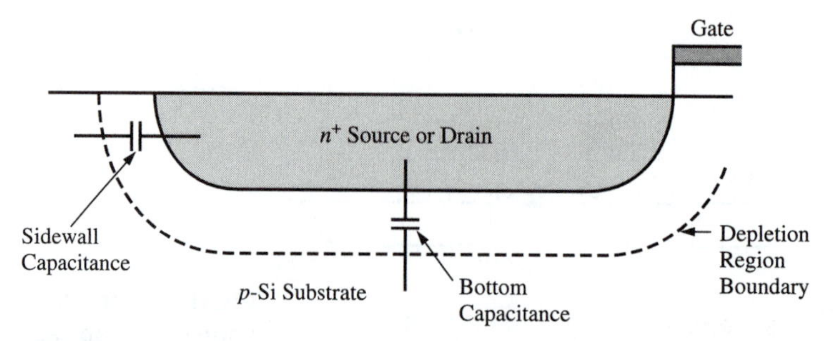

FIGURE D.5 Bottom and sidewall capacitance of the source and drain junctions.[31]

C_{GBO} is obtained in SPICE by multiplying `CGBO` by the gate length L in meters which is specified in the element line.

source-substrate and drain-substrate capacitance

channel stop

The source-substrate and drain-substrate p-n^+ junction capacitance is separated into two components for SPICE. Both the source and drain capacitances have a bottom and a sidewall part as shown in Fig. D.5. The reason for this separation is that during circuit fabrication the impurity concentration of the p substrate outside the gate area is often increased under the field oxide to form a *channel stop* to increase the threshold voltage in these areas to prevent inversion under the circuit interconnections. Thus, the sidewall capacitance per unit area may be higher (by a factor of 2 to 10) than for the bottom capacitance per unit area. As given in Sec. 5.4.1, the depletion capacitance C_d for an abrupt p-n^+ junction is given by

$$C_d = A\sqrt{\frac{q\epsilon_{\text{Si}}N_a^-}{2(V_{bi}-V_a)}} = A\sqrt{\frac{q\epsilon_{\text{Si}}N_a^+}{2V_{bi}(1-V_a/V_{bi})}} = \frac{C_{j0}}{(1-V_a/V_{bi})^{0.5}}, \tag{D.6}$$

where C_{j0} is the zero-bias depletion capacitance per unit area,

`CJ`

$$C_{j0} = \sqrt{\frac{q\epsilon_{\text{Si}}N_a^-}{2V_{bi}}}, \tag{D.7}$$

`MJ`

`CJSW`

`MJSW`

and is parameter no. 32, `CJ`, for the bottom capacitance. The built-in potential V_{bi} was given as parameter no. 22 in Table 8.1 and repeated from Chapter 4 as Eq. (8.54). The grading coefficient (parameter no. 33) for the bottom capacitance is given by the SPICE parameter `MJ` and is 0.5 for an abrupt junction and 0.333 for a linearly graded junction. The zero-bias depletion capacitance per unit perimeter length C_{jsw0} is given by the parameter `CJSW`, parameter no. 34, and the sidewall capacitance grading parameter is given by M_{jsw}.

The total capacitance of the source region C_{BS} is the sum of the area and perimeter capacitance:

C_{BS}

$$C_{BS} = \underbrace{A_S C_{j0} V_{bsj}}_{\text{bottom cap.}} + \underbrace{P_S C_{jsw0} V_{bss}}_{\text{perimeter cap.}}, \tag{D.8}$$

AS and PS

where A_S is the source area, and P_S is the source perimeter, which excludes the side adjacent to the gate, since that side contains the conducting channel and not a channel stop area. Both A_S and P_S are given in the element line. The corresponding total capacitance of the drain region C_{BD} is

C_{BD}

$$C_{BD} = A_D C_{j0} V_{bdj} + P_D C_{jsw0} V_{bds}, \tag{D.9}$$

where A_D is drain area and P_D is the drain perimeter, which also excludes the side adjacent to the conducting channel.

FC

The expressions for V_{bsj}, V_{bss}, V_{bdj}, and V_{bds} depend on the substrate-source voltage V_{BS} and the substrate-drain voltage V_{BD}. These conditions for the source are specified as $V_{BS} \leq \texttt{FC} \times V_{bi}$, and $V_{BS} > \texttt{FC} \times V_{bi}$, where FC is parameter no. 36. A typical value for FC is 0.5. For $V_{BS} \leq \texttt{FC} \times V_{bi}$, the voltage dependences in Eq. (D.8) for the source-region capacitance become

$$V_{bsj} = (1 - V_{BS}/V_{bi})^{-M_j}, \tag{D.10}$$

and

$$V_{bss} = (1 - V_{BS}/V_{bi})^{-M_{jsw}}. \tag{D.11}$$

For $V_{BS} > \texttt{FC} \times V_{bi}$, the voltage dependences in Eq. (D.8) for the source-region capacitance become

$$V_{bsj} = (1 - \texttt{FC})^{-(1+M_j)}[1 - \texttt{FC}(1 + M_j) + M_j V_{BS}/V_{bi}] \tag{D.12}$$

and

$$V_{bss} = (1 - \texttt{FC})^{-(1+M_{jsw})}[1 - \texttt{FC}(1 + M_{jsw}) + M_{jsw} V_{BS}/V_{bi}]. \tag{D.13}$$

CBS
CBD

Similar expressions may be written for V_{bdj} and V_{bds} with V_{BS} replaced with V_{BD}. These source and drain capacitances may also be specified as the zero-bias source capacitance CBS and the zero-bias drain capacitance CBD. These capacitances are parameters no. 37 and no. 38 in Table D.1.

JS

The source and drain reverse saturation currents may be specified as the current density in A/m^2 as parameter JS, parameter no. 39, and the total source and drain currents are obtained as $J_S A_S$ and $J_S A_D$, where the source area A_S and drain area A_D are specified in the element line. The current may also be specified as the current parameter

IS

IS, which is parameter no. 40 and is given in A.

D.2 NOISE PARAMETERS

The noise parameters used in SPICE are listed in Table D.2 and are used to represent the flicker noise. A value of noise at a given time t is unpredictable so that noise is

TABLE D.2 SPICE MOSFET Noise Parameters.

No.	Text Symbol	SPICE Keyword	Level	Parameter Name	Default Value	Units
41	k_f	KF	1–3	Flicker-noise coefficient	0.0	—
42	a_f	AF	1–3	Flicker-noise exponent	1.0	—

characterized as the mean squared value $\overline{i^2}$. Usual measurement involves a narrow bandwidth Δf. Flicker noise ($1/f$ noise) is found in all types of active devices and is associated with the flow of direct current and displays a spectral density of the form

KF and AF

$$\overline{i_{D_F}^2} = k_f \frac{I_D^{a_f}}{f} \Delta f, \tag{D.14}$$

where Δf is a small bandwidth at the frequency f, I_D is the DC drain current, k_f is the flicker-noise coefficient (values range from 0.5 to 2), parameter KF in Table D.2, and a_f is the flicker-noise exponent, parameter AF in Table D.2. The flicker noise is due mainly to random fluctuations of carriers in the surface states and is the dominant noise at low frequencies.

The conducting channel is resistive and exhibits thermal noise which arises mainly from the random thermal motion of the channel charge carriers. Thermal noise is important at high frequencies and is given by

$$\overline{i_{D_T}^2} = \frac{8kT}{3} g_m \Delta f, \tag{D.15}$$

where g_m is the transconductance at the operating point. No SPICE parameters are involved with the thermal noise.

Tables 8.1, D.1, and D.2 have listed 42 parameters for the description of the MOSFET, and parameters must be entered into SPICE2 for every operating temperature because there is no provision for including temperature. Hodges and Jackson[74] have discussed the application of SPICE to digital circuits and present simplifications which can be made to permit hand analysis. For accurate circuit simulation, the SPICE2 parameters would have to be specified for a particular fabrication process.

D.3 ADDITIONAL PSPICE MOSFET MODEL PARAMETERS

PSpice has both additional parameters applicable to the level 1–3 models as well as the level 4 and 5 models. The additional PSpice parameters summarized by Banzhaf[75] are given in Table D.3. The level 4 and 5 models are based on the Berkeley short-channel-IGFET Model (BSIM) and are introduced in Sec. 8.4.3.

TABLE D.3 Additional PSpice MOSFET Parameters.

No.	Text Symbol	SPICE Keyword	Level	Parameter Name	Default Value	Units
43	R_G	RG	1–3	Gate Ohmic resistance	0	Ω
44	R_B	RB	1–3	Bulk Ohmic resistance	0	Ω
45	R_{DS}	RDS	1–3	Drain-source shunt resistance	∞	Ω
46	τ_D	TT	1–3	Source and drain *p-n* junction transit time	0	sec
47	n	N	1–3	Source and drain *p-n* junction ideality factor	1.0	—
48	J_{ssw}	JSSW	1–3	Source and drain *p-n* junction sidewall current/length	0	A/m
49	$V_{bi\,sw}$	PBSW	1–3	Source and drain *p-n* junction sidewall built-in potential	V_{bi}	V

RG
RB
RDS

Parameters nos. 43–45 give additional parasitic parameters. The gate resistance RG enters as part of the resistance in the *RC* time constant between devices. The bulk resistance RB is the resistance of the bulk semiconductor as a resistance in series with the source *p-n* junction and the substrate. The drain-source shunt resistance RDS represents leakage paths between the source and drain. The source and drain transit time TT (parameter no. 46) has been given for the *p-n* junction in Table 5.1 and appears as an additive term to the source and drain capacitance. If the source-region capacitance with the contribution with TT is represented by C^*_{BS}, then

TT

$$C^*_{BS} = C_{BS} + \mathrm{TT}\, g_{BS}, \tag{D.16}$$

where C_{BS} was given in Eq. (D.8) and g_{BS} is the bulk-source conductance given by

$$g_{BS} = \frac{\partial I_{BS}}{\partial V_{BS}}. \tag{D.17}$$

Similar expressions can be written for the bulk-drain capacitance.

The remaining three parameters (nos. 47–49) are used to represent the source-to-bulk and drain-to-bulk currents. The source-to-bulk leakage current is given by

$$I_{BS} = I_{ss}[\exp(qV_{BS}/\mathrm{n}kT) - 1], \tag{D.18}$$

n

where n is the usual *p-n* junction ideality factor, which ranges from 1.0 to 2.0. The saturation current is given by

$$I_{ss} = A_S \times J_s + P_S \times J_{ssw}, \tag{D.19}$$

JSSW

where A_S is the source area, P_S is the source perimeter, which excludes the gate side, J_S is the current density given in parameter no. 39, and J_{ssw} is the sidewall current per

PBSW

unit length. The last parameter no. 49 is the source and drain sidewall built-in potential, which is used in Eqs. (D.10)–(D.13) for the source- or drain-region capacitance.

temperature

In PSpice, the normal room temperature is 300 K and is referred to as TNOM, but may be changed with the .OPTIONS control line, such as

```
.OPTIONS TNOM = 297
```

for the case of a reference temperature of 297 K.

CHAPTER 9

BIPOLAR TRANSISTORS

9.1 INTRODUCTION

William Shockley at Bell Labs submitted his patent request for the junction transistor on June 26, 1948, and the junction transistor patent no. 2,569,347 was issued on September 25, 1951.[1] One configuration of the junction transistor given in this patent is shown in Fig. 9.1; it is very similar to the "prototype" structure used in this chapter to describe transistor action. Semiconductor devices in which both electrons and holes participate in the conduction are termed bipolar devices and for this reason the junction transistor is now more commonly called the bipolar-junction transistor (BJT) or the bipolar transistor. Shockley gave January 23, 1948 as the date for the completion of the conception of the junction transistor based on minority-carrier injection.[1] He had been close to the concept of minority-carrier injection for more than six months, and he described his difficulty[1] to emphasize that "the path to creativity may be neither easy nor direct." His junction transistor patent included heavy doping near the contacts as well as heterojunctions with wide energy-gap emitters to enhance injection efficiency (Sec. 5.6.4), even though no junction transistor had yet been fabricated and demonstrated. The "existence proof" for the junction transistor was made on April 7, 1949, at Bell Labs with a Ge structure fabricated by Bob Mikulyak. The steps involved in the evolution of the Ge junction transistor into the Si integrated circuit were described in Sec. 1.3.

bipolar devices

The metal-oxide semiconductor-based field-effect transistors (Chapter 8) now dominate integrated circuits because of factors such as lower cost and power dissipation, but bipolar transistors have continued to be important where the ability to drive capacitive loads is required and in some families of high-performance logic circuits as well as in analog integrated circuits. An understanding of junction transistor action demonstrates many of the useful properties of semiconductor behavior.

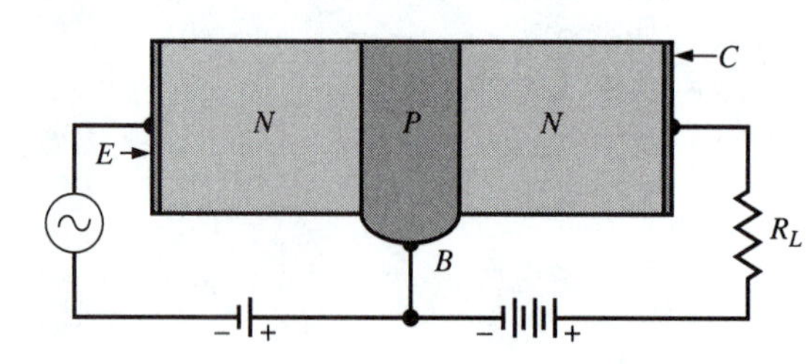

FIGURE 9.1 Representation of the junction (bipolar) transistor in Shockley's transistor patent (2,569,347), which was filed on June 26, 1948. This representation has the *n*-type emitter labeled by *E*, the *p*-type base labeled by *B*, and the *n*-type collector labeled by *C*. The load resistor is labeled by R_L. The emitter-base and collector-base bias voltages are also shown.[1]

Therefore, the goal of this chapter is to illustrate these bipolar concepts rather than to provide an exhaustive analysis of the bipolar transistor.

Section 9.2 begins with the schematic representation of the bipolar transistor. The analysis uses a one-dimensional *pnp* prototype structure similar to the *npn* structure illustrated in Fig. 9.1. A *pnp* structure is used to illustrate transistor action because hole current flow and conventional current flow are in the same direction. However, the more commonly used transistor is the *npn* structure because the electron mobility exceeds the hole mobility (see Fig. 3.3) and *npn* transistors will have higher frequency response.

common-base current gain

The current gain is determined for the common-base configuration to illustrate the basic concepts of transistor action such as injection of minority carriers from a more heavily doped emitter region into the base region. The injected minority carriers diffuse across the base region and flow through the collector-base junction. In the common-base configuration, the current gain is less than unity, but the voltage and power gain are greater than unity because the emitter-base junction is a low-impedance forward-biased *p-n* junction, while the collector-base junction is a high-impedance reverse-biased *p-n* junction.

common-base I–V characteristics

The current–voltage (*I–V*) characteristics for a *pnp* bipolar transistor for the common-base configuration are derived in Sec. 9.3 and begin with the injected hole concentration distribution in the base. This minority-carrier distribution was given in Sec. 4.7.3 for the case when the layer thickness is much less than a diffusion length and with the boundary condition of a zero minority carrier concentration at the base-collector junction. This carrier distribution leads to an expression for emitter-base diffusion current. The recombination of holes in the base is taken into account to determine the collector current. Other current components are the current due to the injection of electrons from the base into the emitter and the space-charge recombination current in the emitter-base junction. These quantities permit determination of the collector current I_C as a function of the collector-base junction voltage V_{CB}. A single curve is given for a fixed emitter current I_E.

common-emitter I–V characteristics

The common-emitter *I–V* characteristics are obtained in Sec. 9.4. An expression for the common-emitter current gain, which is represented by β, is obtained as the ratio of the change in collector current for a change in base current and can be much larger than unity. This β utilizes the expression for the common-base current gain. Other behavior such as the conditions for saturation of the base region when both the emitter-base junction and the collector-base junction are forward biased are also given in this section.

Ebers-Moll model

small-signal AC model

SPICE parameters

emitter-coupled differential amplifier example

heterojunction transistors

Next, in Sec. 9.5, the basic equations which describe the collector current in terms of the emitter and base currents are used to obtain an equivalent circuit. This equivalent-circuit model is known as the Ebers-Moll model,[2] and this model is used in the SPICE representation of the bipolar transistor. The small-signal AC model is introduced in Sec. 9.6, and the high-frequency figure of merit represented by the cutoff frequency f_T is derived. The application of SPICE to represent the bipolar transistor is given in Sec. 9.7. There are 38 SPICE2 and PSpice parameters which are related to the expressions used to represent the *I–V* behavior and the junction capacitances. The junction depletion and diffusion capacitances were given in Sec. 5.4. Two additional SPICE2 and PSpice parameters are used to represent noise. An additional 19 model parameters are used in PSpice which are also summarized in Sec. 9.7. To illustrate an integrated-circuit application, the two transistor emitter-coupled differential amplifier is introduced, and results obtained with PSpice are given in Sec. 9.8. In Sec. 9.9, the use of heterojunctions for the emitter-base junction to improve the high-frequency performance is presented. Bipolar transistors with heterojunctions are called heterojunction bipolar transistors (HBTs). The concepts introduced in this chapter are summarized in Sec. 9.10 and the expressions used to represent the BJT are listed.

9.2 COMMON-BASE CONFIGURATION

9.2.1 Basic Operation

A representative *npn* planar bipolar transistor is illustrated in Fig. 9.2 (*a*). The n^+ region is heavily doped to inject electrons into the *p*-type base region. The p^+ region provides ohmic contact to the *p*-base region, and the n^+-collector region provides a low-resistance contact to the *n*-collector region. The commonly used circuit symbol for the *npn* bipolar transistor is shown in part (*b*), and the *pnp* circuit symbol is shown in part (*c*). The arrow on the emitter gives the direction of conventional current flow. Currents into the transistor will be taken as positive, and currents flowing out of the

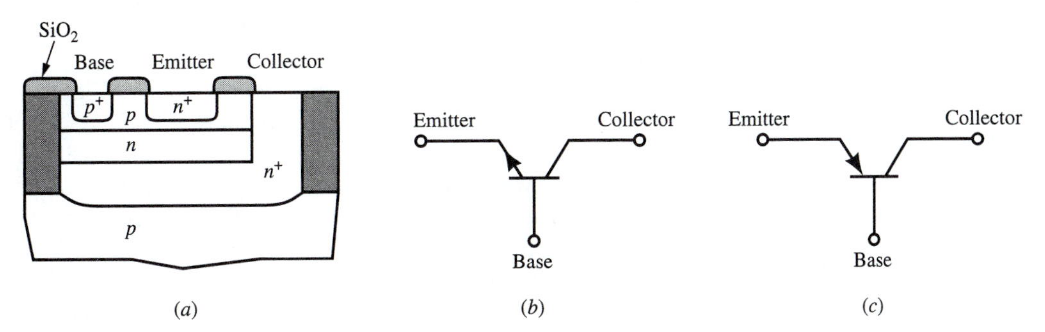

FIGURE 9.2 (*a*) Planar *npn* bipolar transistor. (*b*) Circuit symbol for the *npn* bipolar transistor. (*c*) Circuit symbol for the *pnp* bipolar transistor.[3]

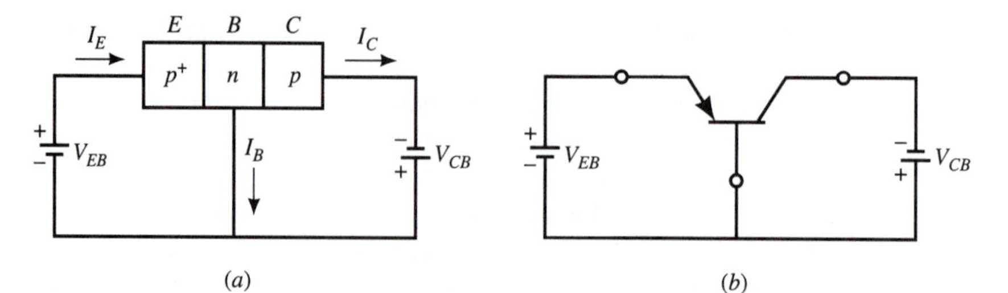

FIGURE 9.3 Common-base configuration. (*a*) One-dimensional representation of the *pnp* bipolar transistor with uniformly doped layers and abrupt step junctions. The appropriate bias polarities are for the normal active mode. (*b*) The *pnp* bipolar transistor in part (*a*) with the circuit symbol for the transistor.

transistor will be taken as negative. To discuss the basic operation of the bipolar transistor, the one-dimensional representation in the common-base configuration for the *pnp* structure shown in Fig. 9.3 (*a*) will be used. The normal bias polarities as well as the normal current flow directions are also shown. All layers are taken as uniformly doped so that the *p-n* junctions are abrupt step junctions as presented in Chapters 4 and 5. The *pnp* transistor with the circuit symbol and the same bias polarities is represented in Fig. 9.3 (*b*). For the *npn* transistor, the bias polarities are opposite to those shown in Fig. 9.3.

In Fig. 9.3 (*a*), the emitter current is represented by I_E and flows into the emitter for the forward bias V_{EB} on the emitter-base junction. No significance for bias polarities will be assigned to the subscripts of the bias polarities, and V_{EB} will be taken as equivalent to V_{BE}. The base current I_B flows out of the base region. The collector-base junction is reverse biased and the collector current I_C flows out of the collection region as shown. In the *active mode,* the emitter-base junction is forward biased and the collector-base junction is reverse biased. It should be noted that by Kirchhoff's law, the sum of the current at a node must sum to zero. Therefore, there are only two independent currents, and if two currents are known, then the third current may be obtained from the other two. The complementary structure is the *npn* transistor, and the bias polarities and current flow directions are reversed. The *pnp* structure will be used to illustrate the basic transistor action, because the minority-carrier current flow (holes) is in the same direction as conventional current flow.

active mode

The *pnp* transistor at thermal equilibrium is represented in Fig. 9.4, with the impurity distribution representing the space-charge regions at each *p-n* junction [part (*b*)], the electric-field variation with distance *x* [part (*c*)], and the potential variation with distance *x* [part (*d*)], as was given previously in Fig. 4.5. The thermal equilibrium energy-band diagram, shown in Fig. 9.4 (*e*), was given for the *p-n* junction in Fig. 4.9 (*b*). Because the emitter is more heavily doped than the collector, the built-in potential, as given in Eq. (4.62), is larger for the emitter-base junction than for the collector-base junction.

For operation in the active mode, the *pnp* transistor has the emitter-base junction forward biased and the collector-base junction reverse biased, as shown in Fig. 9.5 (*a*).

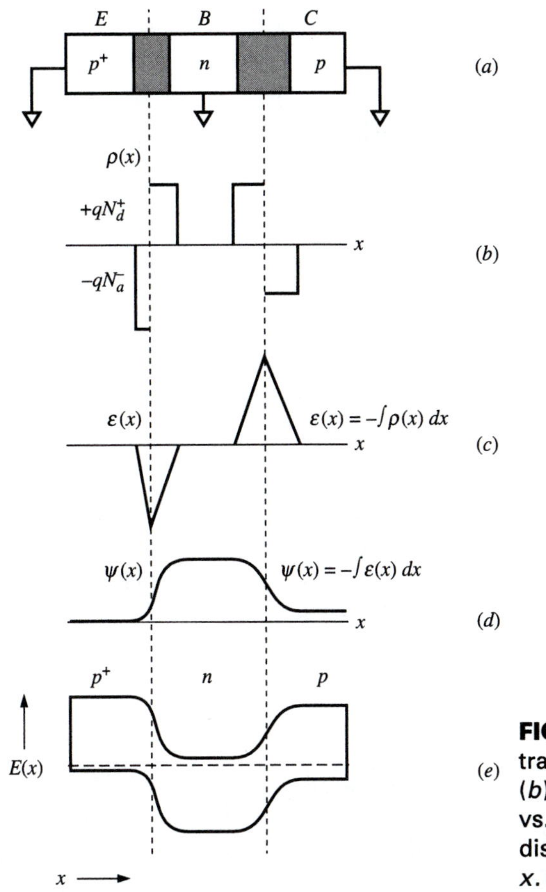

FIGURE 9.4 (*a*) Abrupt *pnp* bipolar transistor at thermal equilibrium. (*b*) Space-charge $-qN_a^-$ and $+N_d^+$ vs. distance x. (*c*) Electric field vs. distance x. (*d*) Potential vs. distance x. (*e*) Energy-band diagram.

An input signal is represented by v_{in}, and the output signal across the load resistor R_L is v_{out}. The energy-band diagram for this bias condition is shown in Fig. 9.5 (*b*). Under these conditions, holes are injected into the base region and diffuse to the collector-base junction. The electric field is positive [see Fig. 9.4 (*c*)], and these injected holes are swept across the collector junction and flow through the external load resistor. The concepts governing this transistor action will be derived in the next part of this section.

9.2.2 Current Gain

The analysis for the current gain of the common-base configuration begins with a schematic representation of the current components for the *pnp* transistor biased in the active mode. These current components are illustrated in Fig. 9.6. The first current component is the hole-diffusion current injected from the emitter into the base and is designated as I_{pE}. The portion of I_{pE} which recombines with electrons in the base region is represented by I_{B_r}, and the difference between I_{pE} and I_{B_r} is I_{pC}. This

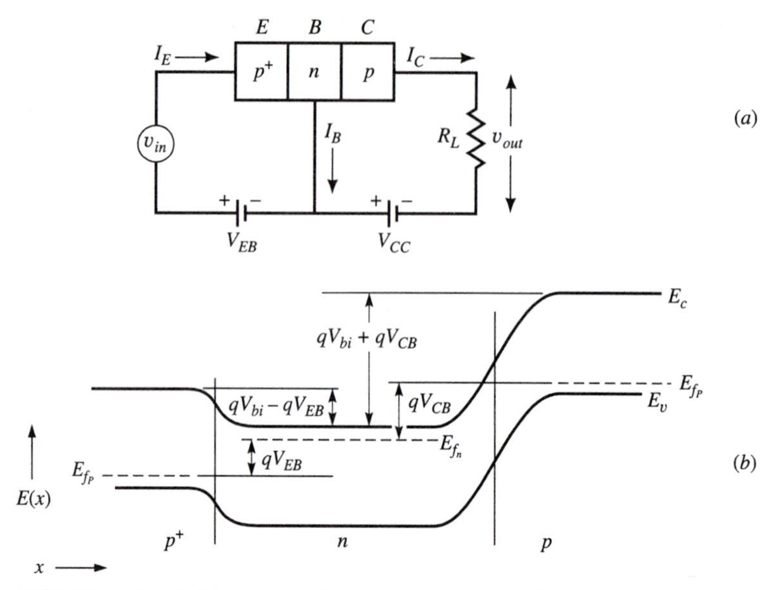

FIGURE 9.5 (*a*) Common-base configuration for an abrupt *pnp* bipolar transistor biased in the active mode with an input signal v_{in} and an output signal v_{out} across the load resistor R_L. (*b*) Energy-band diagram for the bias conditions in part (*a*).

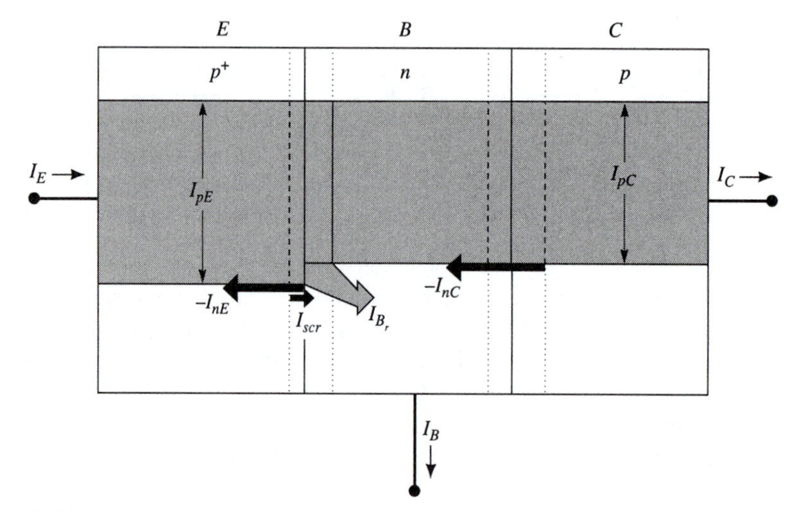

FIGURE 9.6 Representation of the current components for the *pnp* transistor. The emitter hole diffusion current injected into the base is I_{pE}, the portion of I_{pE} lost to recombination in the base region is I_{B_r}, and the portion of the hole emitter current flowing in the collector region is I_{pC}. Additional current components are represented by the collector-base reverse saturation current as $-I_{nC}$, the electron current injected into the emitter from the base is $-I_{nE}$, and the emitter-base junction space-charge (or surface) recombination current is I_{scr}.

hole current I_{pC} is the current in the collector flowing from the emitter to the collector and also flows in the external collector circuit represented by R_L in Fig. 9.5 (*a*). The current represented by $-I_{nE}$ is the electron-diffusion current injected into the emitter from the base. An additional current due to space-charge recombination (and/or surface recombination) in the emitter-base junction is given as I_{scr}. The last current component represented in Fig. 9.6 is the reverse saturation current $-I_{nC}$, which is the current for the reverse-biased collector-base junction with the emitter-base junction open. These current components permit writing expressions for the emitter, base, and collector currents.

From Fig. 9.6, the total emitter current may be seen to be given by

$$I_E = I_{pE} + I_{nE} + I_{scr}. \tag{9.1}$$

Note that the electron current injected into the emitter from the base is a flow of electrons to the left, but a positive conventional current. The collector current is given by

$$I_C = I_{pC} + I_{nC}, \tag{9.2}$$

where the collector electrons entering the base also result in a positive conventional current. Generally, the collector region is more lightly doped than the base region, and as described in Sec. 5.2, the minority carriers from the more lightly doped side dominate the reverse-bias saturation current. The base current is now simply the difference:

$$I_B = I_E - I_C = (I_{pE} + I_{nE} + I_{scr}) - (I_{pC} + I_{nC}), \tag{9.3}$$

where the expressions for I_E and I_C have been used to represent the base current.

The common-base current gain, represented by α_o, is defined as the ratio of the hole current flowing in the collector to the total emitter current:

common-base current gain α_o

$$\alpha_o \equiv \frac{I_{pC}}{I_E} = \frac{I_{pC}}{I_{pE} + I_{nE} + I_{scr}} = \left[\frac{I_{pE}}{I_{pE} + I_{nE} + I_{scr}}\right]\left[\frac{I_{pC}}{I_{pE}}\right]. \tag{9.4}$$

The right-hand side of Eq. (9.4) was written in a form to identify two separate ratios which are commonly used to describe transistor behavior.

The first ratio on the right in Eq. (9.4) gives the fraction of the emitter current injected into the base to the total emitter current. This ratio is called the emitter efficiency and is represented by γ as

emitter efficiency γ

$$\boxed{\gamma = \frac{I_{pE}}{I_E} = \frac{I_{pE}}{I_{pE} + I_{nE} + I_{scr}} = \frac{1}{1 + I_{nE}/I_{pE} + I_{scr}/I_{pE}}}\,. \tag{9.5}$$

Equation (9.5) emphasizes that a high emitter efficiency requires that the diffusion current injected into the emitter from the base and the space-charge recombination current be small compared to the diffusion current injected into the base from the emitter. The other ratio on the right in Eq. (9.4) is the fraction of the minority carriers entering the base which reach the collector. This ratio is called the base transport factor and is represented by α_T as

base transport factor α_T

$$\boxed{\alpha_T = \frac{I_{pC}}{I_{pE}} = \frac{I_{pE} - I_{B_r}}{I_{pE}} = 1 - \frac{I_{B_r}}{I_{pE}}}. \tag{9.6}$$

Equation (9.6) emphasizes that for a base transport factor approaching unity, the recombination in the base region must be small. The common-base current gain given by Eq. (9.4) may be written with Eqs. (9.5) and (9.6) as

common-base current gain α_o

$$\boxed{\alpha_o = \gamma \alpha_T}. \tag{9.7}$$

The collector current I_C in Eq. (9.2) may be written with Eq. (9.6) for I_{pC} to give

$$I_C = I_{pC} + I_{nC} = \alpha_T I_{pE} + I_{nC}. \tag{9.8}$$

With Eq. (9.5) for I_{pE}, Eq. (9.8) may be written as

$$I_C = \gamma \alpha_T I_E + I_{nC} = \alpha_o I_E + I_{nC}, \tag{9.9}$$

where Eq. (9.7) is used to replace $\gamma \alpha_T$ with α_o. The quantity I_{nC} is specified as the collector-base saturation current when the emitter base is open, so that $I_E = 0$, and I_{nC} is represented as I_{CBO} to give the collector current for the common-base configuration as

common-base collector current I_C

$$\boxed{I_C = \alpha_o I_E + I_{CBO}}. \tag{9.10}$$

9.3 CURRENTS IN THE COMMON-BASE CONFIGURATION

9.3.1 Carrier Distribution in Base

The minority-carrier distribution in the base of a transistor biased in the active mode has the boundary conditions as given in Sec. 4.7.3 where the base width W_B is less than the minority-carrier diffusion length, which is L_p for the base of a *pnp* transistor ($W_B < 0.1L_p$). As was shown in Sec. 4.7.8, the minority carrier concentration goes to zero at the edges of the depletion region for the base-collector junction with reverse bias. The geometry of the base region is illustrated in Fig. 9.7 with the depletion width on the emitter side of the base given by x_{n_E} and the depletion width on the collector side of the base given by x_{n_C}. For the present analysis, the depletion widths will be ignored and the neutral region will be taken as W_B rather than the actual thickness of $W_B - (x_{n_E} + x_{n_C})$.

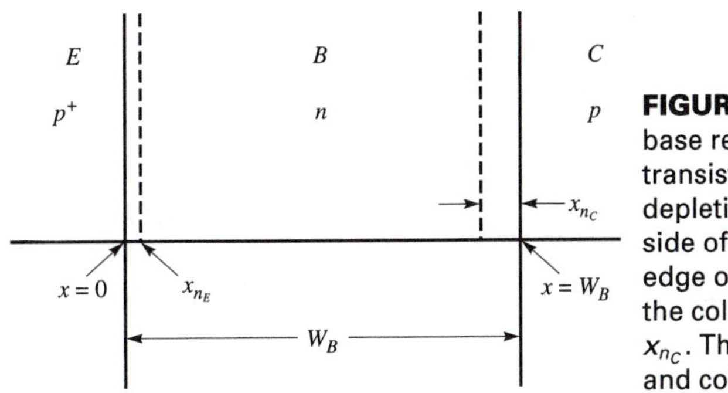

FIGURE 9.7 Geometry of the base region for a *pnp* bipolar transistor. The edge of the depletion region on the emitter side of the base is x_{n_E} and the edge of the depletion region on the collector side of the base is x_{n_C}. The separation of the emitter and collector junctions is W_B.

When the emitter-base junction is forward biased, the minority carrier concentration at x_{n_E} was given in Eq. (4.84) as

$$p_n(0) = p_{no}\exp(qV_{EB}/kT) \qquad \text{at } x = x_{n_E} \simeq 0. \tag{9.11}$$

At the collector-base junction for reverse bias, all minority carriers which reach the edge of the depletion region (Fig. 4.24) will be swept across and the concentration goes to zero at W_B:

$$p_n(W_B) = 0 \qquad \text{at } x = x_{n_C} \simeq W_B. \tag{9.12}$$

For these boundary conditions, the minority carrier concentration was shown in Eq. (4.138) to vary linearly as

$$\boxed{p_n(x) = p_{no}\exp(qV_{EB}/kT)(1 - x/W_B)}\,. \tag{9.13}$$

This linear minority carrier concentration variation in the base region is illustrated in Fig. 9.8 (*a*). If the base layer thickness is not small compared to the diffusion length, the minority-carrier distribution is given by hyperbolic functions.[4] The minority carrier concentrations in the emitter, base, and collector regions are shown in Fig. 9.8 (*b*). The electrons injected into the emitter region from the base are represented by $n_p(x)$ and decrease as $\exp(-x/L_n)$ as given by Eq. (4.96) when the thickness of the emitter region is larger than the diffusion length. In an integrated-circuit transistor, the emitter layer thickness could be less than the diffusion length ($W_E < 0.1L_n$), and $n_p(x)$ would have a linear variation in the emitter. The minority carriers in the collector vary as $[1-\exp(-x/L_n)]$, as given in Eq. (4.167) for the reverse-biased collector-base junction, or a linear variation of $n_p(x)$, if the collector layer thickness is less than the diffusion length ($W_C < 0.1L_n$).

The stored charge in the base Q_B is represented by the shaded area in Fig. 9.8 (*b*) and is readily obtained as the area

$$Q_B = qA\int_0^{W_B} p_n(x)\,dx = qAp_{no}\exp(qV_{EB}/kT) \times W_B/2, \tag{9.14}$$

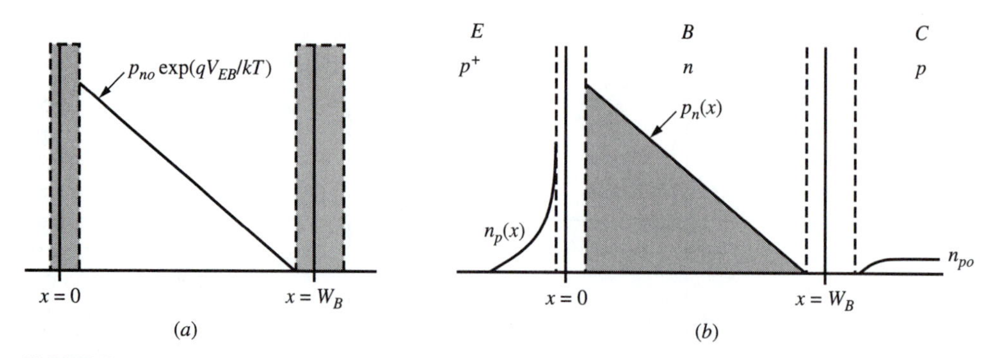

FIGURE 9.8 (*a*) Variation of the minority carrier concentration $p_n(x)$ in the *n*-type base region of a *pnp* bipolar transistor which is biased in the active mode with a base width of W_B. The shaded areas are the depletion regions. (*b*) Variation of the electrons injected into the emitter from the base $[n_p(x)]$, the hole variation in the base $[p_n(x)]$, and the depletion of minority-carrier electrons due to reverse bias of the collector-base junction. The minority-carrier holes in the base are represented by the shaded area. The concentrations are not drawn to scale and $p_n(0) >> n_p(x) >> n_{po}$.

where A is the cross-sectional area of the base region. It is more convenient to replace p_{no} with $n_i^2/N_{d_B}^+$, where $N_{d_B}^+$ is the donor concentration in the base region, to give

base stored charge Q_B

$$\boxed{Q_B = \frac{qAW_B n_i^2 \exp(qV_{EB}/kT)}{2N_{d_B}^+}}. \tag{9.15}$$

The base stored charge Q_B will be used in several expressions for various currents in the bipolar transistor. This expression in Eq. (9.15) is the same expression as was obtained in Eq. (5.60) for the thin-diode case.

9.3.2 Emitter-base Current

The emitter-base diffusion current may readily be found from Eq. (4.106) with Eq. (9.13) for $p_n(x)$ as

$$I_{pE} = -qAD_p \left.\frac{dp}{dx}\right|_{x=x_n} = -qAD_p p_{no} \exp(qV_{EB}/kT)(-1/W_B), \tag{9.16}$$

or with p_{no} replaced with $n_i^2/N_{d_B}^+$ gives

current injected into base from emitter I_{pE}

$$\boxed{I_{pE} = \frac{qAD_p n_i^2 \exp(qV_{EB}/kT)}{N_{d_B}^+ W_B}}. \tag{9.17}$$

Note that the diffusion current is the same as the expression given in Eq. (4.139) for the case when the n-type layer thickness is much less than the diffusion length.

The electron current injected into the p^+ emitter from the n-base is given by Eq. (4.141) when multiplied by the cross-sectional area,

current injected into emitter from base I_{nE}

$$\boxed{I_{nE} = \frac{qAD_n n_i^2}{W_E N_{a_E}^-}[\exp(qV_{EB}/kT) - 1]}, \tag{9.18}$$

when the emitter layer thickness W_E is much less than the minority-carrier diffusion length, as expected for a BJT in an integrated circuit. If $W_E > L_n$, then W_E in the denominator would be replaced by L_n.

9.3.3 Base Recombination Current

The next current component to be considered is the base current due to recombination of the injected minority carriers in the base region which is I_{B_r}. This current may be obtained from the continuity equation for holes, which was given in Eq. (3.51) as

$$\frac{\partial p}{\partial t} = 0 = -\frac{1}{q}\frac{\partial J_p}{\partial x} + (G_p - R_p) \tag{9.19}$$

for steady state. With no external generation, the G_p term is zero and the continuity equation becomes

$$-\frac{\partial J_p}{\partial x} = qR_p = q\frac{\Delta p_n(x)}{\tau_p}. \tag{9.20}$$

Integration of Eq. (9.20) gives

$$-\int_{J_0}^{J_{W_B}} dJ_p = \frac{q}{\tau_p}\int_0^{W_B} \Delta p_n(x)\,dx. \tag{9.21}$$

The integral on the left-hand side of Eq. (9.21) gives I_{B_r} and $\Delta p_n(x)$ may be represented by Eq. (9.13) to give

$$I_{B_r} = -A(\underbrace{J_{W_B}}_{J_{pC}} - \underbrace{J_0}_{J_{pE}}) = \frac{qA}{\tau_p}\int_0^{W_B} p_{no}\exp(qV_{EB}/kT)(1 - x/W_B)\,dx. \tag{9.22}$$

Integration now gives

$$I_{B_r} = I_{pE} - I_{pC} = \frac{qA}{\tau_p}\left[p_{no}\exp(qV_{EB}/kT)\left(x - \frac{x^2}{2W_B}\right)\right]_0^{W_B}. \tag{9.23}$$

Evaluation of the limits at 0 and W_B and replacement of p_{no} by $n_i^2/N_{d_B}^+$ gives

base recombination current I_{B_r}

$$\boxed{I_{B_r} = \frac{qAW_Bn_i^2}{2\tau_pN_{d_B}^+}\exp(qV_{EB}/kT) = \frac{Q_B}{\tau_p}} . \tag{9.24}$$

By comparison of the expression for I_{B_r} with Eq. (9.15) for the base stored charge, the base recombination current becomes Q_B/τ_p.

9.3.4 Collector Current

The collector hole current is the difference of the hole current injected into the base and the base recombination current:

collector hole current I_{pC}

$$\boxed{I_{pC} = I_{pE} - I_{B_r}} . \tag{9.25}$$

The hole current I_{pE} injected into the base was given in Eq. (9.17) and the base recombination current was given in Eq. (9.24).

The collector current for the emitter open and the collector-base junction reverse-biased is the reverse saturation current given by Eq. (4.169) with $V_a = -V_{CB}$:

$$I_{nC} = -\frac{qAD_nn_i^2}{W_CN_{a_C}^-}, \tag{9.26}$$

where W_C is the thickness of the collector region and $N_{a_C}^-$ is the acceptor concentration in the collector. Consideration of only the electron current for the reverse-bias saturation current assumes that the collector is more lightly doped than the base. This current is a positive current because the conventional current is opposite to the direction of the electron flow.

The total collector current is the sum of I_{pC} and I_{nC} given in Eqs. (9.25) and (9.26), and may be written as

$$I_C = I_{pC} + I_{nC} = I_{pE} - I_{B_r} + I_{nC} \simeq I_{pE} \tag{9.27}$$

when $I_{pE} > I_{B_r} > I_{nC}$. With Eq. (9.17) for I_{pE}, I_C may be written as

$$I_C \simeq I_{pE} = \frac{qAD_pn_i^2\exp(qV_{EB}/kT)}{N_{d_B}^+W_B} \times \frac{2W_B}{2W_B}. \tag{9.28}$$

Rewriting Eq. (9.28) as

$$I_C \simeq \underbrace{\frac{qAW_Bn_i^2\exp(qV_{EB}/kT)}{2N_{d_B}^+}}_{Q_B}\frac{2D_p}{W_B^2} \tag{9.29}$$

permits writing I_C as

$$I_C \simeq Q_B\frac{2D_p}{W_B^2} \simeq \frac{Q_B}{\tau_D}, \tag{9.30}$$

where τ_D is the base transit time given in Eq. (5.65). Equation (9.30) emphasizes that the collector current is directly proportional to the stored base charge.

9.3.5 Gummel Number

With the approximation for the collector current given in Eq. (9.28) as $I_C \simeq I_{pE}$, the collector current may be written as

$$I_C \simeq \frac{qAD_pn_i^2}{N_{d_B}^+W_B}\exp(qV_{EB}/kT). \tag{9.31}$$

The denominator of Eq. (9.31) is the integrated doping in the base per unit area:

$$G_N \equiv \int_0^{W_B} N_B(x)\,dx = N_{d_B}^+W_B, \tag{9.32}$$

and then

$$I_C \simeq \frac{qAD_pn_i^2}{G_N}\exp(qV_{EB}/kT), \tag{9.33}$$

where G_N is termed the Gummel number[5] and $N_B(x)$ represents the impurity profile in the base region. Equation (9.33) demonstrates that in the absence of recombination in the base region, the collector current is inversely proportional to the integrated base doping, and the smaller the Gummel number, the higher the collector current for a given V_{EB}. Typical values of G_N in a high performance bipolar transistor range from 10^{12} cm^{-2} to 10^{13} cm^{-2}. A representative impurity profile for high-speed bipolar transistor is shown in Fig. 9.9.

9.3.6 Evaluation of α_o

The expressions for the current components derived in this section may be used to evaluate the expressions given for the common-base current gain in Sec. 9.2.2. In Eq. (9.5)

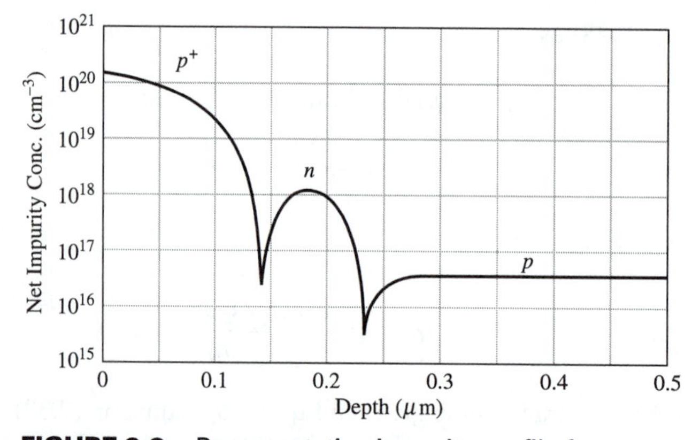

FIGURE 9.9 Representative impurity profile for a high-speed *pnp* transistor.[6]

for the emitter efficiency, I_{nE} was given in Eq. (9.18), I_{pE} was given in Eq. (9.17), and I_{scr} was given in Eq. (4.159). For $V_{EB} > 3kT/q$, the -1 term in the diffusion current expressions may be neglected. By cancellation of common terms qA, the emitter efficiency for the bipolar transistor becomes

emitter efficiency γ

$$\gamma = \frac{1}{1 + I_{nE}/I_{pE} + I_{scr}/I_{pE}}$$

$$\gamma = \frac{1}{1 + \dfrac{D_n N_{d_B}^+ W_B}{D_p N_{a_E}^- W_E} + \dfrac{N_{d_B}^+ W_B x_d \exp(qV_{EB}/2kT)}{2n_i \underbrace{D_p \tau_p}_{L_p^2} \exp(qV_{EB}/kT)}}. \tag{9.34}$$

Equation (9.34) shows that the heavy emitter doping $N_{a_E}^- >> N_{d_B}^+$ *decreases* the electron current from the base into the emitter so that γ approaches unity. The term due to space-charge recombination has a voltage dependence as $\exp(-qV_{EB}/2kT)$ when the $\exp(qV_{EB}/2kT)$ and the $\exp(qV_{EB}/kT)$ terms are combined. Therefore, as V_{EB} increases, the effect of space-charge recombination on γ will be diminished.

The base transport factor given in Eq. (9.6) may be evaluated with Eq. (9.24) for I_{B_r} and Eq. (9.17) for I_{pE} to give

$$\alpha_T = 1 - \frac{I_{B_r}}{I_{pE}} = 1 - \frac{qAW_B n_i^2}{2\tau_p N_{d_B}^+} \frac{N_{d_B}^+ W_B}{qAD_p n_i^2}, \tag{9.35}$$

or by canceling common terms and replacing $D_p \tau_p$ by L_p^2,

base transport factor α_T

$$\boxed{\alpha_T = 1 - \frac{W_B^2}{2L_p^2}}. \tag{9.36}$$

Equation (9.36) shows that the base transport factor approaches unity for $W_B \ll L_p$.

Equations (9.34) and (9.36) give a physical representation of the factors which affect γ and α_T, and therefore α_o. The expression for the diffusion current as well as Eq. (9.34) for γ contains the diffusion coefficients for electrons and holes as minority carriers. In Sec. 3.2, the mobilities for electrons and holes as majority carriers were given. The minority-carrier mobility was considered in Sec. 4.7.4. For both Si and GaAs, the best present choice is to use μ_n(minority carrier) $=$ μ_n(majority carrier) and μ_p(minority carrier) $=$ μ_p(majority carrier). The diffusivity is related to the mobility by the Einstein relation given in Eq. (3.43) as $D = \mu kT/q$.

9.3.7 Modes of Operation

It is helpful in describing the *I–V* behavior of the common-base bipolar transistor to consider operation for the four possible junction bias conditions. The four modes of operation for the *pnp* transistor are shown in Fig. 9.10 for forward or reverse bias on the emitter-base (V_{EB}) and collector-base (V_{CB}) junctions together with the corresponding minority-carrier variations. The emitter and collector regions are considered to have thicknesses much larger than the minority-carrier diffusion lengths so that n_p varies as $\exp(-x/L_n)$ for a forward-biased junction and as $1 - \exp(-x/L_n)$ for a reverse-biased

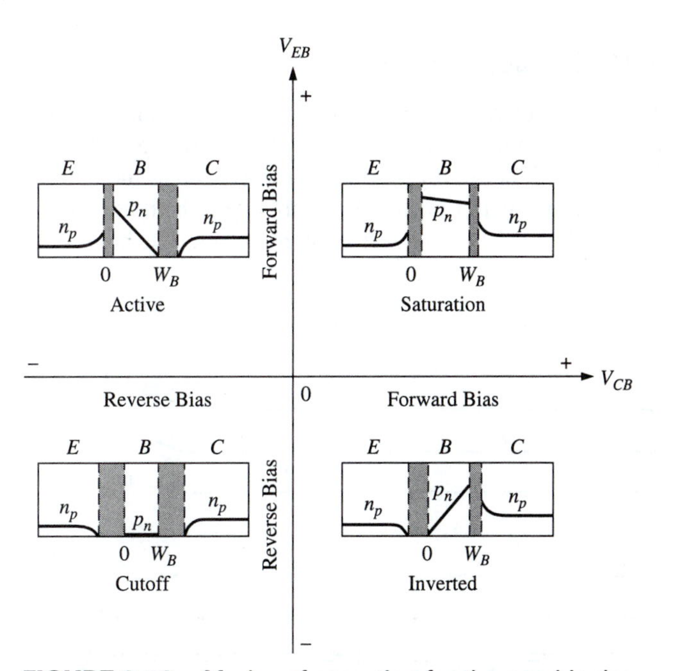

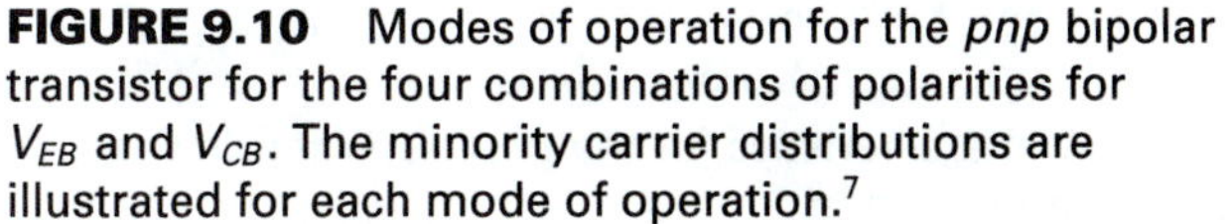

FIGURE 9.10 Modes of operation for the *pnp* bipolar transistor for the four combinations of polarities for V_{EB} and V_{CB}. The minority carrier distributions are illustrated for each mode of operation.[7]

junction. The base region is taken to be much thinner than the minority-carrier diffusion length so that p_n varies linearly with distance. The active mode was illustrated previously in Fig. 9.8 (*b*) for forward bias on the emitter-base junction and reverse bias on the collector-base junction.

In the inverted mode, the emitter-base junction is reverse biased and the collector-base junction is forward biased. For this condition, the collector serves to inject holes from the collector into the base. However, the injection efficiency is poor because the collector is generally more lightly doped than the base.

In the cutoff mode, both the emitter-base and the collector-base junctions are reverse biased. The boundary condition on the emitter side of the base is $p_n(0) = 0$, and the boundary condition on the collector side of the base is $p_n(W_B) = 0$. Therefore, there is no charge in the base region and the transistor is *off* and the collector current approaches zero.

In the saturation mode, both the emitter-base junction and the collector-base junction are forward biased. The boundary condition on the emitter side of the base is $p_n(0) = p_{no} \exp(qV_{EB}/kT)$, and the boundary condition on the collector side of the base is $p_n(W_B) = p_{no} \exp(qV_{CB}/kT)$. For this mode of operation, a large collector current will flow for a small forward bias on the collector-base junction, and the large minority-carrier charge in the base region results in slow turn-off switching times.

9.3.8 *I–V* Characteristics

The *I–V* characteristics for a representative *pnp* transistor are shown in Fig. 9.11 for the common-base configuration.[8,9] The collector saturation current I_{CBO} is the collector current when the emitter circuit is open. The presence of the emitter-base junction influences I_{CBO} because with the emitter circuit open, the hole gradient dp_n/dx must be zero, as shown in Fig. 9.12 (*a*) for zero emitter current. The hole concentration variation

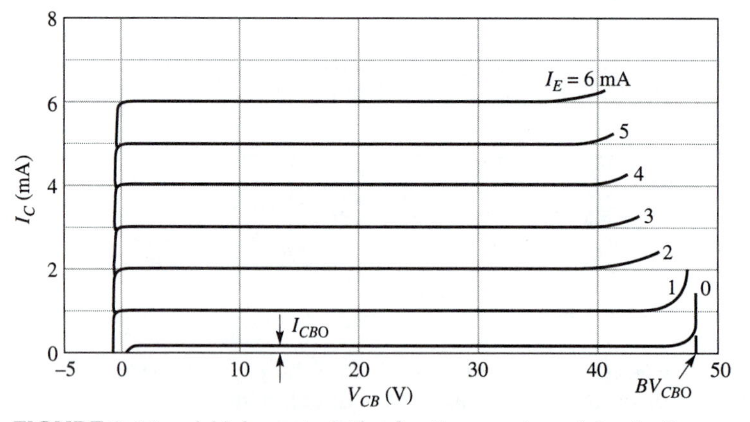

FIGURE 9.11 *I–V* characteristics for the *pnp* transistor in the common-base configuration.[8,9]

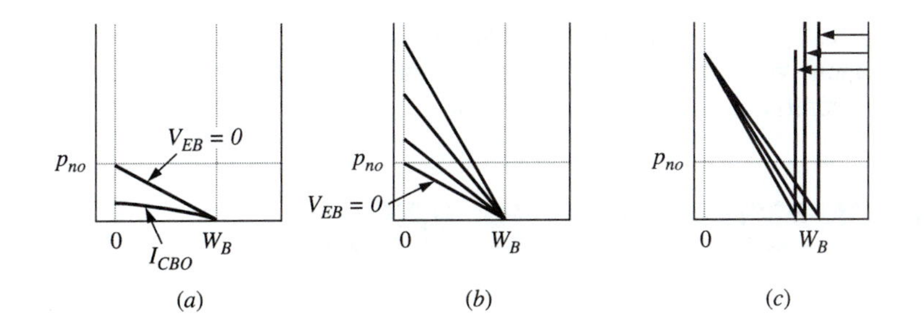

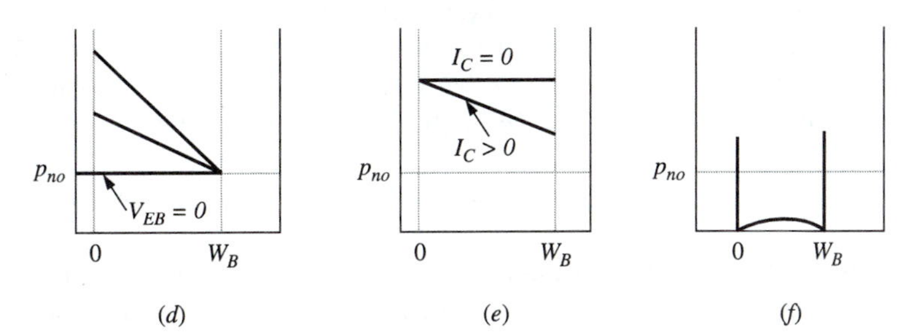

(d) (e) (f)

FIGURE 9.12 Minority-carrier hole concentrations $p_n(x)$ in the base of a *pnp* transistor. (*a*) $p_n(x)$ with the emitter open to give I_{CBO}, and $p_n(x)$ for the emitter-base shorted to give $V_{EB} = 0$. (*b*) $p_n(x)$ for the active mode with V_{EB} positive and V_{CB} negative. (*c*) Variation of $p_n(x)$ in the active mode with a fixed V_{EB} and W_B varying with changes in $-V_{CB}$. (*d*) $p_n(x)$ for V_{EB} positive, but $V_{CB} = 0$. (*e*) Saturation with both junctions forward biased. (*f*) Both the emitter-base and collector-base junctions reverse biased.[8,9]

in the base is also shown in Fig. 9.12 (*a*) for the emitter-base junction short-circuited to give $V_{EB} = 0$ so that $p_n(0) = p_{no}$. As V_{CB} is increased, I_{CBO} begins to increase rapidly at BV_{CBO}. This rapid increase in current may be due to avalanche breakdown, as was presented for the *p-n* junction in Sec. 5.2, or the neutral base width has been reduced to zero and the collector-base depletion region merges with the emitter-base depletion region due to the reverse bias on the collector-base junction. Breakdown by this process is known as punch-through.

For the active mode, the collector current is given by Eq. (9.10), and $I_C \simeq I_E$ when $\alpha_o \approx 1$ and I_{CBO} is small. The hole concentration variation in the base for the active mode is shown in Fig. 9.12 (*b*). Variation of the neutral base width is shown in Fig. 9.12 (*c*) for a fixed $+V_{EB}$, while the reverse bias $-V_{CB}$ is varied. As shown in Fig. 9.11, there is collector current even when $V_{CB} = 0$. The corresponding hole concentration variation is shown in Fig. 9.12 (*d*) to illustrate the hole gradient which gives the collector current even though there is no collector-base voltage. When the collector-base junction is forward biased, both junctions inject holes into the base region, as shown in Fig. 9.12 (*e*), and when V_{CB} is large enough to make the hole gradient go to zero, the collector current goes to zero. The condition for both junctions forward biased is the saturation region. When both junctions are reverse biased, the transistor is cut off and the hole concentration in the base is less than p_{no}, as shown in Fig. 9.12 (*f*).

9.4 COMMON-EMITTER CONFIGURATION

9.4.1 Transistor β

Although the common-base configuration is very useful in describing transistor action, the relationship in Eq. (9.10) between the output collector current I_C and the input emitter current I_E, as $I_C \simeq \alpha_o I_E$ with $\alpha_o \simeq 1$ illustrates the absence of current gain. However, examination of Eq. (9.3) for the base current gives

$$I_B = I_E - I_C = (I_{pE} + I_{nE} + I_{scr}) - (I_{pC} + I_{nC}). \tag{9.37}$$

With Eq. (9.25) for the hole collector current I_{pC} and neglecting the collector-base reverse bias saturation current I_{nC}, which is very small for Si transistors, Eq. (9.37) becomes

$$I_B = (I_{pE} + I_{nE} + I_{scr}) - (I_{pE} - I_{B_r}) = I_{nE} + I_{scr} + I_{B_r}. \tag{9.38}$$

In Eq. (9.38), all the currents represent small currents which are minimized to give γ close to 1 (with $I_{nE} + I_{scr} < I_{pE}$) and to give α_T close to 1 (with $I_{B_r} < I_{pE}$). These currents were represented in Fig. 9.6. Therefore, by arranging the transistor in the common-emitter configuration as shown in Fig. 9.13 for the *pnp* transistor, the output current I_C

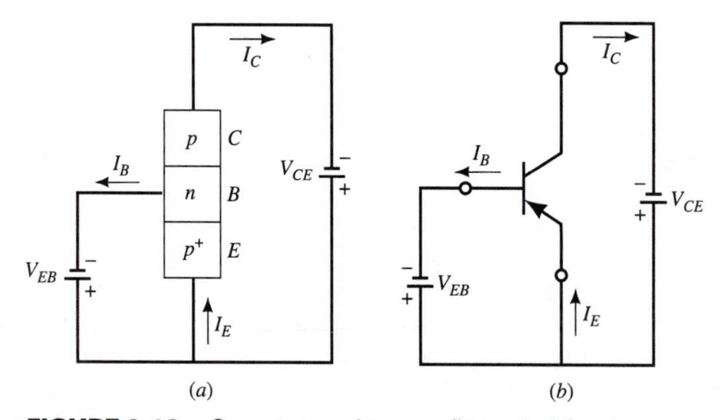

FIGURE 9.13 Common-emitter configuration for the *pnp* transistor. (*a*) One-dimensional representation of the *pnp* transistor with uniformly doped layers and abrupt-step junctions. (*b*) The *pnp* transistor in part (*a*) with the circuit symbol for the transistor.

is much larger than the input current I_B. The normal bias polarities as well as the normal current directions are also shown. For the *npn* transistor, the bias polarities are opposite to those shown in Fig. 9.13. All layers are taken as uniformly doped so that the *p-n* junctions are abrupt step junctions as for the common-base configuration.

The collector current for the common-emitter configuration is the same as given for the common-base configurations, as given in Eqs. (9.8) and (9.10) as

$$I_C = I_{pC} + I_{nC} = \alpha_T I_{pE} + I_{nC} = \alpha_o I_E + I_{CBO}. \tag{9.39}$$

The emitter current is the same as given in Eq. (9.3):

$$I_E = I_B + I_C. \tag{9.40}$$

Note that in the common-base configuration, the input current is I_E, and in the common-emitter configuration, the input current current is I_B. Next, eliminate I_E in Eq. (9.39) with I_E from Eq. (9.40) to give

$$I_C = \alpha_o(I_B + I_C) + I_{CBO}, \tag{9.41}$$

or

collector current I_C

$$I_C(1 - \alpha_o) = \alpha_o I_B + I_{CBO} \tag{9.42}$$

and

$$\boxed{I_C = \frac{\alpha_o}{(1 - \alpha_o)} I_B + \frac{I_{CBO}}{(1 - \alpha_o)}}. \tag{9.43}$$

The current gain for an incremental change in output current (ΔI_C) with respect to an incremental change in the input current (ΔI_B) is known as the common-emitter current gain and is represented by β:

common-emitter current gain β

$$\boxed{\beta \equiv \frac{\Delta I_C}{\Delta I_B} = \frac{\alpha_o}{(1 - \alpha_o)}}. \tag{9.44}$$

The collector-emitter leakage current for $I_B = 0$ is represented by the second term in Eq. (9.43) as

$$I_{CEO} \equiv \frac{I_{CBO}}{(1 - \alpha_o)}. \tag{9.45}$$

In most bipolar transistors, α_o approaches unity, which means that β can be much greater than 1, and I_{CEO} will be much larger than I_{CBO}. For example, when $\alpha = 0.98$,

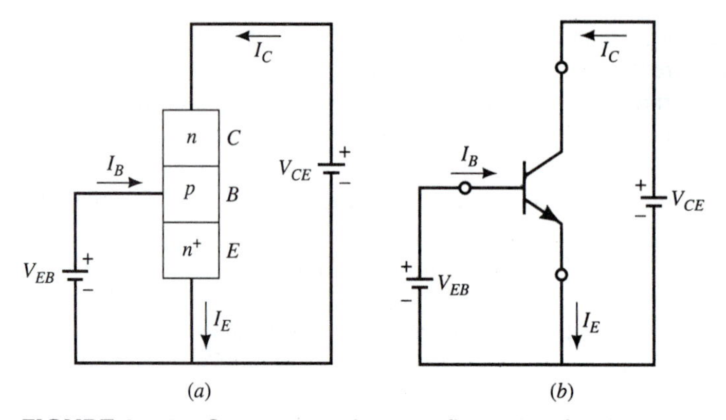

FIGURE 9.14 Common-emitter configuration for the *npn* transistor. (*a*) One-dimensional representation of the *npn* transistor with uniformly doped layers and abrupt-step junctions. (*b*) The *npn* transistor in part (*a*) with the circuit symbol for the transistor.

then $\beta = 49.0$. The usefulness of the bipolar transistor in the common-emitter configuration is that a small base current can control a much larger collector current.

The normal bias polarities and current directions for the *npn* transistor are shown in Fig. 9.14. Equations similar to Eqs. (9.39)–(9.45) may be written for the *npn* transistor.

9.4.2 *I–V* Characteristics

A representative plot of typical collector and base currents as a function of the emitter-base bias is shown in Fig. 9.15. The plot of I_C and I_B vs. V_{EB} has become known as a Gummel plot.[5] At low currents in Fig. 9.15, the base current I_B varies as $\exp(qV_{EB}/2kT)$ due to space-charge recombination current and/or surface recombination current, and then at higher currents, I_B is due to recombination in the base, which was given in Eq. (9.24) as $I_{B_r} \propto \exp(qV_{EB}/kT)$. At high base currents, I_B departs from the $\exp(qV_{EB}/kT)$ dependence due to ohmic resistance of the base region, as introduced in Sec. 4.7.7, and is represented by an $I_B R_B$ voltage drop. This base resistance is generally due to the base region being thinner and more lightly doped than the emitter. The collector current results from the emitter-base diffusion current, which varies as $\exp(qV_{EB}/kT)$ except at high currents where the injected minority carrier concentrations approach the majority carrier concentration in the base and the I_E begins to vary as $\exp(qV_{EB}/2kT)$, as discussed in Sec. 4.7.7. The variation of I_E is related to the collector current as $I_C = \alpha_o I_E$, and therefore, I_C has the same $\exp(qV_{EB}/nkT)$ variation as I_E.

The variation of β, as obtained from the Gummel plot in Fig. 9.15, is shown in Fig. 9.16. At small values of I_C, β does not become relatively constant until the space-charge recombination and/or surface recombination current are much less than

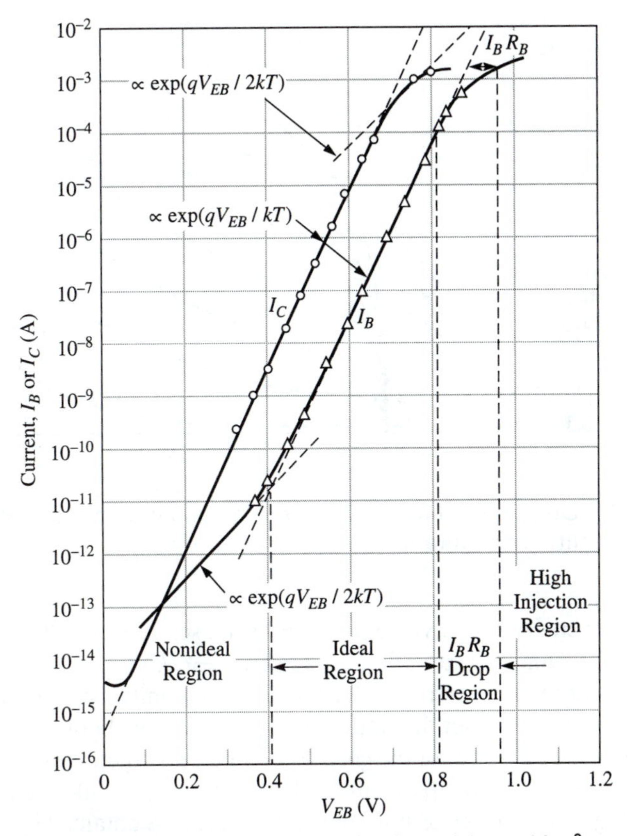

FIGURE 9.15 Gummel plot for I_C and I_B vs. V_{EB}.[9]

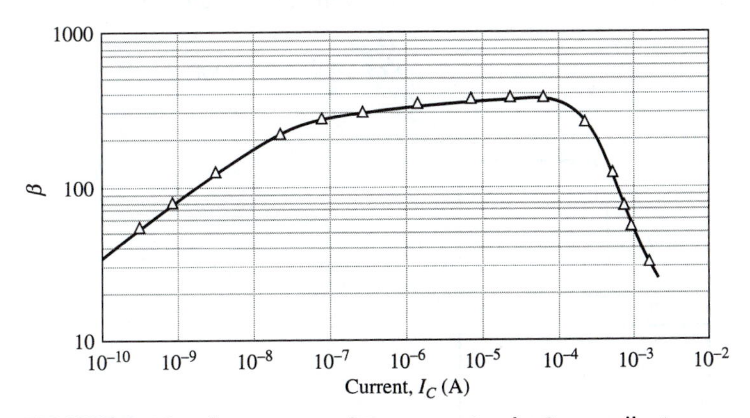

FIGURE 9.16 Common-emitter current gain β vs. collector current I_C.[9]

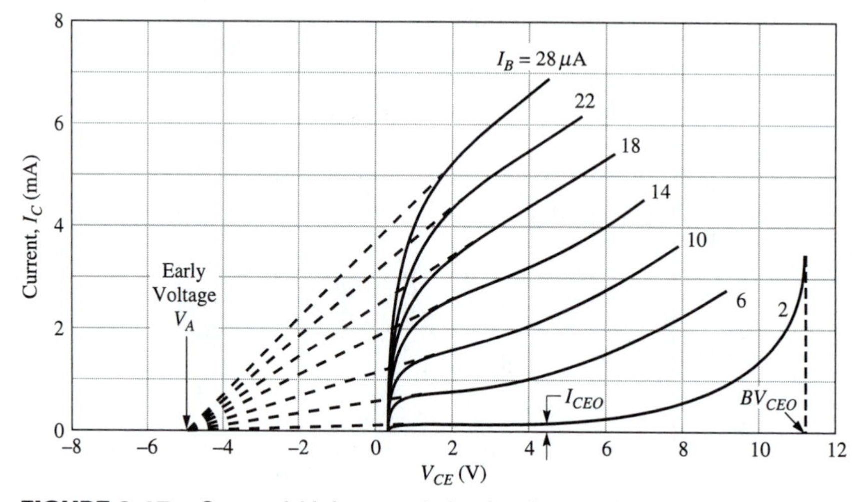

FIGURE 9.17 Output *I–V* characteristics for the *pnp* bipolar transistor in the common-emitter configuration.[8,9]

the diffusion current. At larger values of I_C, β decreases due to the high injection effects which give a diffusion current variation as $\exp(qV_{EB}/2kT)$.

The *I–V* output characteristics for the common-emitter configuration are shown in Fig. 9.17. The condition for the saturation condition when both junctions are forward biased is somewhat different for the common-emitter configuration, because saturation is now determined by the relative values of V_{EB} and V_{CE}, while the polarities remain as shown in Fig. 9.13. The condition for saturation can be obtained by representing the transistor by the circuit shown in Fig. 9.18. Let $I_C \simeq I_E$ and $I_E >> I_B$. Then the current flow I_C through the two resistors results in the voltage drops of V_{EB} and V_{CB}, as shown. From Kirchhoff's law, the sum of the voltages around a closed loop is zero:

$$V_{CE} = V_{CB} + V_{EB} \tag{9.46}$$

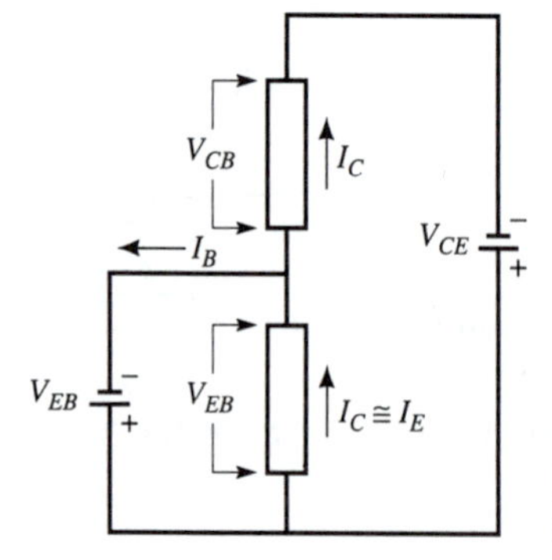

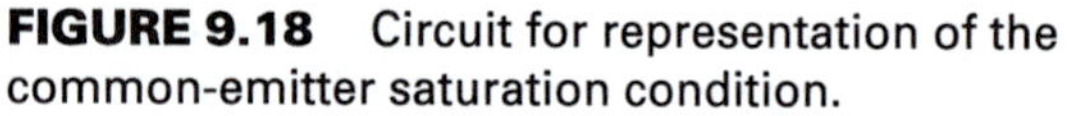

FIGURE 9.18 Circuit for representation of the common-emitter saturation condition.

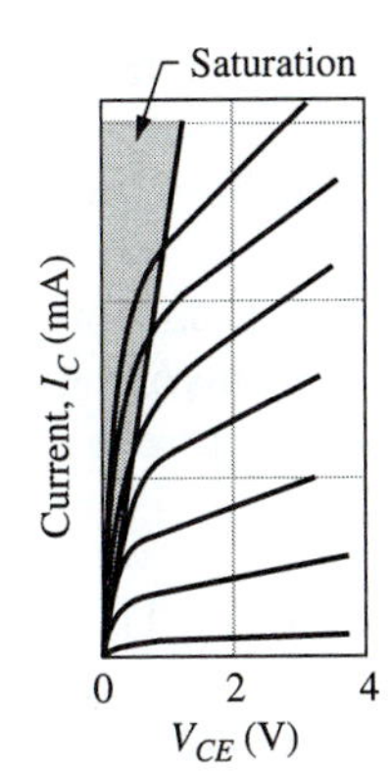

FIGURE 9.19 Representation of the saturation region for the common-emitter configuration.

or

$$V_{CB} = V_{CE} - V_{EB}. \tag{9.47}$$

saturation condition

When $V_{CE} < V_{EB}$, V_{CB} goes negative and the collector-base junction is forward biased. This condition is known as saturation. This saturation region is illustrated in Fig. 9.19, and occurs in Si transistors for V_{CE} of less than about 1 V. The minority carrier concentration variation when both junctions are forward biased was illustrated in Fig. 9.12 (*e*). Because the hole gradient becomes smaller for this condition, I_C approaches zero as shown. Saturation results in large stored-base charge, and therefore, this condition must be avoided for high-speed operation.

The reverse saturation current I_{CEO} shown in Fig. 9.17 was given in Eq. (9.45) as $I_{CBO}/(1 - \alpha_o)$ and is the collector current for the open emitter-base junction. Also, in Fig. 9.17, the collector breakdown voltage is represented by BV_{CEO}.

Another feature in Fig. 9.17 is the Early voltage designated by V_A. Modulation of the neutral base width by V_{CE} was illustrated in Fig. 9.12 (*c*). Because the neutral base width W_B becomes smaller as the reverse bias V_{CE} is increased, α_o will become closer to unity because the base transport factor α_T, which was given in Eq. (9.36) as $(1 - W_B^2/2L_n^2)$, depends on W_B^2. This increasing α_o causes large increases in β and gives significant slope to I_C as V_{CE} becomes larger. Extrapolation of I_C to zero is designated the Early voltage[10] V_A, as illustrated in Fig. 9.17.

9.5 EBERS-MOLL MODEL

9.5.1 Basic Model

For circuit analysis, the previously derived equations which describe transistor action are used to develop circuit models. The most common DC and large-signal, nonlinear model, and the one used in SPICE, is the Ebers-Moll model.[2] This model relates the large signal behavior to the commonly used small-signal parameters, which are derived in the next section. Space-charge and generation currents are not included, but these currents, as well as the junction capacitances and series resistances, are readily added. Recombination in the base region, as represented by the base transport factor α_T, is included. A very good summary was given by Getreu.[11]

The Ebers-Moll model represents two interacting *p-n* junctions by giving I_E and I_C in terms of V_{EB} and V_{CB}. All currents are expressed in terms of the reference currents I_F and I_R. For the *npn* transistor, the reference forward current I_F is given for the emitter-base junction as

$$I_F = I_{nE} + I_{pE} = I_{EBS}[\exp(qV_{EB}/kT) - 1], \tag{9.48}$$

where I_{EBS} is the emitter-base junction saturation current, and V_{EB} is the emitter-base voltage. In Eq. (9.48), I_F is the current that would flow across the emitter-base junction for a given V_{EB} if the collector region was replaced by an ohmic contact. The reference reverse-diode current I_R is given for the collector-base junction as

$$I_R = I_{nC} + I_{pC} = I_{CBS}[\exp(qV_{CB}/kT) - 1], \tag{9.49}$$

where I_{CBS} is the collector-base saturation current, and V_{CB} is the collector-base voltage. In Eq. (9.49), I_R is the current that would flow across the collector-base junction if the emitter region was replaced with an ohmic contact. Use of the *npn* transistor permits writing many of the previous equations which were given for the *pnp* transistor.

The two diodes separately do not represent the transistor which has coupling by the thin-base region. This coupling was represented by the common-base current gain given in Eq. (9.7) as α_o, which is the fraction of the emitter current reaching the collector. In the Ebers-Moll model, α_o for the base-emitter junction is taken as α_F. Also, for the collector-base junction, α_R is the fraction of the collector current reaching the emitter when the collector-base junction is forward biased. The collector current can be expressed in terms of I_F and I_R as

$$I_C = \alpha_F I_F - I_R. \tag{9.50}$$

In Eq. (9.50), where $\alpha_F I_F$ is the collector current due to the emitter-base junction and I_R is the collector-base current. Note that for active bias, the collector-base junction would be reverse biased and $I_R = -I_{CBS}$. Similarly, the emitter current can be expressed as

$$I_E = I_F - \alpha_R I_R, \tag{9.51}$$

where I_F represents the emitter-base current which flows out of the emitter for the active bias mode. The emitter current due to the collector-base junction is represented by $\alpha_R I_R$. The base current is

$$I_B = (1 - \alpha_F)I_F + (1 - \alpha_R)I_R. \tag{9.52}$$

Equations (9.50)–(9.52) give the Ebers-Moll model for the *npn* transistor, which is shown in Fig. 9.20, and this version is called the injection model.

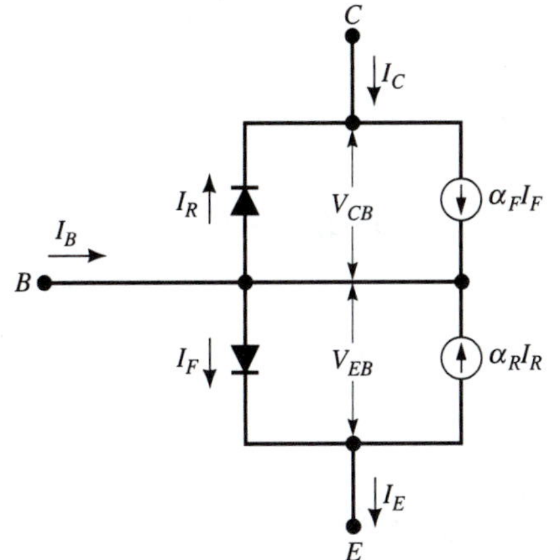

FIGURE 9.20 Basic Ebers-Moll model for the *npn* transistor.

In Eqs. (9.50)–(9.52), four parameters are used to represent I_C and I_E: I_{EBS}, I_{CBS}, α_F, and α_R. Ebers and Moll[2] showed that these four parameters reduce to three parameters. Consider an *npn* transistor where all the layer thicknesses are less than the diffusion lengths, and with the emitter-base area given by A_E and the collector-base area given by A_C. From Eq. (9.17), but for the *npn* transistor, the saturation electron current I_{nES} for the current injected into the base from the emitter is

$$I_{nES} = \frac{qA_E D_n n_i^2}{N_{a_B}^- W_B}. \tag{9.53}$$

Similarly, the saturation electron current I_{nCS} for the electron current injected into the base from the collector is

$$I_{nCS} = \frac{qA_C D_n n_i^2}{N_{a_B}^- W_B}. \tag{9.54}$$

Note that in Eqs. (9.53) and (9.54), $I_{nES} = I_{nCS}$ if $A_E = A_C$. These saturation currents only depend on the base doping, and not the emitter or collector doping.

From Eq. (9.18), but for the *npn* transistor, the saturation hole current I_{pES} for the hole current injected into the n^+ emitter is

$$I_{pES} = \frac{qA_E D_p n_i^2}{N_{d_E}^+ W_E}. \tag{9.55}$$

A similar equation can be written for the saturation hole current for the hole current injected into the collector from the base, which is

$$I_{pCS} = \frac{qA_C D_p n_i^2}{N_{d_C}^+ W_C}. \tag{9.56}$$

In this case, $I_{pES} \neq I_{pCS}$ because the emitter and collector dopings are very different: $N_{d_E}^{+} >> N_{d_C}^{+}$.

The product $\alpha_F I_{EBS}$, with α_F given by Eq. (9.7) as $\gamma\alpha_T$, becomes

$$\alpha_F I_{EBS} = \underbrace{\left(\frac{I_{nES}}{I_{nES} + I_{pES}}\right)}_{\gamma} \alpha_T (I_{nES} + I_{pES}) = \alpha_T I_{nES}. \tag{9.57}$$

The product $\alpha_R I_{CBS}$ becomes

$$\alpha_R I_{EBS} = \underbrace{\left(\frac{I_{nCS}}{I_{nCS} + I_{pCS}}\right)}_{\gamma} \alpha_T (I_{nCS} + I_{pCS}) = \alpha_T I_{nCS}. \tag{9.58}$$

By comparing the expressions for I_{nES} and I_{nCS} given in Eqs. (9.53) and (9.54) for $A_E = A_C$, then

$$\boxed{\alpha_F I_{EBS} = \alpha_R I_{CBS} \equiv I_S}. \tag{9.59}$$

Ebers and Moll,[2] with a three-dimensional analysis and Green's theorem, proved that this relation in Eq. (9.59) holds with no restrictions for the emitter and collector shapes (areas).

9.5.2 SPICE Configuration

To obtain the form of the Ebers-Moll model used in SPICE, the currents flowing through the current sources in Fig. 9.20, $\alpha_F I_F$ and $\alpha_R I_R$, are written as

$$I_{CC} = \alpha_F I_F = \underbrace{\alpha_F I_{EBS}}_{I_S}[\exp(qV_{EB}/kT) - 1] \tag{9.60}$$

and

$$I_{EC} = \alpha_R I_R = \underbrace{\alpha_R I_{CBS}}_{I_S}[\exp(qV_{CB}/kT) - 1], \tag{9.61}$$

where Eq. (9.59) has been used for I_S. The current I_F becomes I_{CC}/α_F and the current I_R becomes I_{EC}/α_R. With these current representations, the transport version of the Ebers-Moll model is shown in Fig. 9.21. This version is readily converted to the most useful form of the Ebers-Moll model, which is known as the nonlinear hybrid-π model.

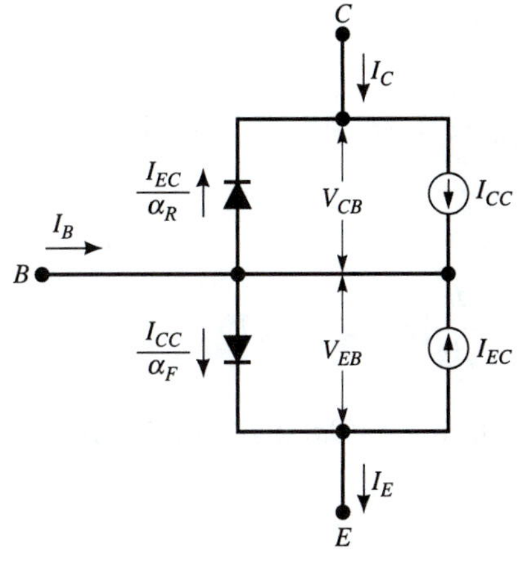

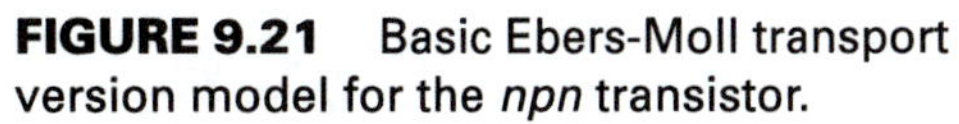

FIGURE 9.21 Basic Ebers-Moll transport version model for the *npn* transistor.

For the transport version shown in Fig. 9.21, the terminal currents are

$$I_C = I_{CC} - \frac{1}{\alpha_R} I_{EC}, \tag{9.62}$$

$$I_E = \frac{1}{\alpha_F} I_{CC} - I_{EC}, \tag{9.63}$$

and

$$I_B = \frac{I_{CC}}{\alpha_F} + \frac{I_{EC}}{\alpha_R} - I_{CC} - I_{EC} = \underbrace{\left(\frac{1}{\alpha_F} - 1\right)}_{(1-\alpha_F)/\alpha_F} I_{CC} + \underbrace{\left(\frac{1}{\alpha_R} - 1\right)}_{(1-\alpha_R)/\alpha_R} I_{EC}. \tag{9.64}$$

In Eq. (9.64),

$$\beta_F \equiv \frac{\alpha_F}{1 - \alpha_F} \tag{9.65}$$

and

$$\beta_R \equiv \frac{\alpha_R}{1 - \alpha_R}. \tag{9.66}$$

Next, the two reference current sources are replaced by a single current source,

$$I_{CT} = I_{CC} - I_{EC} = I_S[\exp(qV_{EB}/kT) - \exp(qV_{CB}/kT)]. \tag{9.67}$$

Then, the two diode currents become

$$\frac{I_{CC}}{\beta_F} = \frac{I_S}{\beta_F}[\exp(qV_{EB}/kT) - 1] \tag{9.68}$$

and

$$\frac{I_{EC}}{\beta_R} = \frac{I_S}{\beta_R}[\exp(qV_{CB}/kT) - 1]. \tag{9.69}$$

Adding and subtracting I_{EC} from I_C in Eq. (9.62) gives the collector current as

$$I_C = I_{CC} - I_{EC} + I_{EC} - \frac{1}{\alpha_R}I_{EC} = (I_{CC} - I_{EC}) - \frac{I_{EC}}{\beta_R}. \tag{9.70}$$

Adding and subtracting I_{CC} from I_E in Eq. (9.63) gives the emitter current as

$$I_E = \frac{1}{\alpha_F}I_{CC} - I_{CC} + I_{CC} - I_{EC} = \frac{I_{CC}}{\beta_F} + (I_{CC} - I_{EC}). \tag{9.71}$$

The nonlinear hybrid-π version of the Ebers-Moll model is shown in Fig. 9.22. This form reduces to the linearized small-signal equivalent circuit for the BJT operating in the forward active region obtained in the next section. The forward-biased emitter-base diode in the Ebers-Moll model becomes the input resistance given as $r_\pi = \beta_F/g_m$ given in the next section. The reverse-biased collector-base diode in the Ebers-Moll model becomes the output resistance r_μ in the small-signal model. Therefore, the version of the Ebers-Moll shown in Fig. 9.22 represents both the large-signal nonlinear and the linear small-signal model, and has become the SPICE BJT model.

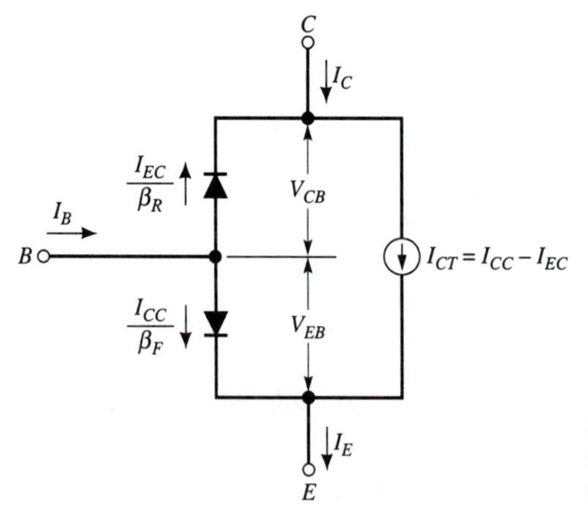

FIGURE 9.22 Nonlinear hybrid-π version of the Ebers-Moll model for the *npn* transistor.

9.6 SMALL-SIGNAL AC MODEL

9.6.1 Equivalent Circuit

Many applications of bipolar transistors are for operation at high frequencies or for high-speed switching in logic circuits. Only the DC characteristics have been considered in the previous sections of this chapter. Small-signal AC means that the AC voltages and currents are small compared to the DC values and the AC output varies linearly with the AC input. The AC signal should be small enough so that the region of the DC characteristic can be approximated by a linear representation. A small-signal equivalent circuit will be obtained in this part of Sec. 9.6. Then, to evaluate the ability of a device to operate at high frequencies, it is helpful to establish characteristic frequencies that are not dependent on specific circuits. The most common figure of merit is the frequency f_T at which the current gain in a common-emitter configuration becomes unity for the output shorted for an AC signal, and f_T will be considered in Sec. 9.6.2.

An amplifying circuit is shown in Fig. 9.23 (*a*) for the *pnp* transistor in the common-emitter configuration with a load resistor R_L and DC supply voltage V_{CC}. The small-signal AC operation of the transistor is illustrated by the load line drawn on the *I–V* output characteristics shown in Fig. 9.23 (*b*). The DC voltage across the transistor V_{CE} and across R_L must equal V_{CC}:

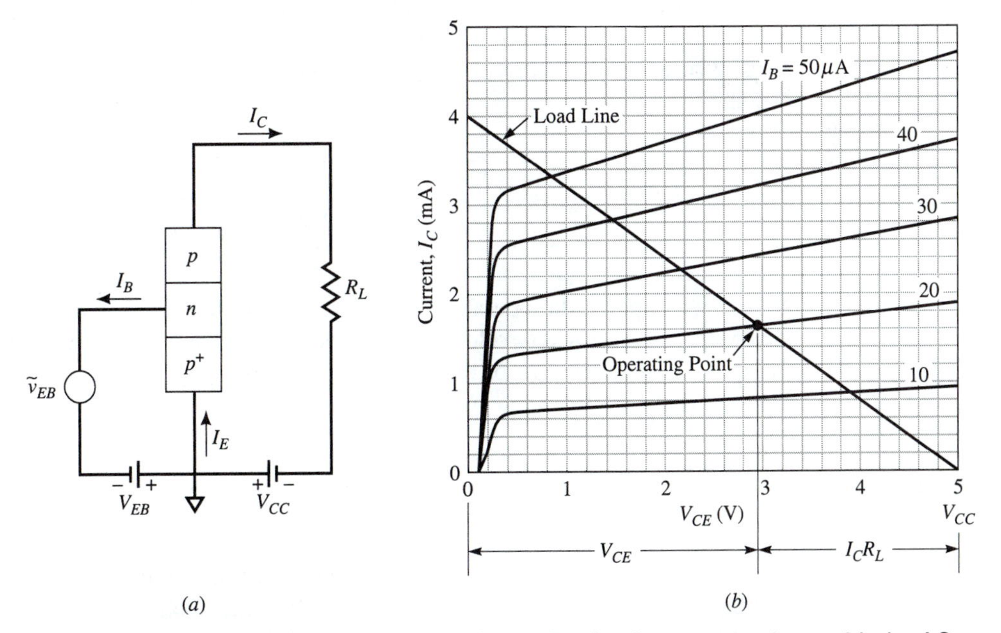

FIGURE 9.23 (*a*) Common-emitter configuration for the *pnp* transistor with the AC input voltage $\tilde{v}_{EB}$. (*b*) Output *I–V* characteristics with the load line and operating point.

$$V_{CC} = V_{CE} + I_C R_L. \tag{9.72}$$

The load line is for $V_{CE} = 0$ and $I_C = V_{CC}/R_L$. The quiescent operating point, or just the operating point, is determined by V_{EB}, which controls I_B. For small variations around the operating point, the nonlinear I–V and C–V characteristics can be linearized, and the incremental currents become proportional to the incremental voltages when the voltage variations are smaller than kT/q. This *small-signal transistor model* is very useful in designing amplifier circuits.

The transconductance (g_m) is the change in output current (I_C) with respect to the change in the input voltage (V_{EB}). With $I_C \simeq I_{pE}$ for the *pnp* BJT, the collector current I_C in Eq. (9.31) may be written as

$$I_C \simeq I_S \exp(qV_{EB}/kT). \tag{9.73}$$

Then, the transconductance becomes

transconductance g_m

$$\boxed{g_m \equiv \left.\frac{\partial I_C}{\partial V_{EB}}\right|_{V_{CE}} = \frac{q}{kT} I_S \exp(qV_{EB}/kT) = \frac{q}{kT} I_C}, \tag{9.74}$$

with Eq. (9.73) for I_C.

The variation of the base current (I_B) with emitter-base voltage (V_{EB}) is defined as g_{EB}, the input conductance. With Eq. (9.44) for the relation between I_B and I_C as $I_B = I_C/\beta$, where the small current $I_{CBO}/(1 - \alpha_o)$ may be neglected, the input conductance becomes

input conductance g_{EB}

$$\boxed{g_{EB} \equiv \left.\frac{\partial I_B}{\partial V_{EB}}\right|_{V_{CE}} = \left.\frac{\partial (I_C/\beta_F)}{\partial V_{EB}}\right|_{V_{CE}} = \frac{g_m}{\beta_F} = g_\pi}. \tag{9.75}$$

Note that at a particular operating point, the forward beta β_F is constant. Because this small-signal model was developed as the linear hybrid π-model, the input conductance in Eq. (9.75) is labeled g_π or the input resistance is

input resistance

$$\boxed{r_\pi = \frac{1}{g_\pi} = \frac{\beta_F}{g_m}}. \tag{9.76}$$

As illustrated in Fig. 9.17, the voltage across the collector-base junction results in a finite slope of the I–V characteristic which is due to base-width modulation and is represented by the Early voltage V_A. In terms of the small-signal parameters, the output resistance r_o is given by

output resistance r_o

$$\boxed{\frac{\Delta V_{CE}}{\Delta I_C} = \frac{|V_A|}{I_C} = \frac{q}{kT}\frac{|V_A|}{g_m} = r_o}, \tag{9.77}$$

where I_C has been represented by Eq. (9.74).

With the emitter-base junction forward biased, there also will be capacitances. The abrupt emitter-base depletion capacitance C_{jEB} may be obtained from Eq. (5.33) and the diffusion (or storage) capacitance C_s (minority-carrier storage in the base) may be obtained from Eq. (5.68). These capacitances with $(q/kT)I_C$ replaced by g_m for the diffusion capacitance have been written as C_π and are

$$C_\pi = A\sqrt{\frac{q\epsilon N_a^- N_d^+}{2V_{bi}(N_d^+ + N_a^-)}}\sqrt{\frac{V_{bi}}{V_{bi} - V_{EB}}} + \tau_D \underbrace{\frac{q}{kT} I_s \exp(qV_{EB}/kT)}_{g_m}. \tag{9.78}$$

The basic *pnp* bipolar transistor small-signal equivalent circuit with these small-signal circuit elements is shown in Fig. 9.24. For the *npn* bipolar transistor, the arrow for g_m would point in the opposite (down) direction.

There are additional circuit elements for the complete small-signal model. The variation of V_{CE} changes the collector depletion-layer width, which results in a change in the base width. The variation in base width results in a change in the minority-carrier charge stored in the base [see Eq. (9.15)] and a change in the base current. A small change ΔV_{CE} in V_{CE} causes a change in ΔI_B, which is represented by a resistance r_μ from the collector to the base of the equivalent circuit shown in Fig. 9.24. This resistance is given by

$$r_\mu = \frac{\Delta V_{CE}}{\Delta I_B} = \underbrace{\frac{\Delta V_{CE}}{\Delta I_C}}_{r_o}\underbrace{\frac{\Delta I_C}{\Delta I_B}}_{\beta_F} = r_o\beta_F. \tag{9.79}$$

Several parasitic elements are also added to the equivalent circuit in Fig. 9.24. One is the depletion capacitance of the collector-base junction, which is designated C_μ, and the other is the the collector-substrate depletion capacitance C_{CS}. The resistive

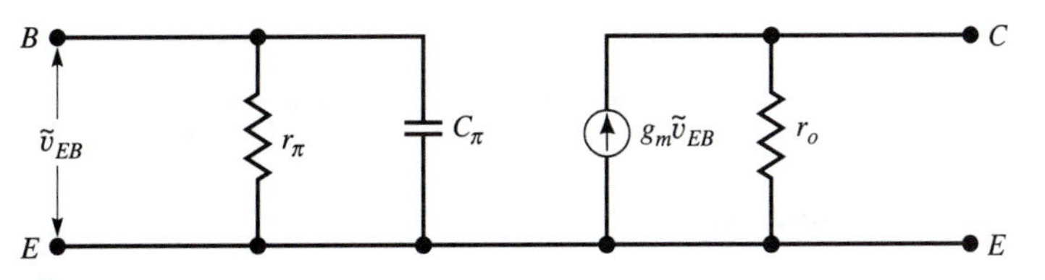

FIGURE 9.24 Basic *pnp* bipolar transistor small-signal equivalent circuit.

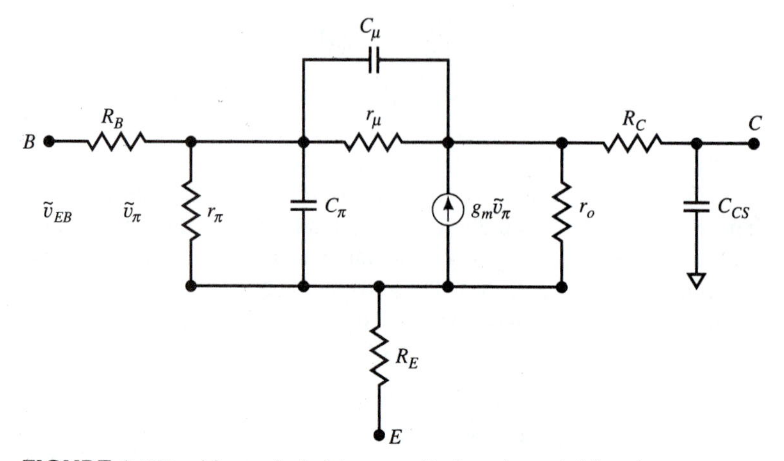

FIGURE 9.25 Linear hybrid-π small-signal model for the *pnp* bipolar transistor. The arrow for g_m points down for the *npn* bipolar transistor.

parasitics are the finite resistances associated with the base, emitter, and collector, which are represented by R_B, R_E, and R_C, respectively. The addition of these resistances and capacitances give the complete small-signal equivalent circuit shown in Fig. 9.25.

9.6.2 Cutoff Frequency f_T

The cutoff frequency f_T is often used to specify the transistor high-frequency performance, and is the frequency at which the current gain in the common-emitter configuration is unity for the output shorted for an AC signal. Measurements of f_T are used to compare the performance of transistors for new Si fabrication technologies and for various combinations of semiconductors used in heterojunction bipolar transistors. The shorted output condition eliminates the output resistance r_o and connects C_μ in parallel with C_π. In this simplified model, R_B, R_E, and R_C are ignored, but a more complete model would generally include R_B. Because r_μ is a large resistance which is now in parallel with the small resistance r_π, r_μ may be neglected. For these conditions, the equivalent circuit shown in Fig. 9.25 for the *pnp* bipolar transisor reduces to the high-frequency equivalent circuit shown in Fig. 9.26.

From Fig. 9.26, the input current is given by

$$\tilde{i}_{in} = [1/r_\pi + j\omega(C_\pi + C_\mu)]\tilde{v}_{EB}, \tag{9.80}$$

and the output current is

$$\tilde{i}_{out} = g_m\tilde{v}_{EB}. \tag{9.81}$$

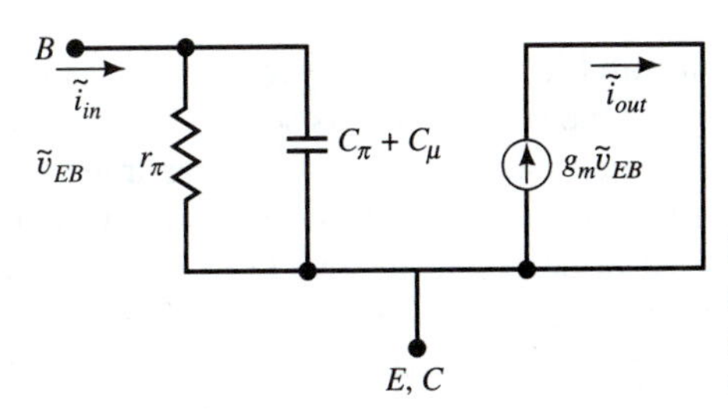

FIGURE 9.26 High-frequency equivalent circuit with the collector shorted to the emitter (output shorted) for the determination of f_T.

The magnitude of the input current is the square root of $\tilde{i}_{in}$ times its complex conjugate:

$$|\tilde{i}_{in}| = \sqrt{1/r_\pi^2 + \omega^2(C_\pi + C_\mu)^2}\,\tilde{v}_{EB}. \tag{9.82}$$

The ratio of the magnitude of the output to the input current is

$$\frac{|\tilde{i}_{out}|}{|\tilde{i}_{in}|} = \frac{g_m}{\sqrt{1/r_\pi^2 + \omega^2(C_\pi + C_\mu)^2}}. \tag{9.83}$$

At high frequencies where $\omega(C_\pi + C_\mu) > 1/r_\pi$, the cutoff frequency f_T is for $|\tilde{i}_{out}|/|\tilde{i}_{in}| = 1$ and is given by Eq. (9.83) as

cutoff frequency f_T

$$\boxed{f_T = \frac{g_m}{2\pi(C_\pi + C_\mu)}}. \tag{9.84}$$

Equation (9.84) is the usual representation of f_T.

Because the diffusion capacitance can be larger than the depletion capacitances, $(C_\pi + C_\mu) \to C_\pi \to C_s$. From Eq. (9.78), $C_s = \tau_D g_m$. Then Eq. (9.84) reduces to

$$f_T \simeq \frac{1}{2\pi\tau_D}. \tag{9.85}$$

With $\tau_D = W_n^2/2D_p$ from Eq. (5.65), f_T may be written as

$$f_T \simeq \frac{D_p}{\pi W_n^2}. \tag{9.86}$$

This expression for f_T emphasizes two factors in the high-frequency performance of bipolar transistors: (1) the cutoff frequency depends directly on the minority-carrier diffusivity in the base region; therefore, minority-carrier electrons will give a higher f_T

because of their higher mobility, and (2) thin base-layer thickness is necessary for large f_T, because f_T varies inversely as the base thickness squared. The role of the minority-carrier diffusivity in the high-speed performance of bipolar transistors is the reason that *npn* transistors are preferred over *pnp* transistors. The cutoff frequency for the metal-oxide semiconductor field-effect transistor (MOSFET) was given in Eq. (8.89) as $f_T \simeq 3\mu_n(V_{GS} - V_T)/(4\pi L^2)$, where L is the gate length. Comparison of Eqs. (9.86) and (8.89) shows that f_T for the MOSFET depends directly on the carrier mobility and inversely on the channel length squared, while f_T for the BJT varies directly with the carrier diffusivity and inversely as the base thickness squared.

The high-frequency behavior of transistors has also been specified by the maximum available power gain at high frequencies.[12] Careful analysis was applied to the transistors as a four-terminal network long before computer simulation was used. For maximum power gain, the load R_L must be matched to the output resistance of the transistor. At high frequencies, the power gain decreases with frequency. When the power gain falls to unity, the maximum frequency of oscillation f_{max} is obtained as

$$f_{max} = \left(\frac{g_m}{16\pi^2 R_B(C_\pi + C_\mu)} \right)^{1/2}. \tag{9.87}$$

Equation (9.87) is derived in Appendix E at the end of this chapter. From Eq. (9.84) with $(C_\pi + C_\mu) \simeq C_\pi \simeq C_s$, g_m/C_s may be replaced by $2\pi f_T$ to give

maximum frequency of oscillation

$$\boxed{f_{max} = \sqrt{\frac{f_T}{8\pi R_B C_s}}}. \tag{9.88}$$

Both f_T and f_{max} are often used as figures of merit for comparison of high-frequency transistors.

9.7 SPICE MODEL FOR BIPOLAR TRANSISTORS

9.7.1 Element and Model Lines

The element and model lines in SPICE for the bipolar transistor have been summarized by Banzhaf.[13] The general form of the *element line* for the bipolar transistor is

bipolar transistor element line

```
QXXXXXX NC NB NE <NS> MODNAME <AREA> <OFF> <IC=VBE,VCE>,
```

where `QXXXXXX` is the name of the bipolar transistor, `NC` is the collector node, `NB` is base node, `NE` is the emitter node, and `MODNAME` is the model name which is used in an associated `.MODEL` control line. These items are required in the bipolar transistor element line. The optional parameters are the quantities in the $< \cdots >$ and an element line may be continued by entering a + sign at the start of the next line. The circuit symbol for the bipolar transistor was shown in Fig. 9.2. The meanings of the optional parameters are:

`NS`	The node of the substrate which defaults to 0.
`AREA`	The area parameter specifies how many of the bipolar transistor model `MODNAME` are connected in parallel to make one `QXXXXXX`.
`OFF`	The initial condition of `QXXXXXX` for DC analysis.
`IC=VBE,VCE`	SPICE will use `VBE` and `VCE` as the initial conditions for the bipolar transistor base-emitter and collector-emitter voltages rather than the quiescent operating point for a transient analysis.

The model form or the *model line* for the bipolar transistor is

bipolar transistor `.MODEL` *line*

```
.MODEL MODNAME NPN<(PAR1=PVAL1 PAR2=PVAL2...)>
.MODEL MODNAME PNP<(PAR1=PVAL1 PAR2=PVAL2...)>,
```

where `MODNAME` is the model name given to a bipolar transistor in the element line, and `NPN` or `PNP` denote that the device is an *npn* or *pnp* transistor; `PAR` is the parameter name of one of the optional parameters listed in Tables 9.1–9.3 for SPICE2 or PSpice, and Table 9.4 for additional optional parameters available in PSpice; and `PVAL` is the value of the designated parameter. Care must be taken to assign the correct units which are also designated in the tables.

9.7.2 SPICE DC Model Parameters

`IS BF BR`

The optional SPICE2 or PSpice DC model parameters, as described by Massobrio and Antognetti,[14] are given in Table 9.1. The first three parameters, I_S, β_F, and β_R represent the bipolar transistor by the Ebers-Moll model given in Sec. 9.5 and provide an accurate DC model. The relationships of these three parameters to I_C and I_B may be illustrated by writing I_C, as represented in Fig. 9.22 and Eq. (9.70), with I_{CT} given by Eq. (9.67). For the active mode, the collector-base junction is reverse biased and the $\exp(-qV_{CB}/kT)$ term in Eq. (9.67) and the I_{EC}/β_R term in Eq. (9.70) may be neglected, and the collector current becomes

$$I_C \simeq I_{CC} \simeq I_S \exp(qV_{EB}/kT). \tag{9.89}$$

For the active mode, the base current I_B given in Eq. (9.64) becomes

$$I_B = \frac{I_{CC}}{\beta_F} = \frac{I_S}{\beta_F}[\exp(qV_{EB}/kT) - 1]. \tag{9.90}$$

The -1 term in Eq. (9.90) may be neglected for $V_{EB} > 3kT/q$, and then with Eq. (9.89) for I_C,

$$\beta_F = I_C/I_B, \tag{9.91}$$

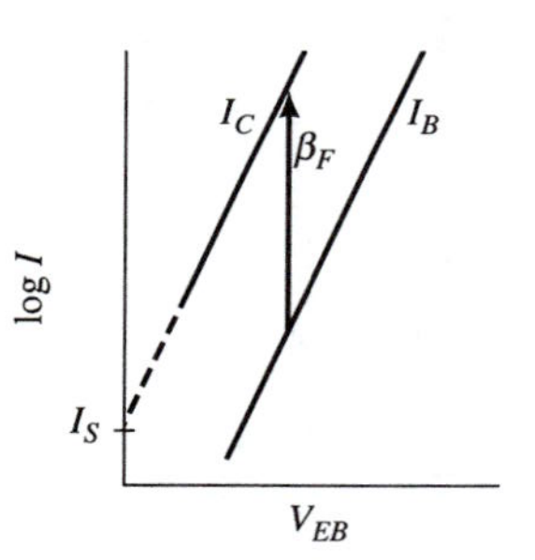

TABLE 9.1 SPICE Bipolar Transistor DC Model Parameters.

No.	Text Symbol	SPICE Keyword	Parameter Name	Default Value	Units
1	I_s	IS	Saturation current	1.0E-14	A
2	β_F	BF	Ideal maximum forward-current gain	100	—
3	β_R	BR	Ideal maximum reverse-current gain	1	—
4	n_F	NF	Forward-current ideality factor (ranges from 1.0 to 2.0)	1.0	—
5	n_R	NR	Reverse-current ideality factor (ranges from 1.0 to 2.0)	1.0	—
6	I_{ssrc}	ISE	Emitter-base space-charge and/or surface-recombination saturation current	1.0E-13	A
7	n_E	NE	Emitter-base ideality factor (ranges from 1.0 to 2.0)	1.0	—
8	I_{ssrc}	ISC	Collector-base space-charge and/or surface-recombination saturation current	1.0E-13	A
9	n_C	NC	Collector-base ideality factor (ranges from 1.0 to 2.0)	1.0	—
10	—	IKF	High-injection "knee" current for β_F roll-off	∞	A
11	—	IKR	High-injection "knee" current for β_R roll-off	∞	A
12	V_A	VAF	Forward Early voltage	∞	V
13	V_B	VAR	Reverse Early voltage	∞	V
14	E_g	EG	Energy gap	1.11	eV
15	—	XTI	Saturation current temperature exponent	3.0	—
16	—	XTB	Forward and reverse β temperature exponent	0	—

as expected from the definition of β in Sec. 9.4.1. As illustrated by the figure in the margin adjacent to Table 9.1, the saturation current I_S represents the extrapolation of I_C to $V_{EB} = 0$. Parameter no. 4 (NF) represents the ideality factor in I_{CC} given in Eq. (9.60) as $I_{CC} = I_S[\exp(qV_{EB}/n_F kT) - 1]$, while parameter No. 5 represents the ideality factor in I_{EC} given in Eq. (9.61) as $I_{EC} = I_S[\exp(qV_{CB}/n_R kT) - 1]$.

NF NR

Several important second-order effects which are present in actual devices are not included in the Ebers-Moll model. These effects include the low base current region where space-charge recombination and/or surface recombination can have an effect on β as well as high-level injection where the injected minority carrier concentration can approach the base majority carrier concentration. Another effect not included in the Ebers-Moll model is base-width modulation as described by the Early effect. These effects are included in the Gummel-Poon model.[15] Parameter no. 6 is the emitter-base junction space-charge recombination saturation current ISE given in Eq. (4.159) and/or the surface-recombination saturation current given by Eq. (4.160). The ideality factor n_E for ISE is parameter no. 7 and generally would be $n_E = 2.0$. This current is in parallel with I_{CC}/β_F as shown in Fig. 9.27. Similarly, parameter no. 8 is the collector-base junction space-charge recombination saturation current ISC given in Eq. (4.159) and/or the surface-recombination saturation current given by Eq. (4.160). The ideality factor n_C for ISC is parameter no. 9 and generally would be $n_C = 2.0$. This current is in parallel with I_{EC}/β_R as shown in Fig. 9.27.

Gummel-Poon model

ISE

NE

ISC

NC

The high-injection "knee" current which is given by parameter no. 10 was represented by Eq. (5.70). This parameter represents the high current injection when the injected minority carrier concentration becomes comparable to the majority carrier

IKF

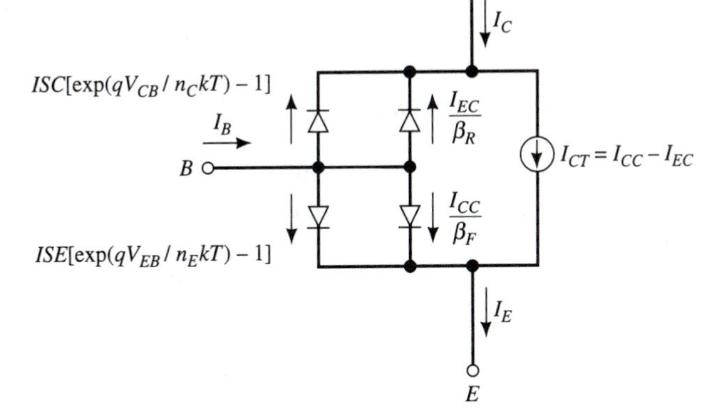

FIGURE 9.27 Gummel-Poon SPICE model for the *npn* transistor with the emitter-base junction space-charge and/or surface recombination current $ISE[\exp(qV_{EB}/n_E kT) - 1]$, and the collector-base junction space-charge and/or surface recombination current $ISC[\exp(qV_{CB}/n_C kT) - 1]$.

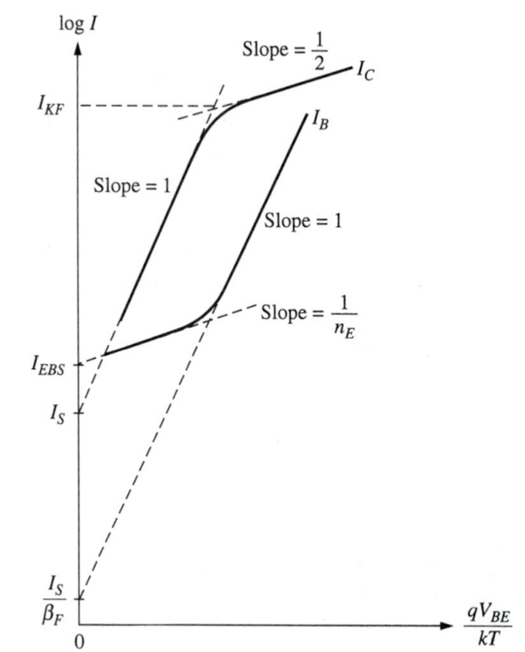

FIGURE 9.28 Gummel plot of I_C and I_B versus qV_{EB}/kT for $V_{CB} = 0$.[16]

concentration. The resulting β_F roll-off was shown in Fig. 9.16. Parameter no. 11 is the same as IKF, but for inverted bias. The effect of these parameters on I_C and I_B is illustrated in the Gummel plot shown in Fig. 9.28.

IKR

The next SPICE parameters, which are also part of the Gummel-Poon model, are for the base-width modulation. Parameter no. 12 is the forward Early voltage V_A, which was illustrated in Fig. 9.17. The Early voltage is related to the slope of the collector current in the output characteristics and is due to the base-width modulation by the collector-base reverse bias. Parameter no. 13 is the Early voltage V_B for inverted bias conditions.

VAF
VAR

The next three parameters represent temperature dependences. Parameter no. 14 is the energy gap E_g with the temperature-dependent expression given in Eq. (2.72) for Si. Parameter no. 15 is the saturation current temperature exponent, and as described in Sec. 5.5.2, is 3.0 for diffusion current. The temperature dependence is represented as

EG

XTI

$$I_S(T) = I_S(300\text{ K})\left(\frac{T}{300\text{ K}}\right)^{\text{XTI}}. \tag{9.92}$$

The temperature dependence of β is given by parameter no. 16 for β_F or β_R as

XTB

$$\beta(T) = \beta(300\text{ K})\left(\frac{T}{300\text{ K}}\right)^{\text{XTB}}. \tag{9.93}$$

EXAMPLE 9.1 Use PSpice to obtain the I_C-V_{CE} output characteristics for a *pnp* transistor in the common-emitter configuration with $I_S = 1.85 \times 10^{-14}$ A, $\beta_F = 65$, $\beta_R = 0.1$, and an Early voltage of 10.0 V. Let I_B vary from 0 to 50 μA in steps of 10 μA and use 0.05 voltage steps for V_{CE} between 0 and 5 V.

Solution (a) First, draw a circuit diagram and number the nodes. (b) Create a file named, for example, bjt-ce.cir.

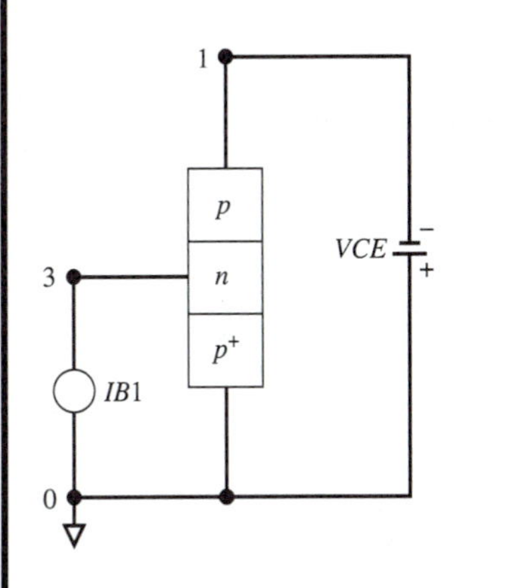


```
bjt pnp common-emitter characteristics bjt-ce.cir
VCE 0 1
IB1 3 0
Q1 1 3 0 Q216
.MODEL Q216 PNP BF=65 BR=0.1 IS=1.85E-14 VAF=10.0
.DC VCE 0 5 0.05 IB1 0.0 50.0U 10.0U
.PROBE
.END
```

The output *I–V* characteristics were shown in Fig. 9.23 (*b*) with a load line.

9.7.3 Resistance and Charge-Storage Parameters

The parasitic resistances and capacitances as well as the base transport time are given in Table 9.2. The base resistance consists of the base contact resistance and the sheet resistance of the base region outside the emitter region, which is parameter no. 17 and is represented by r_B. Parameter no. 18 is represented by r_{BM} and is the resistance of the base directly under the emitter. In SPICE, the total base resistance R_B is expressed as[17]

RB

RBM

$$R_B = r_{BM} + 3(r_B - r_{BM})\left(\frac{\tan z - z}{z \tan^2 z}\right), \tag{9.94}$$

TABLE 9.2 SPICE2 Bipolar Transistor Parasitic Parameters.

No.	Text Symbol	SPICE Keyword	Parameter Name	Default Value	Units
17	r_B	RB	Zero-bias base resistance	0	Ω
18	r_{BM}	RBM	Minimum base resistance at high currents	RB	Ω
19	I_{rB}	IRB	Current where base resistance falls to half its minimum value	∞	A
20	R_E	RE	Emitter resistance	0	Ω
21	R_C	RC	Collector resistance	0	Ω
22	$C_{jE}(0)$	CJE	Zero-bias emitter-base depletion capacitance	0	F
23	$V_{bi_{EB}}$	VJE	Emitter-base built-in voltage	0.75	V
24	m_E	MJE	Emitter-base grading coefficient	0.33	—
25	—	FC	Coefficient for forward-bias depletion capacitance expression	0.5	—
26	$C_{jC}(0)$	CJC	Zero-bias collector-base depletion capacitance	0	F
27	$V_{bi_{CB}}$	VJC	Collector-base built-in voltage	0.75	V
28	m_C	MJC	Collector-base grading coefficient	0.33	—
29	$C_{jS}(0)$	CJS	Zero-bias collector-substrate depletion capacitance	0	F
30	V_{bi_S}	VJS	Collector-substrate built-in voltage	0.75	V
31	m_S	MJS	Collector-substrate grading coefficient	0	—
32	X_{CJC}	XCJC	Fraction of collector-base depletion capacitance connected to internal base node	1	—
33	τ_F	TF	Ideal forward transit time	0	s
34	τ_R	TR	Ideal reverse transit time	0	s
35	X_{tF}	STF	Coefficient for the bias dependence of TF	0	—
36	V_{tF}	VTF	Voltage describing V_{CB} dependence of TF	∞	V
37	I_{tF}	ITF	High-current parameter for effect on TF	0	A
38	$P_{\tau F}$	PTF	Excess phase at $f = 1/2\pi\tau_D$	0	°

where

$$z = \frac{-1 + \sqrt{1 + 144 I_B/\pi^2 I_{rB}}}{(24/\pi^2)\sqrt{I_B/I_{rB}}}. \tag{9.95}$$

IRB
RE
RC

In Eq. (9.95), I_{rB} is the current where the base resistance falls to half of its minimum value and is parameter no. 19. The emitter resistance R_E is given by parameter no. 20, and the collector resistance R_C is given by parameter no. 21.

The capacitances for each junction are modeled in the same manner as was given in Sec. 5.5.2 for the *p-n* junction. The total emitter-base capacitance C_{EB} and the total collector-base capacitance C_{CB} are the the sum of the depletion capacitance C_j given in Eq. (5.34) or Eq. (5.47), and the diffusion capacitance C_s given in Eq. (5.68). These capacitances were illustrated in Fig. 5.16 and for reverse bias only C_j applies, but for forward bias both C_j and C_s are important.

The emitter-base depletion capacitance is given as

CJE

$$C_{jE}(V_{EB}) = \frac{C_{jE}(0)}{(1 - V_{EB}/V_{bi_{EB}})^{m_E}}, \tag{9.96}$$

VJE
MJE

where the zero-bias capacitance $C_{jE}(0)$ is parameter no. 22, the emitter-base built-in potential $V_{bi_{EB}}$ is parameter no. 23, and the emitter-base grading coefficient m_E is parameter no. 24. SPICE uses several different expressions for the capacitance.[18] One expression is used for $V_{EB} = 0$ to a voltage which is a fraction of the built-in potential.

FC
CJC, VJC, MJC
CJS, VJS, MJS

This fraction of the built-in potential is represented by `FC` and is generally 0.5, which is a voltage of $0.5V_{bi_{EB}}$. Parameters nos. 26, 27, and 28 give an equation similar to Eq. (9.96) for the collector-base junction. Parameters nos. 29, 30, and 31 give an equation similar to Eq. (9.96) for the collector-substrate junction.

A better approximation to the distributed resistance and capacitance at the collector-base junction is made by dividing the junction capacitance into two parts. Parameter no. 32 is the fraction of the collector-base depletion capacitance and is

XCJC

represented by `XCJC`.[19] The capacitance $X_{CJC}C_{jC}$ represents the portion of the capacitance between the internal base node and the collector, and $(1 - X_{CJC})C_{jC}$ is the portion of the capacitance between the external base and the collector. This partition is generally only important at high frequencies.

The diffusion (or storage) capacitances are related to the emitter-base current I_{EB} and the collector-base current I_{CB}, as illustrated in Sec. 5.4.3. For the stored charge due to I_{EB}, the emitter-base is assumed to be forward biased and $V_{CB} = 0$. The total forward base-collector transit time τ_F is parameter no. 33 and represents the mean base-collector transit time for minority carriers to cross the neutral base region. The diffusion capacitance was given by Eq. (5.68), and for the emitter-base current becomes

$$C_{SE} = \frac{q}{kT}\tau_F I_{EB}. \tag{9.97}$$

In a similar manner, the diffusion capacitance for the collector-base current becomes

$$C_{SC} = \frac{q}{kT}\tau_R I_{CB}, \tag{9.98}$$

where τ_R is the reverse base-collector transit time given by parameter no. 34. For saturation conditions (both V_{EB} and V_{CB} forward biased), C_{SE} and C_{SC} are assumed to act independently. The parasitic resistances, the depletion capacitances, together with the diffusion capacitances, are shown in Fig. 9.29. This model is called the Ebers-Moll large-signal model.

The variation of τ_F at high collector currents is represented by an empirical equation derived from the variation of f_T, with collector current I_C at various collector-emitter voltages V_{CE}.[20] The cutoff frequency f_T was related to τ_F in Eq. (9.85) as $f_T \simeq 1/2\pi\tau_F$. These effects are modeled by the empirical function [21]

$$\mathsf{ATF} = 1 + X_{tF}\exp(V_{CB}/1.44V_{tF})\left(\frac{I_{CC}}{I_{CC}+I_{tF}}\right)^2, \tag{9.99}$$

STF
VTF

where X_{tF} (parameter no. 35) is the coefficient for the bias voltage dependence of τ_F, V_{tF} (parameter no. 36) is a voltage describing the V_{CB} dependence of f_T, and I_{tF} (parameter

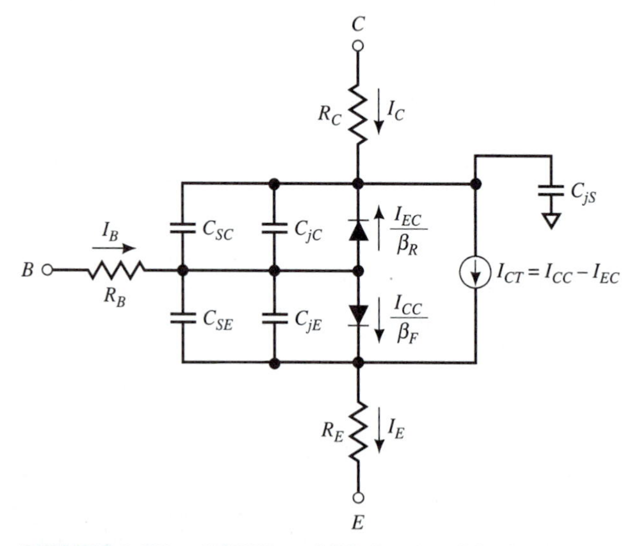

FIGURE 9.29 SPICE2 and PSpice *npn* bipolar transistor Ebers-Moll large-signal model with the parasitic resistances and capacitances. The emitter-base junction capacitances C_{jE} and C_{SE} are the depletion and diffusion capacitances, respectively. The collector-base junction capacitances C_{jC} and C_{SC} are the depletion and diffusion capacitances, respectively.

ITF

no. 37) is a parameter for the high-current effect on f_T. The high-frequency cutoff is given by[21]

$$f_T = \frac{1}{2\pi\tau_F\left[\mathtt{ATF} + \dfrac{2(\mathtt{ATF}-1)I_{tF}}{I_{CC}+I_{tF}} + \dfrac{kT}{q}n_F\dfrac{\mathtt{ATF}-1}{1.44V_{tF}}\right]}. \tag{9.100}$$

PTF

Parameter no. 38, which is given as $P_{\tau F}$, takes into account the measured phase shift which often exceeds the phase shift given by the lumped-parameter equivalent circuit,[22] and is due to distributed effects in the base region. A parameter ω_o is given as

$$\omega_o = 1/P_{\tau F}\tau_F, \tag{9.101}$$

where $P_{\tau F}$ is the phase delay at the idealized maximum bandwidth ($\omega = 1/\tau_F$).

9.7.4 Noise Parameters

Flicker noise in a bipolar transistor has been found experimentally to be represented by a current generator across the internal emitter-base junction. Parameters no. 39 and 40 are listed in Table 9.3 and are used to represent the flicker noise. Flicker noise ($1/f$ noise) is found in all types of active devices and is associated with the flow of direct current and displays a spectral density for the base current of the form

KF &AF

$$\overline{i_{B_F}^2} = k_f \frac{I_B^{a_f}}{f}\Delta f, \tag{9.102}$$

where Δf is a small bandwidth at the frequency f, I_D is the DC drain current, k_f is the flicker-noise coefficient (values range from 0.5 to 2), parameter KF, and a_f is the flicker-noise exponent, parameter AF. The flicker noise is due mainly to random fluctuations of carriers. Massobrio and Antognetti provide further discussion of noise in the bipolar transistor.[23]

9.7.5 Additional PSpice Bipolar Transistor Model Parameters

The additional PSpice parameters are listed in Table 9.4.[13,24] Parameter no. 41, given by NK, is used with parameter no. 10 to represent the emitter-base current for high-level

NK

TABLE 9.3 SPICE Bipolar Transistor Noise Parameters.

No.	Text Symbol	SPICE Keyword	Parameter Name	Default Value	Units
39	k_f	KF	Flicker-noise coefficient	0.0	—
40	a_f	AF	Flicker-noise exponent	1.0	—

TABLE 9.4 Additional PSpice Bipolar Transistor Parameters.

No.	Text Symbol	SPICE Keyword	Parameter Name	Default Value	Units
41	—	NK	High-current roll-off coefficient	0.5	—
42	I_{SS}	ISS	Substrate *p-n* junction saturation current		A
43	n_S	NS	Substrate *p-n* junction ideality factor	1.0	—
44	R_{C0}	RCO	Epitaxial region resistance	0	Ω
45	V_0	VO	Carrier mobility "knee" voltage	10	V
46	Q_{C0}	QCO	Epitaxial region charge factor	0	C
47	γ	GAMMA	Epitaxial doping factor	1E-11	—
48	—	TRE1	RE resistance linear temperature coefficient	0	$°C^{-1}$
49	—	TRE2	RE resistance quadratic temperature coefficient	0	$°C^{-2}$
50	—	TRB1	RB resistance linear temperature coefficient	0	$°C^{-1}$
51	—	TRB2	RB resistance quadratic temperature coefficient	0	$°C^{-2}$
52	—	TRM1	RBM resistance linear temperature coefficient	0	$°C^{-1}$
53	—	TRM2	RBM resistance quadratic temperature coefficient	0	$°C^{-2}$
54	—	TRC1	RC resistance linear temperature coefficient	0	$°C^{-1}$
55	—	TRC2	RC resistance quadratic temperature coefficient	0	$°C^{-2}$

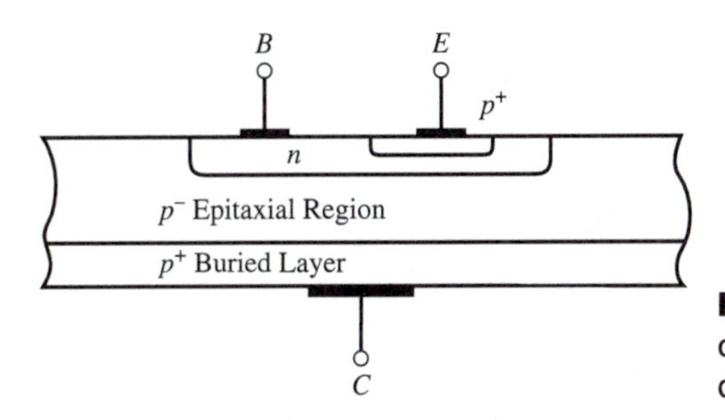

FIGURE 9.30 Schematic representation of a bipolar transistor with a lightly doped p^- epitaxial collector region.[25]

injection conditions. This diffusion current was given in Eq. (9.60) and has the form: $I_{CC} = [\cdots]^{\mathtt{NK}}\mathtt{IS}\exp(qV_{EB}/kT)$, where `NK` is generally 0.5. The next two parameters (nos. 42 and 43) represent a leakage current into the substrate and are represented by the saturation current I_{SS} and the substrate *p-n* junction ideality factor n_S.

`ISS`
`NS`

The next four parameters model what is called quasi saturation.[25] A schematic representation of the structure is shown in Fig. 9.30 where the lightly doped epitaxial collector region is represented by p^-. When a bipolar transistor is operated at high injection levels from the base into the collector region, DC current gain β and cutoff frequency f_T fall rapidly from their maximum values under conditions of quasi saturation. For this condition, the internal collector-base junction is forward biased, while the external collector-base terminals remain reverse biased. Because the collector-base region becomes forward biased rather than the normal reverse biased, the neutral base region widens and results in a smaller β, and the increased collector charge reduces f_T. This effect modifies the common-emitter I_C vs. V_{CE} output characteristics.

`RCO`

Parameter no. 44 is the resistance of the epitaxial region at low-level current flow and is represented by $R_{C0} = \rho w_C/A$, where ρ is the resistivity at thermal equilibrium, w_C is the collector layer thickness, and A is the cross-sectional area. At high collector currents, the quasi neutrality in the epitaxial collector region is represented by $p = n + N_a^-$, where the nonthermal-equilibrium hole concentration p exceeds the thermal-equilibrium hole concentration p_o. When $p > p_o$, the collector conductivity is modulated by the current.

`VO`
`QCO, GAMMA`

The next parameter (no. 45) is used to take into account the minority-carrier velocity saturation at high electric fields, with V_0 representing the drift velocity "knee" voltage. The next two parameters are used to model the excess charge in the epitaxial collector layer. Parameter no. 46 is $Q_{C0} = qAw_CN_a^-/4$, and parameter no. 47 is $\gamma = (2n_i/N_a^-)^2$. The expression used to calculate the collector current was given by Kull et al.[25]

Parameters nos. 48 through 55 give the temperature dependences of the transistor parasitic resistances. In PSpice, the normal room temperature of 300 K is referred to as `TNOM`. Use the `.OPTIONS` control line

```
.OPTIONSTNOM=297
```

to change the reference temperature to 297 K. The temperature variation of the parasitic emitter resistance given as parameter no. 20 is represented by

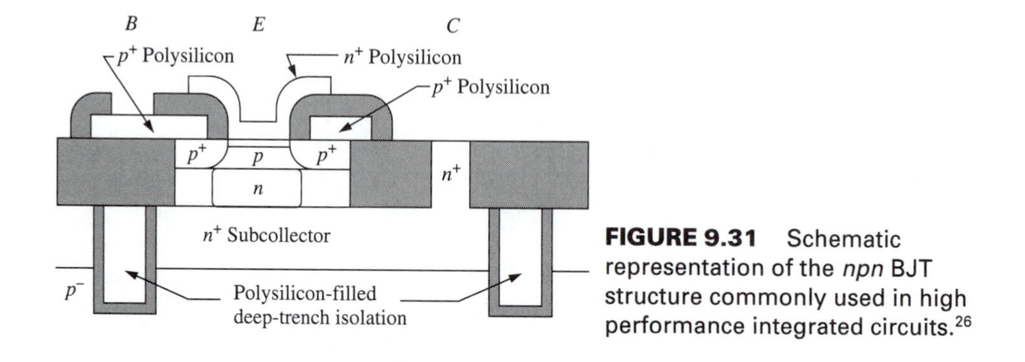

FIGURE 9.31 Schematic representation of the *npn* BJT structure commonly used in high performance integrated circuits.[26]

$$R_E(T) = \texttt{RE}[1 + \texttt{TRE1}(T - \texttt{TNOM}) + \texttt{TRE2}(T - \texttt{TNOM})^2], \tag{9.103}$$

TRE1, TRE2

where TR1 is the linear temperature coefficient for R_E and TR2 is the quadratic temperature coefficient for R_E. The remaining parameters in Table 9.4 are the linear and quadratic temperature coefficients for the other parasitic resistances in Table 9.2.

9.8 APPLICATIONS

9.8.1 Bipolar Transistor Structure

The most widely used bipolar technology is probably the deep-trench, double-polysilicon, self-aligned bipolar technology illustrated in Fig. 9.31,[26] and variations of this structure. This *npn* bipolar transistor structure uses (1) deep-trench isolation, (2) polysilicon emitter, (3) polysilicon base contact, which is self-aligned to the emitter, and (4) the collector region (pedestal collector) is directly underneath the emitter and is more heavily doped than its surrounding regions.

9.8.2 Differential-amplifier Pair Example

The emitter-coupled pair, as illustrated in Fig. 9.32, is a widely used two-transistor subcircuit because cascades of emitter-coupled pairs can be directly coupled without interstage coupling capacitors.[27] These emitter-coupled pairs are called differential amplifiers for analog applications and are also used for very fast digital logic called emitter-coupled logic.[27] The qualitative description of the differential amplifier given in the next part of this section illustrates the basic behavior of this circuit as a small-signal amplifier. Then, Example 9.2 will use PSpice to illustrate the frequency response of the gain and phase.

9.8.3 Qualitative Description of Circuit Operation

As shown in Fig. 9.32, the differential-pair amplifier has two identical transistors Q1 and Q2 whose emitters are connected to the constant current source I_{EE}. Let the input

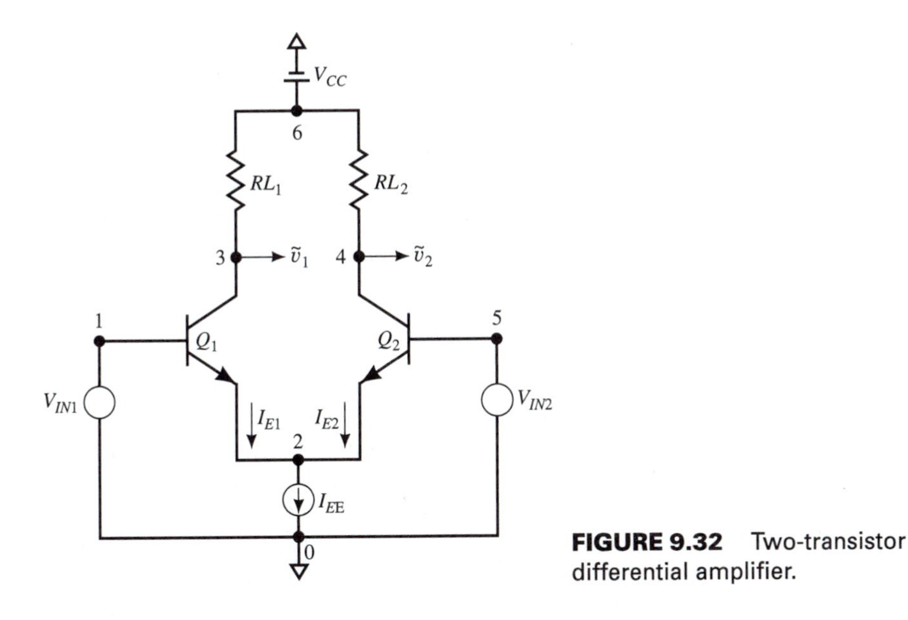

FIGURE 9.32 Two-transistor differential amplifier.

signals V_{IN1} and V_{IN2} be zero so that I_{EE} divides equally between Q_1 and Q_2:

$$I_{E1} = I_{E2} = I_{EE}/2. \tag{9.104}$$

The emitter-base voltage V_{EB} is found from $I_{E1} = I_S \exp(qV_{EB}/kT)$ as

$$V_{EB} = \frac{kT}{q} \ln \frac{I_{EE}/2}{I_S}. \tag{9.105}$$

For bipolar transistors with $I_S = 6.8 \times 10^{-15}$ A and with a constant current source $I_{EE} = 2 \times 10^{-4}$ A, the emitter-base bias for Q_1 and Q_2 is $V_{EB} = 0.609$ V.

With relatively small differences in V_{IN_1} and V_{IN_2}, the current can be increased in one transistor while the emitter current decreases in the other with the restraint that $I_{EE} = I_{E1} + I_{E2}$. The emitter currents given by Eq. (9.60) are

$$I_{E1} = I_S \exp[q(V_{EB} + V_{IN1})/kT] \tag{9.106}$$

and

$$I_{E2} = I_S \exp[q(V_{EB} + V_{IN2})/kT]. \tag{9.107}$$

The ratio of I_{E1} and I_{E2} from Eqs. (9.106) and (9.107) gives

$$\frac{I_{E1}}{I_{E2}} = \exp[q(V_{IN1} - V_{IN2})/kT]. \tag{9.108}$$

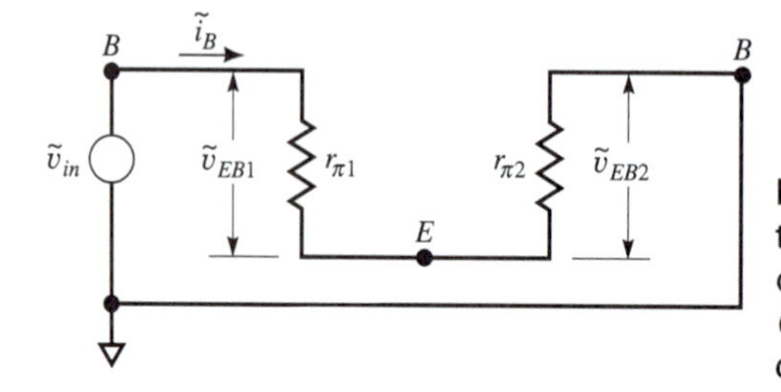

FIGURE 9.33 Representation of the small-signal input ciruit for the differential amplifier with the base of Q_2 grounded and the input signal $\tilde{v}_{in}$ connected to the base of Q_1.

This equation shows that the ratios of the emitter currents is related to the difference of the input voltages to the two transistors.

The amplifier gain G will now be considered for the case when the base of Q_2 is grounded. The gain will be taken as the ratio of the AC output voltage $\tilde{v}_2$ at the collector of Q_2 to the AC input signal $\tilde{v}_{in}$ connected between the base of Q_1 and ground. The basic bipolar transistor small-signal equivalent circuit was shown in Fig. 9.24. At low frequency where the capacitances can be ignored, the small-signal circuit can be simplified as shown in Fig. 9.33. The constant current source I_{EE} is high resistance and will be neglected in this simplified analysis.

For identical transistors, the input resistances, $r_{\pi 1} = r_{\pi 2}$, and the AC signal across the base of Q_1, $\tilde{v}_{EB1}$, is also across the base of Q_2. The AC signal is

$$\tilde{v}_{EB1} = \tilde{v}_{EB2} = \tilde{i}_B r_{\pi 1} = \frac{\tilde{v}_{in}}{2r_{\pi 1}} r_{\pi 1} = \frac{\tilde{v}_{in}}{2}, \tag{9.109}$$

where $\tilde{i}_B$ is the small-signal base current as shown in Fig. 9.33. At the collector of transistor Q_2, the current through R_L is $g_m \tilde{v}_{EB1}$ (see Fig. 9.24 for output current), and

$$\tilde{v}_2 = \tilde{v}_{out} = g_m \frac{\tilde{v}_{in}}{2} R_L. \tag{9.110}$$

The transconductance was given in Eq. (9.74) as $g_m = (q/kT)I_C$. Then with $I_C \simeq I_E = I_{EE}/2$ the AC output signal becomes

$$\tilde{v}_{out} = \frac{q}{kT} \frac{I_{EE}}{2} \frac{\tilde{v}_{in}}{2} R_L, \tag{9.111}$$

and the gain $G = \tilde{v}_{out}/\tilde{v}_{in}$ is

$$\boxed{G = \frac{\tilde{v}_{out}}{\tilde{v}_{in}} = \frac{I_{EE}}{4} \frac{q}{kT} R_L}. \tag{9.112}$$

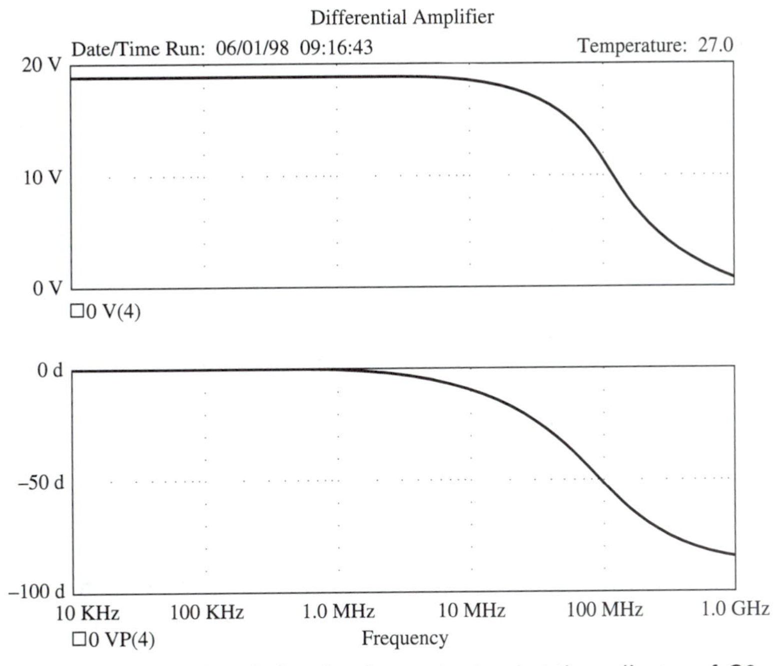

FIGURE 9.34 Plot of signal voltage obtained at the collector of $Q2$ (top plot) and the phase of this voltage (bottom plot) for the SPICE input file given in Example 9.2.

With $q/4kT = 9.7\ \text{V}^{-1}$ at room temperature,

$$\boxed{G \simeq 10 I_{EE} R_L}\,. \tag{9.113}$$

A PSpice example can be used to determine how well the gain is represented by Eq. (9.113) from this analysis.

EXAMPLE 9.2 Selected bipolar transistor parameters specified by MOSIS (MOS Implementation Services) for a particular bipolar fabrication process are: `BF=82`, `IS=6.8E-15`, `NF=1.0`, `VAF=59`, `CJE=0.78E-12`, `VJE=0.85`, `MJE=0.50`, `CJC=0.63E-12`, `MJC=0.50`, and `VJC=0.80`. Fabrication services offered by MOSIS are managed by the Information Sciences Institute of the University of Southern California.

(a) Find the output signal voltage at node 4 over a frequency range of 10 kHz to 1 GHz for the differential amplifier shown in Fig. 9.32 for $I_{EE} = 0.2$ mA, $RL1 = RL2 = 10\ \text{k}\Omega$, and an input signal of 1 mV.

(b) How does the low-frequency gain compare with Eq. (9.113)?

Solution A similar example has been given by Banzhaf.[29] The PSpice input file named dif-amp.cir is

```
DIFFERENTIAL-AMPLIFIER dif-amp.cir
VIN1 1 0 AC 0.001
VIN2 5 0
VCC 6 0 DC 6
IEE 2 0 DC 2E-4
Q1 3 1 2 DIF
Q2 4 5 2 DIF
.MODEL DIF NPN(BF=82 IS=6.8E-15 NF=1.0 VAF=59 CJE=0.78E-12
+VJE=0.85 MJE=0.50 CJC=0.63E-12 MJC=0.50 VJC=0.80)
RL1 6 3 10K
RL2 6 4 10K
.AC DEC 5 10K 1G
.PROBE
.END
```

The gain and phase as plotted with PSpice Probe are shown in Fig. 9.34. The low-frequency gain is 18.9, which is close to the value of 20 predicted by Eq. (9.113).

9.9 HETEROJUNCTION BIPOLAR TRANSISTORS

As bipolar transistor technology has progressed to higher frequency operation, techniques which increase the cutoff frequency f_T and the maximum frequency of oscillation f_{max} have been investigated. The cutoff frequency was derived in Sec. 9.6.2 for a *pnp* transistor and can be represented by the approximate relationship given in Eq. (9.86) as $f_T \simeq D_p/\pi W_n^2$, where D_p is the hole diffusivity in the base and W_n is the base thickness. Therefore, as might be expected, the base thickness is decreased to obtain higher frequency operation. As W_n becomes thinner, the base resistance R_B for a given base acceptor concentration becomes larger. It was shown in Appendix E at the end of this chapter that the maximum frequency of oscillation is taken as the frequency for a maximum power gain of unity and was given in Eq. (E.14) as $f_{max} = \sqrt{f_T/8\pi R_B C_\mu}$, where C_μ is the emitter-base capacitance. To lower the base resistance for thin-base regions, the acceptor concentration is increased. As illustrated in Eq. (9.34), the emitter efficiency (and the common emitter β) degrades as the base doping is increased.

In Sec. 5.6.4, it was shown in Eq. (5.100) for a heterojunction diode that the wide energy-gap semiconductor injects minority carriers into the more heavily doped, but narrower energy-gap semiconductor, while suppressing injection of minority carriers from the more heavily doped and narrower energy-gap base. The use of a wider energy-gap emitter for injection into a more heavily doped, but narrower energy-gap base region forms the heterojunction bipolar transistor (HBT). The basic concepts for HBTs were thoroughly investigated with $Al_xGa_{1-x}As$ emitter regions and heavily doped GaAs base regions. The collector region generally is also GaAs. The variation of the energy gap for $Al_xGa_{1-x}As$ was shown in Fig. 2.30. Recently, several approaches to form Si HBTs have been demonstrated.

A technique for Si-based HBTs is to use a Si emitter with a solid solution of Si_xGe_{1-x} to give a narrower-energy-gap base region.[30] This combination of Si/Si_xGe_{1-x} provides the advantages of HBTs while retaining the well-established Si-based technology. The introduction of Ge in the base provides an additional degree of freedom for higher transistor performance, although the integrated-circuit processing becomes more complex.

Detailed analysis for HBTs has been given by Tiwari,[31] and the most extensive work has been on HBTs based on III-V compound semiconductors such as $Al_xGa_{1-x}As/GaAs$ and $InP/Ga_{0.47}In_{0.53}As$. The example considered here is based on results for a HBT with an InP emitter and a $Ga_{0.47}In_{0.53}As$ base region.[32]

The energy-band diagram for the $InP/Ga_{0.47}In_{0.53}As$ HBT with the emitter-base junction forward biased and the collector-base junction reverse biased is shown in Fig. 9.35. The energy gap of InP at room temperature is 1.351 eV, and the energy gap of $Ga_{0.47}In_{0.53}As$ is 0.860 eV. This composition for $Ga_xIn_{1-x}As$ lattice matches InP. This

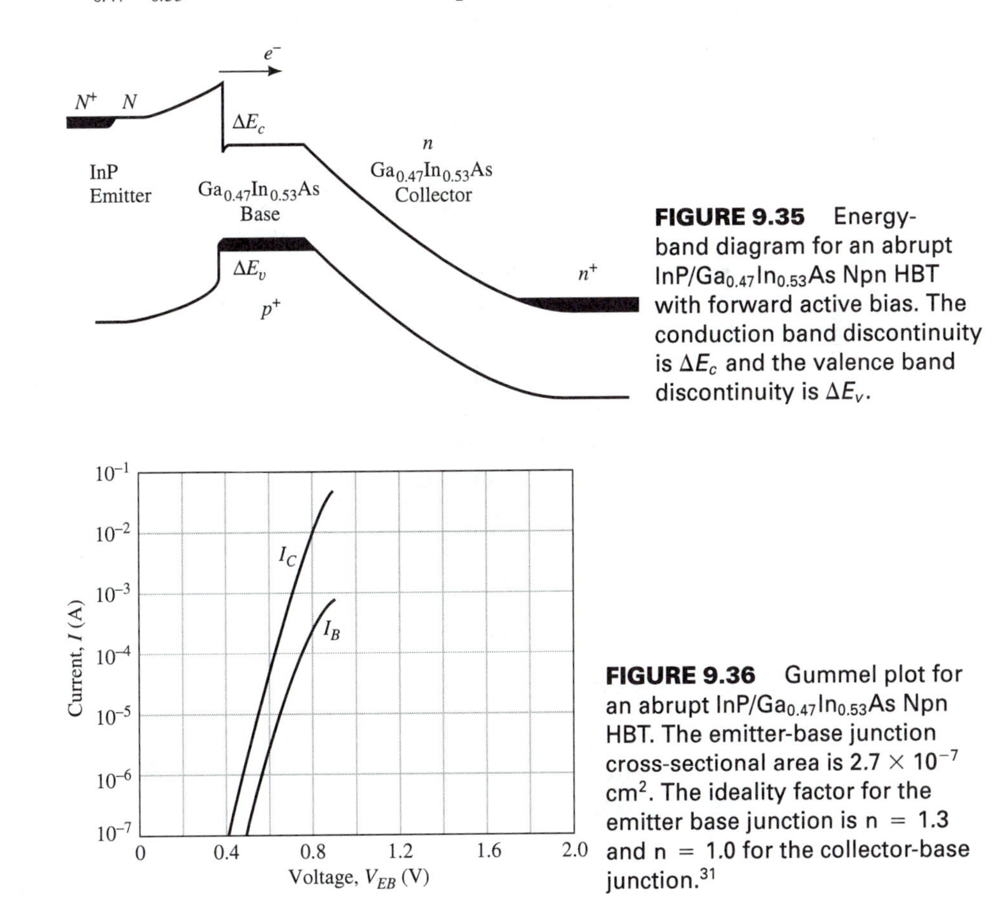

FIGURE 9.35 Energy-band diagram for an abrupt $InP/Ga_{0.47}In_{0.53}As$ Npn HBT with forward active bias. The conduction band discontinuity is ΔE_c and the valence band discontinuity is ΔE_v.

FIGURE 9.36 Gummel plot for an abrupt $InP/Ga_{0.47}In_{0.53}As$ Npn HBT. The emitter-base junction cross-sectional area is 2.7×10^{-7} cm^2. The ideality factor for the emitter base junction is $n = 1.3$ and $n = 1.0$ for the collector-base junction.[31]

HBT had a unity current-gain cutoff frequency $f_T = 300$ GHz. The Gummel plot of the collector current I_C and the base current I_B as a function of the emitter-base bias voltage V_{EB} is shown in Fig. 9.36.

9.10 SUMMARY AND USEFUL EXPRESSIONS

This chapter extended the two-terminal *p-n* junction to the three-terminal bipolar-junction transistor (BJT).

- The key concept in the operation of the bipolar junction transistor is that a low-input resistance forward-biased emitter-base junction injects minority carriers into the base region which diffuse to the high-output resistance reverse-biased base-collector junction.
- Injection efficiency γ is the ratio of the diffusion current from the emitter to the base to the total emitter-base current.
- Base transport factor α_T is the fraction of minority carriers injected from the emitter into the base which reach the collector.
- Common-base current gain is $\alpha_o = \gamma\alpha_T$.
- Common-emitter current gain $\beta = \alpha_o/(1 - \alpha_o) = I_C/I_B$.
- At low base currents, β decreases due to space-charge recombination and/or surface-recombination current at the emitter-base junction.
- At high base current, β decreases as the injected minority carrier concentration becomes comparable to the base-region majority carrier concentration.
- For circuit analysis, the Ebers-Moll model gives the equivalent circuit with three parameters: (1) the saturation current I_S, (2) the forward current gain β_F, and (3) the reverse current gain β_R.
- The Gummel-Poon model adds the emitter-base space-charge recombination and/or surface-recombination current as well as the base-width modulation represented by the Early voltage to the Ebers-Moll model and is used for SPICE.
- Unity current-gain in a common-emitter configuration with the output shorted is given by $f_T \simeq g_m/2\pi C_{S_{EB}} \simeq D_{n \text{ or } p}/\pi W_B^2$, where g_m is the transconductance, $C_{S_{EB}}$ is the emitter-base diffusion capacitance, $D_{n \text{ or } p}$ is the minority-carrier diffusivity in the base, and W_B is the neutral base width.

The following expressions for the BJT were obtained in this chapter.

***pnp* Common-base Current Gain**

$$\gamma \equiv \frac{I_{pE}}{I_E} \qquad \alpha_T \equiv \frac{I_{pC}}{I_{pE}} \qquad \alpha_o \equiv \frac{I_{pC}}{I_E} = \gamma\alpha_T$$

***npn* Common-base Current Gain**

$$\gamma \equiv \frac{I_{nE}}{I_E} \qquad \alpha_T \equiv \frac{I_{nC}}{I_{nE}} \qquad \alpha_o \equiv \frac{I_{nC}}{I_E} = \gamma\alpha_T$$

Common-emitter Current Gain $\beta = \alpha_o/(1 - \alpha_o)$

Base Recombination Current $I_{B_r} = Q_B/\tau_p$

Gummel Number $G_N \equiv \int_0^{W_B} N_B(x)dx$

Minority-carrier Transit Time $\tau_D = W_B^2/2D_{n \text{ or } p}$

Minority-carrier Mobility $\mu_n(p\text{-Si}) = \mu_n(n\text{-Si})$

Ebers-Moll Model

$I_{CC} \equiv I_S[\exp(qV_{EB}/kT) - 1] \qquad I_{EC} \equiv I_S[\exp(qV_{CB}/kT) - 1]$

$I_{CT} \equiv I_{CC} - I_{EC}$

$$I_E = \frac{I_{CC}}{\beta_F} + I_{CT} \qquad I_C = I_{CT} - \frac{I_{EC}}{\beta_R}$$

Cutoff Frequency $f_T \simeq g_m/(2\pi C_{S_{EB}}) \simeq 1/(2\pi\tau_p)$

Bipolar-transistor Element Line

```
QXXXXXX NC NB NE <NS> MODNAME <AREA> <OFF> <IC=VBE>,VCE>
```

Bipolar-transistor Model Line

```
.MODEL MODNAME NPN<(PAR1=PVAL1 PAR2=PVAL2...)>
.MODEL MODNAME PNP<(PAR1=PVAL1 PAR2=PVAL2...)>
```

REFERENCES

1. W. Shockley, "The Path to the Conception of the Junction Transistor," IEEE Trans. Electron. Devices **ED-23**, 597 (1976).
2. J. J. Ebers and J. L. Moll, "Large-Signal Behavior of Junction Transistors," Proc. IRE **42**, 1761 (1954).
3. K. Ng, *Complete Guide to Semiconductor Devices* (McGraw-Hill, New York, 1995), p. 240.
4. S. M. Sze, *Semiconductor Devices: Physics and Technology* (Wiley, New York, 1985), p. 116.
5. H. K. Gummel, "Measurement of the Number of Impurities in the Base Layer of a Transistor," Proc. IRE **49**, 834 (1977).
6. Ng, *Complete Guide,* p. 243.
7. Sze, *Semiconductor Devices,* p. 122.
8. M. J. Morant, *Introduction to Semiconductor Devices* (Addison-Wesley, Reading, Mass., 1964).
9. S. M. Sze, *Physics of Semiconductor Devices,* 2nd ed. (Wiley, New York, 1981), p. 148.

10. J. M. Early, "Effects of Space-Charge Layer Widening in Junction Transistors," Proc. IRE **43**, 1761 (1952).
11. I. Getreu, *Modeling the Bipolar Transistor* (Tektronix, Beaverton, OR, 1976), pp. 10–18.
12. R. L. Pritchard, "High-Frequency Power Gain of Junction Transistors," Proc. IRE **43**, 1075 (1955).
13. W. Banzhaf, *Computer-Aided Circuit Analysis using PSpice,* 2nd ed. (Regents/Prentice Hall, Englewood Cliffs, N.J., 1992) pp. 112–114.
14. G. Massobrio and P. Antognetti, Eds., *Semiconductor Device Modeling with SPICE,* 2nd ed. (McGraw-Hill, New York, 1993) pp. 45–130.
15. H. K. Gummel and H. C. Poon, "An Integral Charge Control Model of Bipolar Transistors," Bell Syst. Tech. J. **49**, 827 (1970).
16. Massobrio and Antognetti, *Semiconductor Device Modeling,* p. 79.
17. Ibid., p. 89.
18. Ibid., p. 69.
19. Ibid., p. 97.
20. Ibid., pp. 97–100.
21. Ibid., p. 99.
22. Ibid., p. 101.
23. Ibid., pp. 299–305, 309.
24. MicroSim Corporation, 20 Fairbanks, Irvine, CA 92718.
25. G. M. Kull, L. W. Nagel, S.-W. Lee, P. Lloyd, E. J. Prendergast, and H. Dirks, "A Unified Circuit Model for Bipolar Transistors Including Quasi-Saturation Effects," IEEE Trans. Electron Devices **ED-32,** 1103 (1985).
26. Y. Taur and T. H. Ning, *Fundamentals of Modern VLSI Devices* (Cambridge University Press, Cambridge, 1998).
27. P. R. Gray and R. G. Meyer, *Analysis and Design of Analog Integrated Circuits,* 3rd ed. (Wiley, New York, 1993), p. 227.
28. A. S. Sedra and K. C. Smith, *Microelectronic Circuits,* 4th ed. (Oxford, New York, 1998), p. 487.
29. Banzhaf, *Computer-Aided Circuit Analysis,* p. 191.
30. D. L. Harame, J. H. Comfort, J. D. Cressler, E. F. Crabbé, J. Y.-C. Sun, B. S. Meyerson, and T. Tice, "Si/SiGe Epitaxial-Base Transistors—Part I: Materials, Physics, and Circuits," IEEE Trans. Electron Devices **42**, 455 (1995).
31. S. Tiwari, *Compound Semiconductor Device Physics* (Academic Press, Boston, 1992), p. 551.
32. J. Laskar, R. N. Nottenburg, J. A. Baquedano, A. F. J. Levi, and J. Kolodzey, "Forward Transit Delay in $In_{0.53}Ga_{0.47}As$ Heterojunction Bipolar Transistors with Nonequilibrium Electron Transport," IEEE Trans. Electron Devices **40**, 1942 (1993).

PROBLEMS

All problems are for room temperature unless another temperature is specified.

9.1 For an *npn* transistor in the common-base configuration and biased in the normal active region, show

(a) The correct bias polarities, and sketch

(b) The charge density vs. distance.

(c) The electric field vs. distance.

(d) The potential vs. distance.

(e) The energy-band diagram.

(f) Why would you expect the built-in potential for the emitter-base junction to be larger than the base-collector junction?

9.2 For an Si *npn* transistor biased in the forward active region, the emitter-base electron-diffusion current is 2.60 mA, the base recombination current is 0.016 mA, the emitter-base hole-diffusion current is 0.061 mA, and the space-charge recombination current is 0.033 mA.

(a) What is the base transport factor?

(b) What is the emitter efficiency?

(c) What is the common emitter β?

9.3 For the *npn* transistor in Problem 9.2, the emitter-base bias is 0.55 V. When the emitter-base bias is increased to 0.65 V, would you expect β to increase or decrease? Please give the reason for your answer.

9.4 For an Si *npn* transistor biased in the forward active region, the base recombination current is 0.012 mA, the space-charge recombination current is 0.025 mA, the emitter-base hole-diffusion current is 0.045 mA, the emitter-base electron-diffusion current is 2.1 mA.

(a) What is the base transport factor?

(b) What is the emitter efficiency?

(c) What is the common emitter β?

9.5 The currents in Problem 9.4 are for $V_{EB} = 0.5$ V. If V_{EB} is increased to 0.6 V, find

(a) The emitter efficiency.

(b) The base transport factor.

(c) β.

Neglect the voltage variation of the neutral base width.

9.6 Show the correct bias polarities for the *pnp* and *npn* transistors in the common-emitter configuration.

9.7 Sketch the minority carrier concentrations for the *pnp* and *npn* transistors for the following bias modes:

(a) Active

(b) Inverted

(c) Cutoff

(d) Saturation.

Let the thickness of the emitter and collector layers be much larger than the diffusion length, but the base thickness be much less than the diffusion length.

9.8 For an *npn* transistor, the emitter-base hole-diffusion current is 0.1 mA, the emitter-base electron-diffusion current is 5.0 mA, the base recombination current is 0.001 mA. What is the transistor β? (Retain quantities to four decimal places.)

9.9 The base region of a Si *npn* transistor is doped with gold to give a recombination center concentration of 5×10^{15} cm^{-3} and a capture cross section of 1×10^{-15} cm^2. The base region is also doped with 3×10^{17} cm^{-3} boron atoms. At this doping level the electron mobility is 500 cm^2/V-s. The thermal velocity is 2×10^7 cm/s. What is the minority-carrier diffusion length in the base region at room temperature?

9.10 Consider an *npn* Si bipolar transistor at room temperature. For the common-emitter configuration, the emitter-collector bias is 5.0 V. It is convenient to use one of the coordinate reading capabilities in .probe in parts (b) and (d). These capabilities are accessed by clicking on one of the icons at the top of the .probe window with the dashed "cross."

(a) The forward beta is 120, the reverse beta is 0.2, and the transport saturation current is 2.65×10^{-17} A. With PSpice, plot the base and collector currents from 1×10^{-9} A to 1×10^{-1} A as a function of the emitter-base voltage. Let the emitter-base voltage vary from 0.4 V to 0.9 V in steps of 0.01 V. This type of plot is called a Gummel plot.

(b) From the plot in part (a), give the transistor beta at 0.70 V.

(c) This next plot takes into account the emitter-base space-charge recombination current and the effects of high-level injection. In addition to the parameters in part (a), the emitter-base leakage saturation current is 6.5×10^{-14} A, the base-emitter leakage emission coefficient **NE** is 2.0, and the corner for forward beta high-current roll-off is 5.0×10^{-3} A. With PSpice, plot the base and collector currents from 1×10^{-9} A to 1×10^{-1} A as a function of the emitter-base voltage. This plot is the same as in part (a), except that space-charge recombination in the emitter-base region and the effects of high-level are taken into account.

(d) From the plot in part (c), obtain beta at $V_{EB} = 0.55$, 0.60, 0.65, 0.70, 0.75, 0.80, 0.85, and 0.90 V. Sketch this variation of beta with the emitter-base voltage.

(e) With .probe, plot β vs. V_{EB} from 0.4 V to 0.9 V.

9.11 **(a)** Draw the Ebers-Moll model for the *pnp* bipolar transistor. Label all the current components in terms of I_S, V_{EB}, V_{CB}, β_F, and β_R.

(b) Write an expression for I_S for the *npn* transistor in the form given in Eq. (9.59) as $I_S = \alpha_F I_{EBS}$ with the expressions similar to Eqs. (9.53) and (9.55) for the emitter-base saturation currents.

9.12 A bipolar transistor has the properties listed in the following table.

Emitter	Base	Collector
$N_d^+ = 5 \times 10^{18}$ cm^{-3}	$N_a^- = 5 \times 10^{16}$ cm^{-3}	$N_d^+ = 2 \times 10^{16}$ cm^{-3}
$\mu_p = 85$ cm^2/V-s	$\mu_p = 370$ cm^2/V-s	$\mu_p = 390$ cm^2/V-s
$\mu_n = 140$ cm^2/V-s	$\mu_n = 1000$ cm^2/V-s	$\mu_n = 1200$ cm^2/V-s
$W_E = 10 \times 10^{-4}$ cm	$W_B = 0.85 \times 10^{-4}$ cm	$W_C = 10 \times 10^{-4}$ cm
$\tau_o = 5 \times 10^{-8}$ s	$\tau_o = 5 \times 10^{-8}$ s	$\tau_o = 5 \times 10^{-8}$ s

(a) What is the minority-carrier diffusion length in the base layer?

(b) At $V_{EB} = 0.6$ V, the emitter-base depletion-layer thickness is 0.1×10^{-4} cm, and at $V_{CB} = -5$ V, the base-collector depletion width is 0.16×10^{-4} cm on the base side and 0.79×10^{-4} cm on the collector side. What is the neutral base width?

(c) What is the ratio of the emitter-base hole-diffusion current to the emitter-base electron-diffusion current for the bias conditions in part (b)?

(d) At the edge of the emitter-base depletion region for the conditions in part (b), what is the recombination rate for injected minority carriers?

9.13 An *npn* transistor has the following properties:

Emitter	Base	Collector
$N_d^+ = 1 \times 10^{19}\ \text{cm}^{-3}$	$N_a^- = 1 \times 10^{17}\ \text{cm}^{-3}$	$N_d^+ = 1 \times 10^{16}\ \text{cm}^{-3}$
$D_{pE} = 1.7\ \text{cm}^2/\text{s}$	$D_{nB} = 20.8\ \text{cm}^2/\text{s}$	$D_{pC} = 10.9\ \text{cm}^2/\text{s}$
$W_E = 1.5 \times 10^{-4}\ \text{cm}$	$W_B = 1.0 \times 10^{-4}\ \text{cm}$	$W_C = 1.8 \times 10^{-4}\ \text{cm}$

Let the emitter-base area and the collector-base area both be 10 μm $\times$10 μm, and the base transport factor $\alpha_T = 0.9953$.

(a) Show that the electron emitter-base saturation current equals the collector-base saturation current.

(b) Show that the forward-diode saturation current, the hole current injected from the base into the emitter, is less than the reverse-diode saturation current, the hole current injected from the base into the collector.

(c) Show that the common-base forward current gain α_F is greater than the reverse current gain α_R.

(d) Show that $\alpha_F I_F = \alpha_R I_R$.

9.14 The *I–V* characteristics for a *pnp* bipolar transistor in the common-base configuration are shown below. For each of the operating points indicated on the *I–V* characteristics, sketch the minority carrier concentrations as a function of distance in the base region. The emitter-base voltage is V_{EB} and the collector-base voltage is V_{CB}. Let the base width be very small compared to the diffusion length. Please label the thermal-equilibrium minority carrier concentration in your sketch as well as nonthermal equilibrium minority carrier concentrations at each depletion region edge in the base.

(a) At the point labeled A.

(b) At the point labeled B.

(c) At the point labeled C.

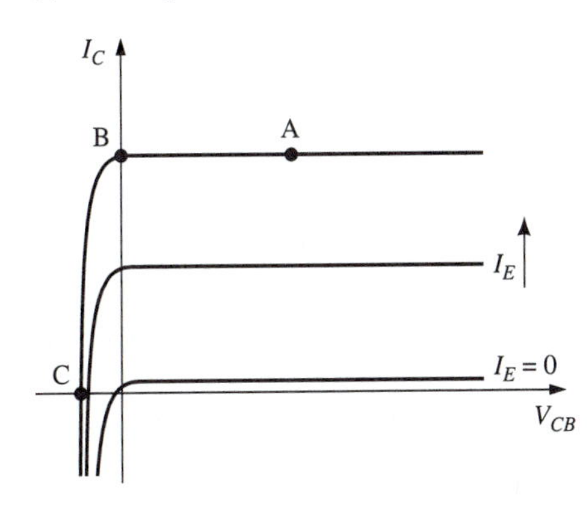

9.15 For a *pnp* transistor in the common-emitter active mode, the ratio of the hole current injected into the base from the emitter to the space-charge recombination current, I_{scr}/I_{pE}, decreases by 25%.

(a) Does the injection efficiency γ change or remain constant and why?

(b) Does the base transport factor α_T change or remain constant and why?

9.16 Draw the Gummel-Poon model for the *pnp* bipolar transistor and label the current components.

9.17 For the bipolar transistor in Problem 9.12 (b),

(a) Find the base transit time.

(b) Find the cutoff frequency f_T.

9.18 For the *npn* transistor with the parameters given in Problem 9.10 (a) and (c), include the Early voltage of 30 V.

(a) Obtain the common-emitter characteristics with a collector voltage of 0 to 5 V and for base currents of 2.5×10^{-8}, 5.0×10^{-8}, 7.5×10^{-8}, and 1.0×10^{-7} A.

(b) At a given collector-to-emitter voltage, why does the spacing between the I_C curves vary for equal changes in I_B?

(c) Obtain the common-emitter characteristics with a collector voltage of 0 to 5 V and for base currents of 2.5×10^{-5}, 5.0×10^{-5}, 7.5×10^{-5}, and 1.0×10^{-4} A.

9.19 The transistor β is shown below as a function of emitter-base bias V_{EB}.

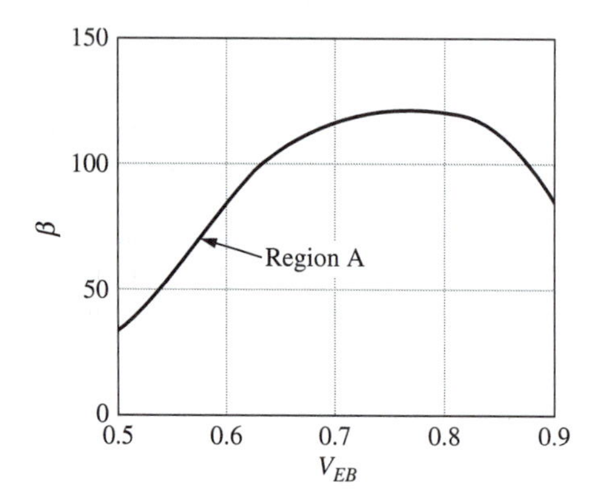

For the region marked by "A" where β increases, explain why the emitter efficiency γ and/or the base transport factor α_T increase.

9.20 For the bipolar transistor in Problem 9.12 (b),

(a) Find the minority carrier concentration at the emitter-base junction.

(b) Find the minority-carrier charge stored in the base for a cross-sectional area of 2×10^{-8} cm^2

(c) Find the ratio of the diffusion capacitance to the depletion capacitance for the emitter-base junction at $V_{EB} = 0.6$ V.

9.21 For the bipolar transistor in Problem 9.12, will avalanche breakdown or punch-through occur first as the collector-base voltage is increased?

9.22 What is the largest donor concentration in the collector region of a Si *npn* transistor which would permit an avalanche breakdown voltage of not less than 25 V?

9.23 Use PSpice to obtain the I_C-V_{CE} output characteristics for the 2N2222A *npn* bipolar transistor in the common-emitter configuration with the following PSpice (student version) parameters: `IS` = 14.3×10^{-15} A, `BF` = 256, `BR` = 6.1, `ISE` = 14.3×10^{-15} A, `NE` = 1.3, `ISC` = 0, `NC` = 2.0, `IKF` = 0.285 A, `IKR` = 0, `VAF` = 74.0 V, `EG` = 1.11 eV, `XTI` = 3.0, `XTB` = 1.5, `RB` = 10Ω, `RC` = 1Ω, `CJE` = 22.0×10^{-12} F, `VJE` = 0.75 V, `MJE` = 0.38, `FC` = 0.5, `CJC` = 7.3×10^{-12} F, `VJC` = 0.75 V, `MJC` = 0.34, `TF` = 0.41 ns, `TR` = 46.9 ns, `XTF` = 3.0, `VTF` = 1.7 V, and `ITF` = 0.6 A. Let I_B vary from 0 to 50 μA in steps of 10 μA and use 0.05 voltage steps for V_{CE} between 0 and 5 V.

9.24 The transfer-function analysis with the `.TF` control line is given by

```
.TF OUTPUTVAR INPUTSRC
```

where `OUTPUTVAR` is a small-signal output variable (voltage or current) and `INPUTSRC` is a small-signal input source (voltage or current). The `.TF` control line will print the input resistance at `INPUTSRC` and the output resistance at `OUTPUTVAR`.

With PSpice, find the input and output resistances for the differential amplifier circuit in Example 9.2.

9.25 Use the parameters given for the differential amplifier in Example 9.2 and vary I_{EE} to investigate whether the gain varies as given by Eq. (9.113) as I_{EE} is increased and decreased by a factor of 10.

9.26 How much does the frequency response change for the differential amplifier in Example 9.2 as *RL*1 and *RL*2 are decreased to 1 kΩ while I_{EE} is increased to 2.0 mA?

9.27 **(a)** For the differential amplifier shown in Fig. 9.32, let V_{IN1} = 2 mV and V_{IN2} = 1 mV. With the transistor and circuit parameters given in Example 9.2, find the output voltage at node 4 from 10 kHz to 1 GHz.

(b) Repeat part (a) with V_{IN1} = 3 mV and V_{IN2} = 1 mV. Does this problem illustrate why this transistor pair is called a differential amplifier?

9.28 The Darlington configuration is illustrated in the figure. This transistor pair permits amplification by direct coupling (eliminates coupling capacitors) and provides high current gain and large input

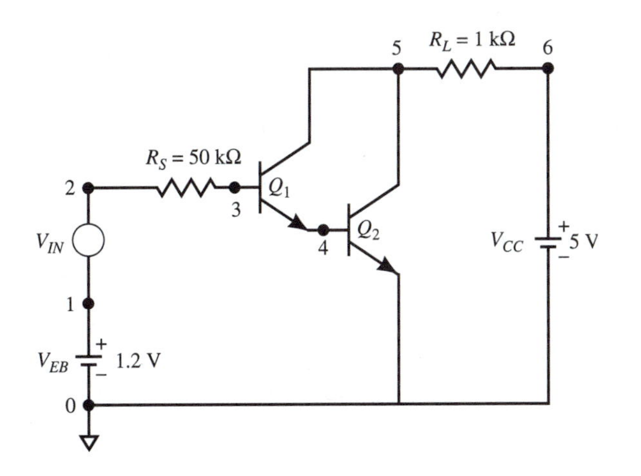

resistance as compared to a single transistor. Note that the emitter of transistor Q1 is directly connected to the base of transistor Q2. For this arrangement, the emitter current and the base current of Q1 are related as $i_e(\text{Q1}) \simeq \beta i_b(\text{Q1}) = i_b(\text{Q2})$. Then, the collector current of transistor Q2 is $i_c(\text{Q2}) \simeq \beta i_b(\text{Q2}) \simeq \beta^2 i_b(\text{Q1})$.

(a) For an input signal of 0.1 mV and a collector voltage of 5 V, use the `.TF` control line (see Problem 9.24) to find the voltage gain (the voltage at node 5 divided by `vin`) and the input and output resistances. Use the transistor parameters given for Example 9.2.

(b) Obtain a representative current gain from the ratio of the output current [i(vcc)] divided by the input current [i(vin)] which is given in the *filename*.out file.

9.29 Compare the voltage at node 4 in Example 9.2 when V_{IN1} is a DC voltage of +1 V and −1 V.

9.30 With the transistor parameters given in Example 9.2 and these additional parameters—`NE=1.32, ISE=4.7E-16, RE=4.8, RC=150, RB=539, RBM=5.0, ISC=1.4E-14`, and `NC=1.04`—plot the output I–V characteristics for the common-emitter configuration with the same base currents and collector-emitter voltage as in Fig. 9.23.

9.31 For the transistor parameters given in Example 9.2 and Problem 9.30, plot β vs. V_{EB} from 0.4 V to 0.9 V.

9.32 For a *pnp* transistor, sketch the minority carrier concentration in the base region for the saturation bias condition.

9.33 This problem illustrates the use of the Schottky diode connected between the collector and base of a bipolar transistor as illustrated earlier in Fig. 6.11. The complete circuit is shown in the accompanying figure. For this problem, use the BJT parameters given in Problem 9.23 for the 2N2222A *npn* transistor. Use the parameters given in Problem 6.22 for the MBD101 Schottky barrier diode. The input voltage `VIN` is a pulse which has a 0 V value at $t = 0$ and a pulsed value of 1.2 V with no time delay. The pulse rise and fall times are 0.5 ns, the pulse width is 10 ns, and the pulse period is 50 ns.

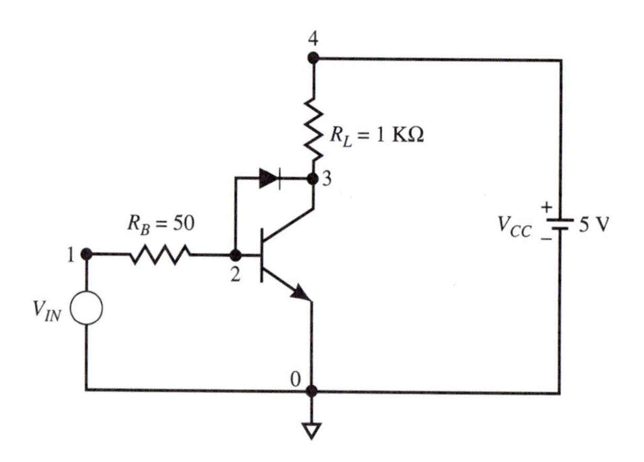

(a) Without the Schottky diode in the circuit, plot the time response of the collector current and the input voltage pulse on the same sheet.

(b) Repeat part (a), with the Schottky diode connected between the base and collector as shown in the figure.

(c) Why is the transistor response faster with the Schottky barrier diode?

9.34 Show that the heavily-doped emitter improves the injection efficiency γ by decreasing the current injected from the base into the emitter, rather than influencing the current injected from the emitter into the base. Use a Si *npn* transistor.

APPENDIX E

E.1 MAXIMUM FREQUENCY OF OSCILLATION f_{max}

The maximum frequency of oscillation f_{max} is found from the maximum available power gain at high frequencies for the equivalent circuit shown in Fig. 9.25 with an input signal current source $\tilde{i}_{in}$ between the emitter and base and with the load resistance R_L between the collector and emitter, as shown in Fig. E.1 (*a*). The output resistance r_o given by Eq. (9.77) is taken as much larger than R_L and may be ignored. Because the emitter-base storage capacitance $C_{S_{EB}}$ is greater than the depletion capacitance $C_{j_{EB}}$, only $C_{S_{EB}}$ is retained for C_π in Fig. E.1 (*a*). The parasitic resistances R_E, R_C, and r_μ, as well as the collector-to-substrate capacitance C_{CS}, are neglected. At high frequencies, the reactance of C_π shorts r_π and reduces the real part of the input impedance to R_B. The high-frequency input power is taken as

$$\tilde{p}_{in} \simeq \tilde{i}_{in}^2 R_B. \tag{E.1}$$

The emitter-base voltage due to $\tilde{i}_{in}$ is

$$\tilde{v}_{EB} = \tilde{i}_{in}/j\omega C_\pi. \tag{E.2}$$

From Eqs. (E.1) and (E.2),

$$\tilde{p}_{in} \simeq \omega^2 C_\pi^2 |\tilde{v}_{EB}|^2. \tag{E.3}$$

To obtain the maximum output power, the value of R_L is matched to the transistor output resistance r_o. The equivalent circuit for obtaining the output resistance with the input open is shown in Fig. E.1 (*b*). With a signal voltage $\tilde{v}_2$ applied at the output as shown, the current $\tilde{i}_2$ is the sum of $g_m \tilde{v}_{EB}$ and $\tilde{i}_3$. The voltage across the two capacitors C_π and C_μ is also $\tilde{v}_2$:

$$\tilde{v}_2 = \tilde{i}_3(1/j\omega C_\pi + 1/j\omega C_\mu), \tag{E.4}$$

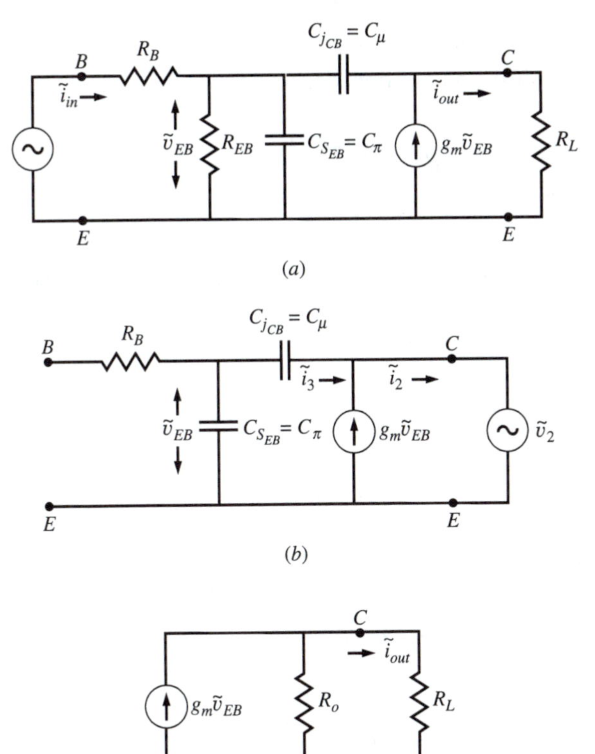

FIGURE E.1 (*a*) High-frequency equivalent circuit for determining maximum available power gain. (*b*) Equivalent circuit for finding the output resistance with the input open. (*c*) Output equivalent circuit with $R_o = R_L$ for maximum available power gain.

and

$$\tilde{i}_3 = j\omega C_\pi \tilde{v}_{EB}. \tag{E.5}$$

With Eq. (E.5) in Eq. (E.4),

$$\tilde{v}_2 = j\omega C_\pi \tilde{v}_{EB}\left[\frac{1}{j\omega C_\pi} + \frac{1}{j\omega C_\mu}\right] = \tilde{v}_{EB}\left[1 + \frac{C_\pi}{C\mu}\right]. \tag{E.6}$$

With $C_\pi > C_\mu$ so that $C_\pi/C_\mu > 1$, Eq. (E.6) may be written as

$$\tilde{v}_2 \simeq \tilde{v}_{EB} C_\pi / C_\mu. \tag{E.7}$$

The current $\tilde{i}_2$ with Eq. (E.5) for $\tilde{i}_3$ is

$$\tilde{i}_2 = g_m \tilde{v}_{EB} + \tilde{i}_3 = g_m \tilde{v}_{EB} + j\omega C_\pi \tilde{v}_{EB}. \tag{E.8}$$

The real part of $\tilde{v}_2/\tilde{i}_2$ with $\tilde{v}_2$ given in (E.7) and the real part of $\tilde{i}_2$ taken from Eq. (E.8) to give the real part of the output impedance R_o:

$$R_o = \frac{\tilde{v}_{EB} C_\pi / C_\mu}{g_m \tilde{v}_{EB}} = \frac{C_\pi}{g_m C_\mu}. \tag{E.9}$$

The output equivalent circuit with the load resistance $R_L = R_o$ as given in Eq. (E.9) is shown in Fig. E.1 (*c*). The current $g_m \tilde{v}_{EB}$ now divides equally between R_o and R_L so that the current through R_L is

$$\tilde{i}_{out} = \frac{1}{2} g_m \tilde{v}_{EB} \tag{E.10}$$

and the output power $\tilde{p}_{out}$ is

$$\tilde{p}_{out} = \tilde{i}_{out}^2 R_L = \left| \frac{1}{2} g_m \tilde{v}_{EB} \right|^2 \frac{C_\pi}{g_m C_\mu} = \frac{1}{4} g_m \frac{C_\pi}{C_\mu} |\tilde{v}_{EB}|^2. \tag{E.11}$$

The maximum available power gain G is

$$G = \frac{\tilde{p}_{out}}{\tilde{p}_{in}} = \frac{1}{4} g_m \frac{C_\pi}{C_\mu} |\tilde{v}_{EB}|^2 \frac{1}{\omega^2 C_\pi^2 |\tilde{v}_{EB}|^2 R_B}, \tag{E.12}$$

or

$$G = \frac{1}{4} \frac{g_m}{C_\pi} \frac{1}{\omega^2 C_\mu^2}. \tag{E.13}$$

The maximum frequency of oscillation is taken for $G = 1$. With Eq. (9.84) for $2\pi f_T = g_m / C_\pi$ with $C_\pi > C_\mu$,

maximum frequency of oscillation

$$\boxed{f_{max} = \sqrt{\frac{f_T}{8\pi R_B C_\mu}}}. \tag{E.14}$$

APPENDIX I

INTRODUCTION TO PSPICE

I.1 GENERAL DESCRIPTION

Numerous books give detailed descriptions for SPICE and PSpice. Examples of these books are:

W. Banzhaf, *Computer-Aided Circuit Analysis Using PSpice, 2nd Ed.* (Prentice-Hall, Englewood Cliffs, NJ, 1992).

M. H. Rashid, *SPICE for Circuits and Electronics Using PSpice, 2nd Ed.* (Prentice-Hall, Englewood Cliffs, NJ, 1995).

G. W. Roberts and A. S. Sedra, *SPICE, 2nd Ed.* (Oxford University Press, 1997).

P. W. Tuinenga, *SPICE: A Guide to Circuit Simulations and Analysis Using PSpice, 3rd Ed.* (Prentice-Hall, Englewood Cliffs, NJ, 1995).

A. Vladimirescu, *The Spice Book* (Wiley, New York, 1994).

Although PSpice has been available for both workstations and personal computers (PCs), only the PC version with *Windows 95* will be presented here. The same procedures apply to both systems. MicroSim Corporation, the developer of PSpice, merged with OrCad in January, 1998. As with most software, PSpice can be expected to continue to be enhanced. The free limited-capability evaluation version of PSpice (release 8.0) is available on the John Wiley World Wide Web site: http://www.wiley.com/college/casey.

A circuit file for PSpice begins with a Title which contains comments such as the name of the circuit file and ends with the `.END` statement. The name of a circuit file is simply a *filename*. A PSpice circuit file must be of the form *filename.cir*. The suffix *.cir* is required in PSpice. Within the PSpice *filename.cir* file, there is no case distinction so that the upper, lower, or mixed cases can be used interchangeably. However, case does matter in filenames (`ex1.cir` is distinct from `EX1.cir`), and the calling names for PSpice and for Probe (the graphics output) are both lower case: `pspice` and `probe`. Output files *filename.out* and *filename.dat* will be generated when PSpice is run. The `filename.out` file contains the circuit description, DC node voltages, voltage source currents, total power dissipated, program run time, or error messages if there is a problem with your *filename.cir* file. After the Title, the Circuit Description is given, then the Analysis Description, and finally the Output Description. The order of the lines between the Title and `.END` statement is not important, but it is helpful to use a designated order to simplify file editing. The order would be:

Title

Circuit Description

Analysis Description

Output Description

.END

A brief description of the three parts of the PSpice file between the Title and .END statement will be given; however, only those parts needed for application of PSpice to devices are included. **It should be noted that no blank lines before the TITLE or after the .END statement are permitted.** General circuit analysis is given in numerous books such as those just cited and will not be considered here.

Numbers for PSpice files may be expressed as integers (5, −5), floating point (5.163), and in scientific notation (5.1E–3). The suffixes for PSpice are:

F = 1E–15 (femto)	P = 1E–12 (pico)	N = 1E–9 (nano)
U = 1E–6 (micro)	M = 1E–3 (milli)	K = 1E3 (Kilo)
MEG = 1E6 (Mega)	G = 1E9 (Giga)	T = 1E12 (Tera)
MIL = 25.4E–6 (0.001 inch)		

The units that are used are: V (Volt), A (Amp), HZ (Hertz), OHM (Ohm), H (Henry), F (Farad) and DEG (degree). Distances are given in meters. Note that F (intended to be Farad) will be interpreted as "femto" if there is no intervening prefix (like U, for micro, for example). In PSpice files, continuation of a line is preceded by a plus (+) sign and comment lines are preceded by an asterisk (*).

I.2 CIRCUIT DESCRIPTION

The circuit description specifies the circuit elements and sources. The nodes of the circuit elements and sources are given by integer numbers with 0 reserved for ground. PSpice identifies a circuit element or source by the first letter on the `element` or `source line`. In PSpice, the element line is often called the General Form. These elements and sources are summarized in Table I.1.

TABLE I.1 First Letter of Elements and Sources

First letter	Circuit element or source
B	GaAs MESFET (not available in SPICE2)
C	capacitor
D	diode
E	voltage-controlled voltage source
F	current-controlled current source
G	voltage-controlled current source
H	current-controlled voltage source
I	independent current source
J	JFET
K	mutual inductors (transformers)
L	inductor
M	MOSFET
Q	bipolar junction transistor
R	resistor
S	voltage-controlled switch (not available in SPICE2)
T	transmission line
V	independent voltage source
W	current-controlled switch (not available in SPICE2)

I.2.1 DC Voltage and Current Sources

The General Form of the `element line` for a DC voltage source is given by

```
VXXXXXX N+ N- DC <value>
```

where `XXXXXX` represents from 1 to 6 alphanumeric characters for the name of the voltage source. SPICE considers that positive current flows from the positive node `N+` to the negative node `N-` within the element. The product of the current (through) and voltage (across) will be positive for a resistor since the power absorbed by a resistor is always positive. However, the current through the DC source will be negative and it supplies (delivers) positive power.

The General Form of the `element line` for a DC current source is given by

```
IXXXXXX N+ N- DC <value>
```

Current flows from the `N+` node through the current source to the `N-` node.

I.2.2 DC Current Measurement

DC currents may be measured with a *dead-voltage source,* which is a DC voltage source with a zero-valued voltage. This voltage source is inserted in the circuit where the current is to be measured. For example,

```
VI1 12 14
```

would measure the current through the nodes 12 to 14. All node currents are obtained in PSpice, but dead-voltage sources may be used to control the sign of a current to be plotted or used in a square-root function.

I.2.3 Piecewise Linear Source

A piecewise linear description of an arbitrary waveform may be represented by pairs of values of time T_j and voltage V_j. The element line is given by

```
VXXXXXX N+ N- PWL(T1 V1 <T2 V2 T3 V3 ...>)
```

where the voltage at T1 is V1. For example, `VGEN 1 0 PWL(0NS 5 10NS 5 10.2NS -5)` is a piecewise linear voltage source between node 1 and ground with an initial voltage of 5 V at $t = 0$ nsec and remains at 5 V at 10 nsec, and then goes from 5 V at 10 nsec to −5 V at 10.2 nsec. Piecewise current sources are given in a similar representation.

I.2.4 Pulse Source

A voltage or current pulse may be specified by the element line:

```
VXXXXXX N+ N- PULSE(V1 V2 TD TR TF PW PER)
```

which can specify a single pulse or a pulse train. A similar expression may be used for a current pulse. The model parameters for pulse sources are given in Table I.2. In Table I.2, TSTEP and TSTOP are the values given in the .TRAN control line (Sec. I.3.2). Figure I.1 illustrates these parameters.

I.2.5 Sinusoidal Source

A sinusoidal voltage waveform is specified by the element line:

```
VXXXXX1 N+ N- SIN(VO VA FREQ TD DF PHASE)
```

where the parameters are given in Table I.3.

TABLE I.2 Model Parameters for Pulse Source.

V1 (initial value)	must specify	Volt or Ampere
V2 (pulsed voltage)	must specify	Volt or Ampere
TD (delay time)	0	sec
TR (rise time)	TSTEP	sec
TF (fall time)	TSTEP	sec
PW (pulse width)	TSTOP	sec
PER (period)	TSTOP	sec

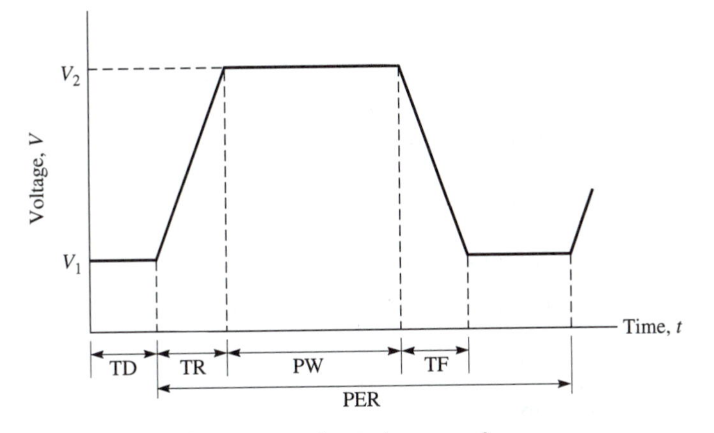

FIGURE I.1 Parameters for pulse waveform.

TABLE I.3 Model Parameters for Sinusoidal Source.

`VO` (offset value)	required	Volt or Ampere
`VA` (amplitude)	required	Volt or Ampere
`FREQ` (frequency)	`1/TSTOP`	Hertz
`TD` (start delay)	0	sec
`DF` (damping factor)	0	1/sec
`PHASE` (phase angle)	0	degree

`TSTOP` refers to the stop time in the transient analysis specified by the `.TRAN` control line (Sec. I.3.2). The sinusoidal start time can occur after time zero, and it may be exponentially damped if desired.

I.3 ANALYSIS DESCRIPTION

I.3.1 DC Analysis

With semiconductor devices, it is often desirable to vary a DC voltage to generate *I–V* characteristics. The .DC control line is given by

```
.DC VXXXXX1 Start Stop Inc <VXXXXX2 Start2 Stop2 inc2>
```

where `VXXXXX1` is the first DC independent voltage source to be varied. The initial voltage is given by `Start`, the final voltage is given by `Stop`, and the increment size in each step is given by `Inc`. A second optional source may be specified by `VXXXXX2`. When two sources are specified, the first will be stepped through from the starting voltage to the final voltage for each value of the second source. Either of these voltage sources could be current sources. For example, the base current IBASE could be swept from 10 μA to 50 μA in 5-μA steps while the collector voltage VCE is swept from 0 to 5 V by

```
.DC IBASE 10U 50U 5U VCE 0 5 0.05,
```

where U represents μA.

I.3.2 Transient Analysis

The transient analysis uses the .TRAN control line, which is given by

```
.TRAN TSTEP TSTOP <TSTART <TMAX>> <UIC>
```

with the analysis performed over the time interval from 0 to TSTOP. The parameter TSTEP is the time interval used with a .PRINT or .PLOT control line and has no effect when the graphics postprocessor probe is used. The default time step used for analysis is TSTOP/50. The default value of TSTART is zero and designation of a value may be omitted. If the PSpice plot is not smooth with the default time step, additional calculated points are needed. The parameter TMAX, which represents the increment between data points, may be used to give a smaller increment between data points. The optional parameter UIC means use initial conditions, which would be specified with a .IC control line. For example,

```
.TRAN 1.0E-5 2.5E-4 0.0 2.0E-6
```

would produce an analysis for $0 \leq t \leq 2.5\times10^{-4}$ s with the largest increment between data points of 2.0×10^{-6} s.

I.3.3 AC Analysis

The AC analysis uses the .AC control line, which is given as

```
.AC DEC ND FSTART FSTOP,
```

where FSTART is the lowest frequency (which may not be negative or 0) and FSTOP is the highest frequency. DEC will make PSpice divide the frequency range into decades, with ND points per decade. Frequencies will be logarithmically spaced.

Additional analysis control lines are: operating-point analysis, sensitivity analysis, noise analysis, and Fourier analysis. These analysis control lines are also described in the books on PSpice.

I.4 .PROBE

PSpice has a graphics postprocessor called Probe, which produces a large variety of graphical results where the x axis is specified by the Analysis Description in the previous section. Various voltages and currents may be selected to be plotted on the y axis.

TABLE I.4 Probe Mathematical Functions.

Function	Description	Function	Description
D(Y)	derivative of Y	S(Y)	integral of Y over the range of the x-axis variable
SQRT(Y)	square root of Y	COS(Y)	cosine of Y(radians)
SIN(Y)	sine of Y(radians)	TAN(Y)	tangent of Y(radians)
ATAN(Y)	inverse tangent, radians	ARCTAN(Y)	inverse tangent, radians
ABS(Y)	absolute value of Y	SGN(Y)	$+1$ if $Y > 0$, 0 if $Y = 0$, -1 if $Y < 0$
EXP(Y)	exp(Y)	LOG(Y)	log base e of Y
LOG10(Y)	log base 10 of Y	AVG(Y)	running average of Y over the range of the x axis variable
PWR(Y,Z)	$\|Y\|$ raised to the Z power	MIN(Y)	minimum value of Y over the range of the x axis
DB(Y)	20 $\log_{10}$ of Y	RMS(Y)	running RMS value of Y over the range of the x axis variable

The free evaluation version of PSpice from MicroSim Corporation, now OrCad, has been provided on a CD for *Windows 95* (or floppy disks for Windows or Macs) which sets up all the necessary files as well as a schematic capture program and a parts library. Setup for the workstation Unix version, which is no longer supported, generally requires additions such as to your .cshrs file.

For the Output Analysis in the circuit file, the command `.probe` or `.PROBE` generates the *filename.dat* file, which contains the graphical data. Probe permits mathematical manipulation of the various voltages and currents. The mathematical operations are given by / for division, * for multiplication, + for addition, and − for subtraction. The most commonly used mathematical functions available in Probe are given in Table I.4. Additional functions are given in the Probe Add Traces Functions or Macros dialog box described in Sec. I.4.1.

I.4.1 Probe on PC with *Windows 95*

To illustrate the use of PSpice and Probe, the diode transient response given in Example 5.1 with the circuit shown in Fig. 5.19 will be used. On a PC with *Windows 95*, the circuit file can be created with the Notepad text editor or with the MicroSim text editor included with the PSpice release 8.0. To open the MicroSim text editor, open the Start menu and choose Programs to open the submenu, which lists MicroSim Eval8. In the menu that comes up for MicroSim Eval8, select Accessories, which brings up the menu with MicroSim TextEdit. Click the left mouse button on MicroSim TextEdit to open. Click on File and click on New in the drop-down list. Enter the text for the circuit file:

```
diode turn-off transit example-appen
vin 1 0 pwl(0ns 5 10ns 5 10.2ns -5)
rl 1 2 100
d1 2 0 ddif
.model ddif d(is=1.1e-15 n=1 tt=1.0e-8 vj=0.986 cjo=5.0e-12)
.tran 0.25ns 50ns 0.0 0.25ns
.probe
.end
```

create circuit file with MicroSim TextEdit

The first line is the title of the circuit file and will appear at the top of the Probe plot. No *, as required for comments, is used. The next line gives the input voltage as a piecewise linear source, which was described in Sec. I.2.3 of this appendix. This voltage source is +5 V at $t = 0$ and at $t = 10$ ns. Then, at $t = 10.2$ ns the voltage switches to −5.0 V. The load resistor `rl` is 100 Ω and is connected between nodes 1 and 2. The diode designated as `d1` in the circuit of Fig. 5.19 is connected between nodes 2 and 0 and has the `MODNAME` of `ddif`. (See Sec. 5.5 for the diode SPICE parameters and for the `.model` line.) As given in Sec. I.3.2, the `.tran` Analysis Control Line gives `TSTEP` = 0.25 ns, `TSTOP` = 50 ns, `TSTART` = 0.0 s, and `TMAX` is the Δt used in the simulation which is 0.25 ns. The `.probe` Output Description runs the graphics postprocessor to permit plotting the output results. The file is entered as shown in Fig. I.2.

To save the file, click on File and click on Save As in the drop-down menu. Enter the file name in File name box as also shown in Fig. I.2. The filename has been selected as `example-appen`. As shown in the Save As type box, the suffix .cir is automatically included. Click on Save. Next, return to the Start menu and select Programs and then select MicroSim Eval8 again. On the menu that comes up, click on PSpiceA_D. In the PSpiceAD window, click on File and then click on Open on the drop-down list. In the Open window, shown in Fig. I.3 with the PSpiceAD window, click on example-appen, which appears in the File name box. Click on Open and PSpice runs. Click on File in the PSpiceAD window and click on Examine Output in the drop-down list to review the results of the PSpice simulation. If there are errors in the circuit file, error information is given. Click on File and then on Exit in the drop-down list to return to the PSpice AD window.

To view the graphical output, click on File and then click on Run Probe in the drop-down list. The Probe window appears as shown in Fig. I.4(a). To assign the quantity to be plotted, either click on Trace or the tenth icon from the right. If Trace is selected, click on Add in the drop-down list, and the Add Traces window opens as shown in Fig. I.4(b). The Simulation Output Variables are listed on the left, while the window on the right lists the Probe mathematical functions, listed in part in Table I.4. Click on I(d1) in the left box to plot the diode current. The I(d1) appears in the Trace Expression box. Click on Ok, and the plot shown in Fig. I.5 appears. The x and y scales may be changed by clicking in Plot and selecting the X Axis settings or the Y Axis settings. To print, click on the third icon from the left (printer icon) in the Standard bar and the plot shown in Fig. I.5 results.

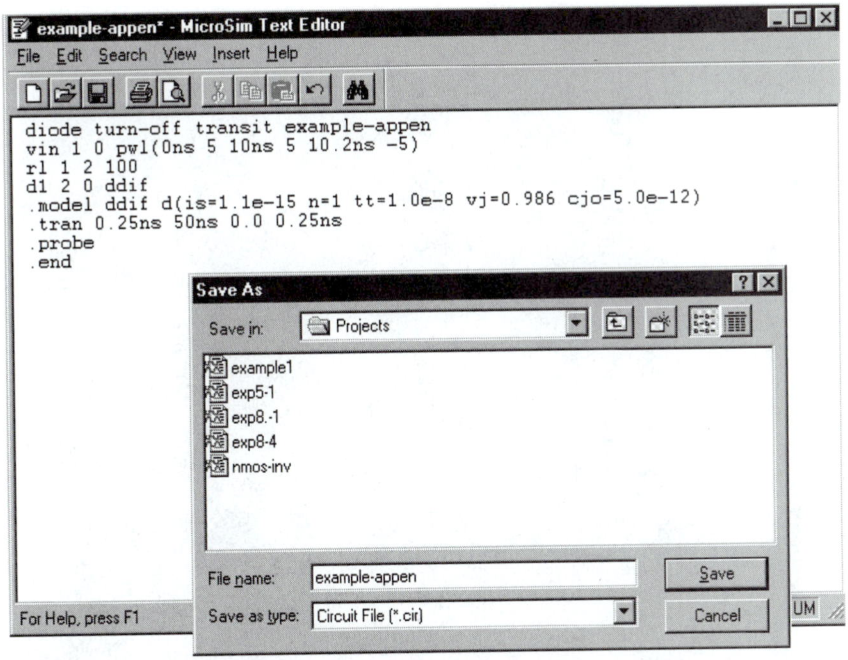

```
diode turn-off transit example-appen
vin 1 0 pwl(0ns 5 10ns 5 10.2ns -5)
rl 1 2 100
d1 2 0 ddif
.model ddif d(is=1.1e-15 n=1 tt=1.0e-8 vj=0.986 cjo=5.0e-12)
.tran 0.25ns 50ns 0.0 0.25ns
.probe
.end
```

FIGURE I.2 The MicroSim Text Editor window with the circuit file named `example-appen` together with the Save As window.

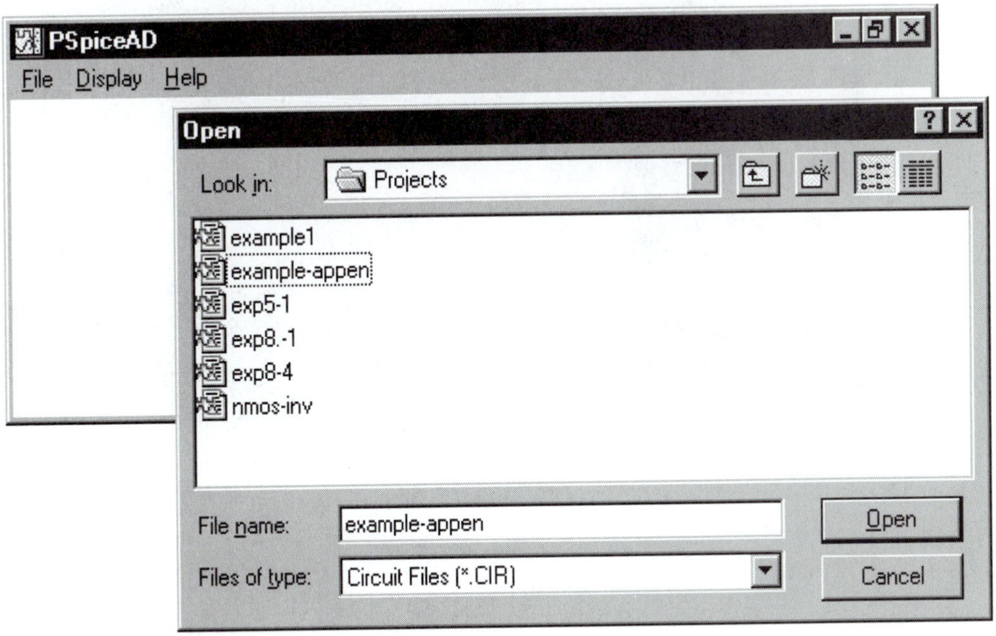

FIGURE I.3 PSpiceAD window with the Open window to designate the file to be run.

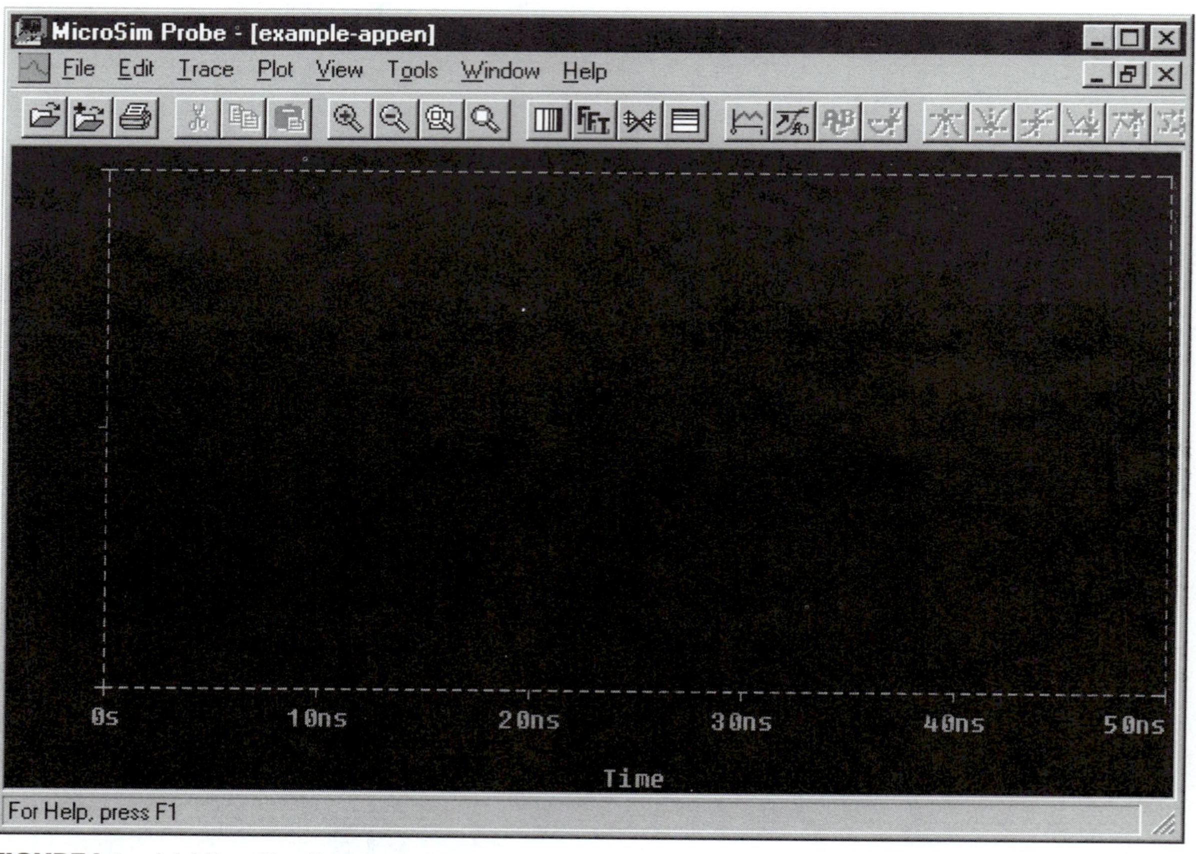

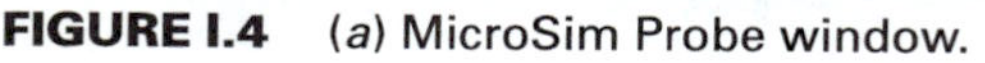

FIGURE I.4 (*a*) MicroSim Probe window.

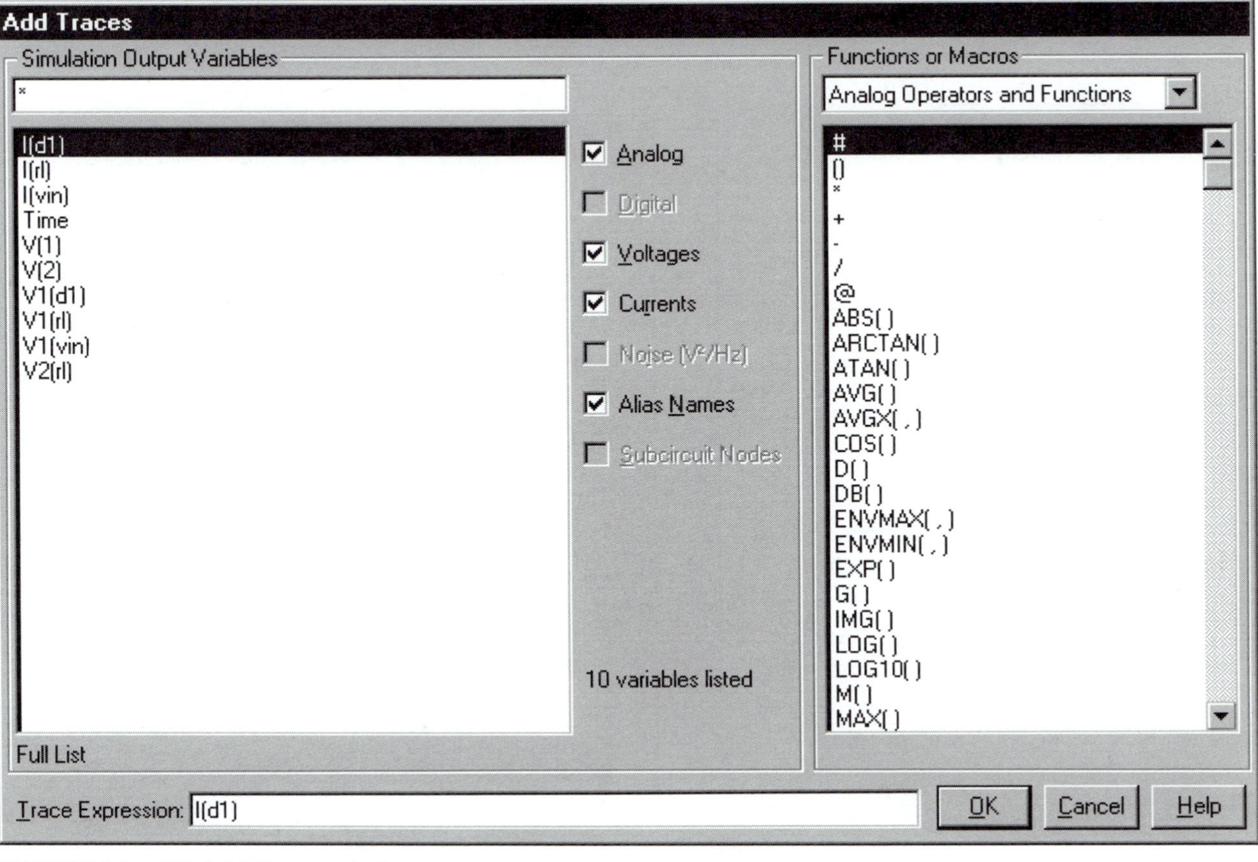

FIGURE I.4 (*b*) Add Traces window.

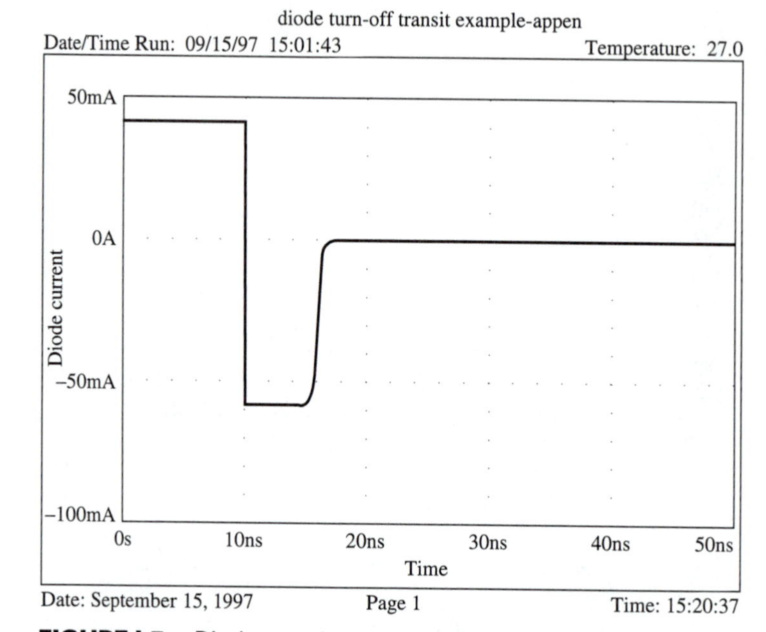

FIGURE I.5 Diode-transient response.

INDEX

N

O

P

Q

R

S

T